Boundary Objects and Beyond

Infrastructures Series

edited by Geoffrey C. Bowker and Paul N. Edwards

Boundary Objects and Beyond

Working with Leigh Star

Edited by Geoffrey C. Bowker, Stefan Timmermans, Adele E. Clarke, and Ellen Balka

The MIT Press
Cambridge, Massachusetts
London, England

This book was set in ITC Stone Serif Std by Toppan Best-set Premedia Limited. Printed and bound in the United States of America.

Library of Congress Cataloging-in-Publication Data is available.

ISBN: 978-0-262-02974-2 (hc); 978-0-262-52808-5 (pb)

10   9   8   7   6   5   4   3   2   1

Based on the Conference: A Celebration of Susan Leigh Star: Her Work and Intellectual Legacy, University of California, San Francisco (UCSF), September 2011, http://www.sscnet.ucla.edu/soc/LeighStar/. **Organized by** Stefan Timmermans (UCLA), Geoffrey C. Bowker (University of Pittsburgh) and Adele E. Clarke (UCSF). **Sponsored by** Science, Technology, and Society Program, Division of Social and Economic Sciences, National Science Foundation; Department of Sociology, UCLA; Department of Social and Behavioral Sciences, UCSF.

Contents

Prologue

Geoffrey C. Bowker

I write this with a heavy heart and tremulous hands; with joy and pleasure and love . . . sharing as ever Leigh's delight in contradiction.

I was graced to weave my life together with hers. She had a soft voice, an impish smile, and a lovely laugh. In pain—and pain was ever present, waxing and waning— her voice would drop and in her eyes I could sense her suffering. In strength—and always was there strength—the beauty of her light shone through. Together we drew down the moon, conjured classifications and discovered poetry abounding.

This volume brings together friends and colleagues who were touched by and who touched her. Not a self-contained set, but offering a flavor of the way she was in the world: intimately, disconcertingly personal; changing lives; generating ideas; thinking the unimaginable with fierce will and phrasing it with delicate touch.

When we made our final move, she had been going through a long and difficult stretch. I am so happy that her last months were full of optimism and renewal. Always a good gauge of her spirit was whether she was writing poetry (or rather, whether she was writing what is formally called poetry): she was, and it was wild and wonderful and copious.

Her next project, which we had played with for years, was to be a work on the poetics of infrastructure. She talked of the orphans of infrastructure. People sequestered away from the ways and wires of the world on an old leper colony island in Hawaii; the Dionn quintuplets whose every evacuation was recorded and whose every moment of idle play was annotated. People who did not fit the units of social programs and software and led strangely beautiful lives. She talked of infrastructure as a thing of beauty . . .

This volume is not a closure, but a living work through which many who do not know her can come to be touched by her magic and those who did can once again feel the beauty of her spirit.

If we collectively have succeeded, then, to quote Leigh, we will "speak Her name clearly into your eyes."

Introduction: Working with Leigh Star

Stefan Timmermans

Once the process gets going it keeps on going, given constant interactions with other people and all kinds of humans and nonhumans in the world. I don't know enough about death to know whether or in exactly what forms it might keep going afterwards, except that the ongoing actions we leave embedded in the world constitute one such action.

(Star 1995a, 20)

Susan Leigh Star (Leigh) became one of the most influential science studies intellectuals of the last decades. Along with other scholars, her work shifted the research agenda from bibliometric and Mertonian preoccupations with the social organization of science (i.e., its reward systems, organization of scientific communities, invisible colleges, dominant values, etc.) to a focus on the production of scientific facts. Summarizing her generation's approach, Star wrote, "Among our common interests and beliefs was the necessity of 'opening up the black box' in order to demystify science and technology, that is, to analyze the process of production as well as the product. The methodological directives here for us were familiar: understand the language and meanings of your respondents, link them with institutional patterns and commitments, and, as Everett Hughes said, remember that 'it could have been otherwise'" (Star 1988, 198). This awareness that "it could have been otherwise" turned the assumption that science reflected "nature" or "facts" into a topic of sociological inquiry.

Drawing from pragmatism and the Chicago school's sociology of work, Star's scholarship highlighted the messy practices of discovering science. Where others conducted ethnographies of the work of laboratory scientists' manipulating tools and data to construct epistemic insights, Star quickly became drawn to the work behind the work: the countless, taken-for-granted and often dismissed practices of assistants, technicians, and students that made scientific breakthroughs possible. One of her goals was to restore agency to scientific work by examining who did the actual work and who received credit for that work. Digging ever deeper into the conditions that make science possible, Star foregrounded the infrastructure of classifications, technologies, paperwork, and regulations that constrained and constituted scientific work. She aimed for a radical symmetry in her work: not only evaluating the winners and losers of science evenhandedly but examining the lives of those caught up against their will in the new globalizing technoscientific regimes. Within the post-Merton sociology of science, her work thus opened up the field of science studies to what others took for granted.

Star was greatly inspired by Everett Hughes's approach to the sociology of work, as passed on by her dissertation advisor Anselm Strauss and intellectual muse Howard Becker—themselves Hughes students. Looking at science as work, a job to be done one task at the time by someone in a particular organizational form demystified the special character of science and offered a vantage point of the practice of doing (rather than talking about) science. For Star, scientists were neither heroes nor villains, but people "doing things together"—the things being an activity they called science. Looking at science as work opened up the laboratory to bread-and-butter sociological questions such as how do people come to believe what they believe about nature? How do scientists find a common language across disciplinary boundaries? How does science contribute to the stratification of society? The emphasis on work drew Star to examine the activities that are ignored, invisible, and deleted. Her methodological mantra was to restore the work: following who did what, when, where, how, and with what kind of consequences.

From the Chicago school of sociology, Star was stimulated by the strong focus on conceptualizing, developing sensitizing notions (Blumer 1954), in the grounded theory tradition (Glaser and Strauss 1967). An accomplished poet and wordsmith in her own right, Star was drawn to the intellectual community of grounded theorists and pragmatists and the creativity ("stretch your imagination" [Star 1991b]) of this theory/methods package (Clarke and Star 2008). Grounded theory did not only stimulate her intellectually but also drew attention to the joys and, occasionally, frustrations of doing scientific research, which considering that her topic was scientific work inevitably drew her to reflect on the commonalities between her own work and the work of scientists (Star 2007). She taught workshops and courses on qualitative methods, showing how well-developed concepts can illuminate hidden dimensions in a research project. Indeed, for Star, pragmatism infused grounded theory with an epistemological compass: "How does one create a life-work and remain open; open to the data, open to being wrong, to redoing one's own work, actively to seek out new views and mistakes? For me, that has come through the privilege of teaching grounded theory, and of collaborating with people who like to work this way. That is, to embrace a continuous, embedded, imbricated, multiple, constantly compared way of making sense of myself" (Star 2007, 90).

With a strong pragmatism at her core, Star was further influenced by feminism, race theories, ecological thinking, symbolic interactionism, actor-network theory, ethnomethodology, linguistics, activity theory, metaphysics, theology, cognitive science, phenomenological psychology, distributed artificial intelligence, and anyone who produced exciting intellectual insights.

Star situated science and technologies within ecologies of knowledge (Rosenberg 1979). As such, she aimed "to understand the systemic properties of science by analogy with an ecosystem, and equally important, all the components that constitute the system" (Star 1995a, 2). Rather than taking the perspective of the university administrator or award-winning professional scientist, Star examined the interrelationships between different layers of a system within an entire ecosphere. Thus, a study of the design of computer chips almost immediately jumped to an analysis of institutions, organization charts, the material characteristics of chips, regulations, and gurus in the field (Star 1995b). And she conceptualized a study of information systems as composed of people and things, in ecological relationships, with representations and signals and fluid ways of working.

Star highlighted how the ecological knowledge perspective eschewed reductionist analyses: "by ecological, I mean refusing social/natural or social/technical dichotomies and inventing systematic and dialectical units of analysis" (Star 1988). Rather than dualism, Star, following Dewey, saw phenomena as continuous (Star 1991b, 277). She encouraged the simultaneous analysis of human and material worlds: "The advantages of such an analysis are that the increased heterogeneity accounts for more of the phenomena observed; one does not draw an arbitrary line between organism and environment, one can empirically 'track' lines of action without stopping at species, mechanical or linguistic boundaries, and especially without invoking a reified conception of society" (Star 1995a, 13). Unlike Latour, Star focused on nonhuman agency in order to highlight how social life is recalibrated and restratified. Her goal was not to democratize the human-nonhuman divide but to analyze the powers of the nonhuman in reshaping a human world.

From pragmatist philosophy, Star took to heart the difficult notion that understanding is based on consequences not antecedents. Indeed, her advisor Anselm Strauss controversially noted that gender and race cannot be presumed a priori to be necessarily relevant but need to "earn" their theoretical relevance (but see Clarke 2008). Yet, schooled as a feminist, Star gravitated to examining the consequences of information technologies for those at the margins, those who are both outsiders and insiders. Studying the outliers and forgotten was theoretically important because the "voices of those suffering from abuses of technological power are among the most powerful analytically" (Star 1991a, 30).

One of Star's goals was to present the process of building big science from the viewpoints of the excluded rather than from those of management or great scientists. "Every enrollment entails both a failure to enroll and a destruction of the world of the non-enrolled" (Star 1991a, 49). She encouraged her students to ask the question: Cui bono? Who benefits? Her point was not to do checklist research in which social scientists check off the "usual suspects" (Shim 2002) of race, class, and gender. Rather, she asked hard questions about the marginalizing as well as liberating powers of technosciences and their stratifying consequences. Star embraced a theory of intersectionality in which marginalities are not pre-given but the systemic consequences of new sociotechnical arrangements that create new—and fluid—forms of exclusion. Every standpoint, every self, every interest counts but they also come with "overhead" (Bowker and Star 1999, 309). Sometimes these overhead costs were deliberately sought out as acts of subversion, at other times the cost was the collateral damage of facing new forms of science and technology. Her project was to reveal the subversion, the collateral damage and suffering as well as the valuable accomplishments of new information regimens.

Restoring the intended and unintended fallout of new information systems and data structures was analytically important because such things are systematically deleted from official records. Hence, Star's interest in the sociology of the invisible and articulation work: "As the work of negotiating, resource management, and many other processes are deleted from representations, one group's interests begin to take precedence over another's, subverting the fundamental pluralism of human interaction. . . . The more the work is rendered deleted and the more invisibles are invoked as explanations, the more suffering there is" (Star 1991b, 278–279).

Articulation work refers to keeping things on track through dispersed bits of local knowledge (Strauss 1988). It is the kind of work done by others that allows an executive or a famous scientist to cross multiple worlds. This is the work of nurses, secretaries, research assistants, copy editors, and repair people.

Among the most celebrated of Leigh's concepts was the notion of boundary object (Star and Griesemer 1989; Star 1989; Star 2010). This concept captures the possibility of cooperative scientific work in the absence of consensus. Cooperation does not always follow from a preexisting consensus but can be achieved with objects flowing through various communities of practice or social worlds. Star used "object" loosely in a Meadian sense as something that people act toward and with; it may be a thing but also, for example, a theory. A boundary object is something that has different and quite specific meanings in intersecting communities but also has a common meaning to facilitate cooperation across communities. "Boundary objects are objects which are both plastic enough to adapt to local needs and the constraints of several parties employing them, yet robust enough to maintain a common identity across sites" (Star and Griesemer 1989, 393). Boundary objects thus satisfy the informational requirements of various communities while still having distinct meanings in each one of them. The boundary part of boundary object referred not necessarily to the periphery of communities but to overlapping areas (as in a Venn diagram) between different communities. Considering that boundary objects emerge through work processes, Star noted that they are themselves in flux. They can easily lose some of their distinct meanings for different communities when they become standardized. Such standardizations, however, may in turn produce more residual categories that defy standardization and stimulate the emergence of new boundary objects (Star 2010).

Star developed this concept in a study of Berkeley's Museum of Vertebrate Zoology. This museum required an extensive collection of natural specimens in order to build up its reputation, and professional biologists were trying to capture California's biodiversity before it would disappear. They enrolled help from amateur collectors who sought legitimacy through their collecting efforts. Other interested parties were trappers, farmers, and traders who often caught rare animals. A key boundary object that facilitated autonomy and communication across these social worlds became the animal specimens. An economy of bartering, trading, and buying sprung up in which each community kept autonomy over the collection of animal specimens, but archival and research science was made possible with the stipulation of minimum standards about the annotation of habitat, conditions, time and geographical location of capture and preservation, and the specimens. Thus the specimens had different identities in each social world but gained a common identity around the scientific documentation of California wildlife.

This boundary object was further related to other boundary objects that regulated communication and relationships between communities. These included standardized information forms for the collection of specimens, repositories of specimens without specification of the ultimate goal of the repository, guidelines that did not capture the granularity to execute tasks but were flexibly adaptable to local needs, and maps of California that gave different social worlds a set of common geographical boundaries but could be filled in differently by each group to suit its needs and goals. The focus on boundary objects allowed Star to examine how objects achieve coherence across social worlds while still having distinct identities within specific communities. Boundary objects reflect neither a consensus nor a top-down imposition of requirements on others: "Rather, boundary objects act as anchors or bridges, however temporary" (Star and Griesemer 1989, 414).

The idea of boundary objects has been extremely influential, giving science, technology, and society (STS, also known as science and technology studies) scholars an entire new approach for studying how communities interact and build science. The perspective of boundary objects offers an engaged counterpoint to the emphasis on translating other people's priorities into scientific ones (translation and *intéressement*), instead highlighting forms of collaboration that preserve the uniqueness of each community of practice (Clarke and Star 2008).

In contrast, working in computer and information science departments and with computer scientists, Star was struck by the tendency of some computer scientists to create formalisms of mathematical models of the world they aimed to capture while others embraced a more empirical approach—trying to collect every data nuance. Social science, especially the qualitative social science she practiced, was often lumped dismissively with an empirical approach and contrasted with the presumed elegance and intellectual superiority of models. Based on the frictions that her presence generated in computer science circles and drawn by both the intellectual and ethical implications of the formalism-specificity issue, Star developed a life-long interest in the trade-offs between formalisms and the messiness of data, work, and life.

She understood that formalization and its related processes of simplification, standardization, universalization, rationalization, objectification, and representation were required to allow technologies to work outside the context where they were created. Yet, formalisms and the technologies that incorporated them often fail to work, requiring what Gasser (1986) called "workarounds" or ad hoc tinkering around formal demands. Rather than seeing this as redemption for the empirical approach, as a pragmatist Star asked under what conditions do questions of formalism arise: "How can something be simultaneously concrete and abstract?" To her question, we can add: Who puts together the database? What is included and what is left out? It was not because people do not do what they are expected to do, or that databases or information technologies have no effect. Again drawing on pragmatist thinking, she saw universalisms as nothing more than standing agreements across communities, not a priori facts. She, along with colleagues in the computer-supported work community, advocated for joining empirical research with formalism, noting that formalism is inevitably locally and temporally situated, distributed, and contingent, and needs to be made to work (Star 1992).

The theme of standards and classification has been present in Star's work from its early beginnings (Star 1983), but reaches its pinnacle in her magnum opus *Sorting Things Out*, coauthored with Geoffrey Bowker (Bowker and Star 1999). In it, Star observes that classification systems and standards tend to blend in with the background infrastructure. Yet, these seemingly boring tools embed deeply social and ethical histories of conflict and compromise in a technical nomenclature that then provides the often-invisible infrastructure for contemporary living.

Classifying is a memory practice to both hold on to certain characteristics and send other elements into oblivion. Star gravitated to an examination of how classification systems cause troubles for individuals, coining the concept of torque to notate situations when a formal classification system is mismatched with an individual's biographical trajectory, membership, or location. Starting from the observation that one person's standard is another's confusion and mess, Bowker and Star articulated the moral dimension of a study of classifications:

The moral questions arise when the categories of the powerful become the taken for granted; when policy decisions are layered into inaccessible technological structures; when one group's invisibility comes at the expense of another's suffering. . . . We need to consistently explore what is left dark by our current classifications ("other" categories) and design classification systems that do not foreclose on rearrangements suggested by new forms of social and natural knowledge. (Bowker and Star 1999, 320–321)

Here we vividly see Star's overarching commitment to social justice through better understanding of the workings of infrastructure.

Political categories become naturalized through standards and classification systems. Bowker and Star link the International Classification of Disease (ICD) to the birth of the welfare state, and note generally how different kinds of political regimes, including the former apartheid regimen in South Africa, depended on various classification systems. With different information requirements and different infrastructures to gather data, each one of the ICD categories constitutes a "frozen policy." It is frozen because the category is reified when introduced in a classification system, and it is policy because it nevertheless feeds relentlessly into further political developments. The ICD data, for example, allowed competing groups of medical specialists, public health officials, and laissez-faire economists to take credit for Britain's decline in mortality rates in the nineteenth century. Bowker and Star significantly note that in the process of making records and producing knowledge, organizations use classification systems to selectively forget the past.

Star and some colleagues joked that their work reads like a manifesto for the Society for the Study of Boring Things (Lampland and Star 2009, preface). What is a dull classification, an annoying information infrastructure, or a set of easy to ignore standards for some, becomes the analytic object for others. For Star, studying these forms of social life meant restoring the work and the political/ethical/social struggles that went into the creation of the formal. She was not against standards or protocols but recognized both the danger of leaving those tools unexamined and the violence they could inflict on those who did not neatly align with them.

References

Blumer, Herbert. 1954. "What Is Wrong with Social Theory?" *American Sociological Review* 18:3–10.

Bowker, Geoffrey, and Susan Leigh Star. 1999. *Sorting Things Out: Classification and Its Consequences*. Cambridge, MA: MIT Press.

Clarke, Adele. 2008. "Sex/Gender and Race/Ethnicity in the Legacy of Anselm Strauss." *Studies in Symbolic Interaction* 32:159–174.

Clarke, Adele, and Susan Leigh Star. 2008. "Social Worlds/Arenas as a Theory-Methods Package." In *Handbook of Science and Technology*, ed. E. J. Hackett, O. Amsterdamska, M. Lynch, and J. Wacjman, 113–139. Cambridge, MA: MIT Press.

Gasser, Les. 1986. "The Integration of Computing and Routine Work." *ACM Transactions on Office Information Systems* 4:205–225.

Glaser, Barney G., and Anselm L. Strauss. 1967. *The Discovery of Grounded Theory*. New York: Aldine Publishing Company.

Lampland, Martha, and Susan Leigh Star. 2009. *Standards and Their Stories: How Quantifying, Classifying, and Formalizing Practices Shape Everyday Life*. Ithaca, NY: Cornell University Press.

Rosenberg, Charles E. 1979. "Towards an Ecology of Knowledge: On Discipline, Contexts and History." In *The Organization of Knowledge in Modern America*, ed. A. Oleson and J. Voss, 440–455. Baltimore: Johns Hopkins University Press.

Shim, Janet K. 2002. "Understanding the Routinized Inclusion of Race, Socioeconomic Status and Sex in Epidemiology: The Utility of Concepts from Technoscience Studies." *Sociology of Health & Illness* 24:129–150.

Star, Susan Leigh. 1983. "Simplification in Scientific Research: An Example from Neuroscience Research." *Social Studies of Science* 13:208–226.

Star, Susan Leigh. 1988. "Introduction: The Sociology of Science and Technology." *Social Problems* 35:197–205.

Star, Susan Leigh. 1989. "The Structure of Ill-Structured Solutions: Boundary Objects and Heterogeneous Distributed Problem Solving." In *Readings in Distributed Artificial Intelligence*, ed. M. Huhns and L. Gasser, 37–53. Menlo Park, CA: Kaufman.

Star, Susan Leigh. 1991a. "Power, Technologies, and the Phenomenology of Conventions: On Being Allergic to Onions." In *A Sociology of Monsters: Essays on Power, Technology and Domination*, ed. J. Law, 26–56. London: Routledge.

Star, Susan Leigh. 1991b. "The Sociology of the Invisible: The Primacy of Work in the Writings of Anselm Strauss." In *Social Organization and Social Process*, ed. D. R. Maines, 265–284. New York: Aldine de Gruyter.

Star, Susan Leigh. 1992. "The Trojan Door: Organizations, Work, and the 'Open Black Box.'" *Systems Practice* 5:395–410.

Star, Susan Leigh. 1995a. "Introduction." In *Ecologies of Knowledge*, ed. S. L. Star, 1–38. Albany: State University of New York Press.

Star, Susan Leigh. 1995b. "The Politics of Formal Representations: Wizards, Gurus, and Organizational Complexity." In *Ecologies of Knowledge: Work and Politics in Science and Technology*, ed. S. L. Star, 88–118. Albany: State University of New York Press.

Star, Susan Leigh. 2007. "Living Grounded Theory: Cognitive and Emotional Forms of Pragmatism." In *Handbook of Grounded Theory*, ed. A. Bryant and K. Charmaz, 75–93. Los Angeles: Sage Publications.

Star, Susan Leigh. 2010. "This Is Not a Boundary Object: Reflections on the Origin of a Concept." *Science, Technology & Human Values* 35:601–617.

Star, Susan Leigh, and James R. Griesemer. 1989. "Institutional Ecology, 'Translations' and Boundary Objects: Amateurs and Professionals in Berkeley's Museum of Vertebrate Zoology, 1907–39." *Social Studies of Science* 19:387–420. [See also chapter 7, this volume.]

Strauss, Anselm L. 1988. "The Articulation of Project Work: An Organizational Process." *Sociological Quarterly* 29:163–178.

I   Ecologies of Knowledge

1   Revisiting *Ecologies of Knowledge: Work and Politics in Science and Technology*

Susan Leigh Star

Editors' Note: This chapter originally appeared as the introduction to a volume Leigh Star edited titled *Ecologies of Knowledge: Work and Politics in Science and Technology* (1995a). It offers an important and still useful overview of both early science and technology studies and her own innovative perspectives on the sociology of knowledge and its production processes, questions at the heart of science and technology studies. The original chapter referred extensively to papers in that volume in ways that would be confusing here. We therefore included citations to clarify where these works appear as [1995].

In 1967 Howard Becker wrote an article that became a clarion call to sociology, his presidential address to the Society for the Study of Social Problems. It was entitled "Whose Side Are We On?" and reminded sociologists that pretensions to value neutrality were themselves value laden. He argued that we must choose to recognize that all perceptions are located in a hierarchy of credibility. In other words, people consider the source of any statement or perception, and discount those produced by lower-status people.

Whose side are we on in social studies of science and technology? What hierarchies of credibility are we tacitly or explicitly assigning? And what language can we invent to investigate these questions honestly?

We could do well to borrow from Patricia Hill Collins's (1986) essay on black feminist thought and its contributions to the structure of sociological knowledge. She argues that African American women's radical explorations of the meaning of self-definition and valuation, the interlocking nature of oppression, and the importance of redefining culture constitute a challenge to sociology's basic beliefs about itself. The challenge takes the form of asking sociology to learn from "the outsider within"—the double glasses of insider and outsider, articulating the tension of both being a sociologist and being excluded by its frame of reference.

The papers in *Ecologies of Knowledge: Work and Politics in Science and Technology* (Star 1995a) are all attempts to frame the question of whose side we are on by examining science as a radically contextual, problematic venture with a very complicated social mandate, if any. Our purpose here is more than polemics; rather than valorizing or denigrating science as a monolith, we are taking an ecological view of work and politics. And we, too, are "outsiders within," as Woolgar's [1995] essay argues—both strangers and intimates in the world of science. Our work challenges the moral order of science and technology making—and in turn places us in a complex, often tense moral position.

Ecologies of knowledge[1] here means trying to understand the systemic properties of science by analogy with an ecosystem, and equally important, all the components that constitute the system. This is not a functional (or functionalist) approach, with a closed-system organic metaphor at its core. As Sal Restivo (1988) notes in his description of science as a social problem, we want to approach science as a set of linked interdependencies inseparable from "personal troubles, public issues and social change agendas," not a social structure with one or more *dysfunctional parts*. Science and technology become monsters when they are exiled from these sorts of questions (Law 1991; Haraway 1992; Star 1991a; Clarke 1995a), just as other symbolic monsters have been borne from the exile of women's strength from the collective conscience, or in the demonizing of people of color. In Michele Wallace's words: "The absence of black images in the reflection of the social mirror, which such programmatic texts (from *Dick and Jane*, to Disney movies, to *The Weekly Reader*) invariably contract, could and did produce the void and the dread of racial questions" (1990). In our exorcism, we simply want to see scientists and technologists as ordinary, as citizens, neither villains nor heroes.

Each of the authors who contributed to *Ecologies of Knowledge: Work and Politics in Science and Technology* (Star 1995) calls in different ways for an ecological analysis, including a restoration of the exiled aspects of science. Thus by *ecological* we mean refusing social/natural or social/technical dichotomies and inventing systematic and dialectical units of analysis. I think this reflects the dissatisfaction with conventional ways of approaching organizational scale and units of analysis, a dissatisfaction brought on no doubt in part by our respondents (scientists and technologists), who themselves are continually plagued by these questions. Croissant and Restivo [1995] examine a large-scale set of relationships, as between science and all other institutions and social arrangements. Kling and Iacono [1995] put the context back into analyses of technological innovation, noting that only such an approach can overcome simpleminded technological determinism or technocracy. Star [1995b] makes a similar point for the class of "artifacts" called formalisms, or formal representations, and their relationship to organizational complexity. Fujimura [1995] looks at the ecology of the workplace in a Hughesian sense, and thus goes beyond the simple adoption of new technologies as a determining factor in scientific social change. Law and Callon [1995], Woolgar [1995], and Latour [1995] each break the traditional boundaries of what can be included in an analysis of technology and social organization, recommending a broader, more democratic kind of analysis that is both moral and deeply ecological. In this, Law and Callon [1995] call attention to the local expertise of the scientist/engineers who, in their eyes, make no distinction between *technae* and *politaea* (Winner 1986). Woolgar [1995] refuses another "great divide" in so doing—that between individual psyche and collective repertoires of behavior. Clarke [1995b] challenges us to examine the stuff of science—the material substrate—instead of ignoring it in the service of idealist theories. Lynch's (1995) work extends this point to a revaluation of the idea of "place" in science, arguing instead for distributed, material "topical contextures" in which to examine scientific genre. Latour [1995] (writing as Johnson) uses the device of analyzing an ordinary, low-tech system—the door—and its surprising complexity to point to the inseparability of the technical and the social.

Our key questions here are those of general political theory and of feminist and third world liberation movements: *Cui bono?* Who is doing the dishes? Where is the garbage going? What is the material basis for practice? Who owns the means of knowledge production? The approach begins in a very plain way with respect to science and technology by first taking it "off the pedestal" (Chubin and Chu 1989)—by treating science as just something that people do together. Some of this means looking at science and technology as the occasion for people to do political work—not necessarily by other means, but fairly directly. Science as a job, science as practice, technology as the means for social movements and political stances, and science itself as a social problem—collectively, these articles take science/technology as the occasion for understanding the political and relational aspects of what we call knowledge. This introduction situates these papers with respect to science and technology studies (STS), and gives a sense of the problems they are addressing in social theory.

Most work in STS has not been seen as general social theory or as contributing in a fundamental way to social science theory. While historians, philosophers, and even computer scientists show a great deal of interest in the new sociology/anthropology of science, most social scientists view it as a kind of luxury, an arcane corner of the discipline offering only specialized insights. One purpose of this book is to demonstrate that social studies of science and technology are addressing a set of questions central to all social science. In this selection of science studies research, readers will see that science and technology are the vehicles for analyzing some very old questions. How do people come to believe what they believe about nature and social order? What are the relationships between work practices and social change? What is the trajectory of social innovations? Who uses them and for what purposes? As people from different worlds meet, how do they find a common language in which to conduct their joint work? How can we study people's work critically, yet as ethnographers or historians respect the categories and meanings they generate in the process? Finally, what are the boundaries between organism and environment; how fixed are they; how can we know them; and are they meaningful a priori?

Perhaps because learning another scientific language is a prerequisite for doing the kind of social studies of science/technology described herein, scholars in science studies tend to be an omnivorous bunch. We read history and philosophy of science as well as scientific tracts in the substantive fields we study, feminist theory, and *Science* magazine. We often borrow models from those writings as well as from other areas of social science. We work across national boundaries in informal groups clustered around an analytic topic: the use of metaphor, for example, or the extent of technological determinism or infrastructural change. The field is small (although growing), lively, and filled with debates, cross-fertilization, and often surprising collaborations. For example, sociologist Michael Lynch [1995], who is interested in visual representation, has collaborated with art historian Samuel Edgerton on analyzing the pictures astronomers create (Lynch and Edgerton 1988). Steve Woolgar became the project manager of an industrial computer development firm in order better to understand the process of technology construction. Diana Forsythe (1992, 1993), a cultural anthropologist by training, has worked for many years in collaboration with computer scientists building medical expert systems, both critiquing the notion of expertise and acting as a designer. Because the field is so interdisciplinary, the term *science studies* often replaces the disciplinary-specific label, such as *sociology of science*.

Historical Review

In the United States, most sociology of science before the late 1970s was dominated by the work of Robert K. Merton and his associates (see, for example, Merton 1973; Zuckerman 1977, 1989) and by a group of researchers conducting bibliographic citation analysis (from which work came the concept of "invisible colleges"; see Crane 1972; Mullins 1973a,b). While there had been much criticism of functionalist sociology on a number of fronts, particularly from symbolic interactionism and Marxism, the sociology of science received scant attention. Symbolic interactionists, for example, had studied work and occupations, medicine, deviance, gender, the family, urban life, and education, critiquing functionalism in each of these areas. Yet they had produced only a few scattered monographs and articles on science (for example, Glaser 1964; Strauss and Rainwater 1962; Marcson 1960; Becker and Carper 1956; Bucher 1962) and had undertaken no large research programs in this area.

Meanwhile, in Europe in the early to mid-1970s,[2] a group of researchers began a series of studies to demonstrate, contra Merton, that science was not "disinterested, communistic, and universal." Many of them had also been deeply influenced by Kuhn's *Structure of Scientific Revolutions* (1970), a book that had questioned the cumulative nature of science and raised the issue of the incomparability of scientific viewpoints or paradigms. They were concerned to show that science was not neutral, that the outcomes and content of science as well as access to it as a profession were determined by structural commitments, political positions, and other institutional considerations. MacKenzie's (1981) work on the interrelationships of statistics and eugenics is a good example of these efforts. These researchers were also concerned to demonstrate the constructed nature of science and its view of nature. Thus, they were strongly antipositivist. Some of this work was done at the University of Edinburgh, and the "interests" model became especially associated with the "Edinburgh School" (see, for example, Barnes 1977). Other important centers with overlapping approaches were in Paris and Amsterdam.

In 1979, Bruno Latour and Steve Woolgar published *Laboratory Life*, probably the book from the field that is most well known outside of science studies. The book was an ethnographic study of a scientific laboratory, and its purpose was to document the creation of a scientific fact. Using a variety of techniques from anthropology, semiotics, and ethnomethodology, they traced the birth of a biological fact in the context of lab work. They concentrated on a process they called "deletion of modalities," a progressive stripping away of contextual information about production, with the end result being a fact bare of its own biographical information. The book was an immediate success and was one of the factors helping spawn a series of laboratory studies and descriptions of act-making, often ethnomethodological in approach.

The combination of fieldwork and antipositivism was familiar to American symbolic interactionists, who welcomed the chance to apply these techniques to science and to learn from colleagues in Europe. A number of collaborations ensued among researchers in America, England, France, and the Netherlands pursuing these viewpoints. (See Fujimura, Star, and Gerson 1987; Clarke and Gerson 1990; Star 1992; and Clarke 1990, 1991 for reviews of this work.) Among our common interests and beliefs was the necessity of "opening up the black box" in order to demystify science and technology; that is, to analyze the process of production as well as the product. Methodological directives for those of us working in the interactionist tradition were familiar: Understand the language and meanings of your respondents, link them with institutional patterns and commitments, and, as Everett Hughes once said, remember that "it could have been otherwise" (Hughes [1959] 1971, 552). Many of our colleagues in Europe held similar views, albeit from very different traditions: Do not accept the current constructed environment as the only possibility; try to understand the processes of inscription, construction, and persuasion entailed in producing any narrative, text, or artifact; try to understand these processes over a long period of time (some of this work is represented in Law 1986; Bijker, Hughes, and Pinch 1987; Callon, Law, and Rip 1986).[3]

There were and are many other groups throughout the world studying science or technology: policy makers, historians, analysts of the impact of technology (particularly computing and automation), and the number is growing rapidly. Another important development began as programs of social science research gelled into science, technology, and society programs at a number of technical institutions and regular universities. New undergraduate and graduate STS programs began to spring up, both within traditional departments and as interdisciplinary programs. Early STS programs represented an amalgam of interests: ethics and values in science and engineering, studies of social impacts of technology, and history of science and technology. They were often an academic home for science criticism, that is, studies that demonstrated scientific bias (racism, sexism, classism) or danger resulting from scientific and technological research and development (nuclear and toxic wastes, recombinant DNA, technological disasters). Criticism of the sacrosanct institution of science and explication of the constructed nature of nature have remained core problems in science studies, and there is currently lively debate about the role of activism in the field, as Croissant and Restivo [1995] indicate.

Questions of Organizational Scale

Questions of organizational scale have always plagued (or some might say, graced) social science. Is social change individual or aggregate? How can we understand the relationship between social facts and individual experience?

These questions appear in science as it is interlaced with presumptions about the nature of scientific inquiry. If researchers accept that nature is simply "out there" waiting to be discovered, they may append to that belief the idea that "anyone can do it, geniuses better than the rest of us." There is nothing that logically ties these two together, but much of the received mythology about science involves great men (*sic*), great moments, great labs, and great accidents of Nature revealing herself. This combination of individualism, positivism, and elitism conspires to confuse the question of the appropriate level of organizational scale at which to conduct inquiry. So, the secondary literature on science is littered with psychologism, reified "societies" that act in mysterious ways on believers, and participant histories that claim exclusive centrality for powerful, rich institutions and people.

Against this trend is a lively debate, partially represented in the pages of this book, about the right unit of analysis for studying science. In escaping from the nasty things mentioned in the previous paragraph, sociologists and anthropologists of science have invented, borrowed, or transformed units of analysis from other parts of the discipline or from science itself: bandwagons, social movements, political economy and large-scale work organization, units of action and activity that cross human/nonhuman boundaries, the taken-for-granted truth about Nature that reflects old and widespread conventions (and superstitions).

How is the little black box of the computer, the test tube, or the door-closer joined with phenomena at larger scales of organization? This is a fundamental question about science and technology, but it appears whenever one questions the nature of local social arrangements as articulated with those at a distance or with considerably more power and purview. All of the articles [in *Ecologies of Knowledge: Work and Politics in Science and Technology* (Star 1995a)] attempt to answer this question ecologically and propose several modes for doing so.

At the largest scale of organization, questions are raised here about the utility and role of science or technology in maintaining or changing the status quo. This is asked not simply in terms of technological determinism, but in terms of larger scale issues, a central one being: Can there be a revolutionary science/technology in the absence of revolutionary social change in other spheres? To the extent that one believes in the interpenetration of spheres and science as a social institution of its historical time and place, the answer must be no. This puts the question of political commitment squarely at the center of science studies. For one thing, it is difficult to escape examining oneself as a scientist while engaged in studying scientists full time. Truly revolutionary science or technology thus means full-scale revolution. The sociology of science might allow us better to understand what that might mean.

Methodological Issues

Interwoven with questions of scale and politics are questions of method. Scientists are very challenging respondents. For one thing, all scientists share with us concerns about reliability and validity of data, robustness of findings, and the meaning of those findings. I never met a scientist who had not thought about the issues raised in this introduction. As a group of respondents, scientists are particularly difficult and rewarding because they have often thought rigorously about the issues we are investigating, and about which we are ourselves uneasy [Woolgar 1995]. So the work of our respondents blends with our own. The meaning of participant observation in this case can begin radically to change.

There are several kinds of work to be juggled in doing research in STS, each of them methodologically challenging. First, there is the map and language that the scientists themselves use in their work. Second, there is our mapping of the work practices and organization. Third, we create maps of the communications between domains. Fourth, a complex "nested" map is generated that shows who answers to whom, and why. It is at this level that questions of unit of analysis, or scope, often show up in force. Reconciling the different maps is a nontrivial methodological problem, again common across many domains of social science and political life.

Why I Am Not a Nazi: Realism and Relativism in Science and Technology Studies

One of the curious things about being in the STS field is that one is immediately plunged into philosophical debates about realism and relativism. Briefly, realism is the position that "there really is a there out there, and it's true in some absolute sense." Relativism holds that truths are relative to a place, time, or person (often a historical situation or geographic/cultural location). All researchers in science studies have had the experience of being challenged about the "underlying truth" of science: What about the scientific method? What about truth? What about the laws of gravity?

I have been involved in science and technology studies for about fifteen years, first as a science critic, then as a historical sociologist and ethnographer. I have given over one hundred talks on various aspects of the sociology of science and technology. In almost every presentation, I have been asked some version of what I now call the "there there" question: But are you saying it's *all* socially constructed? Doesn't that mean anything could be true? Isn't there anything out there? Are you saying that scientists are making it all up? Are you saying germs don't really make you sick, or gravity doesn't really make things fall down?

It is indicative of the central place of science in mainstream Western belief systems that *merely to imply* that the acquisition of scientific knowledge is work, not revelation, seems to involve the kind of radical idealism (if not radical autism) alluded to above. But this is not necessarily the case.

To say, as Hughes ([1959] 1971, 552) did about social order, that "it could have been otherwise" is not to say that it *is*. And to say that the conditions of nature or science are the result of collective enterprise that includes humans and nonhumans is not to imply that the merest whim on the part of an individual could overturn them. Rather, as social scientists, let's ask: Under what conditions do such questions about reality routinely get raised?

First, the term *socially constructed* is a reformist term, inserted into titles in sociology/anthropology of knowledge and science. Its initial purpose was to demonstrate that the reports of science that had been stripped of production history were missing important historical and situated accounts; second, to restore accounts of the actual work and its organization to those reports. Furthermore, if one takes "society" as the scientific problem, then the image of a society "out there" structuring an experience that is then entered into the canon of research doesn't make any sense either.

I call the idea that cumulative collective action is flimsy the "mere society" argument. It is paralleled by simplistic perceptions about, for example, socialization and gender. The argument goes something like this: "So, she's been socialized as a girl. Well, let's just let her into the corridors of power and de-socialize her, and then everything will be ok." Such a statement depends on a trivial and reified conception of both socialization and gender. Whatever bundle of actions, past and present, we might think of as "socialization" here is far more complex and durable than most of us realized in the early days of feminism. Similarly, the notion of "institutionalized racism" has been crucial in understanding that racism is not simply a matter of people not being nice to each other, nor necessarily to be found in a single set of micro-interactions—rather, it is a web of racist discourse and practices that extends through and informs all human practice—and cannot be simply transcended (hooks 1990). The durable bundle of actions and experience that comprise "science" has a similar sturdy complexity. This complexity does not defy its ontological status as "created," however. The constructivist or relativist schools in science studies (and I will not explicate the subtle differences between them here) have often been accused of flimsiness or mentalism on grounds that deeply confuse epistemology (how do you know what you know) with ontology (how are you what you are).

Scholars in science studies have disagreed about this issue and will continue to do so for some time to come. Yet a thread runs throughout the work of the groups represented in this issue: Let's replace the either/or dichotomy of constructed versus real with more useful concepts. Concepts such as workplace ecology, *irréductions* (Latour 1987), sociological imagination, networks and translations, and boundary objects (Star and Griesemer 1989) are important here. Wimsatt's (1980) concept of "robustness" has similarly been an important replacement for more restrictive concepts of reliability and validity. He, borrowing from biologist Richard Levins, defines it as "the intersection of independent lies," or more sociologically, the durability of collective action despite the fragility of any one instance.

During the 1980s there were scores of articles and books addressing this class of questions in science and technology studies. They have important links as well with earlier work in other parts of sociology and anthropology. For example, the debate in the 1960s about labeling deviance asked whether some things aren't *really* sick (or unnatural). The sociology of art has been concerned with the question of whether some things aren't *really* just beautiful (in a timeless or transcendental fashion).[4] The enduring concern with ethnocentrism in anthropology has recently exploded in debates about the place of the anthropologist and whether the knowledge constructed by anthropology is rightly seen as a jointly created fiction. In sociology and anthropology of medicine, the debate occurs as a question about whether one can differentiate physiological disease from "illness behavior"—aren't some things *really* just germs?

But the analytic trick in each of these cases is to raise the concept of "really" to the status of rigorous, reflexive inquiry and ask: *Under what conditions does the question get raised?*

One of the difficult things about trying to analyze an institution as central as science is that one challenges the received views of things for audiences and respondents. In giving talks that defend the above position, I have sometimes been called a Nazi, or parallels have been drawn between the social construction of science and Nazi science. It took me a while to figure out what people were talking about in these accusations, since being a Nazi is anathema to me.

If one takes the point of view that fascism requires a kind of situation ethics and requires that one redefine the situation according to opportunism or a kind of distorted view of science and nature, then any attempt to make relative any situations (especially natural or scientific ones) becomes morally threatening. This is so because one antidote to fascist ideology is to affirm an overriding value in human life, a universal value that cannot be distorted by the monstrosities informed by local, parochial ideologies of racism and genocide. Ethicists often base their arguments on this presumption. The worst thing for an ethicist is to hear arguments that plead "special circumstances"—the name of the game is finding good universals.[5] Yet this criticism of relativists as Nazis shows another kind of confusion, which again relies on a separation of the social and the natural and a separation between the conditions of production and the product. If the relative ontological status of a phenomenon is inextricably embedded in the conditions of production, then it's not a question of an analyst legitimating genocide or situation ethics. Rather, the question on a meta-level becomes: How can we make a revolution that will be ontologically and epistemologically pluralist yet morally responsible? Can we be both pluralist and constructivist, hold strong values and leave room for sovereign constructions of viewpoints? These are not new questions, either; both the French and American Revolutions were fueled by them. I would claim that there is stronger evidence for Nazism arising from ignorance of the conditions of production of knowledge than from exploring the relative configurations of these conditions in different times and places; more oppression from the appeal to absolute natural law than from negotiations about findings. While I'm not implying here that science studies is the best weapon against totalitarianism, the fact that this question arises so frequently in so many different contexts is to me indicative of the fact that science has been such an inviolate institution, certainly in academia.

Current Intellectual Development in STS

Taking on science as a social construction grew beyond either interest explanations or laboratory ethnographies by the end of the 1980s. Science and technology studies has over the past several years worked hard at two central intellectual currents, both of which are at the core of an ecological analysis of science (or perhaps, in some sense prior to it). The first is the establishment of science as materially based (see especially Clarke and Fujimura 1992a,b; Haraway 1989; Clarke [1995b]; Lynch [1995]); the second is science as a form of practice (see especially Star 1989; Pickering 1992).

It is remarkable for how long accounts of science (in history, philosophy, and sociology/anthropology of science) neglected to notice that much of the activity we call science consists of people manipulating materials, including specimens, media and cultures, breeding colonies, and display items. This material culture of science is important not just as another form of exoticism, but for the ways in which it is constituent of scientific findings and constraining of the ways we perceive scientific meaning.

Of Humans and Nonhumans

One of the issues that appears in different ways in the papers in *Ecologies of Knowledge: Work and Politics in Science and Technology* (Star 1995a) is the issue of "where to draw the line" in analyzing science and technology. Traditional studies usually drew the line at the edge of the black box, whatever it might be: the computer, the laboratory, the closed scientific work group. The argument in *Ecologies of Knowledge* is that science studies in the past have left out some of the most important actors, the "nonhuman" ones. Many of the new sociologists of science are engaged in a kind of democratization of this analysis, as the papers here demonstrate. If one adopts an ecological position, then one should include all elements of the ecosphere: bugs, germs, computers, wires, animal colonies, and buildings, as well as scientists, administrators, and clients or consumers (see Clarke and Fujimura 1992b and Latour 1987 for an analysis of this). The advantages of such an analysis are that the increased heterogeneity accounts for more of the phenomena observed; one does not draw an arbitrary line between organism and environment, one can empirically "track" lines of action without stopping at species, mechanical or linguistic boundaries, and especially without invoking a reified conception of society.

On the other hand, this kind of analysis presents some serious ethical problems—on both sides (Singleton and Michael 1993). For many years feminists, radical ecologists, and pantheists have recommended a kind of analysis that does not exclude anything from the natural world. The exclusion of animals, the biological environment, and other parts of the natural context has been one of the major sources of alienation under patriarchy, late capitalism, or religions that are antinature (Griffin 1978; Merchant 1980; Harding 1991). Restoring the natural world to the research context would be an ethical and political advance. On yet another hand, the papers by Kling and Iacono [1995] and that by Star [1995b] are written from within a research context in which it is not humans who have been privileged at the expense of nonhumans, but vice versa. It is computers and automation that have occupied a privileged position vis-à-vis human beings, often because of the inadequate social analysis held by computer movement advocates. An ethical social problems position in this case most likely involves checking the power of nonhumans and their advocates and seeing that humans understand it contextually, not democratizing the nonhuman position. Thus where Latour [1995] is concerned to restore ecology from one side, Kling and Iacono [1995] and Star [1995b] are concerned to restore it from another.

What are the moral values invoked by such analyses? I think that there are no simple answers. The dividing lines should not really be between advocates of humans and advocates of nonhumans, but between ecologists and reductionists. In furthering the cause of an ecologically responsible, socially and philosophically sophisticated analysis of science and technology, we need to confront head-on questions of scale, of boundary drawing, and of mystifying science and technology, *as well as* questions of race, sex, and class. To do that, recursively and reflexively, we need an ecological approach.

A recent collection of papers in the sociology of science highlighted this debate in the field (Pickering 1992). Collins and Yearley's (1992) paper in that volume accuses Latour and Callon of playing "epistemological chicken" in the interests of advancing the nonhuman analysis at the expense of the human. Callon and Latour (1992) respond with a defense of their position, claiming it as a heuristic analytic device that pushes the boundaries of science studies beyond reified sociological categories. Fujimura's (1992) paper in that volume, taking a symbolic interactionist/pragmatist and feminist perspective, rejoins with the claim that neither side has comprehended the human stakes involved, and that when the debate is phrased as humanists versus poststructuralists, once again concerns of all women and men of color, as well as other minorities, are ignored.

The debate between the British and the French, on the one hand, and Fujimura's (1992) claim that from an American pragmatist perspective the issues are misframed are important for the ecological analyses presented here. If we take ecological to mean treating a situation (an organization or a country or interactions and actions) in its entirety looking for relationships, and eschewing either reductionist analyses or those that draw false boundaries between organism and environment, then indeed the human/nonhuman question is reframed. The axes within the ecological space are four:

1. continuity versus discontinuity
2. pluralism versus elitism
3. work practice versus reified theory
4. relativity versus absolutism

On the left-hand, radical side go continuity, pluralism, and relative ecologies of work practice; the reactionary side is discontinuity (or divides great and small), elitism (or pretensions to a single voice), reified theory (or deletions of the work in representations of it), and absolutism (or "there really is a little bit of determinism").

A central fight within American sociology, and subsequently within sociology of science as discussed above, has been against functionalism, that school of thought which sees a closed-world, top-down, organism-like social order that draws its imperative from an imputed physiology-writ-large. The sociological field in America is mined with "hot spots" that come from the scars of these historical battles. From the pragmatist side are the words *consensus*, *boundary maintaining*, *natural*, and *obvious*.

The fights between the British and the French resonate along these axes in three ways. First, the relativism of both schools satisfies the pragmatist concern with pluralism. They both imply that there are frames of meaning, definitions of situations, different perspectives based on experience. Due to the long history of fighting that pseudosingular voice, such pluralism became the salient relativist dimension for pragmatists/interactionists in STS. The fact that it also has profound resonance along the axis of nature/society, people/things, and so on doesn't carry as much historical weight, for the reason that pragmatists never believed in that divide in the first place. John Dewey and Arthur Bentley, for example, spent their entire careers fighting these notions in analytic philosophy; they appear as similar fights in the work of symbolic interactionist sociologists influenced by them, such as Howard Becker and Anselm Strauss. So the work of Collins and Pinch (1982) on the different valuations placed on parapsychology and psychology resonates with that commitment to pluralism, too, seeming to restore the voice of the underdog to scientific debates and balance it out— to even out the "hierarchy of credibility" discussed above.

The French actor-network theories, and their emphasis on the inclusion of nonhumans [see Latour 1995], find a match in the pragmatist concerns about continuity and process. Because the mandate of the pragmatist research program since the 1920s has been to "follow the actors," it is not surprising that there have developed strong ties between French researchers and American symbolic interactionists.

A central tenet of the pragmatist work in STS has been to think of scientists as people who are doing a certain kind of work. Among other things, science is a job. It's a very interesting one, because it turns out that even *calling* it a job invokes the wrath of American functionalists, most philosophers, many deans and administrators, and most computer scientists. Simultaneously, this vision of science as work invokes the appreciation and support of many historians.

As a symbolic *interactionist*, I agree with Callon and Latour and Lynch and Woolgar that new methods that will lead us across traditionally accepted boundaries are crucial, and those that will help manage the "rich confusion" of things and people are absolutely critical for science studies at this time. Because of my pragmatist concerns with work, I would add work itself to the rich confusion, in the form of activity, practice, and/or work organization. I would also like to emphasize a neglected dimension in the worlds of STS and science/technology: the great divide between formal and empirical.

Work, Formalisms, and Divides

One of the most confusing actants in this complex ecology is the work of scientists and technologists who create formalisms,[6] including those working with information technologies. The impact of STS in these spheres is not so much limited by concerns with relativism/realism—indeed, here Latour's early point about scientists not being naive realists is absolutely true. But many scientists I've known are naive formalists, especially those in information technology and computer science. One thing about computing technology is that it allows one, paradoxically, to use a very concrete thing to manipulate representations that are quite formal. The great divide that computer science itself then produces is between the formal and the empirical—this is reproduced many times across the sciences, including social sciences. On the one side are formalists who believe that computers are embodied mathematical theories—theories come to life. (I do not exaggerate here; if anything, I am understating the case.) On the other side are engineers who argue that *only* making things (e.g., machines, programs, new speed records) really counts for technological advance. And on a third (much, much smaller) side are a few brave souls who argue (mostly for reasons of safety or ethics) that empirical studies must join together inseparably with formal models. They are as yet few in number, and have suffered enormous academic stigmatization for their stance. Some details of the debate in computer science itself can be found in Newell and Card (1985, 1987) and Carroll and Campbell (1986). These pieces and Fetzer (1988) give a sense of the vituperative flavor of the debate. Star (1992) reviews the work of people in several disciplines working in this "third force" (see also Star 1995c).

The formal/empirical canyon is a complex and compelling great divide (see Bowker and Sankey 1993/1994 for a discussion of the issues), one that is only beginning to receive attention in STS. And here are the stakes. A formalist would argue that when building air traffic control systems, human fallibility and bias is such that we are virtually killing people to rely on it. The computations are so many and so intricately interconnected that only a machine can be smart enough (statistically, and there's the rub) to do them fast enough before the (now much more complex) airplanes fall out of the sky. The empiricist argues that all such mathematical systems are unprovably fallible (using strong formal allies like Gödel and Schrodinger), too big or too dangerous to test, and that we'd better stop the reliance on formal testing or we will all (literally) be blown sky high by Star Wars or a series of high-tech accidents.

A significant divide indeed, since it captures all life forms in its technological threads. . . .

The formal/empirical divide is also richly represented in social sciences. On the one side, in American social science especially, there are formalist fundamentalists who believe that life "really is" a mathematical model, and that empirical data axe incidental at best to its representation. Only quantitative truth matters. On the other are empiricist fundamentalists who sneer at numbers, algorithms, or other sorts of formal models.[7] Star's [1995b] and Woolgar's [1995] papers explore ways to refuse this great divide of formal versus empirical in computer science, psychology, and in social theory more generally.

As both a scientist and a citizen, I have a great stake in closing this divide. I do believe that the stakes are as high as the empiricists in computer science claim, although I don't think they are as centralized as many of them envision. Rather, I think the consequences of maintaining the formal/empirical divide are highly distributed, and reside as much in forms of bureaucracy, education, and exchange as in bombs and air traffic control systems.

The Status of Matter and the Absolute

I have on occasion taught Latour's *Science in Action* (1987) to scientists. The book's central tenet is that "nature" is nowhere to be found apart from the web of work and inquiry constituting the relations of science. To my initial surprise, the class discussions often became theological in nature. It seems my students who are in the sciences aren't afraid to use words like *God* or *soul*. However, among social scientists such discussions seem to be completely taboo. It's easier to talk about sex or excrement or almost anything than to talk about one's deepest spiritual or metaphysical beliefs. (I suppose in the United States at least it's because religious fundamentalists conjure up images of antiscience, antifeminist fanatics.)

There are a set of questions in STS of science that resemble metaphysical and theological questions throughout the ages. Beyond the questions of humans and nonhumans, and of formal and empirical knowledge, many of the deeper issues in the debate seem to reflect divergent opinions about the status of matter itself. The questions go something like this: A couple of years ago I gave a talk at UCLA about my work in collaborating with computer scientists in artificial intelligence. I talked among other things about the primacy of distributed (cross-personal, organizational, or community) cognition. Harry Collins was also there, and as "devil's advocate" asked me a question: "Agreeing that cognition is social, isn't there a limit to that? How do you explain, for example, that you could wake up in the morning speaking English, go alone into a locked room, and come out at night still speaking English? Doesn't that imply that there's some cognition only in the head?" At the time, I didn't have much of an answer for him, but in thinking through the human/nonhuman debate for *Ecologies of Knowledge: Work and Politics in Science and Technology*, I now venture a bit of translation.

First of all, when speaking of the brain in that fashion, one is implicitly speaking about matter, about physicality. Traditionally, people have had a difficult time modeling or speaking of "brain" and "thought" without invoking one of the original great divides, that between mind and matter. If we want to refuse that great divide, we must thus carefully examine how we think about matter itself. This is perhaps the microarchitecture of an ecological analysis. If we are to go beyond the current debates, STS researchers must confront this basic dichotomy and not allow in "a little bit of determinism" here and "a little bit of realism" there.

Think of matter as composed of arrangements of space and time—some very, very fast, as with light, some very, very slow, as with rocks. In an Einsteinian/quantum mechanics fashion, this matter has no absolute speed or rhythm. Rather, its rhythm and speed derive from its context. The rearranging of space-time configurations is a constant, never-stopping process, although some speeds are too slow for us to perceive as anything but stopped.

Another way in which this rearrangement works is as a relative location. Analytically, it is extremely useful to think of human beings as *locations* in space-time. We are relatively localized for many bodily functions and for some kinds of tasks we perform alone. But for many other kinds of tasks we are highly distributed—remembering, for example (Middleton and Edwards 1990). So much of our memory is in other people, libraries, and our homes. But we are used to rather carelessly localizing what we mean by a person as bounded by one's skin. Pragmatist philosopher Arthur Bentley cautions against the philosophical contradictions this brings about in his brilliant essay, "The Human Skin: Philosophy's Last Line of Defense" ([1954] 1975). The skin may be a boundary, but it also can be seen as a borderland, a living entity, and as part of the system of person-environment. Where the skin is, indeed, under some conditions becomes a very interesting question. But as an unthinking, linelike division between inside and outside, where "self" is on the inside, it makes no sense philosophically. Parts of our selves extend beyond the skin in every imaginable way, convenient as it is to bound ourselves that way in conversational shorthand. Our memories are in families and libraries as well as inside our skins; our perceptions are extended and fragmented by technologies of every sort.

All the matter in our body can be thought of this way, including the brain. In 1896, John Dewey wrote a critique of reflex psychology that still stands today ([1896] 1981). He noted that the common image of psychology was that a stimulus would happen, "go in" to the brain, stop there, be processed, and something would come back out. This was complete nonsense, said Dewey. It doesn't "go in" through nothingness . . . there is an event that changes the air, interacts with skin, with nerves. It is continuous, and there is never a time when it "stops." The arc is a convenient notation for a dualist, reductionist psychology, however, and makes certain things amenable to quantification.

I reiterate Dewey's critique with respect to cognition and the individual, and recommend it to researchers in STS. Learning English is a series of continuous events, of changes, rearrangements in the space-time of your body. Once the process gets going it keeps on going, given constant interactions with other people and all kinds of humans and nonhumans in the world. I don't know enough about death to know whether or in exactly what forms it might keep going afterwards, except that the ongoing actions we leave embedded in the world constitute one such action; for example, the books we write may be read after our deaths.

So the alone person in the room speaking English before and after is one case of a time of aloneness (a typically very short time, otherwise it would be solitary confinement, the ultimate penalty in most cultures). The ecological image for the aloneness of that learning is analogous to holding your breath—you still need oxygen, but you take your lungs "out of play," or put them on hold, for a moment. So the alone person is aside in the sense of not being together with others. Aloneness seen this way is not a vindication for mentalism or for the primacy of the individual; it's just a special case of relocating.[8]

There are historical and contemporary neurophysiologists who view the brain this way, as well as the brain-body-environment. The images have a wide range. Some have seen the brain and cognitive functioning as a "re-entrant, emergent process," where that which is sensed keeps circulating in the brain forever. Others find that we can't process sound without hearing what is before and after it—perception is entirely relative with respect to context. Once something is perceived, the action of perception continues indefinitely, changing and being changed by other events near it, sometimes resonating, sometimes clotting up or clumping up, sometimes fading into background noise.

To think this way we must vastly complexify the way we think and talk about matter. The brain is not a lump of meat with a few electric channels strung though it. The body/brain of any one person is a location of dense rearrangements, nested in like others. When we use the shorthand "individual" or "individual cognition," we are thus only pointing to a *density*.

Thus, in speaking of aggregates of peopled, material ecologies (including in them things, built environments, the natural world), we have a basis of resolution of the realism/relativism dichotomy and of the formal/empirical divide. An ecological analysis refuses to ground beliefs, including scientific beliefs, in something outside of this webbing location.

Moral Implications

What are the moral implications of this view? Scientists certainly don't hesitate to address such concerns in bringing up these issues, and neither should we in STS. Sociology and economics began as moral ventures, addressing this class of problems with respect to the place of human beings on earth, and of the prescriptive nature of our relationships with each other. From this evolved a long investigation of those relationships, in the middle of which we seem to have forgotten that the whole purpose of the enterprise was to speak responsibly, in a disciplined and collective fashion, to Very Big Questions. What is a moral order? What are values and passions, and how are they arranged? What do we owe tradition, and what do we owe innovation? In the case of STS, this includes the question: What is matter?

Activity theorist Yrjö Engeström (1990) argues that technology occurs when joint activity between two actors is articulated. It does not preexist such action, and as the tool occurs, it comes to form part of the subsequent material conditions mediating further action. Material conditions are not only such things as stuff, money, climate, and bodies, but also refer to *durable arrangements* that have consequences on the trajectory of action as material conditions (e.g., Butler 1990).

In order for this not to be read as an idealist statement, one must firmly resist great divides between individual and environment, between technology and knowledge, and between language and thought. *The emphasis is on use and consequences, not antecedents*. And the result is an inversion of everyday thought. It requires resistance to the functionalist defaults of presumption. So ethnomethodologists say that "constraints are also resources"—or feminists observe that structurelessness can be tyrannical (Freeman 1975)—or pragmatists that "things perceived as real are real in their consequences" (Thomas and Thomas [1917] 1970). Let me say it another way: If you think of matter (including people) as a space-time arrangement in the way I've described it, and you also then think of rearrangements or reconfigurations in those arrangements as having consequences, you can easily come to a very political but nonreductionist and nonpositivist account of moral orders. There are some very slow moving/ very large-scale, quite general standing arrangements and settled questions constituting multiple moral orders that, taken ecologically, constitute what we think of as "societies."

I think Collins and Yearley (1992) have a quite legitimate fear that including nonhumans in an undifferentiated way threatens our moral order (and in particular our moral order as social scientists). The very real image behind the passion in their critique of Callon and Latour is, well, does a cat have just as much right as a human being? Are we going to anthropomorphize machines in a nonchalant way so as to render our moral critiques worthless? Aren't we either like silly pantheists on the one hand running around in meadows all day worshiping daisies, or like grim mechanists on the other, giving primacy to machines and their attendant constraints?

I think we need to say that such criticisms occur in the presence of deeply anti-ecological power structures in academia and government, and they are the "third force" or silent partner in the argument coming from Collins and Yearley (1992) against Callon and Latour (1992). They seem to be saying, If you let anything into the analysis, and give away the important differences between humans and nonhumans, you are throwing away our birthright as discussants of moral order—our "birthright as social scientists." The debate becomes confounded at precisely this juncture. Callon and Latour (1992) insist that they are not "leveling" between humans and nonhumans, merely including. From the pragmatist point of view I must ask: Under what conditions, and for whom, does this inclusion imply leveling, nihilism, and claims to a lost birthright?

One of those conditions does have to do with giving up a privileged position, as indeed Latour [1995; see also Callon and Latour 1992] states. And perhaps this is another place where pragmatist feminists make a contribution. We didn't come into the academy, or science, or social science, as insiders. We come from a numerically small, "holdout" tradition in American sociology that has had pluralism as its deepest commitment, as that from which all sorts of relativisms derive. We were among the first generation with more than a token number of women in our positions. Intellectually, our tradition never thought it had a better way of looking at things than our respondents; our whole social science was fashioned on the premise that *they* were making the news, and we were only reporting it. As long as we were talking about drug addicts or prostitutes, such sentiments seemed acceptable for many liberals. But now that we're talking about scientists, we threaten some basic theological commitments. We honestly believe that there are no positions that are epistemologically superior to any others. But I do at the same time argue with and try to overthrow those I don't agree with! Relativism in this sense does *not* imply neutrality—rather, it implies forswearing claims to absolute epistemological authority. This is quite different from abandoning moral commitments.

We're Scientists Too

This brings me to two last notes about the relationship between scientists and STS. First, I have said above that I am a scientist. I notice that the way we talk (I, too) about scientists is usually as "them." Why? We share the same fate, use their results, enter into dialogue with each other, and even go to each other's conferences and try to live in each other's departments.[9] Others have noticed this in other ways—in STS the reflexivists, and in anthropology the so-called new ethnographers. Fujimura (1991) has written an analysis of this movement from the points of view of pragmatism, feminism, and antiracism, emphasizing the importance of multivocality (pluralism) that is grounded in experience, practice, and identities.

I think the reflexivists and new ethnographers are moving in an important direction, which is to see ourselves as cocreators of our scientific narratives, and to place ourselves actively in those texts (see Denzin 1989 for a methodological explication of this in social science). But it's a difficult undertaking. In analyzing reflexively, they/we are addressing, among other things, the conditions of their-our own work. And that is dangerous in several ways. The first is simply in overestimating the impact of academic politics. Fujimura (1992) notes that such reflexivity can become "academic politics" veiled as moral reform. Another "silent partner" in this argument, which she makes explicit, is feminism. For those of us who "grew up" intellectually in feminism, it is a little bewildering to encounter in science studies the attitude that gender is one of those boring, reified categories like "society" itself, which must be endured but not mined. Feminist theory is not about civil liberties, or rather, only in the sense that it completely redefines the notion of civil liberties. It began in its recent Western incarnation as a movement about exclusion—"we need more women in *x*"—or about barriers. But the central, exciting parts of feminist theory concern exactly those issues raised here. We have an example of a personal, often private, pervasively and somewhat unevenly distributed phenomenon: the oppression of women.

Just documenting this was not an easy task. However, when done to the exclusion of other kinds of analysis, especially in the academy, it can indeed be boring. But the exciting part of feminism has been to invent ways to *see* this phenomenon: to understand the role of invisible work, to articulate the asymmetries between listening and speaking, between hearing and agreeing, to form a political reform program that was simultaneously collective and private (consciousness raising and action), to struggle with extant categories and define the subtle ways in which coalition becomes co-optation.

There are two major parts of feminist scholarship and activism that have been largely absent from science studies: community and spirituality. For feminists, the community aspect was especially important for the early years of theory formation as well as for mobilizing social movements. The result was a highly interdisciplinary development, women's studies. At first, feminist scholarship was so beleaguered, so new, and so dependent on emerging community that people from all fields were welcome. So the analysis of poets was equal to if not greater than that of scientists; the experience of an eighty-year-old woman as important for the critique as that of a twenty-five-year-old (although this certainly not without its own struggle), and there was an incredible heterogeneity in the sorts of analysis brought to bear on issues. There was an openness in figuring out the *questions* that was very important for maintaining the scope and vision of the movement as these questions shifted into scholarly arguments. This was coupled with an important inclusion and participation of feminist theologians—Mary Daly, Carol Christ, Rosemary Ruether, Nelle Morton, and others—who were not afraid to tackle questions of God/dess, the Absolute, power, imagination, and so on. These were taken up seriously as questions, discussed and debated; feminist scholars read all sorts of writing. We attempted simultaneously to redefine knowledge, politics, family, race, community, nature, and spirituality. We created experimental forms of writing and of worship, and seriously rethought our commitments to the most basic of questions.

There were, of course, large parts of the women's movement that converted these questions into silly mysticism, or called them silly mysticism. And there were scholars who went about the business of converting gender into just one more variable in a positivist pantheon; "Women in X" became a favorite career-building strategy for some. But for many others the scope of the questions remained undiminished. I think that even for women who thought "feminist spirituality" was nonsense and a detraction from bigger issues, such as those of class and capitalism, the dialogue about ultimate questions remained alive, and long-term survivors from both sides came to appreciate the power of the questioning as the "daily-ness of moral, ethical and political conundrums became clearer to us"[10] (see Anzaldúa 1987).

What does it mean, then, to include yourself in debates about such things? In the feminist movement, answers emerged in the form of communities, committees, commitments, and combats, as well as a new field of scholarship. When we say "self-inclusion" in STS, what does it mean? We are tackling one of the most important, widespread institutions of our time—science and technology. And if this doesn't mean social reform of some sort, what, by our own analysis, *does* it mean? (By social reform here, by the way, I include reforming the division of labor in the academy—more on that below.) Where are our communities and commitments? What sort of moral order are we building or trying to reform? I think the problematic of including ourselves in the analysis can be viewed as a problem of work organization, too. Why not just include it as another kind of observation? Why not adopt the French stance of including everything democratically (which to an American most emphatically also does *not* mean without difference)?

One of the most important philosophical tenets of pragmatism is that it is the consequences of an action that constitute logic and belief, not putative antecedents (which are anyway unknowable). So from my point of view, we can't know about the consequences of including ourselves in the analysis until we try. But reflexivity does not work at arm's length. Rather, it implies radical change quite close to home (even if not *in* the home), and the consequences of working for that are serious.

At the same time, for us as feminists, differences such as between human and nonhuman, race and ethnicity are not abstract. The *lived*, experienced differences as embodied in specific locales and moments, and communities, are central (Star 1994).

Studying Nonhumans: A Professionalization Movement in STS?

With respect to nonhumans, there are similarly large questions that may be viewed politically and pragmatically. Some people have full-time jobs observing or tending nonhumans. We call these people scientists, laboratory technicians, data entry clerks, various kinds of technical monitors, some computer scientists, and still-life painters. Some people have full-time jobs observing or tending humans. We call these people baby-sitters, parents, teachers, attendants, some kinds of servants, and old-fashioned sociologists. Perhaps most people have full-time jobs observing a "rich confusion" of humans and nonhumans: some computer scientists, dressmakers and tailors, bus drivers, doctors and nurses, most scientists, police, literary critics, some of the new sociologists of science, and so on. I confuse the points here in order to make a point. As Latour has beautifully argued in *Science in Action* (1987), we *all* have full-time work that is interacting both with humans and nonhumans—the mingling is inescapable.

The constitution of anyone's work is a mixture of human and nonhuman which can be analyzed ecologically. But the nature and quality of that composition will reflect back on the organization of the work itself in important ways. So in this, I consider myself an amateur nonhuman-watcher as compared with my license and mandate as a sociologist to be a people-watcher. To change the ecological mix with respect to my work organization means changing the organization in which I work. It is not merely an exercise of imagination, but a real political risk. And that brings us back to self-in-the-study: it's pretty dangerous to do so. You walk across so many great divides that your feet develop webs . . .

But let us consider the movement to include nonhumans as a case of amateurs in a line of work trying to professionalize. We have as a group rather casually (even as passionate amateurs, compared with professional scientists) observed nonhumans. Some of us are making ready to go from amateur to professional status. In so doing, we should draw on lessons from other professionalizing groups, within and outside of academia.

And like all professionalizing movements, we face obstacles in the form of those who already claim the full-time territory in our own organizations (in this case, the university). But you could look at all forms of moving from low-percentage nonhuman watching to high-percentage in these terms—for example, the early years of computer science and the automation of various kinds of work.

What happens in a professionalization movement? I'll be brief: To establish a profession you need several things. First is a license and mandate from those whom you serve (Hughes [1959] 1971). You need to have research to legitimate the movement and professional societies to secure it. You need to seize the means of evaluation for yourself. And you need to be in control of methods. Typically, resistance from the old full-timers who control the domain comes in the form of ridicule, gatekeeping of positions and other resources, the invocation of various great divides, and so on. The key to success in professionalization movements is to develop good infrastructure—training programs, methods, technologies—and to offer a new kind of service. I think, being a pragmatist and a little reflective, that we might even be able to improve on the nasty aspects of professionalization—for example, use the phenomenon of professionalization in order to understand the reaction to our proposed new organizational order, and eschew the elitism and secrecy that too often go with professions. In our studies of science, we try to include the voices of people not traditionally heard in accounts of science: lab technicians, sponsors, administrators, spouses (usually wives), and consumers. In our studies of science we are also trying to observe and account for the nonhumans not traditionally heard, in ways not often practiced by those in power. Following Abbott (1988), as well, we can speak of a *system of professions*[11] rather than stand-alone entities. Sciences and technologies cannot be separated analytically from professional governments, from medicine, or from any other profession.

Let's adopt a stance about this: Rather than simply creating a professionalization movement with respect to nonhumans, and thus becoming scientists in a way that often makes us uneasy, let us change the way science is organized. And it is here that the authors in *Ecologies of Knowledge* have the most to say. Perhaps one way to characterize their voices is as a reclaiming of the term *network* from some of its unfortunate discontinuous connotations and affiliations. A web is composed of filaments, and a seamless web should be an oxymoronic term. There's no empty space in a seamless web, but our image of *network* is that it is filaments with space between. For this reason I prefer ecology. Let's use networks- without-voids for an ecological analysis. And in that ecology, let us be epistemologically democratic, including toward our own work organization.

There's so much to be done.

Information Technology: Hope and Fun, Hype and Danger

I live in the summers in Silicon Valley, California. My closest friends for years have been "hackers," many of whom occupy highly paid positions as systems programmers in computer design firms. In the 1980s they rode the remarkable growth boom of The Valley, switching jobs frequently for higher paid and more scientifically interesting situations. By the beginning of the 1990s, the recession had cut deeply into California life. Companies like Apple and IBM are firing vast numbers of people. Like so many others in the United States, even the top programmers and systems developers in Silicon Valley these days are often "working scared."

Outside work, my friends join progressive organizations, working for feminist, gay rights, and antiracist causes, and embrace ecology as both movement and fundamental ideology. They opposed Star Wars in the 1980s and continue to work with organizations such as Computer Professionals for Social Responsibility. They are all avid readers of science fiction, and some are active participants in science fiction fandom. In more than the economic sense, they are working scared. As much as any group I know, they grasp the double-edged significance of the technologies they are making and how embedded they are in life in the United States these days—the visionary possibilities and the fun, on the one hand, and the dangers and violence on the other. All the computers in the world will not clean up the terrible pollution in California. We know now that dreams of telecommuting as panacea were flawed in many ways [see Kling and Iacono 1995]. The "neat toys" that my friends create for peaceful, medical, or leisure purposes may come to form the substrate for surveillance or weaponry—or for world peace. They continue to dream and to work and to hope, living with the contradictions as do we all.

My good friends—the Tuesday dinner crew—have taught me more than all the laboratory studies in the world about the links between politics and playing and working in science and technology. I would like to thank them for that (and for the good food, Linnea), with the hope that we will all live to see our better fantasies realized, and not our worst nightmares.

The Cost of Opposing Hype
In boom times there is a terrible cost to opposing a bandwagon, the mirror image of Fujimura's analysis of the blandishments and career advancement prospects for joining

one (1987). Silencing and isolation are among them. This summer I was interviewed by *Science* magazine about some studies I had done of computer systems currently being designed for scientists. I had explained some of the difficulties of responsible design of such systems, and that many scientists, had told me that they did not like or use such computer systems, even though they liked our carefully designed system. I spoke to the reporter of potential problems, and of my fears that the system I was evaluating might increase structural disparities between rich and poor labs, even while I hoped that it would level the differences. The resulting story was published with no mention of my part of the study or my comments; an apologetic e-mail note to me from the author recently said that there had been no room, and that perhaps some of my words would appear in a later article on computing and the humanities. The whole tone of the article was celebratory of such systems, full of visionary language of the sort criticized by Kling and Iacono [1995] and by the women of the Women, Information Technology and Scholarship group at the University of Illinois (Taylor, Ebben and Kramarae 1993).

In an atmosphere of expansion and bandwagons, there is little room for complexity and caution. Perhaps science fiction, poetry, and film *are* the best possible subversive vehicles for countering the tsunami of computer hype—for the sociological imagination (Star 1995c). I wrote the following poem after reading an adulatory article on virtual reality.

The Net
"we are part of the records we keep"
—Gayatri Spivak

i
network
and the word flares trumpets
shining webs
connect me
dissolving time and space
network
soaked with information
all there is to know
the little wire
next to my bed
network
the net
work
feeding the teenage son
waiting in line for the pass
word

shaking the numbness from the shoulders
the arms

ii

my best friend lives two thousand miles away
and every day
my fingertips bleed distilled intimacy
trapped Pavlovas
dance, I curse, dance
bring her to me
the bandwidth of her smell

iii

years ago I lay twisted below the terminal
the keyboard my only hope for work
for continuity
my stubborn shoulders
my ruined spine
my aching arms
suspended above my head soft green letters
reflect back:
Chapter One
no one can see you
Chapter Two
your body is filtered here
Chapter Three
you are not alone

iv

oh seductive metaphor
network flung over reality
filaments spun from the body
connections of magic
extend
extend
extend
who will see the spaces between?
the thread trails in front of me
imagine a network with no spaces between
fat as air
as talk
this morning in the cold Illinois winter sun
an old man, or perhaps not so old
made his way in front of a bus his aluminum canes inviting spider thoughts

a slow, a pregnant spider
the bus lumbering stopped
and in the warm cafe I read of networks and cyborgs
the clean highways of data
the swift sure knowing
that comes with power
who will smell the factory will measure the crossroads
will lift his heavy coat from his shoulders
will he sit before
the terminal

v

it's too late for romance
the chestnut tree blooms no more
the com and pigs in this vast flat place
travel the network too
their genes secure in stock indexes
it's too late for bitterness
but still there is a space
in the net
a choice of cyborgs
oh brave new world
for the courage to choose the mundane
the rough wool of a winter coat
draped over an old back
a smell, a feel of her hair
the unfamiliar intimacy of the dancing letters
literacy
or survival

vi

am I the only one who strokes the scars
the Frankenstein neck
who wonders
when the stitches will come
out

Conclusion

Each paper in *Ecologies of Knowledge: Work and Politics in Science and Technology* (Star 1995a) addresses issues common across a variety of domains in social science and for wider questions of social order. Fujimura [1995] extends Hughes's ([1959] 1971) concept of the ecology of the workplace to study the creation of standardized "packages" in scientific work and in biotechnology research. Woolgar [1995] raises the notion of the nature of moral order in the doing of our own work, and the threats posed by crossing traditional boundaries. Lynch [1995] defines genre and representation with respect to a context of textures, conventions, and work practices. Star [1995b] treats formalisms as a form of technical artifact, and raises ethical and methodological questions about their impact in very large systems.

Kling and Iacono [1995], drawing on literature from social, professional, and scientific movements, look at the symbolic and political uses of computers to create "computer-based social movements." Several of the papers (theirs, Croissant and Restivo's [1995], and Star's [1995b] ask: Are science and technology themselves social problems? All eschew scientific or technological determinism and look at how science and technologies are being used in the service of various social movements and structures. All find a conservative effect. Neither the so-called computer revolution nor the scientific revolution has been very revolutionary for many of our lives, certainly not in the sense of moral and political order. Restivo (1988) uses this observation to propose a return to Mills's sociological imagination, and to ask important questions about widely held implicit assumptions, including those held by researchers in STS.

Law and Callon [1995] place the analysis of technological innovation and its attendant failure or success at the heart of social change as well. They concern themselves with fidelity to the categories of the actors; this includes a methodological demand for "heterogeneity." We should understand scientists and engineers on their own terms, as engineers of the social as well as of the technical. The failure of the group they studied to create an "obligatory point of passage" within their organization raises a larger question: What does it take to become a gatekeeper? What kinds of autonomy can be negotiated in technological design and implementation? Bruno Latour's (aka Jim Johnson's) piece [1995], for all its humor, presents a serious philosophical challenge: Where are the boundaries of technical phenomena? What happens when we draw analytic boundaries in unconventional places or when we take the methodological mandate of "following the actors" to include everything in the site chosen, including mundane things? The processes of inscription and pre-inscription that he describes are important tools for understanding technological determinism and social change.

Adele Clarke [1995b] as well follows all of the actors in the setting of reproductive research, describing a regime of materiality embedded in the very stuff of scientific work. The mundane materials of physiological work become the key pivots upon which findings revolve; only by understanding this previously ignored concrete "substrate" may we get at abstract theoretical developments.

Many great divides have been rejected, crossed, and perhaps will be closed by STS research. I hope that *Ecologies of Knowledge: Work and Politics in Science and Technology* (Star 1995a) will contribute to that process—that our hopes and our fun, not our fears and violence, will inform the closure.

Notes

1. After writing this piece Adele Clarke brought to my attention an article by Charles Rosenberg (1979b) with a similar title—and I trust a complementary approach!

2. I am loosely grouping apples and oranges here for the purpose of creating a coherent narrative. There were functionalists and antifunctionalists in both Europe and America, of course, and many different schools of thought represented by this brief description. I do not have space here to detail the contributions of American ethnomethodology, French history and philosophy of science, or Marxist and anarchist analyses, among others. See Bowker and Latour 1987; Lynch [1995]; and Croissant and Restivo [1995].

3. I am grateful to Françoise Bastide, Geof Bowker, and Bruno Latour for discussions of these issues.

4. I am grateful to Howie Becker for pointing this out.

5. I should note that it is precisely this quest, and these terms, that some feminist ethicists are attempting to revolutionize. See Addelson 1991.

6. That is, artifacts such as formal models, algorithms, formulae, quantitative simulations, and so forth.

7. It is important to note here that formal models need not be mathematical. Rather, they are an attempt to explicate the interacting constraints of a set of conditions. In that sense, the work of the American pragmatist sociologists such as Hughes, Blumer, Becker, and Strauss, and their students, has some extremely formal elements. However, this has been combined with a reliance on empirical data—the pragmatist attempt at refusing this particular great divide. (See Glaser and Strauss 1967 for the first volley in this particular version of the war.)

8. We have a paucity of language for this which does not recommit some version of the great divides. The only somewhat adequate language for this I have heard is theological analysis, which would refute the distinction between the immanence and transcendence of God/Goddess.

9. Here I must note, with varying degrees of exhaustion and success, but that is another story.

10. Thanks to Adele Clarke for these words.

11. It is important to note that *system* here does not mean a functional system, or a closed system of any sort. Rather, *system* means a set of interrelated contingencies, whose scope is unknown and that is quite open, both analytically and with respect to other contingencies entering in. *Ecology*, again, in this sense, is not a closed ecosystem, but rather a term that emphasizes the open interdependence of ongoing processes. It is not, in other words, a functionalist description.

References

Abbott, Andrew. 1988. *The System of Professions: An Essay on the Division of Expert Labour*. Chicago: University of Chicago Press.

Addelson, Kathryn Pyne. 1991. *Impure Thoughts: Essays on Philosophy, Feminism and Ethics*. Philadelphia: Temple University Press.

Anzaldúa, Gloria. 1987. *Borderlands/La Frontera: The New Mestiza*. San Francisco: Aunt Lute Books.

Barnes, Barry. 1977. *Interests and the Growth of Knowledge*. London: Routledge and Kegan Paul.

Becker, H. S. [1967] 1970. "Whose Side Are We On?" In *Sociological Work: Method and Substance*, ed. H. S. Becker, 123–134. New Brunswick, NJ: Transaction Books.

Becker, Howard, and James Carper. 1956. "The Development of Identification with an Occupation." *American Journal of Sociology* 61:289–298.

Bentley, Arthur. [1954] 1975. "The Human Skin: Philosophy's Last Line of Defense." In *Inquiry into Inquiries: Essays in Social Theory*, ed. S. Ratner, 195–211. Westport, CT: Greenwood Press.

Bijker, Wiebe, Thomas Hughes, and Trevor Pinch, eds. 1987. *TheSocial Construction of Technological Systems: New Directions in the Social Study of Technology*. Cambridge, MA: MIT Press.

Bowker, Geof, and Bruno Latour. 1987. "A Booming Discipline Short of Discipline: (Social) Studies of Science in France." *Social Studies of Science* 17:715–748.

Bowker, Geoffrey C., and Howard Sankey. 1993/1994. "Truth and Reality in Social Constructivism." *Arena Journal* 2:233–252 [online @ http://arena.org.au/j2/].

Bucher, Rue. 1962. "Pathology: A Study of Social Movement within a Profession." *Social Problems* 10:40–51.

Butler, Judith. 1990. *Gender Trouble: Feminism and the Subversion of Identity*. New York: Routledge.

Callon, Michel, and Bruno Latour. 1992. "Don't Throw the Baby Out with the Bath School! A Reply to Collins and Yearly." In *Science as Practice and Culture*, ed. Andrew Pickering, 343–368. Chicago: University of Chicago Press.

Callon, Michel, and John Law. 1995. "Engineering and Sociology in a Military Aircraft Project: A Network Analysis of Technological Change." In *Ecologies of Knowledge: Work and Politics in Science and Technology*, ed. Susan Leigh Star, 281–301. Albany: State University of New York Press.

Callon, Michel, John Law, and Ari Rip, eds. 1986. *Mapping the Dynamics of Science and Technology*. London: Macmillan.

Carroll, John M., and Robert L. Campbell. 1986. "Softening Up Hard Science: Reply to Newell and Card." *Human–Computer Interaction* 2:227–249.

Chubin, D., and E. Chu. 1989. *Science off the Pedestal: Social Perspectives on Science and Technology*. Belmont: Wadsworth Publishing.

Clarke, Adele E. 1990. "Controversy and the Development of Reproductive Sciences." *Social Problems* 37:18–37.

Clarke, Adele E. 1991. "Social Worlds/Arenas Theory as Organizational Theory." In *Social Organization and Social Process: Essays in Honor of Anselm Strauss*, ed. David Maines, 119–158. Hawthorne, NY: Aldine de Gruyter.

Clarke, Adele E. 1995a. "Modernity, Postmodernity and Reproduction, 1890–1993, or 'Mommy, Where Do Cyborgs Come from Anyway?'" In *The Cyborg Handbook*, ed. C. H. Gray, H. Figueroa-Sarriera and S. Mentor, 139–156. New York: Routledge.

Clarke, Adele E. 1995b. "Research Materials and Reproductive Science in the United States, 1910–1940." In *Ecologies of Knowledge: Work and Politics in Science and Technology*, ed. Susan Leigh Star, 183–225. Albany: State University of New York Press.

Clarke, Adele E., and Elihu M. Gerson. 1990. "Symbolic Interactionism in Science Studies." In *Symbolic Interactionism and Cultural Studies*, ed. Howard S. Becker and Michal M. McCall, 179–214. Chicago: University of Chicago Press.

Clarke, Adele E., and Joan H. Fujimura, eds. 1992a. *The Right Tools for the Job: At Work in the 20th Century Life Sciences*. Princeton, NJ: Princeton University Press.

Clarke, Adele E., and Joan H. Fujimura. 1992b. "What Tools? Which Jobs? Why Right?" In *The Right Tools for the Job: At Work in the 20th Century Life Sciences*, ed. Adele E. Clarke and Joan H. Fujimura, 3–44. Princeton, NJ: Princeton University Press.

Collins, Harry M., and Trevor Pinch. 1982. *Frames of Meaning: The Social Construction of Extraordinary Science*. London: Routledge and Kegan Paul.

Collins, Harry M., and Steven Yearley. 1992. "Epistemological Chicken." In *Science as Practice and Culture*, ed. Andrew Pickering, 301–326. Chicago: University of Chicago Press.

Collins, P. H. 1986. "Learning from the Outsider Within: The Sociological Significance of Black Feminist Thought." *Social Problems* (Special Theory Issue 6) 33:14–32.

Crane, D. 1972. *Invisible Colleges: Diffusion of Knowledge in Scientific Communities*. Chicago: University of Chicago Press.

Croissant, Jennifer, and Sal Restivo. 1995. "Science, Social Problems, and Progressive Thought: Essays on the Tyranny of Science." In *Ecologies of Knowledge: Work and Politics in Science and Technology*, ed. Susan Leigh Star, 88–118. Albany: State of University New York Press.

Denzin, Norman. 1989. *Interpretive Interactionism*. Newbury Park, CA: Sage.

Dewey, John. [1896] 1981. "The Reflex Arc Concept in Psychology." In *The Philosophy of John Dewey*, ed. J. J. McDermott, 136–148. Chicago: University of Chicago Press.

Engeström, Yrjö. 1990. "What Is a Tool? Multiple Meanings of Artifacts in Human Activity." In *Learning, Working, and Imagining*, 171–195. Helsinki: Orienta-Konsultit Oy.

Fetzer, James. 1988. "Program Verification: The Very Idea." *Communications of the ACM* 31:1048–1063.

Forsythe, Diana. 1992. "Blaming the User in Medical Informatics: The Cultural Nature of Scientific Practice." *Knowledge in Society* 9:95–111.

Forsythe, Diana. 1993. "Engineering Knowledge: The Construction of Knowledge in Artificial Intelligence." *Social Studies of Science* 23:445–477.

Freeman, Jo. 1975. *The Politics of Women's Liberation: A Case Study of an Emerging Social Movement and Its Relation to the Policy Process*. New York: McKay.

Fujimura, Joan H. 1987. "Construction 'Do-able' Problems in Cancer Research: Articulating Alignment." *Social Studies of Science* 17:257–293.

Fujimura, Joan H. 1991. "On Methods, Ontologies, and Representation in the Sociology of Science: Where Do We Stand?" In *Social Organization and Social Process: Essays in Honor of Anselm L. Strauss*, ed. David Maines, 207–248. Hawthorne, NY: Aldine de Gruyter.

Fujimura, Joan H. 1992. "Crafting Science: Standardized Packages, Boundary Objects, and 'Translation." In *Science as Practice and Culture*, ed. Andrew Pickering, 168–214. Chicago: University of Chicago Press.

Fujimura, Joan H. 1995. "Ecologies of Action: Recombining Genes, Molecularizing Cancer, and Transforming Biology." In *Ecologies of Knowledge: Work and Politics in Science and Technology*, ed. Susan Leigh Star, 302–346. Albany: State University of New York Press.

Fujimura, Joan H., Susan Leigh Star, and Elihu M. Gerson. 1987. "Methodes de Recherche en Sociologie des Sciences: Travail Pragmatisme et Interactionisme Symbolique." *Cahiers de Recherche Sociologique* 5 (2): 65–85.

Glaser, Barney. 1964. *Organizational Scientists: Their Professional Careers*. Indianapolis: Bobbs-Merrill.

Glaser, Barney, and Anselm Strauss. 1967. *The Discovery of Grounded Theory*. Chicago: Aldine.

Griffin, Susan. 1978. *Woman and Nature: The Roaring Inside Her*. New York: Harper & Row.

Haraway, Donna. 1989. *Primate Visions: Gender, Race, and Nature in the World of Modern Science*. New York: Routledge.

Haraway, Donna. 1992. "The Promises of Monsters: A Regenerative Politics for Inappropriate/d Others." In *Cultural Studies*, ed. Lawrence Grossberg, Cary Nelson, and Paula Treichler, 295–337. New York: Routledge.

Harding, Sandra. 1991. *Whose Science? Whose Knowledge? Thinking from Women's Lives*. Ithaca, NY: Cornell University Press.

hooks, bell. 1990. *Yearning: Race, Gender, and Cultural Practices*. Boston: South End Press.

Hughes, Everett C. [1959] 1971. *The Sociological Eye: Vol. 2 Selected Papers on Work, Self and the Study of Society*. Chicago: Aldine.

Kling, Rob, and C. Suzanne Iacono. 1995. "Computerization Movements and the Mobilization of Support for Computerization." In *Ecologies of Knowledge: Work and Politics in Science and Technology*, ed. Susan Leigh Star, 119–153. Albany: State University of New York Press.

Kuhn, Thomas S. [1962] 1970. *The Structure of Scientific Revolutions*. 2nd ed. Chicago: University of Chicago.

Latour, Bruno. 1987. *Science in Action: How to Follow Scientists and Engineers through Society*. Cambridge, MA: Harvard University Press.

Latour, Bruno (Jim Johnson). 1995. "Mixing Humans and Nonhumans Together: The Sociology of a Door-Closer." In *Ecologies of Knowledge: Work and Politics in Science and Technology*, ed. Susan Leigh Star, 257–277. Albany: State University of New York Press.

Latour, Bruno, and Steve Woolgar. 1979. *Laboratory Life: The Social Construction of Scientific Facts*. Beverly Hills: Sage.

Law, John. 1986. "On the Methods of Long-Distance Control: Vessels, Navigation and the Portuguese Route to India." In *Power, Action and Belief: A New Sociology of Knowledge?*, ed. John Law, 234–263. A Sociological Review Monograph. London: Routledge.

Law, John. 1991. *A Sociology of Monsters: Essays on Power, Technology, and Domination*. New York; London: Routledge.

Law, John, and Michel Callon. 1995. "Engineering and Sociology in a Military Aircraft Project: A Network Analysis of Technological Change." In *Ecologies of Knowledge: Work and Politics in Science and Technology*, ed. Susan Leigh Star, 281–301. Albany: State University of New York Press.

Lynch, Michael. 1995. "Laboratory Space and the Technological Complex: An Investigation of Topical Contextures." In *Ecologies of Knowledge: Work and Politics in Science and Technology*, ed. Susan Leigh Star, 226–256. Albany: State University of New York Press.

Lynch, Michael, and Samuel Y. Edgerton. 1988. "Aesthetics and Digital Image Processing." In *Picturing Power: Visual Depiction and Social Relation*, ed. G. Fyfe and John Law, 184–220. Sociological Review Monograph 35. London: Routledge and Kegan Paul.

MacKenzie, Donald. 1981. "Notes on the Science and Social Relations Debate." *Capital and Class* 14:47–60.

Marcson, Simon. 1960. *The Scientist in American Industry*. New York: Harper.

Merchant, Carolyn. 1980. *The Death of Nature: Women, Ecology and the Scientific Revolution*. San Francisco: Harper and Row.

Merton, Robert K. 1973. *The Sociology of Science: Theoretical and Empirical Investigations*, ed. Norman W. Storer. Chicago: University of Chicago Press.

Middleton, David, and Derek Edwards, eds. 1990. *Collective Remembering*. London: Sage.

Mullins, Nicholas. 1973a. *Theory and Theory Groups in Contemporary American Sociology*. New York: Harper & Row.

Mullins, Nicholas C. 1973b. "The Development of Specialties in Social Science: The Case of Ethnomethodology." *Science Studies* 3:245–273.

Newell, Allen, and Stuart Card. 1985. "The Prospects for Psychological Science in Human–Computer Interaction, Part 1." *Human–Computer Interaction* 1:209–242.

Newell, Allen, and Stuart Card. 1987. "Straightening Out Softening Up: Response to Carroll and Campbell, Part 2." *Human–Computer Interaction* 2:251–267.

Pickering, Andrew, ed. 1992. *Science as Practice and Culture*. Chicago: University of Chicago Press.

Restivo, Sal. 1988. "Modern Science as a Social Problem." *Social Problems* 35:206–225.

Rosenberg, C. 1979. "Toward an Ecology of Knowledge: On Discipline, Context, and History." In *The Organization of Knowledge in Modern America, 1867–1920*, ed. Alexandra Oleson and John Voss, 440–455. Baltimore: Johns Hopkins University Press.

Singleton, Vicky, and Mike Michael. 1993. "Actor-Networks and Ambivalence: General Practioners in the UK Cervical Screening Program." *Social Studies of Science* 23:227–264.

Star, Susan Leigh. 1989. *Regions of the Mind: Brain Research and the Quest for Scientific Certainty*. Stanford: Stanford University Press.

Star, Susan Leigh. 1992. "The Trojan Door: Organizations, Work, and the 'Open Black Box.'" *Systems Practice* 5:395–410.

Star, Susan Leigh. 1994. *Misplaced Concretism and Concrete Situations: Feminism, Method and Information Technology*. Odense, Denmark: Gender-Nature-Culture Feminist Research Network.

Star, Susan Leigh, ed. 1995a. *Ecologies of Knowledge: Work and Politics in Science and Technology*. Albany: State University of New York Press.

Star, Susan Leigh. 1995b. "The Politics of Formal Representations: Wizards, Gurus, and Organizational Complexity." In *Ecologies of Knowledge: Work and Politics in Science and Technology*, ed. Susan Leigh Star, 88–118. Albany: State University of New York Press.

Star, Susan Leigh. 1995c. *The Cultures of Computing*. Oxford: Blackwell Publishers, Inc.

Star, Susan Leigh, and James R. Griesemer. 1989. "Institutional Ecology, 'Translations' and Boundary Objects: Amateurs and Professionals in Berkeley's Museum of Vertebrate Zoology, 1907–39." *Social Studies of Science* 19:387–420. [See also chapter 7, this volume.]

Strauss, Anselm L., and Lee Rainwater. 1962. *The Professional Scientist: A Study of American Chemists*. Chicago: Aldine.

Taylor, H. Jeanie, Maureen Ebben, and Cheris Kramarae, eds. 1993. *WITS: Women, Information Technology, and Scholarship Colloquium*. Urbana, IL: Center for Advanced Study.

Thomas, W. I., and Dorothy Swaine Thomas. [1917] 1970. "Situations Defined as Real Are Real in Their Consequences." In *Social Psychology through Symbolic Interaction*, ed. Gregory P. Stone and Harvey A. Farberman, 54–155. Waltham, MA: Xerox Publishers.

Wallace, Michele. 1990. *Invisibility Blues: From Pop to Theory*. London: Verso.

Wimsatt, William. 1980. "Reductionist Research Strategies and Their Biases in the Units of Selection Controversy." In *Scientific Discoveries: Case Studies*, ed. T. Nickles, 213–259. Boston: D. Reidel.

Winner, Langdon. 1986. *The Whale and the Reactor: A Search for the Limits in the Age of High Technology*. Chicago: University of Chicago Press.

Woolgar, S. 1995. "Representation, Cognition, and Self: What Hope for an Integration of Psychology and Sociology?" In *Ecologies of Knowledge: Work and Politics in Science and Technology*, ed. Susan Leigh Star, 154–179. Albany: State University of New York Press.

Zuckerman, Harriet. 1977. *Scientific Elites: Nobel Laureates in the United States*. New York: Free Press.

Zuckerman, Harriet. 1989. "The Sociology of Science." In *Handbook of Sociology*, ed. Neil Smelser, 511–574. Newbury Park, CA: Sage.

2 Ecological Thinking, Material Spirituality, and the Poetics of Infrastructure

Maria Puig de la Bellacasa

When a scholar captures your affections, what is passed on to us is hard to fathom. Beyond theories, concepts, and methods, we inherit gestures and ways of doing. This has been my feeling since joining the collective homage and mourning for Susan Leigh Star. Star's work has opened unusual ways of thinking and knowing for many of us. When I was a young feminist philosopher trying to understand the singular power of the sciences I first encountered the commonly named "Onions Paper" (Star 1991, 50; and see chapter 13, this volume) and the introduction to her edited volume *Ecologies of Knowledge* (Star 1995; and see chapter 1 this volume). These texts became foundational for me. More than a decade after discovering her work and only after some years of enjoying the privilege of her friendship I can hardly accept she's gone and that I am participating in posthumous celebrations of her. Rereading those writings, I longed to talk to her again, to tell her how astonished I was to realize I had forgotten how much was in them, the many lines of thought and seminal questions she had opened in early science and technology studies that are far from being closed, and mostly, how much her ways, and some of her insistent yearnings, have unwittingly marked me, shaped my own work, become part of the infrastructure of my soul. To put it in her own words when paying tribute to her teacher Mary Daly: within a "subjective-collective," thinking is *weaving* (Star 2009, 335) and the threads we weave with rarely belong only to us. But Star's influential gestures and ways of doing go well beyond what we usually call scholarly or intellectual knowing. She fostered particularly caring "modes of attention"[1] to marginalized experiences that, as she always made explicit in her work in science and technology studies, were marked by radical feminist thinking. Part of the magic of her work and scholarship is thus an affectively charged politics of knowledge fuelled with love and care for seemingly unimportant but very tangible experiences, bodies, and relations. And so I thought that in writing to celebrate her contributions I would engage with how Star's very singular knowledge politics had shape-shifting effects in our ways of knowing. I would explore how this might happen through her untamed and imaginative thinking, working in mysteriously affective ways. But in the process, shape-shifting happened again, bringing to me notions I had not expected.

Reading some of Star's writings which I was unfamiliar with, those that focus on "infra-structure" (Star 1999; Star and Ruhleder 1996), moved me in unexpected ways of think-ing about my current topic of research fixation: the contemporary collective reclaiming of the soil under our feet as a vital ecology. This essay speaks of how this led me to engage with relatively subtle dimensions of Star's scholarship: her attention to ecology, to metaphor, and the relation of both to spirituality. What unfolds is a particularly powerful entanglement of ecological thinking and material spirituality, a form of "eco-logical poetics" (Selby 1997) at the heart of contemporary reclamations of the soil.

The essay starts by addressing Star's commitment to attention to marginalized expe-riences and how this relates to her preference for ecologies in contrast with the notion of networks. This lays out a questioning of the predominant atmosphere where ecology appears as a "background" against which technoscience develops—allowing planetary ecologies to serve as field for network extension rather than obliging a responsive rela-tion. A second section engages with alternative ways of thinking revealed by envision-ing soil ecology as an infrastructure, and reflects on Star's practice of metaphor. Finally engaging with Star's call to integrate spirituality and community in science and tech-nology studies, I focus on spiritual meanings of soil. The chapter concludes with a reflection on ecological poetics as a possible naming for reimagining naturecultural spiritualities.

I "Spaces Between"

oh seductive metaphor
network flung over reality
filaments spun from the body
connections of magic
extend
extend
extend
who will see the spaces between?
(Star 1995, 32; Star, chapter 1, this volume)

Star often interrupted the flow of thought and ideas with poems. "The Net," from which the verses above are extracted, was included in her introduction to *Ecologies of Knowledge*, a foundational collection for the field of science and technology studies. These verses disrupt celebrations of virtual reality, the language of computer hype, a pervasive discourse at the time they were published. The poem slows down the trium-phalism surrounding networking successes, expressing both fascination and hesitation about metaphors promising flights of spreading extension to our corporeal finitudes. It does so with a question: *who will see the spaces between?* Star's efforts to look after spaces between are well known. In spaces between she saw the violence and pain in the lives of the forgotten, the silenced, the erased, the invisible worker and her "deleted" work, all residual categories like the "none of the above" or not elsewhere classified; the anomalies and all the singular experiences destroyed by bureaucratic mechanistic arrangements that privilege the norm. "The Net" calls for attention to those spaces between, to interruptions in overconnected networks that reveal fissures in technoscientific cultures.

There is a commitment to justice in Star's work.[2] She knew what it means to be an outsider, however *within*, but she coupled this awareness with a sturdy sense of humor, never situating her outsiderness on a higher moral ground. So though her particular path into the thinking of injustice is well revealed by the way she persistently asked *cui bono?* [to whose benefit?], we should remember that Star's *cui bono* is very different from that of a justifier, or a righter of wrongs, very different too from a *cui bono* satisfied by suspicious debunking—*who benefits from the crime?*—driven by desire to reveal the hidden motives indicating that something may not be what it appears at first to be. I believe she would have identified the guilt and boredom at stake in these satisfactions—something to remember when we see moral punishment replace social justice. Star's commitment to justice has nothing to do with adjustment; on the contrary she'd rather question the bases of rule as, for instance, the constitution of categories (Bowker and Star 1999). In other words, Star's pragmatics of investigating consequences at the margins is absolutely not reducible to exposing the benefits of those who have something to gain. Rather, if there is a moral calculus here it would direct us to realize the loss which comes upon us when we overlook other possible worlds hidden or silenced in marginalized spaces—including the fun of them! Further, this is not about a balance sheet between gains and losses; there is no equivalence.

This is because for Star, spaces between are not only about pain, violence, and survival: these fissures are also about possibility. Spaces between are created when things fall off established charts, creating split-selves; multiple memberships in different communities of practice: new forms of living are born in spaces between. In negotiations within, and with these uncharted belongings, we create possibility. Attending to spaces between is also about fostering these possibilities. The feminist Chicana poet-thinker Gloria Anzaldúa reclaimed the Aztec word *Nepantla* to name a state of political and spiritual in-betweenness in which those marginalized and oppressed by existing regimes engage in strategies of survival (Anzaldúa 1987). Both Star and Anzaldúa were *nepantleras*, dwelling at the borderlands of existing categories, fostering a poetics of creative resistance in the middle worlds. This work is as important as attention to violence and pain because the interstices of alternative possibility are also obscured by seamless accounts of both technological progress and doom.

So to say that Star had a commitment to justice wouldn't by itself capture her mode of attention to marginalities. I have used in another context a notion of "speculative commitment"—borrowing from Isabelle Stengers's work on speculative philosophy (Stengers 2002)—to speak of Star's and other feminist scholars' singular contributions to science and technology studies (Puig de la Bellacasa 2011). Trying to think how things would be different if we saw the world through *spaces between* indeed reveals a strong attachment to situated and positioned visions of what a liveable and caring world could be: Star did take sides (Star 1991). But her commitment is speculative because she would let neither a given situation nor even her own position—that is, an acute awareness of pervasive dominations and exclusions—confirm in advance what *is* or *could* be. I can imagine her telling us how tedious that would be. The question "who will see the spaces between?" is a yearning to which no definitive answer can be given. We do not know, however good our intentions, whether we will be able to see spaces between. But also, perhaps more importantly, we do not know what spaces between can become. The fact that her hesitations about the liberatory feel of the network metaphor expressed in "The Net" comes as a question rather than a judgment exposes a singular carefulness in Star's commitment, an assumed vulnerability that contributes to the creative force of her affective knowledge politics. Rereading her renowned "Onions Paper," I realized that, of course, Star had thought about this indeterminacy of commitment. Thinking with Howard Becker and with John Dewey,[3] she affirmed that "we involve ourselves in many potential actions; *these become meaningful in the light of collective consequence*s, jointly negotiated" (Star 1991, 50; emphasis added).

One of these crucial collective negotiations of still indeterminate commitments is happening today around multiple sites of ecological breakdown faced by Earth's people—i.e., not only humans—raising questions of naturecultural justice rather than just social, read "humanist." In the light of these collective consequences I want to explore the challenges to speculative inquiry that Star's interpellation toward network thinking might pose. What *spaces between* are being created by the drive for extension of the very ontology of technoscientific networks? To that question we might need to add this one: What creases and cracks are being created by the tensions between the power of creative networks and ecological relationships on Earth? It is Star's preference to speak of ecologies rather than networks that triggered this wondering.

II Ecologies—"Networks-without-voids"

A web is composed of filaments,
and a seamless web should be an oxymoronic term.
There is no empty space in a seamless web,
but our image of *network* is that it is filaments with space between.
For this reason I prefer ecology.
(Star 1995, 27)

Coming from a scholar whose work developed at the beating heart of the network kingdom—information and communication technologies—the persisting character of this ecological stance across her work has an enigmatic character. Less if we know how important was the metaphor of ecology for early twentieth-century Chicago sociology, the pragmatist philosophical roots of symbolic interactionist sociology in which Star was trained (see chapter 2 in Clarke 2005; Clarke and Star 2007).[4] Of course, here ecology does not refer specifically to "natural" ecosystems but to a particular form of relating. An ontology, with its correspondent mode of thought and attention to ecological thinking, is here distinguished from network living and thinking. However the terms of distinction in the quote starting this section are somehow enigmatic, they sound more like a poetic riddle hidden in an academic piece. This style works to ensure that the terms of reappropriation of this subtle critique of networks are left open. In other words, to turn this into a blanket judgment against the idea of networks (or of Actor Network Theory, or ANT) would be as much to blatantly misread as it would be to ignore the importance of the distinction at a moment when Star and other fellow (feminist) thinkers (such as Adele E. Clarke and Joan Fujimura) were voicing concerns about network thinking (*and* ANT); not the least because Star's argument here was embedded in her work of gathering a volume with seminal interventions in STS at a time when the concept of network was gaining importance in the field. While she indeed was making a point by asking the shape-shifters of the then-emerging discipline not to "simply create a professionalization movement with respect to nonhumans and thus becoming scientists in a way that often makes us uneasy," the invitation is radically open: "let us change the way science is organized." One way, she said, would be to engage in a "reclaiming of the term network of some of its unfortunate discontinuous connotations and affiliations" to go for "networks-without-voids for an ecological analysis" that would also be "epistemologically democratic, including toward our own work organization" (Star 1995, 27).

Much could be unpacked from those words. In continuity with the attention to spaces-between, an ecological analysis of networks would involve imagining "networks-without-voids." In that sense we can see that ecological thinking was for Star deeply grounded in the attention to that which escapes dominant visions but is still vital for the living of a world. Together with this ontological point goes the fact that ecological thinking commits to foreground these spaces. Again this is illuminated by how, in her "Onions Paper" ecological analysis appears in contrast with network thinking and early ANT vocabulary: "Every enrollment entails both failure to enroll and a destruction of the world of the non-enrolled. Pasteur's success meant simultaneously failure for those working in similar areas, and a loss and world-destruction for those outside the germ theory altogether. (. . .) A stabilized network—and thus a 'successful' one—is only stable for those who are 'members'; and this involves the private suffering of those who are not standard members and the nonmembers" (Star 1991, 49).

Taking up Bruno Latour's example of the pasteurization of France, Star reminds us how Pasteur's networking success meant also failure and loss. It involved a "world-destruction" that didn't have to do with the scientific controversy between Pasteur and Pochet, but with the "ecological effects of Pasteurism and its enrollment." Star mentioned bodies of knowledge destroyed by germ theory and that are being reclaimed only now: "immunology, herbal wisdom, acupuncture, the relationship between ecology and health": We can also take from here that one of the roles of the epistemological democracy she was calling STS to foster includes paying attention to the ecological effects of network enrollment that can be perceived in spaces between.

Putting together Star's hesitations about the seductive metaphor of networks and her preference for ecologies is an invitation to think speculatively about what more could be said about the contrast between these ontologies/modes of thought. Together with other thinkers (feminist and beyond) in STS, Star was attentive to the exclusions produced by stabilized networks and their seductive extensiveness. *The Net* manifested how network thinking is in love with extension and circulation. Of course, stabilization is key for the success of extending networks, but only until the next opening. The word of network is alliance, the provisional, the switchboard. Dimitris Papadopoulos has drawn my attention to the inherent productivism in the term network (Papadopoulos 2011). Networks induce an addictive fantasy of freedom—well attuned to the "free" global marketplace. Networks speak of particular ways of relating, of drawing lines of connection between one point and other, leaving behind countless spaces as background for their agency flows. Agency and power are indeed distributed, yet never evenly. We must always remember, in Star's words, that a stabilized network—and thus a "successful" one—is only stable for those who are members—involving often the privatization of pain for those who are not standard members, or are not members at all.

An ecology, by contrast, evokes a site of intensities, synergies, and symbiotic processes within relational compounds. Ecological circulation functions in cyclic interdependent ways rather than by extension. In contrast with the distribution and circulation of agency between multifarious agents that define a network, what characterizes an ecology is that the world it composes is inseparable from a certain durability of the ethos and practices at stake in a relatively stable territory (Deleuze and Guattari 1987; Stengers 2010). In other words, the dominant existential drive of an ecology is not so much to extend itself but to hold together resilient relationships—that are, of course, not necessarily pleasant, one need only to think of the intimate and interdependent relations between predators and prey (Stengers 2010). Ecological thinking is attentive to the capacity of relation-creation, to how different beings affect each other, to what they do to each other, the internal "poiesis" of a particular configuration. This is not to say that a focus on networks necessarily overlooks relations that hold together as stable but that extension remains the beating heart of network thinking. The very existence of the verb "networking" exposes the particular quality of this dynamic mode of thought, as does the interest in alliances and connections (usually strategic). Finally, though relational thinking is common to both network and ecological thinking there is an important difference consistent with the attention to what holds together: speaking of ecology inevitably invokes life and death. Ecological thinking involves the acknowledgment of finitude (and renewal) and therefore a certain resistance to the deliriums of infinitude of extension metaphors. That is also why ecological thinking cannot avoid ethical and political thinking of consequences of world-destruction and, as a corollary, of the possibilities of regeneration.

Star had the art of "staying with the trouble" as Donna Haraway would say (Haraway 2010). While the task she gave to herself was to look out for the spaces between, and she never held back from expressing her views about injustice, she always spoke of our ambivalent relations to technodreams rather than purifying them into right and wrong. *The Net*, for instance, also speaks of yearnings for connections at a distance that information and communication technologies (ICT) could fulfill. In this spirit, the point couldn't be to infer that networks cannot be ecological or that ecologies cannot benefit from networks. Thinking further, this contrast between network thinking and ecological thinking shouldn't lead to create a new binary but rather invite engagement with the historicity of our concepts and imaginaries.[5] The question is rather: Can the contrast between ecologies and networks speak of different collective consequences of scientific and technological developments in the current state of ecological breakdown? The need to think the problem strikes me when ecological concerns become so easily translated into a lure compatible to network extension and its perpetual futurosophy. One can think of the boom of "sustainable" business. Look at how, today, the prevalent solution to ecological breakdowns is an extension of greentech markets. Here is green policy: does your house have a high carbon footprint? Install triple glazing; we will lend you the money—never mind the tons of PVC tripling toxic trash. And what will be our stance when most expressions of doubt and concern about the desirability of a technology are easily dismissed as backward or simplistic, when reminders of the possible ecological destructions by extensions of a network are silenced by the orders of economic realism? And what happens to "progressive" network thinking when we bring to the forefront the "background stage" of technoscientific networks, that is, increasingly crushed planetary ecologies?

I have no answer to such questions in general, and in order to avoid moralistic binarism (its *either* networks *or* ecologies) I want to stay with how Star's work fosters modes of attention to think of networks in an ecological rather than a "connectionist" way; that is, one that tries to avoid instituting irrelevant "background stages." We can do so by following a line of thought in which Star's ecological thinking manifests fully: the ontological and ethical attentions she gave to "infrastructure," intimately coupled with her care for the "residual" categories and labors hidden (Star and Bowker 2007). The next section thus continues this exploration of Star's ecological thinking with an exploration of soil ecology as infrastructure.

III   Soil as Bioinfrastructure—The Power of Metaphor

Biology is an inexhaustible source of troping. It is certainly full of metaphor, but it is more than metaphor.

(Haraway 2000, 82)

Experts in infrastructural analysis please forgive me for displacing this complex notion into an unlikely terrain: the soil below our feet. Yet I have to bear witness to the fact that "Steps for an Ecology of Infrastructure" (Star and Ruhleder 1996) as well as the methodological text "The Ethnography of Infrastructure" (Star 1999) provide a beautiful way to unpack the specific wonders of soil ecology. Thinking with these texts the final home of all residues appears as more than just an accumulation of dirt, more than the solid ground upon which we circulate: it emerges as one of our most vital infrastructures, a *bioinfrastructure* (Puig de la Bellacasa 2014). The story can be told as what Geoffrey Bowker (quoted in Star 1999, 380) has named an "infrastructural inversion"— or should I say a conversion? It could be entitled: "How a City Dweller Discovered the Magic of Dirt between the Pavements." As Star and Rudheler note, people "do not necessarily distinguish the several coordinated aspects of infrastructure" (Star 1999, 381). This accounts for the "embeddedness" of infrastructure, for the fact that it is "sunk into and inside of other structures, social arrangements and technologies." The ecology of soil is embedded in the food we eat, the air we breathe, the water we drink. Most children of the scientific-industrial age rarely notice this. But lately more of us do as ecological activists have started talking of "peak soil," that is a moment of global "resource collapse" of this limited essential compound of lively matter, a far more looming devastation than any other "peak" (Shiva 2008; Wild 2010).

So the personal infrastructural revelation was not that personal after all. First, again, soil ecology is learned "through membership"—another characteristic of infrastructure according to Star and Ruhleder. For me the discovery of the importance of soil ecology that made soil pass "from background to topic" came through encounters with a community, a "social world" (Clarke and Star 2007) of knowledge makers across multiple memberships and for whom soil has long been the topic: soil scientists, gardeners, ecological activists, and even indigenous people fighting the destruction of their land. Another characteristic of infrastructure: its particular "reach or scope" always "goes beyond a single event or one-site practice" (Star 1999, 381). Infrastructure manifests its existence locally, through our material everyday relationships with it: though soil is everywhere, and the breakdown has become planetary, we can only engage locally with it. The calls for planetary awareness are thus often starting from the local level, creating interdependent discussions. Second, the infrastructural inversion happened at "the moment of breakdown" identified by Star and Rudheler as the moment when: "The normally invisible quality of working infrastructure becomes visible." Something is not working well in our soil ecologies. As Star and Rudheler show, even the use of "back-up mechanisms" further emphasizes the occurrence of an infrastructural breakdown. We can think of the fertilizers industrially produced to repair poor soil, unable to nourish, revealing not only that something is gone wrong but also further affecting infrastructural arrangement by their very intervention (i.e., further destroying soil's ecological capacities for regeneration). In other words, if soil ecology was working and healthy, most of us could go around without really noticing it. Noticing it is an event. What we do with that event is what matters: our responses are part of the relational infrastructural arrangement.

Within this general breakdown is a shift that is rearticulating, however slightly, a sense of community. A re-learning of soil as infrastructure is being fostered by scientists and ecologists (including "lay" gardeners and farmers) for whom knowledge of the soil involves attending carefully to its inhabitants and its ecology.

Good organic farmers, and a few conventional ones, are acutely protective of their soil, treating it with the commitment, concern and empathy normally reserved for close family members. I have seen organic farmers sniffing and even tasting their soil, and disrobing its virtues with familiarity and affection. A handful of healthy soil can contain millions of life forms from tens of thousands of different species [. . .]. Pesticides, fertilisers, animal wormers [. . .] can all drastically reduce these populations, not by just a few percent but by 10 or even 100 per cent. Imagine the outcry from WWF if anyone could see the carnage.

So if you can't see the fungi, bacteria and invertebrates and you don't feel inclined or qualified to taste your soil, how do you know it is healthy? (Watson and Baxter 2008, 14)

Depending on who is looking at soil and for whom that work is done some things are considered important and others not. Soil doesn't reveal the same aspects whether it is a scientist who looks at it through her microscope or a gardener who digs her plot—for a living or for fun. But there is something that joins them all. Neither sees soil as an irrelevant backdrop for burying the debris that technoscientific networks leave behind while they reconnect to find "solutions" (e.g., fertilizers). Their work is drawing attention to spaces between the pavements, bringing the "backstage of elements of work practice" to the forefront. This perception relates too with awareness brought in by ecologies of "managing" excess. It is only relatively recently, since the late twentieth century, that waste nonsusceptible to decay has become a highly ethically charged category of matter. If it cannot become soil, we have a problem. But also, most interestingly, the meaningfulness is intensified of calls to care for the invisible, nonhuman, laborers of the soil that make it possible (Lowenfels and Lewis 2006). What is being indicated is that this mostly invisible infrastructure only lives and works well if we humans work and live with proper attention to processes of soil renewal, if we live together well with its invisible workers, such as worms and microbial communities (i.e., by not pouring fertilizers and pesticides into their habitat).

The lessons emerging from the reclaiming of soil are in tune with Star's call for networks-without-voids: there cannot be such thing as an irrelevant background in an ecological worldview. Soil ecology is not an extensible network, an infinite resource: it is an infrastructural arrangement essential to life on this planet. Looking at this ecology as infrastructure involves thinking about modes of relationalities in interdependency, that is, some kind of community. Understanding the place of humans in soil ecology as a form of belonging might change the meaning of community itself. But to say so also reveals something about Star's approach to infrastructure: that it is not only about materials but also about meanings that are neither separable from, nor reducible to, what we usually conceive as materiality (Barad 2007).

Before immersing into exploring these meanings, a word should be said about the effects of this somehow displaced appropriation of Star's understanding of infrastructure. Initially conceived to analyze the relational arrangements of human actions and technologies, it reveals their potential to engage with a nonhuman-made world. But every good Western thinker indoctrinated into secular science has been warned against anthropomorphism—or other such things as *hylozoism*, meaning "an attribution of purpose, will and life to inanimate matter, and of human interests to the nonhuman" (Schaffer 1991, 182). Indeed I can be held accountable here for something that we could call *anthroporganizationalism*: the attribution of human organizational practices to natural worlds. And it will not be enough of a justification to claim that STS has widely questioned the human/nonhuman distinction both for analytical and political reasons. Indeed, there is still some trouble in compounding together without distinction all-things-nonhuman (Papadopoulos 2010). So what work is being done by the breach of specificity of this displaced engagement with infrastructure?

Maybe there is something in Star's work that presages why I have felt somehow authorized to engage her notion of infrastructure in this way, to stretch it as a form of analysis into a naturecultural ecology. In displacing infrastructure, I could have also been taken by a quality of writing that makes of theory also a trope. In other words, Star's way of speaking of infrastructure has impressed me as metaphor. Before examining further the consequences of this metaphorical use, we can look at what it involves in terms of elucidating her modes of attention. *Star-the-poet* belongs to a feminist clan of inspiring godmothers (among whom are also Haraway, Anzaldúa, Barad) exploring alternative ways of knowing our world, among which is a materialist understanding of meaning. Material worlds can be source of metaphor or trope that doesn't means, as Haraway warns, that they are *only* metaphor.

Star's poetics were not just contained in the poems she included in her writings. I have two very vivid impressions of gestures that affected me beyond the reading her work. One was how she had presented herself in a bio accompanying the program of a talk she was giving. I was fairly struck by the fact that she presented herself first as a poet, and then as an academic. As a person struggling with my own academic identity—fearing its fairly isolating effects—I was inspired, feeling some relief that such a position was possible. Another vivid memory is that of her presidential lecture at the annual meeting of the Society for Social Studies of Science in 2006. There she told the story of a startled and enchanted encounter with a bobcat near her home. She described this moment of recognition between two earthy beings sharing the same grounds of life at the edge of radical alterity as the kind of interruption that she invited us to weave into the thinking of technology. She didn't tell us how. The call was compelling in more mysterious ways. When telling the story in her soft voice, she interrupted the flow; one could have heard a feather fall. She had that power.

But more explicitly it is her work that affirmed metaphor as a powerful force, one that can build community by transgressing across splits. In her words:

Because we are all members of more than one community of practice and thus of many networks, at the moment of action we draw together repertoires mixed from different worlds. Among other things, we create metaphors—bridges between those worlds.

Power is about whose metaphor brings worlds together, and holds them there. It may be a power of the zero-point or a power of discipline; of enrollment or affinity; it maybe the collective power of not-splitting. Metaphors may heal or create, erase or violate, impose a voice or embody more than one voice. (Star 1991, 52)

What bridgings are produced by engaging metaphorically with infrastructure? Perhaps an ecological, relational-intensive, way of thinking of infrastructural arrangements in naturecultures could include in the same breath ecologies and networks, relations and connections. But mostly, for me, what defying the perils of metaphorical anthroporganizationalism does to the ecologies of soil is to push us to acknowledge their importance as something that is being *made* (not given), not a passive "resource" but an active organization involving humans and nonhumans that requires our participation for its renewal and liveliness, not only for its use and consumption. Moreover, this bioinfrastructure processes many meanings that most of us only have human ways of accessing: scientific research, gardening, agriculture, eating, composting. Other possible bridgings could be happening in the reclaiming of soil between ecological movements (understood in the largest sense possible), and the sciences of soil through what soil can mean across these worlds. These involve some cosmopolitical recompositions (to borrow Isabelle Stengers's notion [Stengers 2005]) by which different fields of practice are working toward a change in how we give care and attention to this overtly present, but relatively ignored world (*cosmos*). It is because these changes in material practices are not separable from changes in consciousness and meanings that in the next section I follow Star's message about the bridging qualities of metaphor into another terrain that she invited us to explore: spirituality. I address the reclamation of soil through a spiritual dimension rooted in its powerful metaphorical-material meanings. In turn, this contributes to a larger notion of a Star-inspired argument for ecological thinking: if ecology is infrastructure, spirituality might also be seen as infrastructural to ecology. And by *thinking spirituality as infrastructural* this ethereal term gains in thickness and flesh, exposing a rather material form of spirituality.

IV   Soil's Messages—a Material Spirituality

We all come from the Goddess
And to her we shall return
Like a drop of rain
Flowing to the ocean
("Chant of the Goddess," reclaiming ritual chant)

A rainbow of soil is under our feet:
Red as a barn & black as a peat.
It's yellow as lemon and white as the snow;
Bluish gray . . .
so many colors below.
Hidden in darkness as thick as the night:
The only rainbow that can form without light.
Dig you a pit, or bore you a hole,
You'll find enough colors to well rest your soul.
(Hole 1985)

The song above is by Francis D. Hole a professor of soil science, renowned also for his lyrical approach to his object of study (Hole 1988). The ritual chant comes from the spiritual-political neopagan movement, Reclaiming, whose members have been involved in ecological activism as a way to restore relations with Earth—their focus of worship, the Goddess. Both songs introduce this section to indicate that there is more than a transformation in sociomaterial and economic practices at play in the shift of awareness with regard to soil ecology. A change in affections and consciousness, a change in the sense of community is occurring in which organized rationalities (sciences of soil, ecological movements) as much as mundane matters (the basic need of humans for food) are rearticulating imaginaries of human belonging. I am mostly thinking with people who attend to the soil as an organic living habitat. However it is more widely that the living aspect of soil manifests a powerful metaphorical edge that mingles with the spiritual, and not only in "Western" culture. This being the culture I know better, and of which much has been told about its compulsive separation of human beings from the "natural," it is the Judeo-Christian theological weight of mud that comes to my mind when seeking to tell a spiritual story about the infrastructural quality of soil for earth's communities.

The Old Testament tells us that God created and shaped man from the earth's matter. Adam, the first man, is named after the masculine derivation of a particular type of earth found in the Middle East, reputed to be "ruddy," recalling the color of blood. Here even the nouns for acre, ground, land take their name from this color: "adhamah" being the feminine root of the word—derived from Akkadian into Hebrew (Botterweck, Ringgren, and Fabry 1997, 76). No comments needed here on how this hierarchical derivation contradicts the storytelling of woman coming from the rib of man. This account attests to what is common among many creation stories in which humans were initially modeled by gods from mud or clay—the very Latin word "human" comes from earth, ground, soil: *humanus*, humus. Soil is where organic life begins—literally—and where all life ends—dust to dust, decay.

Soil rhymes with soul (McIntosh 2004). That might say it all. Some basic material truth lies in this metaphorical derivation into the spiritual dimension of soil: the fact is that we humans, like most other complex living organisms on this planet, live on the food we grow from the soil. This organic spiritual meaning of soil is explicit in Western movements of neopagan spirituality for which working with the soil is a basic practice of caring for the Earth Mother (Puig de la Bellacasa 2010), attending to a body to which all things return. Here worship involves a form of everyday ecological activism, developing material modes of attention to current breakdowns in the webs of life. Caring properly for the soil requires relearning to know it as a living community: "Earth-honoring agriculture would generate abundance, but its primary intention would be not to grow profits, but rather *to grow soil*—living, healthy, complex soil—as a fertile matrix for living, vital, health-sustaining food. To grow soil, we need to appreciate and understand that soil is a living matrix of incredible complexity, the product of immense cycles and great generative processes" (Starhawk 2004, 161; emphasis added).

Starhawk is a spiritual neopagan activist deeply committed to organic forms of working with soil such as permaculture as a coherent engagement with Mother Earth, not just as a symbol for ecofeminist spirituality but also as a material entity of which humans are part. Something these eco-activists voices are insisting on is the value of a very active community laboring hard in the growing of soil rather than only consuming (from) it, and they are connected with the sciences of soil ecology to do this. The spiritual and scientific "messages" of soil meet here to encourage a different relation with soil, as a living world rather than merely a resource (Ingham 2004). Thickening the Judeo-Christian tale, we could then add that the scientific meaning of *humus* is not strictly synonymous with soil, but only one component of soil, a most nourishing component, the sturdy stable end product of laborious processes of decomposition and decay made possible by lively soil communities. Could this knowledge invite acknowledgment of worms et al. as cocreators of human's very matter? The point here is that in these forms of neopagan spiritualities, the binary spiritual/material is radically put into question. Soil is not seen as dust, or dirt, to receive humans after death, nor is soulless inert matter shaped by god and infused with spirit to create a soulful form (humans): soil is in itself part of a living organic web of being of which many creatures including humans are working for its ongoing (re)creation.

We live on the invisible work happening in soil as infrastructure. But soil workers being mostly very small nonhumans, most of us can only hear their messages through specialist spokespersons. One of the common difficulties in researching the two overlapping worlds of interest in science and technology studies—technocultures and naturecultures—is that the category of "nonhumans" is highly differentiated. And so are their spokespersons: whether for pieces of software (technologists) or for biological beings (scientists). Who is bringing up the messages from the soil workers? Who is giving voice to the breakdown of soil's nourishing capacities? Strong voices seem to be coming from a reappropriation of soil sciences by (everyday) ecological activists and growers (Carlsson 2008; Lowenfels and Lewis 2006). They are intervening in the material-meanings of soil, seeking for more liveable spaces for humans and nonhumans at the heart of technoscientific productivism.

I have come to write about this cosmopolitical recomposing of naturecultural relations through the spiritual aspects of soil. One reason for this is that these aspects constituted an important element of the training in permaculture practices that played an important role in my personal experience of infrastructural inversion.[6] But another reason is that I am trying to find ways to listen to Star's call in her introduction to *Ecologies of Knowledge*, to reintegrate two aspects into the thinking of science and technology: *spirituality* and *community*. In secular academia, pronouncing the word "spirituality" triggers all sorts of red lights and security alarms. It is my tribute to Star's own call to bring up the word here, but also to bear witness that hers was a very material spirituality, noncontradictory with her profound empiricism. It would be a lie to say that STS analysis is soulless (Latour 1996). But paradoxically it looks as if we are often more at ease with recognizing "souls" and other spiritual manifestations in technoscientific networks and artifacts—the ghosts in the machines—than we are with other living actors in our ecologies. Maybe this is because "we" create them? The fact is that in order to engage with some of the ongoing shifts in human-nonhuman relations for the sake of healthier ecologies of soil, we might need to stop separating these material practices from what transcends them. By this I mean not a supernatural world above "reality," but what we could call, inspired by Star's approach to infrastructure, an *infranatural* dimension: something that exceeds us individually and collectively, *but from within*. The question remains open about how to prolong Star's call to reintegrate spirituality and the meanings we will give to it.

V   Cues toward an Ecological Poetics

I went out to the hazel wood
Because a fire was in my head.
(W. B. Yeats, "The Song of Wandering Aengus")

And the birds of the air
And the beasts of the soil
And the fishes of the desperate seas
Will know who I am
And our substance will expand
As part of everything
(Jolie Holland, "Goodbye California")

Yet if metaphorical work is about the power to bridge, the word spirituality might not work particularly well in the context of the communities of membership on which I am speculating here. This doesn't mean it is out of the landscape but that it might produce adverse effects for the project. After all, how many soil scientists would recognize themselves in the statement that their sciences are feeding into forms of Earth worshipping? But more conceptually speaking, "spirituality," as Donna Haraway has reminded me, is probably still bound to be an ill-suited word for materialist meaning practices. The very word "spirit" is entrenched in a tradition of separation of spirit and body, mind and matter. By the way, she suggested, only half joking, and quoting her partner, the physics teacher, ecologist, and radio director Rusten Hogness, we could just call it *compost*. And we can then, as David Abram suggests, also return to the root of the word spirituality, spirit, related to *spirare*, to breath, as a reminder that every living being needs this breathing—including good compost.[7] Whatever path we might choose, Star's call does not settle into a single answer. Yet we can keep the question: How can the openings she produced by pronouncing the word "spirituality" be included in our analysis?

I will not try to provide an answer but I can share some cues. One is in the very way Star articulated the question at the heart of the STS community. In *Ecologies of Knowledge*, the question of spirituality appears through her acknowledgment of belonging to and early training in a community of knowing:

At first, feminist scholarship was so beleaguered, so new, and so dependent on emerging community that people from all fields were welcome. So the analysis of poets was equal to if not greater that of scientists; the experience of an eighty-year-old woman as important for the critique as that of a twenty-five-year-old (although this certainly not without its own struggle, and there was an incredibly heterogeneity in the sorts of analysis brought to bear on issues . . .). This was coupled with an important inclusion and participation of feminist theologians—Mary Daly, Carol Christ, Rosemary Ruether, Nelle Morton, and others—who were not afraid to tackle questions of God/dess, the Absolute, power, Imagination, and so on. (Star 1995, 24)

Star admitted the presence of "silly mysticism" or that silly mysticism became an easy accusation to dismiss the potential of these experiences. But she insisted on the joys and troubles of "not splitting": that is, in not making any of these questions and experiences irrelevant a priori. We can also read here Star's commitment to pluralism, epistemological democracy, and an enduring belief in the fertility of relating multiple marginalities and memberships. And thus we see that the thought of spirituality appears here with the spirit of a community becoming possible also because it is somehow invisible. A *space between*.

This leads me to a second cue suggested by how "spirituality" appears framed in an early special issue of the feminist journal *Quest*, from the very time Star is describing above:

The contemporary women's movement has created space for women to begin to perceive reality with a clarity that seeks to encompass many complexities. This perception has been trivialized by male dominated cultures that present the world in primarily rational terms. *Reality is not only rational, linear, and categorized into either/or—it is also irrational and superrational. Because we do not have a new word for this struggle to comprehend this totality and incorporate that understanding into our action, we are calling it spirituality.* (Davis and Weaver 1975, 2; emphasis added)

This second cue also leads us to a community open to multiple memberships and an experimentation with the arts of "not splitting." And it can be said that this kind of nonsplitting practice by which a community is created is based on a form of *faith*, a suspension if not of doubt at least of skepticism: *because we do not have a new word for this struggle*. Because, again, we do not know if we will become able to speak for the spaces-between we are articulating. That which Judy Davis and Juanita Weaver called "spirituality" is also that mysterious, unexplainable, and even irrational something because it is unknown, a something that always dwells in the possibility of a community to come. Because the world as we remake it, even with very old pieces, is always somehow new.

A third cue when looking for meanings and effects of spiritual openings was given me at an event in celebration and honor of Star.[8] I went there with the same question in mind: how to prolong her calls for spirituality? I thought I'd be walking on delicate ground, but I had got it wrong. I was far from being the only one bringing the infranatural into the picture. In her opening address to a panel exploring the meanings of *cui bono*, Karen Barad started by engaging with the material practicalities of her Jewish practice; Donna Haraway offered a view of the cosmic and artistic meanings of her favorite metaphor of *terrapolitical* relationality of the cat's cradle game as used in Native American myth creation; through the miracles of audiovisual technologies Jake Metcalf's presentation made us shiver at the sounds of howling coyotes in the night, recorded in Star and Bowker's home in the Santa Cruz mountains. . . . There is definitively something in Star's call that has not been left unheard. A very material spirituality, or that-for-which-we-do-not-have-a-final-word, was definitively present at that gathering. And it was Bowker who gave me a good cue to a word that might do, for now, for the something I'm trying to get at in the ongoing collective remaking of soil ecology, when he evoked the work that he and Star had been doing lately on *the poetics of infrastructure*. I do not know in what way this work would have unfolded and will still unfold. But the very mentioning of that phrasing made me return to how Star's modes of attention found their expression through poetry. As I mentioned earlier, her poetry is not only in her poems but also in how, always, an awareness of something that exceeds us finds a pathway through her work.

Poetics go beyond the art of writing poetry and the theory of it. *Poesis* is about making creatively. The quality of making speaks to the fact that poetry is not only *about* the world—or its metaphor—it is about *making* worlds. Poets who were dear to Star such as Adrienne Rich are celebrated for grounding their poetry in this belief. Also, and correlatively, poetics disturbs the contradiction between metaphor and truth. Characteristically, one of the words Star kept pronouncing, without irony, when writing at the midst of the most poststructuralist—or postmodern—times was truth: "How can we say the truth about our lives?" (Star 2007) This commitment to something that could be called truth marks one of the singularities of her work, however it is the embodiment of a particular form of knowing in which truth has nothing to do with the equivalence or congruence with a prefigured reality but more to do with the contribution to its making. Such is "the pragmatism of truth" she advocated: a truth that is about effects rather than just interpretation. Her pragmatic of consequences is based on a pluralism of truths as the condition of community.

The intrinsically entangled power of truth—as metaphor—and poesis—as making worlds—are aspects the theologian and ecological activist Alastair McIntosh discusses wonderfully in his book *Soil and Soul*. One of the ecological struggles he recounts is the resistance to final naturecultural annihilation that accompanied the progression of the British interior empire into the Scottish ecology and economy through "clearing the land" of its people and severing relations among soil communities. But he also explores the surviving remnants of this oral culture, and its modern reclaiming, through the figure and the words of the *bards*. This brings us back to the poet's craft. McIntosh reminds that the historical figure of the bard was the vehicle of recording and transmitting history. These public poets were respected scholars and somehow politicians—their spoken knowledge of truth disturbed the distinctions of scholarship, politics, and poetry. But most important, the bard, like the shaman in other cultures, was deeply embedded in a connection with a community of beings that goes beyond ourselves:

To the poet, historical truth could not be separated from representation in the language of metaphor. As such it is a qualitative reality rather than a black-and-white absolute. History, to the mythopoetic mind, was not just literal; more importantly, it was also a very psychological and spiritual reality (. . .) the bare bones of historical fact had to be fleshed out with illustrative meaning. In this way not just the truth but also the whole truth would be told. . . . To such a mind, nature itself has a life beyond the one-eyed seeing that allows perception only of its mundane facets. . . . Such, then is the (. . .) "otherworld," existing not necessarily as some distant Eden . . . but interpenetrating the world all around us. (McIntosh 2004, 72–73)

Of course McIntosh's use of an almost banished culture to understand the present is in itself a *mythopoetic* endeavor, rooted in a process theology of perpetual recreation, in which there is no original or final act of creation, an ongoing process involving all beings. And this recreation is intrinsically word metaphor and matter. This is the type of creation that we can read as poesis: encompassing spirituality *and making* beyond the divide between spirit and matter. The poetics of infrastructure is *in* the world rather than about/above the world. Coming back to the breakdown of ecological futures, and thinking of the kind of membership and commitment that we can hopefully glimpse in the reclaiming of soil as a living ecology, we can think about contemporary attempts to explore an "ecological poetics" as ways of expressing-recreating the infranatural.

Giving the final word to the poets seems most appropriate here. Stuart Cooke, a researcher in ecological humanities and poet writes: "I am looking at ways of *articulating ecological networks*, or of allowing energy to flow through language(s) in the same way that it flows constantly between different forms and materials in an ecosystem. Ideally, I'd like my language to graft onto the ecology in question. In order to do so, poetry must become as surprising, as fractured and as many-sided as any other member of a complex ecology" (Cooke n.d.; emphasis added).

This project involves a rearticulation between very different types of work.[9] Nick Selby, speaking of Gary Snyder's poetics, finds meanings for an "ecological poetics" in the "striving for a visionary integration with the land," which for him means necessarily "to mark the divorce between nature and culture . . . the split between word and world that is exposed in our work of reading these poems, and which can be read as a product of a capitalist economy of exchange" and of Western culture. He affirms that to "see the poem as work-place is to expose the workings of language, and to make fraught our relationship to the object world. *The ecological lesson . . . lies, finally, in an attending to the fracture in the very guts of the real*' (Selby 1997; emphasis added).

I will miss Leigh till the day I join the worms. Yet her uniquely creative legacy is at work. We need her to help us read the cues toward an ecological poetics (or an "ecopoetics" (Gander and Kinsella 2012, 11)) that could *attend to the fractures*—spaces between—as a possibility for reinfusing spiritual-material meanings in our thinking and making of naturecultural infrastructures. My way of continuing to talk to her is to engage with her inspirational voice through her writings, to trust that she is still speaking through our weaving with her work. The integration of modes of attention that her poesis inspires between ecological thought, metaphorical bridging, and material spirituality, coupled with her care for avoiding easy purifications, has much to offer to the naturecultural collective negotiations of ecological commitments at the heart of technoscience's networks. And it starts with challenging a human-centered sense of community.

Notes

1. I owe encountering the phrasing "modes of attention" to Natasha Myers and Joe Dumit's invitation to think the "Lag" in the panel they organized for the 2011 annual meeting of the Society for Social Studies of Science. This invoked both an embodied sense of modes of perception and the quality of care, of paying attention and listening, that were so characteristic of Leigh Star.

2. This commitment was explored and celebrated in the conference "The State of Science and Justice: Conversations in Honor of Susan Leigh Star" organized by the Science and Justice Working group, June 2–3, 2011, at the University of California, Santa Cruz.

3. I thank Adele Clarke for reminding me of Star's longstanding interest in John Dewey's philosophy.

4. Thanks again to Adele Clarke for making me aware of this tradition and of the fact that Star borrowed the phrase "ecologies of knowledge" from Charles Rosenberg (Rosenberg 1979).

5. It would be relevant to see, though too lengthy to discuss here, how this speculative discussion of ecology sits with the current astonishing proliferation of the ecological metaphor in numerous spheres of thought and practice. Ecology is used indiscriminately to refer to any type of relational clusters and processes without any discussion of what is implied by the notion in contrast to the previously prevalent "network" (examples of this gleaned here and there are: "data ecologies," "media ecologies," "industrial ecologies"). It does seem that ecology has become a predominant turn of speech—without necessarily implying eco-ethical regimes of living but maybe, rather sadly, naturalized relations.

6. See http://www.earthactivisttraining.org/ (accessed April 2, 2015).

7. Thanks to Isabelle Stengers for reminding me about Abram's take on the word "spirituality." Stengers herself explores novel relations between modes of thought and experience—spiritual, scientific, and philosophical. See her text "Reclaiming Animism" (Stengers 2012). See also Belgian anthropologist Benedikte Zitouni's insightful statement on "ensoulment" as a way to enrich the social sciences (Zitouni 2012).

8. See conference mentioned in note 2.

9. "Articulation" was a dear concept to Star and her community. The concept of "articulation work" in particular was developed by Anselm Strauss, with whom Star collaborated and who had a strong influence in her work (Strauss 1988).

References

Anzaldúa, G. E. 1987. "The Creation of Man Meister Bertram—Grabow Altarpiece." In *Borderlands/La Frontera: The New Mestiza*, 1375–1383. San Francisco: Aunt Lute Books.

Barad, K. M. 2007. *Meeting the Universe Halfway: Quantum Physics and the Entanglement of Matter and Meaning*. Durham, NC: Duke University Press.

Botterweck, J. G., Ringgren, H., and H.-J. Fabry. 1997. *Theological Dictionary of the Old Testament*. Grand Rapids, MI: Wm. B. Eerdmans.

Bowker, G. C., and S. L. Star. 1999. *Sorting Things Out: Classification and Its Consequences*. Cambridge, MA: MIT Press.

Carlsson, C. 2008. *Nowtopia: How Pirate Programmers, Outlow Bicyclists, and Vacant-Lot Gardeners Are Inventing the Future Today!* Edinburgh: AK Press.

Clarke, A. E. 2005. *Situational Analysis: Grounded Theory after the Postmodern Turn*. Thousand Oaks, CA: Sage.

Clarke, A. E., and S. L. Star. 2007. "The Social Worlds Framework as a Theory-Methods Package." In *The Handbook of Science and Technology Studies*, ed. E. J. Hackett, O. Amsterdamska, M. Lynch, and J. Wajcman, 113–137. Cambridge, MA: MIT Press.

Cooke, S. n.d. Personal webpage. Ecological Humanities. http://www.ecologicalhumanities.org/cooke.html (accessed December 3, 2012).

Davis, J., and Weaver, J. 1975. "Dimensions of Spirituality." *Quest: A Feminist Quarterly* 1 (4, Women and Spirituality): 2–6.

Deleuze, G., and F. Guattari. 1987. *A Thousand Plateaus: Capitalism and Schizophrenia*. Minneapolis: University of Minnesota Press.

Gander, F., and J. Kinsella. 2012. *Redstart: An Ecological Poetics*. Iowa City: University of Iowa Press.

Haraway, Donna J. 2000. *How Like a Leaf: An Interview with Thyrza Nichols Goodeve*. New York; London: Routledge.

Haraway, Donna J. 2010. "Staying with the Trouble: Becoming Worldly with Companion Species." Paper presented at the Colloquium Series of the Center for Cultural Studies. University of California, Santa Cruz, October 10.

Hole, F. D. 1985. "A Rainbow of Soil." http://extension.illinois.edu/soil/songs/songs/htm (accessed July 22, 2015).

Hole, F. D. 1988. "The Pleasures of Soil Watching." *Orion Nature Quarterly* (Spring): 6–11

Ingham, E. R. 2004. "The Soil Foodweb: Its Role in Ecosystems Health." In *The Overstory Book Cultivating Connections with Trees*, ed. C. R. Elevitch, 62–65. Holualoa, HI: Permanent Agriculture Resources.

Latour, B. 1996. *Aramis, or the Love of Technology*. Cambridge, MA: Harvard University Press.

Lowenfels, J., and W. Lewis. 2006. *Teaming with Microbes: A Gardener's Guide to the Soil Food Web*. Portland, OR: Timber Press.

McIntosh, A. 2004. *Soil and Soul: People versus Corporate Power*. Rev. ed. London: Aurum.

Papadopoulos, D. 2010. "Insurgent Posthumanism." *Ephemera: Theory & Politics in Organization* 10 (2): 134–151.

Papadopoulos, D. 2011. "Alter-ontologies: Towards a Constituent Politics in Technoscience." *Social Studies of Science* 41 (2): 177–201.

Puig de la Bellacasa, M. 2010. "Ethical Doings in Naturecultures." *Ethics, Place and Environment: A Journal of Philosophy and Geography* 13 (2): 151–169.

Puig de la Bellacasa, M. 2011. "Matters of Care in Technoscience: Assembling Neglected Things." *Social Studies of Science* 41 (1): 85–106.

Puig de la Bellacasa, Maria. 2014. "Encountering Bionfrastructure of Bios: Ecological Movements and the Sciences of Soil." *Social Epistemology* 28 (1): 26–40.

Rosenberg, C. E. 1979. "Toward an Ecology of Knowledge: On Discipline, Context and History." In *The Organization of Knowledge in Modern America*, ed. A. Oleson and J. Voss, 440–455. Baltimore: Johns Hopkins University Press.

Schaffer, S. 1991. "The Eighteenth Brumaire of Bruno Latour." *Studies in History and Philosophy of Science* 22 (1): 174–192.

Selby, N. 1997. "Poem as Work-Place: Gary Snyder's Ecological Poetics." *Sycamore: A Journal of American Culture* 1 (4): 1–19. http://www.english.illinois.edu/maps/poets/s_z/snyder/selby.htm (accessed September 4, 2015).

Shiva, V. 2008. *Soil Not Oil: Environmental Justice in a Time of Climate Crisis*. Cambridge, MA: South End Press.

Star, S. L. 1991. "Power, Technologies and the Phenomenology of Conventions: On Being Allergic to Onions." In *A Sociology of Monsters: Essays on Power, Technology and Domination*, ed. J. Law, 26–56. London: Routledge.

Star, S. L., ed. 1995. *Ecologies of Knowledge. Work and Politics in Science and Technology*. Albany: State of New York University Press.

Star, S. L. 1999. "The Ethnography of Infrastructure." *American Behavioral Scientist* 43 (3): 377–391.

Star, S. L. 2007. "Interview." In *Philosophy of Technology*, ed. J. K.-B. Olsen and Evan Selinger, 223–231. New York: Automatic Press.

Star, S. L. 2009. "Weaving as Method in Feminist Science Studies: The Subjective Collective." *Subjectivity* 28 (1): 335–336.

Star, S. L., and G. C. Bowker. 2007. "Enacting Silence: Residual Categories as a Challenge for Ethics, Information Systems and Communication." *Ethics and Information Technology* 9: 273–280.

Star, S. L., and K. Ruhleder. 1996. "Steps toward an Ecology of Infrastructure: Design and Access for Large Information Spaces." *Information Systems Research* 7:111–134.

Starhawk. 2004. *The Earth Path: Grounding Your Spirit in the Rhythms of Nature*. San Francisco: Harper.

Stengers, I. 2002. *Penser avec Whitehead: Une libre et sauvage création de concepts*. Paris: Editions du Seuil.

Stengers, I. 2005. "The Cosmopolitical Proposal." In *Making Things Public: Atmospheres of Democracy*, ed. B. Latour and P. Weibel, 994–1003. Karlsruhe: ZKM/ Center for Art and Media and Cambridge, MA: MIT Press.

Stengers, I. 2010. *Cosmopolitics I*. Minneapolis: University of Minnesota Press.

Stengers, I. 2012. "Reclaiming Animism" [electronic version]. *e-flux*. http://www.e-flux.com/journal/reclaiming-animism/ (accessed December 18, 2012).

Strauss, A. L. 1988. The Articulation of Project Work: An Organizational Process. *Sociological Quarterly* 29:163–178.

Watson, G., and J. Baxter. 2008. *Riverford Farm Cook Book*. London: Fourth Estate.

Wild, M. 2010. "Peak Soil: It's Like Peak Oil, Only Worse" [electronic version]. *Energy Bulletin*. http://www.energybulletin.net/52788 (accessed January 2012).

Zitouni, B. 2012. "Ensoulment or How Panpsychism and Animism Can Enhance Social Sciences" [electronic version]. Paper presented at the Conference "What's New about New Materialisms?" Center for Science, Technology, Medicine & Society (CSTMS), University of California, Berkeley, May 4–5, 2012.

3   Don't Go All the Way: Revisiting "Misplaced Concretism"

Nina Wakeford

"Don't go all the way."

This was my mother's response upon my announcement, at the age of nineteen, that I had become a feminist and was studying feminist philosophy at university. Of course I knew exactly what she meant, just as I had known what she meant when she left me with the injunction "just don't kill us" as her last words to me before I left for university. "Not going all the way" meant:

- don't become a lesbian, and therefore in her eyes
- don't become a man, therefore
- don't also become some sort of weird transsexual, therefore
- don't shame us with your monstrosity, and give up all hope for us having grandchildren.

"Not killing us" [sic] simply meant don't get pregnant if you're not married. The year was 1971 when I left for university; I became a feminist in 1973.

(Star 1994; chapter 6, this volume)

Of course Leigh *did* much of what her mother feared she might, and immersed herself in the lesbian communities of San Francisco. And with these experiences she began the working paper "Misplaced Concretism and Concrete Situations."

(Star 1994; chapter 6, this volume)

How can we remember Leigh's queerness? I sense the legacy of these years reverberates through her work, not just the impact of feminism, but also a recognition of a particular kind of ideological struggle, and the tensions within communities trying to dismantle the infrastructure of the patriarchy. It is work that often meant doing difficult feelings differently. Kathleen Stewart has invoked the term "atmospheric attunement" as a way to talk about the labor of becoming sentient to bodies, rhythms; a practice of being in noise and light and space (2011). Stewart asks:

What happens if we approach worlds not as the dead or reeling effects of distant systems but as lived affects with tempos, sensory knowledges, orientations, transmutations, habits, rogue force fields? What might we do with the proliferation of little worlds of all kinds that form up around conditions, practices, manias, pacings, scenes of absorption, styles of living, forms of attachment

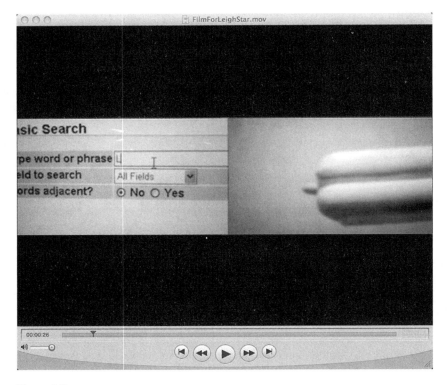

Figure 3.1
Still from *Misplaced Concretism/Film for Leigh Star*
Source: Wakeford and Keal 2010.

(or detachment) identities and imaginaries, or some publically circulating strategy for self-transformation? (Stewart 2011)

As she does in "Misplaced Concretism," Leigh often began her work by sharing her sense of affective attunement, a strategy far removed from many of the other worlds in which she operated, but familiar to much of queer theory, which has tackled the politics of feelings—whether in debates about the limits of intimacy, or battles about recognition and misrecognition. The now-intense interest among queer theorists in matters of public affect and intimacy in public (e.g., Berlant) finds itself already prefigured in Leigh's interweaving of stories of struggles, personal-political, with accounts of institutions, organizations, and systems. So it is in this spirit that here I want to revisit "Misplaced Concretism" by building up a new kind of project of misplacement and concretism—a project which has been built in the spirit of other feminists questioning the development of "the empirical." It is called "inventive" sociology and one of its

Figure 3.2
Still from *Misplaced Concretism/Film for Leigh Star*
Source: Wakeford and Keal 2010.

operating sites is not the office but the studio. Patricia Clough (2009) has called for sociology to open its doors to an empiricism that is at the very limit of the phenomenal. I want to propose activities in this vein. Clough says: "We can speak of an empiricism of sensation, not an empiricism of the senses, not the sense knowledge underpinning methodological positivism, but an empiricism of the 'in-experience' of affect at the very limit of the phenomenal" (51).

Please Sing This Before Continuing

Lavender Jane: View from Gay Head, Alix Dobkin, 1973[1]

[SING!] I heard Cheryl and Mary say
There are two kinds of people in the world today
One or the other a person must be
The men are them, the women are we

And they agree it's a pleasure to be
A lesbian, lesbian
Let's-be-in no man's land
Lesbian, lesbian
Any woman can be a lesbian

Carol is tired of being nice
A sweet smile, a pretty face, submissive device
To pacify the people for they won't defend
A woman who's indifferent to men
She's my friend, she's a lesbian

And Liza wishes the library
Had men and women placed separately
Ah, but theirs is the kingdom
She knows who she'll find
In the HIStory of MANkind
But then she's inclined to be ahead of her time
She's a lesbian, lesbian
Let's-be-in no man's land
Lesbian, lesbian
Any woman can be a lesbian

And women's anger Louise explains
On a million second places in the master's games
It's real as a mountain, it's strong as the sea
Besides, an angry woman is a beauty
She's chosen to be a dyke like me
She's a lesbian, lesbian
Let's-be-in no man's land
Lesbian, lesbian
Any woman can be a lesbian

So the sexes do battle, they batter about
The men-s are the sexes I will do without
I'll return to the bosom where my journey ends
Where there's no penis between us friends
Will I see you again
When you're a lesbian, lesbian
Let's-be-in no man's land
Lesbian, lesbian
Any woman can be lesbian
Every woman can be a lesbian

For me Leigh always had, and I wanted to come up with a term that I think she would have enjoyed, a *frisson of queerness*. Perhaps that is due to the fact that the first article I read of hers was not anything to do with boundary objects or standards, but rather was written as a lesbian feminist in a volume called *Against Sadomasochism: A Radical Feminist Analysis*, published in 1982, which she also coedited. For this volume Leigh wrote an essay entitled "Swastikas: The Street and the University" and also interviewed Audre Lorde. The volume also includes an article "Lesbian S & M: The Politics of Dis-illusion" by Judy (later Judith) Butler. In her own article Leigh noted, with distress, the adoption of storm trooper outfits, military boots, and swastikas—or full Nazi regalia—of gay men in her neighborhood of the Castro. She and Lorde discussed the position that sadomasochism constitutes the "institutionalized celebration of dominant/subordinate relationships"—the political claim of the book as a whole. At one point in the interview, Star comments to Lorde:

Star: I often feel that there's a kind of tyranny about the whole concept of *feelings*, as though, if you feel something then you must act on it.
Lorde: You don't *feel* a tank or a war—you feel hate or love. Feelings are not wrong, but you are accountable for the behavior you use to satisfy those feelings.

In terms of addressing, let alone *satisfying* our feelings, I've never been quite sure of the capacity of science and technology studies (STS), although Leigh's contributions to the field's professional meetings were always an exception. I made my own way into STS via an interest in non-normative identities and social movements and how they were being facilitated by the Internet in the late 1980s. It was these concerns—specifically how lesbian communities were being created online (which sounds anachronistic now)—that originally led me to arrive in California, and to meet Leigh and Geof, facilitated by another "famous"-to-me lesbian feminist, the novelist Anna Livia, who taught alongside them at the University of Illinois. One of my driving lessons—offered by Anna—was en route to Leigh and Geof's BonnyDoon residence where we were greeted with dinner and lively conversation, not least about the consequences of getting in a redwood hot tub that had only just been installed. Having recently arrived in California, all three indulged me in my "I'm new in town" stories—Leigh asked me if I'd seen Steve Martin's film *LA Story*—advising that it would help explain Californian culture!

I still recommend *LA Story* to others—not just to understand aspects of the Californian mindset, but also for its invocation of a whole range of knotty issues about insiderness and outsiderness, and its astute satire on the creation of affective atmospheres—most notably in a scene in which Steve Martin's girlfriend Sandee, played by Sarah Jessica Parker, shows off her "spokesmodeling" skills honed in on a course where she is learning to speak toward objects while pointing at them.

Figure 3.3
Still from *Misplaced Concretism/Film for Leigh Star*
Source: Wakeford and Keal 2010.

Sandee's display of how she is learning to channel attention toward objects functions as a form of cheerleading—an aspiration to changing the affective atmosphere around a product. Its contemporary equivalent is the genre of "unwrapping videos" on YouTube, in which people share recording of themselves taking the packaging off their latest technological gadget, all the while narrating the process.

The full title of Working Paper 11 of the Danish Feminist Research Network reads "Misplaced Concretism and Concrete Situations: Feminism, Method and Information Technology." Published in 1994 it began as a lecture given to the Department of Information and Media Sciences at Aarhus University—part of a conference entitled "Exploring Cyborgs: Feminism and Shifts of Paradigms."

Summarizing the paper's argument provides an opportunity to remember the political urgency of its arguments. However, in this unfolding and remembering of the content itself, we could also open ourselves to an affective attunement—an alignment with

Figure 3.4
Still from *Misplaced Concretism/Film for Leigh Star*
Source: Wakeford and Keal 2010.

a time when the idea of cyborgs was fresh and newly generative, a time when the politics of identity had not been jettisoned but was still useful politically and intellectually.

My own return to this work has happened more experimentally, through the making of a four-minute digital video produced for the workshop that was the origin point of many contributions in this volume. This video is neither an illustration of the arguments nor a distillation of any key points—although all the words in the film do come from the essay. Rather, in the spirit of recent attempts to challenge the form of doing STS, this is a more-than-representational engagement-with, a becoming-with, and features a young London radical filmmaker (Lily Keal) encountering Leigh's thoughts and words for the first time through the working paper.

This switching of modes, from a more traditional "reading" to a becoming-with, is part of an exploration of the politics of method which I think of as "studio sociology." Studio sociology attempts to take on a new affective register of method—building on much that Leigh said about the impossibilities of re-representing multiplicity, margin-ality, and suffering, but, at the same time, the absolute necessity of making them if not visible, then sensible. "One of the things which is important here is honoring (I won't say capturing) the work involved in borderlands and boundary objects" (Star 1994; chapter 6, this volume).

Leigh set us all a challenge—to honor and make visible, without entrapment—ambi-guity, and even to have an *ethics of ambiguity*, while keeping in mind the politics of method, the urgencies of social movements and our own bodies.

In the United Kingdom, sociology has recently been swept up in a debate about new data sources, their ownership, and the consequences for the future of the discipline. In brief, the massive datasets owned by Google had led some to suggest that we need to return to a more descriptive sociology—which doesn't just rely on narrative, but rather seeks to link narrative number and images. In science and technology studies, this attempt has been filtered through John Law's injunctions that we should recognize the importance of mess—both in the world and in our own methods.

In an attempt to be more inventive with our methodological repertoire, I've turned to the studio—both literally immersing myself in studio practice as part of a research process, and more theoretically analyzing what a studio might "do." for STS. For many artists a studio holds a process and makes multiple things present. As the artist Rebecca Fortnum says:

The studio "holds" the process non-sequentially and makes it "present." A plethora of stuff—evi-dence of initial explorations, unfocused desires, material experiments, blind alleys, slight asides as well as labors of love, moments of sheer boredom and utter playfulness—can be simultaneously present. Artists will often collect objects and images not knowing why and "live" with them for a while before utilizing them or not. The studio can afford the artist an encounter with this ac-cumulated material that, initially at least, may not be articulated. (Fortnum 2013)

I should say straight away that I don't think what emerges from studio sociology is "art"—but rather a different kind of sociological engagement, an "inventive sociology." As Celia Lury and I have suggested: "It is not possible to apply a method as if it were indifferent or external to the problem it seeks to address, but that method must rather be made specific or relevant to the problem. . . . Inventive methods are ways to intro-duce answerability into a problem . . . if methods are to be inventive, they should not leave that problem untouched (Lury and Wakeford 2012, 2–3).

The first premise of this practice is to claim that our data may arrive in our studio site as both a traditional product of empirically driven data collection *and* as material for transformation. The traditional interview would enter the studio with an expanded set of affordances—including, for example, the many properties of recorded sound—volume, distortion, background interferences, and so on. With these affordances put centre stage, interviews can be used for novel kind of aesthetic arrangements—subject to exploration and experimentation resulting in new kinds of methodological

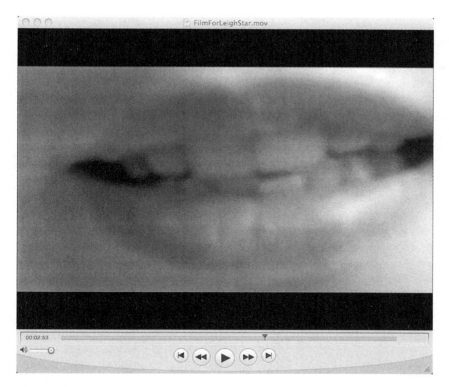

Figure 3.5
Still from *Misplaced Concretism/Film for Leigh Star*
Source: Wakeford and Keal 2010.

engagements. Of course we will have to be careful that we aren't too nostalgic about the capacities of the studio. However if our hopes are a bit embarrassing, let's integrate that too into our expanded sense of affective politics.

The second premise of studio sociology is that we acknowledge and deploy the *multiple* temporalities that may accompany studio work. "Studio time is defined by a mobile cluster of tenses, quotes of past embodied in completed works, some abandoned, others waiting for resurrection, at least one in process occupying a nervous present through which the future plunges . . . a future exerting on the present the pressure of unborn ideas" (Fortnum 2013).

Law (2004) appeals for slow and uncertain methods—in the studio we might encounter processes of both slowing down and speeding up. The corporate and policy-making worlds that attempt to incorporate social science insights and methods sometimes admonish academically focused researchers for not being fast enough—for not being able to keep up with "innovation speed"—a part of our involvement in the

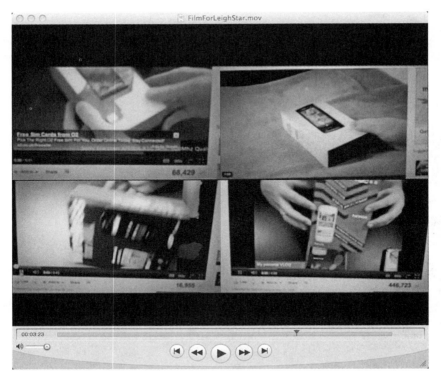

Figure 3.6
Still from *Misplaced Concretism/Film for Leigh Star*
Source: Wakeford and Keal 2010.

anticipatory regime with its associated valences of speculation and enforced preparedness for the future.

The recalcitrance of studio processes—and specific materials—might slow our processes down—but more stimulating is to imagine what might happen to the *forms* that
emerge from studio sociology if they are subject to some version of studio time. This
would bring sociology face to face with where in the future it situates itself—whether it
reaches to some future dates, or rather just to the period we might think of as the
"next." Design studios are quite good at thinking "nextness" —for example, through
the device of the prototype. Studio sociology might expand sociology's techniques for
next-making—perhaps even by taking up some vocabulary of prototyping through
materials.

The third premise of studio sociology is that it would take on board the making and
remaking of the boundaries of the studio as a physical place or metaphor—and become
implicated in (and take responsibility for) what happens when materials cross this

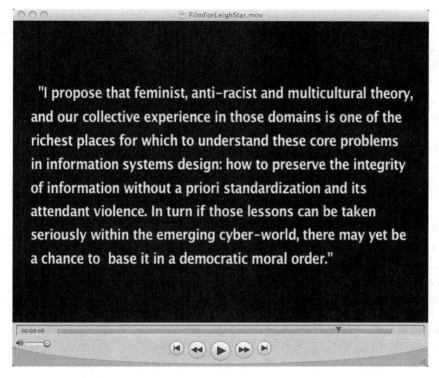

Figure 3.7
Still from *Misplaced Concretism/Film for Leigh Star*
Source: Wakeford and Keal 2010.

boundary. For artists, the moment at which objects leave the studio for the gallery is the moment they "become art"—a particularly transformative moment for contemporary art when the object was Duchamp's urinal. Bruno Latour's exhibition "Making Things Public" has suggested that sociologists might participate in exhibitions to create gatherings and assemblies of objects. Studio sociology might take on exhibitions as an obligatory point of passage. More radical perhaps would be to do so by using the exhibition as a material setting in which attempts might be made to transform materials into practices of involvement—or engagement devices.

"My best hope for this cyborg essay," Leigh says, "is that it will produce *in you* a gestalt switch where multiple memberships and multiple meanings come to the foreground; and it will become *purity* and *universality* that routinely bear the burden of proof" (Star 1994, italics added; chapter 6, this volume).

The working paper (Star 1994; chapter 6, this volume) is an attempt, as she puts it, to begin to define a place where feminism, critical theory, multiculturalism, and information science might meet. We are still, I would argue, in urgent need of more of these meeting places.

In the introduction Leigh then jumps directly into her personal experience—encountering San Francisco in the 1970s, with its culture of radical lesbian feminism.

These were years of "painful heady explorations"—but also of revelations—

This was, for those of you who remember the working paper, the moment of the "turkey baster disaster"—let me summarize Leigh's account.

San Francisco and Oakland contained thriving communities of lesbians, some of whom were trying to have children. The efforts of lesbians to have children, often using the sperm of gay men in San Francisco that was driven across the Bay Bridge to Oakland for insemination, provoked a fundamental rethinking of categories and politics.

This was due to the arrival of men—rumor had it that most of the children born from such insemination were male—some said as many as 85 percent, due apparently to the cooling down of the sperm as it was driven in traffic across the Bay Bridge. The longer the drive the more likely it was to be a boy. Male babies provoked a crisis for a social movement that had such essentialist politics, Leigh argues. Before, men were mutants. After the birth of boy children—"'Men' become modified by 'our sons'; 'mutants' were newly domesticated and intimate" (Star 1994; chapter 6, this volume). "The anecdote is absolutely emblematic of how Leigh works with the category of personal experience. Leigh remembers 'we took the misplaced concretism of sex and re-situated it within the concrete experience of gender and relationships'" (Star 1994; chapter 6, this volume).

Presenting this argument in the Nordic context, the audience must have been baffled by the nature of this object—this turkey baster. For in the working paper a charming hand-drawn sketch of a turkey baster appears on page 9, helpfully labeled to indicate rubber bulb and shaft. And if ever you are looking for a meticulous description of the use of a turkey baster—conventionally vis-à-vis turkeys that is, rather than in its alternative use—I thoroughly recommend the description that goes alongside it on page (Star 1994, 9, and see figure 6.1, this volume)—which is both generous to an audience for whom this object is strange and very funny.

Leigh's first point here is *not just* about the material conditions of experience but *also*, and more importantly, for thinking about marginality, how feminism offers a way to do contradiction, for thinking about many things all at once. She says: "I hold that it is the *all at once-ness* that is at the core of feminist survival, and as a consequence at the core of our relationship to science and technology. The power to hold multiple, contradictory views in a moral collective is necessary in shaping the divergence between Big Brother and a positive cyborg-inhabited multiverse" (Star 1994; chapter 6, this volume).

So from this non-normative personal experience, the working paper begins to elaborate a way of thinking in which an emphasis on method is central. Ever homosocial, Leigh draws on the work of Isabelle Stengers and Donna Haraway writing on science and technology, bell hooks writing on counter-hegemonic discourse, Gloria Anzaldùa "La conscientia de la mestizo," as well as Patti Lather's (1993) attempts to construct alternative validities in the aftermath of poststructuralism. In all of these feminist authors she finds examples of how feminism as a method might create "robust findings" through the articulation of multiplicity, contradiction, and partiality while standing in a politically situated, moral collective.

Star (1994) uses a formal list to reinforce the following attributes of feminist method:

- an experiential and collective basis
- having a processual nature
- honoring contradiction and partialness
- situating historicity with attention to specificity
- the simultaneous application of all these points. (15)

One problem that she picked up on in 1994 was that information systems design had very few ways of understanding or being able to theorize multiplicity, yet alone being able to fully incorporate the demands of feminist method which she was proposing.

So the following sections try to forge a theoretical toolkit for the development of a method of multiplicity for information systems, in an attempt to resist the language and practice of standardization.

Bringing the paper back to an audience of information system designers, she begins to move to more generalized questions that pertain to this field:

1. How can objects inhabit multiple contexts at once, and have both local and shared meaning?
2. How can people living in one community, and drawing their meanings from other people and objects situated there, communicate in another?
3. What is the relationship between the two?
4. What range of solutions to these three questions are possible, and what consequences attend each of them? (Star 1994; chapter 6, this volume)

Leigh's response, of course, is to point to feminist and other allied theories as one of the richest places for understanding the core problems of information systems design. This is why, she says, in my favorite provocation of the pamphlet, computer scientists should read African American poets.

Indicating her debt to her sociological roots, Leigh points out that she understands marginality in what she calls the "old fashioned sense" of Robert Park's "marginal man"—the one who has double vision by virtue of having more than one identity to negotiate—rather than marginality as embedded in a center/margin model.

But she goes beyond saying we should just understand the world from the point of view of the margins—rather this becomes just part of a potential method.

The method toolkit incorporates other elements, woven into an argument in which understandings of marginality and multiplicity function together.

First, we must think of "communities of practice" in so far as they link together actions, people and artifacts. Second, we have to find the "boundary objects" and the work of translation and communication between communities of practice. Third, we should consider "borderlands" and "monsters." She offers the following explanation: "A monster occurs when an object refuses to be naturalized. A borderland occurs when two communities of practice co-exist in one person."

The argument is explicitly about relationality. If information systems design wishes to understand from this a model, it should be one of many-to-many relational mapping—between the multiple marginality of people and the multiple naturalizations of objects through boundaries and standards. This map must point simultaneously to the articulation of selves *and* to the naturalization of objects.

Of course all this articulation is not a trivial matter—it constitutes work, or as Leigh introduces it here, "overhead."

"Every community of practice has its overhead: paying your dues, being regular, hangin', being cool, being professional, people like us, conduct becoming, getting it, catching on . . ."

She puts the concepts of articulation work and invisible work on the table, not just to highlight that they exist, but also to propose that "*modeling* articulation work is one of the key challenges in the design of co-operative and complex computer and information systems." She calls on Lucy Suchman's (2007) work on situated actions to talk about real-time contingencies. A connection that takes her back to the lesbians. "*Driving across the Bay Bridge with the Turkey Baster; what happens when there's a traffic jam?*" (Star 1994, 32)

We can't dismiss anomalies, Leigh argues—for "monsters arise when the legitimacy of multiplicity is denied." There are risks as systems become integrated—as objects or ideas become naturalized in more than one community of practice.

The working paper concludes by directly addressing this issue of naturalization and power. Ideals of generalization and transparency are dismissed. Rather Leigh calls for an ethics of ambiguity—a way of handling multiplicity and marginality. The stress is on recognizing and working with membership of multiple communities of practice, and noting that the objective voice creates the "suffering of monsters in borderlands."

As information systems design was turning its attention to cooperation across heterogeneous worlds, in this working paper Leigh suggested another possibility as method in this field. Drawing on experience—her concrete situations—rather than formal systems with their misplaced concretism, and integrating feminist authors, she poses strong challenges to conventional information design—politically and practically. As the final paragraph makes clear, these ideas have traveled through struggle and anger in social movements just as much as more institutionalized academic knowledges. Revisiting the struggles—our own as well as those of others—enables an opening out of the debates on marginality and multiplicity.

Note

1. A video of this poem/song being sung is available at https://www.youtube.com/watch?v=jfPUELVSLvE.

References

Clough, P. 2009. "The New Empiricism: Affect and Sociological Method." *European Journal of Social Theory* 12 (1): 43–61.

Fortnum, R., ed. 2013. *On Not Knowing*. London: Black Dog Publishing.

Lather, P. 1993. "Fertile Obsession: Validity after Poststructuralism." *Sociological Quarterly* 34 (4): 673–693.

Law, J. 2004. *After Method: Mess in Social Science Research*. London: Routledge.

Lury, Celia, and Nina Wakeford, eds. 2012. *Inventive Methods: The Happening of the Social*. London: Routledge.

Savage, M., and R. Burrows. 2007. "The Coming Crisis of Empirical Sociology." *Sociology* 41:885–899.

Star, Susan Leigh. 1994. "Misplaced Concretism and Concrete Situations: Feminism, Method, and Information Technology." Gender-Nature-Culture Feminist Research Network Working Paper 11. Odense University, Denmark. [See also chapter 6, this volume.]

Stewart, K. 2011. "Atmospheric Attunements." *Environment and Planning. D, Society & Space* 29 (3): 445–453.

Suchman, L. 2007. *Human-Machine Reconfigurations: Plans and Situated Actions*. 2nd ed. Cambridge: Cambridge University Press.

Wakeford, Nina, and Lily Keal. 2010. *Misplaced Concretism/Film for Leigh Star*. YouTube.

4 Anticipation Work: Abduction, Simplification, Hope

Adele E. Clarke

This chapter offers a meditation on some overlapping, nonfungible, and messy work processes of anticipation, intended as a series of provocations to honor the intellectual legacies of Susan Leigh Star, known widely as Leigh.[1] Drawing deeply on her work, I seek a useful vocabulary for some of the *processes of the labor of anticipation, the work of anticipating.* Perhaps most significantly, that work may or may not be visible, may or may not be taken into account and credited, may or may not be valued (at times, like other work, depending upon the doer and/or the legitimacy of that which is antici- pated), and so on (Star 1985, 1991a,b; Star and Strauss 1999). Like other work, anticipa- tion work may be stratified (raced, classed, gendered, masculinized, feminized) and may be glorified or marginalized, outsourced and/or peripheralized. It may be profes- sionalized (expertified, scientized, [bio]medicalized, legalized), venture or labor or other capitalized, technologized and/or deskilled. My goal for this chapter is that antic- ipation work be *recognized* as work, as labor, as effortful, and as potentially fraught, however hopefully it may be undertaken.

In writing on Anselm Strauss's sociology of the invisible, Leigh (Star 1991a, 265; emphasis in original) argued that understanding invisible work involves:

an old sociological/philosophical question: the relationship between the empirical/material on the one hand, and the theoretical/abstract on the other . . . I call this the relationship between the visible and the invisible. . . . The visible things are actions, stuff, bodies, machines, buildings . . . they are many . . . they are . . . everywhere. In facing the tyranny of blind empiricism, however, we temper the clutter of the visible by creating invisibles: abstractions that will stand quietly, cleanly, and docilely. . . . The central insight of Anselm's research is not that the invisible can be restored—for it was never lost—but that the visible and the invisible are dialectically inseparable. *Work is the link between the visible and the invisible.*

Invisibles are routinely created and reified, especially perhaps in the creation of theo- ries, but also in the production of classifications, standards, and organizing systems of many sorts (e.g., Bowker and Star 1999; Lampland and Star 2009; Busch 2012), includ- ing the "educated guesses" so characteristic of anticipation. This is accomplished by what Leigh (Star 1983) wrote about as *simplification*: deleting discussions of the concrete work that went into their production, who did it, the politics of the division of labor, and so on. She wrote: "*Deleting the work* can be understood by listening to another kind of pattern—patterns in the invisibles, such as silences, omissions, areas of neglect" (Star 1991a, 266). Citing two feminists, poet Adrienne Rich (1978) and sociologist Dorothy Smith (1987), Leigh (Star 1991a, 267; emphasis added) sought to "restore agency" to explore the "double understanding of both the organization of the work *and* the organization of the deleted work."[2]

Sustaining Leigh's commitments here, I begin with some historical background on underlying questions which riveted her across her career concerning how uncertainties are managed in different situations.[3] What can happen/is allowed to happen in moments when "things could be otherwise" (Hughes 1971, 552)? Across her life, the frameworks for taking up such questions of uncertainty shifted from the closed- versus open-systems debates that initially engaged her in the 1980s, to attempted management of uncertainties through anticipation work which intrigued her toward the end of her life. I then dwell on three forms of the labor of anticipation which I believe are often (though not always) invisibled/deleted. First I take up *abduction*, the feedback loopings from empirical elements to conceptualizations of them through which we produce and perform anticipation. My focus here is on some of its roots in pragmatist philosophy, especially how "educated" guessing involves considerable work. I then discuss Leigh's (Star 1983) key processes of the work of *simplification* in science. Leigh's major contribution in this early paper concerns how the actual doing of various kinds of work "disappears" through simplifying strategies manifest especially in publications. She carefully noted the loss of complexities and how the important and requisite *work of taking responsibility* for doing anticipatory labor is rendered invisible. Here I try to extend the applicability of Leigh's key processes of simplification into nonscientific spaces and places.

Anticipation (as opposed to dread with which it shares a future orientation) also evokes what Seigworth and Gregg (2010, 9) call "bloom-Spaces." Here hope as promise is a possible angle of vision, even if threat is imminent and hope may also be commoditized (Zournazi 2002a; Brown 2003). Last I turn to *hope*, attempting to specify how it liminally weaves itself into possible futures in relation to anticipation work as fuel, as energy source, as drive, as process, as product (Star 2009). To understand anticipation, we need to grasp how hope(s) for a future are conceptualized, produced, distributed, consumed, *and* invisibled.

These three processes of anticipation work—abducting, simplifying and hoping—may occur in heterogeneous orderings and combinations, often simultaneously. Hope, for example, can be *both* engine and outcome. There are different kinds of tasks involved in each, some of which will be detailed here. They are often held together by what Anselm Strauss (1988) called *articulation work*. Such work organizes the doing of a particular task or set of tasks that require cooperation and coordination to accomplish properly, and that may require efforts to get "things back 'on track' in the face of the unexpected, and modifies action to accommodate unanticipated contingencies" (Star 1991a, 275). Think, for example, of what goes into getting a nonroutine MRI done on a hospital patient, from doctor's orders to nurse's arrangements, to orderlies to-ing and fro-ing, the MRI techs, the radiologists and their department, hospital transport regulations, and the PACS (electronic means of image sharing) which ultimately deliver the goods to radiologists and clinicians, hopefully in a timely fashion (Tillack 2012). Leigh emphasized: *"The important thing about articulation work is that it is invisible to rationalized models of work"* (Star 1991a, 275; emphasis in original).[4] It is a form of labor not usually taken into account, as it often invisibly manages uncertainties in a situation.[5] Yet that management is requisite. In conclusion, I discuss the shifting politics of responsibility for doing anticipatory labor especially in relation to its visibility.

From Closed vs. Open Systems to Anticipation

My intuition is . . . that the world is largely messy. It is also that contemporary social science methods are hopelessly bad at knowing that mess. Indeed it is that dominant approaches to method work with some success to repress the very possibility of mess. They cannot know mess, except in their aporias, as they try to make the world clean and neat. (Law 2007, 595)

Every age has its image of the future (Polak 1961). During the decades after World War II, the concept of "future orientation," a core trait of *The Organizational Man* and the bourgeoisie more generally became widely known (Whyte 1956).[6] Manifest across the social sciences, military sciences, computer and information sciences, international politics and development, it was deeply linked to looking and planning ahead, especially long-term planning. This concept appeared simultaneously with existentialist philosophical assumptions about the relentless neutrality of the universe, sequillae of the atom bomb, and early theoretical precursors to poststructuralism.

In computer and information sciences, where Leigh found herself participating actively as a qualitative sociologist in the early to mid-1980s, quite intense debates in the discipline centered on the usefulness of different metaphors for future orientation. Competition was between metaphors of closed and open systems. Open systems should have the flexibility to deal with uncertainties by their openness to new forms of data, information, all kinds of change, be capable of self-restructuring, and so on. They not only tolerate but engage messiness. In sharp contrast, formalist closed systems were often linked to computer modeling where data were bounded or pregiven and formal models based upon that data were to be imposed. Star's most famous publication of that era was a manifesto for open systems titled "The Structure of Ill-Structured Solutions: Heterogeneous Problem-Solving, Boundary Objects and Distributed Artificial Intelligence" (Star 1988). Her engagement with these questions endured. Her last book (Lampland and Star 2009) was titled *Standards and Their Stories: How Quantifying,*

Classifying and Formalizing Practices Shape Everyday Life. The book she was working on at her death was to be titled *Against Formalisms*. In short, much if not most of her scholarly life was spent in engaging and challenging closed-systems approaches to managing uncertainties, and in trying to specify the kinds of losses and missed alternative solutions which such systems engender, sometimes tragically.[7]

Very similar debates happened across many other disciplines. In economics, closed systems forms of modeling and formalisms won the day. Praxeology, central to the Austrian economics of von Mises ([1962] 2006), Hayek ([1944] 2007) and Friedman ([1962] 2002), adheres to this approach, so important to the transnational rise of neoliberalism, especially in its Chicago economics framing.[8] In Leigh's and my discipline of sociology, since World War II these debates have centered on the validity and utility of quantitative versus qualitative methods. Initially, quantitative survey research, the empirical and ideally predictive arm of functionalist social theory was dominant nationally and exported transnationally, especially through demography. More recently, this evolved into rational choice theory and various forms of quantitative modeling (e.g., Eriksson 2011). In the late 1960s, qualitative approaches were resurrected from their prewar slump by a groundbreaking and discipline-challenging book, *The Discovery of Grounded Theory* by Barney Glaser and Anselm Strauss (Glaser and Strauss 1967). Both had been Star's teachers in graduate school. This book was a founding manifesto of what has since become a transnational and transdisciplinary renaissance in qualitative approaches to understanding social life not only across the social sciences and humanities but also in professional education as well.[9]

Symbolic interactionist sociologist Herbert Blumer captured the closed versus open systems debate in sociology during the late 1960s in his contrast between definitive versus sensitizing concepts:

[T]he concepts of our discipline are fundamentally sensitizing instruments. Hence, I call them "sensitizing concepts" and put them in contrast with definitive concepts. . . . A *definitive concept* refers precisely to what is *common* to a class of objects, and by the aid of a clear definition in terms of attributes or fixed bench marks. . . . A *sensitizing concept* lacks such specification. . . . Instead, it gives the user a general sense of reference and guidance in approaching empirical instances. Whereas definitive concepts provide prescriptions of what to see, *sensitizing concepts merely suggest directions along which to look.* (Blumer 1969, 147–148; emphasis added)

The implication of Blumer's distinction is that the precision of definitive concepts does not work well—is not as helpful—in analyzing the messy world we see about us. In a very recent contribution to this ongoing debate John Law (2007, 595), quoted in the epigraph above, brilliantly calls methods that simplify and tidy up life "forms of hygiene" (see also Taylor 2005). Such methods manage the uncertainties of complexity by building "great walls" to exclude them and focusing solely on what is inside. While such methods may protect users from the mess largely by deleting it and other uncertainties from consideration, they do little to analyze it.

It can be argued that today anticipation and related concepts such as promissory futures are displacing "future orientation." One interesting new feature is that anticipation explicitly connotes affect while sustaining a temporal orientation (e.g., Adams, Murphy, and Clarke 2009). Anticipation is an affective state highly congruent with the demands of neoliberal capitalism and its requisite expansions (Lave, Mirowski, and Randalls 2010). Further, anticipation is especially characteristic of biocapitalism with its deeply uncertain and "promissory" properties (e.g., Thompson 2005; Sunder Rajan 2006, 2012), and of biomedicalization (Clarke et al. 2010), the often-dramatic expansions of biotechnological and other technoscientific approaches in biomedicine and public health (Mackenzie 2013). As Sunder Rajan (2006, 14) noted, there is "a shifting grammar of life, toward a future tense." Anticipation can thus be linked to the emerging "sociology of expectations" that centers on scientific and technological innovation (e.g., Brown 2003, 2005; Brown and Michael 2003; Brown, Rappert, and Webster 2000; Pollock and Williams 2010). Here "expectations can be seen to be fundamentally generative," the shared energizing impetus toward innovation despite uncertainties (Borup et al. 2006, 285).

There is also, however, an essential and sustained tension. Hope is paradoxical (Mattingly 2010) and its flip side is dread. Hope can no longer pretend to be naive. For example, failed expectations haunt contemporary biomedicine (e.g., Nerlich and Halliday 2007), where unfulfilled promises of the past such as gene therapies, often prematurely touted in the media, are increasingly tracked (e.g., Morrison 2012; Prainsack, Geesink, and Franklin 2008). Recently Tutton (2012) analyzed the manifestations of the reverse of hope in biotech. Investment materials are now also "promising pessimism," offering explicit articulations of risks and potential down sides of biotech research as well as the up sides. Thus these materials can themselves demonstrate *both* "the volatility and the promise of the life sciences in the 21st century." In its explicit recognition of uncertainties, its appropriation of uncertainty, anticipation is always already haunted (Gordon 1997; Murphy 2013; Novas 2006).

In this chapter, however, I am especially interested in anticipation as affect and potential in Brian Massumi's more Deleuzian sense: "I use the concept of 'affect' as a way of talking about that margin of maneuverability, the 'where we might be able to go and what we might be able to do' in every present situation" (Zournazi 2002c, 212). It is this margin of possibility, of potential movement, that Vincanne Adams, Michelle Murphy, and I argued is anticipation as characteristic of the contemporary moment, pervasive in discourse *and* action, however haunted and rife with contradiction and loss or risk of loss (Adams, Murphy, and Clarke 2009, 246):

One defining quality of our current moment is its characteristic state of anticipation, of thinking and living toward the future. Anticipation has epistemic value, a virtue emerging through actuarial saturation as sciences of the actual are displaced by speculative forecast. It is a politics of temporality and affect. Key dimensions are: injunction as the moral imperative to characterize and inhabit states of uncertainty; abduction as requisite tacking back and forth between futures, pasts and presents, framing templates for producing the future; optimization as the moral responsibility of citizens to secure their "best possible futures"; preparedness as living in "preparation for" potential trauma; and possibility as "ratcheting up" hopefulness, especially through technoscience. Anticipation is the palpable sense that things could be (all) right if we leverage new spaces of opportunity, reconfiguring "the possible."

Anticipation as this "palpable sense that things could be (all) right *if* . . ." we can just figure out how to act in Masumi's "margin of maneuverability" is the grounding of this chapter. It was this edginess that intrigued Star.

One largely ignored facet of anticipation is how much work it actually takes to "optimize," to "live in preparation," to anticipate—to gather information, calculate, consider, plan, foresee, decide, act, and so on. The pragmatist symbolic interactionist sociology in which Star and I were steeped has long focused on work, occupations, and professions—from the factory floor to the scientific laboratory and especially the clinic and hospital—and the management of heterogeneous uncertainties in those settings (Shaffir and Pawluch 2003). Focusing (usually through fieldwork) on *what people were actually doing* (rather than what they *said* they were doing) allowed early scholars to see tasks as multifaceted, collaboration as complex and stratified, and the work per se as quite heterogeneous, both under- and overvalued, and so on (e.g., Becker 1970, 1986; Strauss [1971] 2001, 1988; Strauss et al. [1985] 1997).[10]

Drawing upon Leigh's and my shared history of involvement with both the interactionist sociology of work and feminist attention to hidden/invisibled work, I next elaborate three of the many forms of work of anticipation: abduction, simplification and hope. For simplification's sake, I discuss them one at a time, but in practice, they are often pursued simultaneously or in varying order, messily munged together. Ultimately, they are co-constitutive. Further, and a point to which I shall return, they are each and all both cognitive and experiential, affective and effective as they facilitate our managing of uncertainties (Star 2007; Adams, Murphy, and Clarke 2009).

Abduction

[A]ll thought would be totally impossible in a universe in which abduction was not expectable. (Bateson 2002, 134)

One key form of anticipation work is abduction. The work of abduction is predicated upon collecting appropriate sorts and amounts of data or information, shaped by the specific problematics of the situation at hand.[11] It then involves tacking back and forth multiple times between the empirical information collected (possibly with great care and at considerable expense) and new theorizings about that data to generate new conceptualizations—and adding a future-orientedness to its utility. Dumit and Meyers wrote similarly of "a positive feedforward cycle" where "attention becomes relatively fixed, as if one may now know what to look at and what to look for."[12]

While coined during the Enlightenment,[13] the concept of abduction has its modern philosophical roots in the work of American pragmatist Charles Sanders Peirce (1839–1914). Abduction is commonly contrasted with induction. For Peirce, "Abduction seeks a theory. Induction seeks for facts" (Peirce quoted in Fann 1970, 35).[14] He argued that the process of abduction—the goings back and forth between information and theorizing through that information—is a form of guessing. And Peirce's guessing is distinctively pragmatist—pragmatic in the sense of being useful rather than expedient (discussed below). He spoke of abduction as a kind of "guessing instinct" (ibid.; Peirce 1929). This chapter could actually be subtitled "The Injunction to Keep on Guessing" as this is the heart of anticipation work.

Abduction is not random, however. It is guessing for very specific reasons. Peirce asserts that abduction "supposes something of a different kind from what we have directly observed, and frequently something which it would be impossible for us to observe directly" (Peirce quoted in Fann 1970, 9). Abductive guessing thus involves going beyond the known. Peirce argued, "An hypothesis . . . has to be adopted which is likely in itself and *renders the facts likely*. This step of adopting a hypothesis as being suggested by the facts, is what I call *abduction*" (Peirce quoted in Fann 1970, 31; emphasis mine).[15] Peirce further asserted that a hypothesis adopted in this way could only be adopted *"on probation"* and *must be tested* (ibid.; emphasis mine).[16] I will return to both the tentativeness and sustained efforts of inquiry implied in Peirce's term "on probation."

Experiential

Abduction is not solely intellectual or cognitive, but also experiential. "The abductive suggestion comes to us like a flash. It is an act of insight, although of extremely fallible insight" (Peirce quoted in Fann 1970, 35). Dewey similarly talked about "suggestions" that "just spring up, flash upon us, occur to us" when we are open to them (quoted in Strübing 2007, 593). One of the German scholars who has recently pursued abduction in its pragmatist formations is Reichertz (2007, 220), who discussed Peirce's ideal outcome of "lightning striking." He finds this process both seductive and exciting: "The secret charm of abduction lies straight in this kind of inference-being: abduction is sensible and scientific as a form of inference, however *it [also] reaches to the sphere of deep insight and new knowledge*" (ibid., 216; emphasis added). In abducting, "one has decided (with whatever degree of awareness and for whatever reason) no longer to adhere to the conventional view of things . . . [seeking instead] . . . a creative outcome which engenders a new idea. This kind of association is not obligatory, and is indeed rather risky" (ibid., 219).

As one work process of anticipation, I am trying here to point to the ongoing loop-ings, the "feedforward loopings" involved in abduction and what they engender.[17] For example, Fann further argues that abduction generates a gestalt that is greater than the sum of its parts: "something not implied in the premises is contained in the conclu-sion" (Fann 1970, 34). Massumi (2002, 211) later asserted that a larger gestalt itself becomes an excess that *enters* the situation. Thus abduction may involve surplus, excess, an affective "charge" that *itself enters the situation in which it is occurring,* propel-ling it. Lightning strikes indeed!

Peirce detailed two quite different "macro-strategies" well-suited to "enticing" abductive processes or creating a "favorable climate" for such lightning strikes to occur (Reichertz 2007, 221). First, "the presence of *genuine doubt* or *uncertainty* or *fear* or *great pressure to act* is a favorable weather situation for abductive lightening to strike" (ibid.; emphasis in original). Second, Peirce recommended "musement" as a strategy to pro-voke successful abduction: "Enter your skiff of musement, push off into the lake of thought, and leave the breath of heaven to swell your sail. With your eyes open awake to what is about or within you, and open conversation with yourself: for such is all meditation" (Peirce quoted in ibid.). Here, instead of near panic, leisure and relative calm are requisite. We can characterize these two strategies as desperation and reflec-tion, and note that there are spaces between as well. Paradoxically, Reichertz (ibid.) remarks that the capacity for abduction is "a state of preparedness for being taken unprepared."

Pragmatic

Pragmatist philosophy provides the ontological and epistemological grounding for symbolic interactionist sociology, one of Leigh's and my traditions, providing the grounding elasticities and flexibilities that allow it to change, to respond to new situa-tions. Significant here, "Peirce considered abduction to be the essence of his pragma-tism. He insisted that it was essential to history, that it constituted the first stage of all inquiries, and that it was a necessary part of perception and memory" (Fann 1970, 5). For Peirce, pragmatist philosophy saw hypotheses "as explanations of phenomena held as *hopeful suggestions*" (Peirce quoted in ibid., 44; emphasis mine).[18]

In sharp contrast with conventional use of the term to indicate pursuit of narrow economic advantage and/or exploitation, pragmatism here is concerned with utility—*usefulness* for the goals at hand—whatever they may be (e.g., James 1921; Maines 1991). An abductively generated "theory" or "suggestion" is allowed to remain in force as long as it is useful, helpful, until we run into its limits and must seek further. "[I]f it shows itself to be useless, it is abandoned" (Reichertz 2007, 222). "To say that a given theory is an interpretation—and therefore fallible—is not at all to deny that judgments can be made about the soundness or probable usefulness of it" (Strauss and Corbin 1994, 279).

The ultimate pragmatist test of a theory is its usefulness—which may, and likely will, end. Anticipation work is never done.

Toward the end of her life, Star (2007) wrote on cognitive and emotional forms of pragmatism. She documented her early lived experiences of seeking a mode of abductive analysis, and the impossibility of separating much less letting go of either the cognitive or emotional aspects of the work: "I wouldn't have known how to say it, exactly, but I was looking for a way simultaneously to incorporate formal and informal understandings of the world. I sought a methodological place that was faithful to human experience, and that would help me sift through the chaos of meanings and produce the eureka of new, powerful explanations" (ibid., 76–77). That method was grounded theory.

Grounded Theory Analysis: An Exemplar of Abduction Work   Abduction has been sufficiently taken up in the interpretive human sciences over the past several decades to constitute an "abductive turn," particularly lively in the literature on qualitative research (Reichertz 2007, 216). There it has been deemed characteristic of grounded theory, the analytic approach generated by Barney Glaser and Anselm Strauss (Glaser and Strauss 1967), in which both Leigh (Star 2007) and I were trained. Glaser and Strauss (ibid., 239) themselves had used the term "induction" to indicate moving from empirical materials to conceptual framings of them—a "grounded theory" of the substantive area of research.[19] More recently, especially vis-à-vis the Straussian form of grounded theory—this process is now seen as "abduction."[20] Philosophically, this tacking back and forth is called "'abductive' reasoning . . . a sort of 'third way' between the Scylla of inductive reasoning and the Charybdis of hypothetico-deductive logic" (Atkinson, Coffey, and Delamont 2003, 149).

Highly exemplary here, Strübing (2007, 595) offers a helpful diagram of evolving grounded theory through the researcher's abductive attitude of pragmatic problem solving (see figure 4.1).[21] But this diagram is too neat. In practice, abduction involves zigs and zags zooming off the page—up, down, sideways, a more three-dimensional topographic orientation. Strübing (ibid., 585) in fact asserts that "theory is process" in pragmatism: "The presumption of the fluid and interactive character of reality results in the need for a similarly processual understanding of theory." To continue to take new data (pertinent information of any kind) into account demands "an openness of the researcher, based on the 'forever' provisional character of every theory" (Strauss and Corbin 1994, 279). Here we can hear echoes of Peirce's adopting a hypothesis only "on probation," testing it repeatedly, and changing it as necessary to maintain its utility.

In grounded theory we also have an exemplar of abduction as a *social* process. A key part of "doing" grounded theory has been doing analytic and other research work in

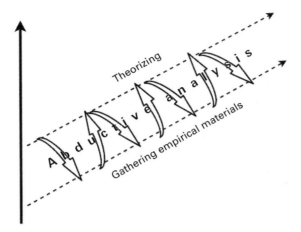

Figure 4.1
Abductive analysis
Source: Based loosely on figure 27.2 in Strübing 2007.

small groups. Glaser and Strauss both created small groups in the classes where grounded theory was initially taught and qualitative research continues to be taught in this fashion at UCSF (Clarke 2012) and elsewhere. Small groups were also full-on working groups for major grounded theory research projects (Lessor 2000; Wiener 2007). Working groups provoke analysis by members having to come to grips again and again with others' perspectives, others' interpretations of data, others' commentaries on preliminary analyses, and so on. Strauss (1987) believed so strongly in such group processes that in one book, he actually provides transcripts of group sessions—literally inscribing the work of abduction—week after week.

Peirce himself had presumed some social grouping of "abductors": "[P]rogress in science depends upon the observation of the right facts by minds *furnished with appropriate ideas*" (Peirce quoted in Fann 1970, 31; emphasis in original). That is, some things have to pave the way for an abductive theory to make sense and those things need to be more or less shared among the abductors. Thus sociality can be both *source and process* of abduction, and certainly a work site.[22]

Doing the work of abduction also involves us in simplification efforts, to which we next turn.

Simplification

Here I seek to elucidate some of the work of simplification: deleting, sorting, rearranging, (re)presenting. My guides are Star's (1983) key paper on simplification in scientific work, and our shared mentor Anselm Strauss. Leigh starts her paper with a quote from pragmatist philosopher William James: "The facts of the world in their sensible diversity are always before us, but our theoretic need is that they should be conceived in a way that reduces their manifoldness to simplicity. . . . The simplified result is handled with far less effort than the original data" (ibid., 205).

Simplification is characteristic of life much more generally, not only the scientific work about which Star was writing, and is requisite for anticipation work. It involves more explicitly situated analytic claims making and the avoidance of both the overgeneralization and overabstraction (e.g., van den Hoonaard 1997) that is such anathema to Law's injunction for us to acknowledge the mess. In Daly's (1997, 360, 353) words, the challenge is to engage the emergent theorizing "not as objective truth but as a located and limited story. . . . [T]o keep theory in play but to redefine theory in a way that keeps the theorist in play." Open systems indeed—with ourselves as active elements!

Star (1983, 206) wrote the following about simplification in scientific work, which I have edited as you can see: "[W]ork is complicated. Any set of ~~scientific~~ tasks involves multiple problems, qualifications, exigencies, demands and audiences. To work without getting lost in endless exigencies, ~~scientists~~ [people] must draw boundaries and exclude some kinds of artifacts and complications from consideration."

Star's paper focused especially on the kinds of constraints scientists experience that push them toward various kinds of simplification. Taking off from Leigh's (Star 1983, esp. 211–212) work, I attempt to render these "pushy constraints" more generically.[23] That is, I think the constraints Leigh specified pertain far beyond formal science to any situation of inquiry. Constraints that may push us to simplify include the following:

• the need or desire to work across multiple communities of practice, disciplines, or social worlds, which promotes simplification of language, preference for using mutually known technologies, etc.;
• the need or desire to generate useful applications (new explanations, clinical alternatives, etc.);
• the need or desire to gather data quickly and/or from a delimited domain and/or with limited technical capacities;
• pressures toward drawing conclusions before sufficient data or analysis has occurred;
• pressures to formulate conclusions in delimiting ways to be comprehensible by particular (and possibly varied) audiences;
• representation rules or limits which demand certain stylistics, deletions, and/or alterations; and
• specializations which demand certain technologies, deletions, and/or alterations.

All these constraints should feel familiar to us, as we live with many of them routinely in academia. Our own publications are exemplars of simplification work. The constraints may push us or push back against us. But pushy they are.[24]

Simplification Work

To live in anticipation as a routine part of life, we do the work of simplification. We may do so before, during, and after abduction, and simplification is often part of hoping as well. As Leigh argued:

> The analytic premises here are that: (1) one does not understand work simply by understanding its products, nor (2) understand it *without* reference to the products; that (3) work involves joint effort over time, and thus is both interactive and processual; and that (4) meaning does not inhere in the nature of ~~scientific~~ work, but is continually renegotiated by workers and consumers. *The units of analysis employed here are tasks and activities*, not individuals or their allegiance to theories [or policies]. (Star 1983, 210; second emphasis added)

Her key point, that the units of analysis in understanding simplification work are tasks and activities and not individuals or their allegiances, is quite radical. It immediately shifts the analysis to concrete practices or processes of the work rather than center on the workers.[25] Leigh's paper was thus prescient of the "turn to practice" in science and technology studies and beyond (e.g., Schatzki et al. 2001).

Simplification may involve many kinds of work processes and practices including (Star 1983):

- breaking problems into pieces to produce manageable, "doable" (Fujimura 1987) projects and tasks;
- simplifying the amounts and kinds of data or information to be collected and limiting sources;
- de- and re-contextualizing as part of the work involved in transporting meaning out of and into different contexts (e.g., Almklov 2008);
- defining and managing "saturation"—when data already gathered are deemed to be sufficient, "enough";[26]
- limiting our theorizing; and
- constraining our representations.

All of these or only some may be used in particular situations. They offer a repertoire of possibilities, with accompanying risks.

Simplification is necessary because of TMI (too much information)—the increasingly common situation in which there is too much information, too much data to manage—or too much affect. Leigh (Star 2007, 89) herself wrote about "flooding" as a state of being overwhelmed by data, codes, comparisons, provisional arrangements, and partial theorizings: too much "seems worthy. How to weed? How to choose?"[27] Such a situation is not, she argues, a romantic interlude, but rather, in Dewey's words: "a call to effort, a challenge to investigation, a potential doom of disaster and death" (Star 2007, 89). In short, inadequate simplification can be deadly and yet doing it most painful. Lather (2001, 202–203) beautifully captures dwelling on the horns of this dilemma as part of the process of doing qualitative research: "Placed outside of mastery and victory narratives, ethnography becomes a kind of self-wounding laboratory for discovering the rules by which truth is produced. Attempting to be accountable to complexity, thinking the limit becomes the task, and much opens up in terms of ways to proceed for those who know both too much and too little."

A key question, of course, is what are the risks of simplification? What is likely to "go missing"? What is likely to be invisible(d)?[28] Drawing again on Star's (1983) paper, these include the following:

- relations *between* pieces of problems;
- material and political elements of the situation;
- technical and theoretical elements of the situation;
- the precise nature of the reconciliations: what was ceded? what was left on the table?;
- discussion of the actual work done in abduction and simplification and hoping;
- an accounting of who actually did the different tasks involved and who did the articulation work—took overall responsibility;
- temporality; etc.

In sum, Leigh argued that what goes missing via simplification are the processes, work, and relationships that went into it. It is their absence, as well as those of the corporeal bodies involved (Casper and Moore 2009), that allows science to appear "objective," "inevitable," the "culture of no culture" (e.g., Haraway 1997; Traweek 1999).

Leigh (Star 1983, 1986, 1989) was particularly concerned with the forms of mobility that "truths" produced through simplification in scientific work gained, allowing them to move across specialties and disciplines, gaining allies and generating successful scientific social worlds, communities of practice. Boundary objects par excellence! Sleek and mutably mobile, such simplifications could evade and dance around rules of evidence, she argued, giving us lessons for analyzing simplification—and not only in scientific work.[29] Wherever it is done, simplification work is always political work, downplaying some uncertainties and featuring others. Grasping the politics involved is itself an intrinsic part of it, commonly erased.

Hope

"I have to say I'm confused about hope, about how it feels and what it can do."
(Duggan and Muñoz 2009, 275)

Anticipation comes pre-wrapped in affect—hopefully inflected. But hope is "a slippery little concept" (Eliott 2005, 29). It is a classic keyword in Williams's ([1976] 1983) sense that it is many things and its various situated genealogies need to be (re)traced across time and space. Public discourses (rather than personal engagements) about hope largely evanesced around spiritual/religious venues or vis-à-vis politics, only expanding in the twentieth century. Here I briefly track public discourse about hope, noting the changing domains of its liveliness. I then discuss hoping as part of anticipation work.

A Conceptual Genealogy of Hope
Hope can be "a stance toward reality that requires careful cultivation."
(Mattingly (2010, 4)

Hope discourse is on the move. A recent history asserts that it fell largely under the aegis of religion until the twentieth century; then, as secular versions were formulated, hope was relocated in the promises of modern science, technology, and medicine (Eliott 2005, 4; see also Hughes 1989). For example, the rise of scientific medicine was distinctively marked as hopeful in what I called "healthscapes," visual cultural manifestations of the institutional displacement of the church by huge cathedral-like hospitals to house these new forms of hope at the turn of the twentieth century (Clarke 2010; Hansen 2009).

Yet as late as mid-twentieth century, hope was still described as a "tabood topic," inappropriate for public discussion (Eliott 2005, 4). Nor was it high on scholarly agendas. But after World War II, philosophers began exploring hope (ibid.), I suspect in response to Sartrian existentialism that questioned the existence of God and hope alike in the aftermath of the Holocaust and atomic warfare. Gradually, beginning in the 1960s, the salience of hope in medical outcomes began to be explicitly explored, soon generating a tidal wave of work into the present moment, including the placebo effect (Eliott 2005, 30; Harrington 1999). A key shift to explicit discourse analyses occurred here with DelVecchio Good and colleagues' (DelVecchio Good et al. 1990) social constructionist framing of the biomedical "discourse on hope," noting too that this discourse had generated a "political economy of hope" in biomedical domains.[30] (I return to this below.)

Perhaps the site of the most activity around hope across the twentieth century and into the present lies in consumption. Advertising and public relations disciplines produce the vast industry of hope discourses so characteristic of "The Century of the Self."[31] Personal identity transformation via consumption became hope incarnate, and hopeful anticipation of the consequences of purchasing are now understood as much more wondrous than actual ownership or use (e.g., Miller and Rose 1997; DeGrazia 1996; Zukin and Maguire 2004).

A "preoccupation with the future" was also central to the rising interest in planning so characteristic of the hope for a better world in the postwar era (Eliott 2005, 16). Planning of all kinds exploded—somehow rescued from its earlier Soviet taint—or perhaps in response to Soviet initiatives linked to the Cold War (Leys 2005; Cooper and Packard 2005). Extensive military/defense planning including major technological initiatives such as the Defense Advanced Research Projects Agency (DARPA), initiated in 1958, that bridged (or blurred) governmental, industry, and academic divides in innovative ways and continues to do so (Belfiore 2009).[32]

While a public discourse of hope has always cathected to politics (e.g., Laclau and Mouffe 1985; Duggan and Muñoz 2009), Zournazi (2002b) has recently noted that the New Right has been especially successful in mobilizing hope via promoting highly simplified versions of both problems and solutions. As Star well knew, if we render uncertainties invisible, hope work is much easier. Alternatively, Papadopoulos and colleagues (Papadopoulos, Stephenson, and Tsianos 2008, 73) argued that the optimizations so smoothly proffered by biocapital may be subverted by what they call "imperceptible politics": "It is only possible to work on the real conditions of the present by invoking imaginaries which take us beyond the present." These can be hopeful imaginaries or what Erik Olin Wright called "real utopias" toward which to work, doable projects to instantiate and generate political hopefulness.[33] They answer "yes" to the question, "Can collective hope without delusion or guarantee generate future possibilities beyond any present expectations?" (Duggan and Muñoz 2009, 276).

In the new millennium, hope has also become a major incitement to discourse in academia. Pragmatism offers renewed possibilities (e.g., Koopman 2009). A key theoretical provocateur about "hope today" was Mary Zournazi (2002a) who interviewed major intellectual figures about their views.[34] As part of "the affective turn" (Clough 2007, 2010; Puar 2009), Seigworth and Gregg (2010) began their edited collection with an essay titled "An Inventory of Shimmers" about elusive but present sites of hope and possibility. In anthropology, Jason Antrosio asserts a "moral optimism" "that what currently exists does not have to be"; moving toward such new spaces involves going beyond critique to explore noncapitalist alternatives (Burke and Shear 2013, 17). As part of "teaching for hope" (Lyon-Callo 2013), academia is actively producing an often-feminist literature on queer as a new hopeful strategy (e.g., Munoz 2009; Puar 2009; Valentine 2007). Anticipating and hoping are "practices which are at the heart of social transformation long before we are able to name it as such" (Papadopoulos, Stephenson, and Tsianos 2008, xii). In short, hope has "come out" and is alive and dwelling in interesting niches.

Hope Work   Here I want to focus on the *work* of hoping as part of anticipation and often intense. Reminiscent of or perhaps extending Foucault's ([1982] 1997) technologies of the self, hope work is often seen as a moral responsibility. Eliott (2005, 28) predicted that as hope becomes "something that the individual can do for themselves and learn to do better," there will be more how-to books and other forms of hope-full injunction. And today we do see many reincarnations of Norman Vincent Peale's *The Power of Positive Thinking* (Peale 1952) and its non-Christian kin, and a plethora of caregivers seeking to instill hope, from therapists to life coaches and personal trainers as Foucault ([1982] 1997) anticipated. Rose's (2007) argument that "optimization"—the increasing legitimacy of securing "the best possible future" that is requisite today—centers on hope work. Dumit (2012) has shown that such optimization may require ingesting drugs for life in hopes of reducing risk. Yet hope work itself is risky as hope usually comes prepackaged with its flip side. Overmedication is one risk (ibid.) and panic is another, as Orr (2006) has exquisitely described. Berlant (2006) warns us about the potential cruelty of optimism.

One space of explicit scholarly encounter between hope and its absence lies in debates about "regimes of truth" versus "regimes of hope" (e.g., Moreira and Palladino 2005; Brown 2003, 2005). These debates have implications about the labor of hope—hope work. Brown and Michael (2003, 4) argued that we "need to shift the analytical angle from *looking into* the future to *looking at* the future, or how the future is mobilized in real time to marshal resources, coordinate activities and manage uncertainty." That is, hope for the future is hope in a future that is envisioned and articulated now—via anticipation and hope work.[35] They characterize a "broader turn in the very basis of the politics of the biosciences . . . a shift from facts and evidence (regimes of truth) toward a conduct of debates through the meta-abstractions of hope, expectations, and the future" (Brown 2005, 339)—a "regime of hope" and its false ally "hype."[36]

In a major response pertinent here, Moreira and Palladino (2005) argue that these regimes are not polarized and instead are so mutually imbricated as to make them mutually parasitic.[37] I next walk you through Moreira and Palladino's definitions that are framed in terms of biomedicine but have more general implications for hope work. They assert the following:

> [T]hese two positions can be understood as deploying more general forms of argument, which, rhetorically, revolve around the tropes of "truth" and "hope" . . . The "regime of hope" is characterized by the view that new and better treatments are always about to come, being tested, "in the pipeline." More specifically, research and development is justified by the promise of finding miraculous cures for debilitating diseases. Such promise entails endless deferrals to stabilize the identity of the therapy, its constituents and effects, deferrals that can be justified in various manners. . . . The following opposition of "truth" and "hope" perhaps best captures the spirit of such deferrals: "We do not know the truth: there is hope."
>
> The "regime of truth," on the other hand, entails an investment in what is positively known, rather than what can be. That is to say, it is characterized by the view that most medical therapies are less effective than claimed, and this involves the constant returning of new and promising approaches to their original claims, their clinical failures and their ethical downfalls. [T]his view . . . can be then opposed to the "regime of hope" by a quite different opposition of "truth" and "hope" . . . "We know the truth: there is no hope." (67)

In terms of anticipation work, I am arguing here that there are commonly "doubled" efforts (Lather 2007) that engage these two regimes of hope and truth, negotiations between some kinds of "coming to terms with the possible" and *at the same time* attempting to generate and produce other kinds of "hopes for the future" regardless of the "truth." The politics of such prospections are intense and ever with us, operating at once at individual, collective, national, and supranational levels: Is the glass half empty or half full? Are the very real miracles of technoscience and biomedicine worth the risks incurred and the doors forever closed by committing to them? Are more direct forms of democracy possible?

Both expectation and anticipation are future-oriented modes of being, with anticipation more cathected to hope. Hope is a primal energy source for action. The work of generating or producing, distributing, and consuming hope is at the heart of anticipation along with abduction and simplification. As Peirce (quoted in Fann 1970, 37) argued, abduction "goes upon the hope that there is sufficient affinity between the reasoner's mind and nature's to render guessing not altogether hopeless."

In the end then, in anticipating, hoping (the gerund connotes the work—the action) must "go beyond hope,"[38] go to the end of hope. Hoping must not only encounter but also engage its risks and possibilities of loss and failure. The very daringness of hope and of hope work is captured in the interesting phrase "hoping against hope."[39]

Exemplars of Anticipation Work

Thus far, my discussion of the processes of anticipation work has been quite abstract. Here I illustrate them in two exemplars from the health field with which I am most familiar. These exemplars themselves foreground different processes, providing some sense of the range of variation in the actual doing of anticipation work.

First, Martine Lappe (2012, 2014) followed an innovative prospective epidemiologic study focused on autism risk factors during fetal development and early childhood designed to investigate some possible environmental causes. She interviewed autism researchers and mothers who had a child already diagnosed with autism and were expecting another child (whose risks for developing autism would be heightened). During that pregnancy, the families agreed to participate in this long-term (multiyear) autism gene-environment interaction study. Lappe argues that the research she followed is itself situated within a larger social and scientific process she calls *anticipating autism*. This captures how autism is increasingly being produced in the United States as a possible future that should be *preemptively* considered, acted upon, and engaged in the present by prospective parents. Anticipation work indeed! Lappe found mothers doing the bulk of the work around planning for a potentially autistic second child, including orchestrating their families serving as research subjects and opening their homes to routine monitoring, in part to access services they might need.

The abduction work that the mothers Lappe studied relentlessly pursued was assessing their new baby's development against both traditional developmental yardsticks and, much more fraught, against the remembered developmental patterns evinced by their older child later diagnosed with autism. This abduction work was both intense and relentless precisely because it is believed that the earlier autism is diagnosed and treated, the better. Windows of opportunity for successful intervention in early child-hood can abruptly close. One form of simplification work that engaged the mothers extensively was learning about and assessing the many forms of environmental risk to which they, their autistic child, and their fetus or newborn were being exposed. Today information about environmental risks in foods, homes, cars, and so on, is in essence infinite and infinitely conflicting, and these mothers had to both limit their gathering and simplify their processing of it. Which information sources were trust-worthy? Which risks mattered more? Less? How could they best be managed? Last, hope work permeated these women's lives, perhaps most especially hoping that their research participation would alleviate the risks of another of their children being diag-nosed with autism. But they also hoped that the environmentally focused research would clarify causality and enable a reduction in autism rates—hoping for other fami-lies as well as their own. Lappe's at-risk and highly motivated participants were con-stantly looking back to move forward, looking ahead to act now, awash in anticipation work.

My second exemplar is also medically oriented but quite different, focusing on the increasing shift of chronic illness management from the clinic into the home, seeking to reduce health care costs by relying more on free labor done largely by family. While such labor shifts occur across the life course, Stefan Timmermans and Betina Freidin (Timmermans and Freidin 2007) studied caretakers of school-aged children with asthma, mostly mothers. They sought to understand how health professionals socialize mothers into intensive caretaking for their children with asthma and how the mothers navigate and negotiate[40] that work. The more effective the mother is in her efforts, the less likely the child will be to have an asthma attack requiring medical intervention, usually in a costly emergency room. Providers "called upon mothers' affection for their children, relying on a mixture of fear, hope, uncertainty and reassurance to emphasize that extensive health interventions [by the mother] were needed to keep the child's asthma under control" (ibid.,1360).

Articulation work—the hidden coordination, timing, checking up, making sure, and overall taking of responsibility across the full trajectory—was found to be central to caring for an asthmatic child: "Care providers assign broad caretaking tasks that require further articulation work to get the job done. . . . Articulation work was linked to the uptake of responsibility through care providers' delegation [largely to mothers] of dif-fuse asthma caretaking tasks. In this process, medical work spills into family life—often with far-reaching consequences" (Timmermans and Freidin 2007, 1351–1352, 1361). For example, to maintain their asthmatic children and be prepared for emergencies, mother must: be sure prescriptions are filled and available at home and in their child's schoolbag, notify teachers of changes in the asthma situation, make and attend medi-cal appointments, keep emergency numbers and cab fare handy, and so on, ad infinitum.

The abduction work done here by mothers is intense and fraught—really, it is hyper-vigilance. It involves ongoing assessments and reassessments to determine what interventions are requisite now or soon to avert or minimize an asthma attack: "Mothers have to figure out for themselves which environmental changes will help their child and decide when to call in professionals when, for instance, a drug's side effects become troublesome" (Timmermans and Freidin 2007, 1361). Their challenging simplification work is the designation and paring away of unnecessary information and tasks so that they may be able to care for their child "well enough." Most were working mothers and also had to do their jobs "well enough" to keep them as well or risk impoverishment! Both the providers and mothers engaged in hope work for different reasons. Here hope work for the mothers is not (or not only) for a wondrous future cure, but for having relatively normal school and work lives—getting through a day or a week or maybe even a month without a life-threatening crisis.

While I have distinguished here among the three kinds of anticipation work—abducting, simplifying, and hoping—in reading these summaries one senses that these are often blurred in concrete practice, flowing into each other and temporally looping back and forth. As well, anticipation work can often be characterized as anxiety ridden and may even require hypervigilance.

Conclusions

Anticipation consists in "the practices employed to navigate daily life and to sustain relations." (Papadopoulos, Stephenson, and Tsianos 2008, xii)

My aim in this chapter has been to elucidate some of the key processes of anticipation so that they may be more commonly recognized as work, as labor, as effortful, and as potentially fraught. I began from Leigh's (e.g., Star 1983) abiding concern with the management of uncertainties in different situations. Across her career, Leigh Star's engagements with these issues shifted from debating closed versus open systems to a serious interest in anticipation work as a strategy for managing uncertainty. (Anticipation can, of course, be seen as an open system par excellence.) I then explicated three key processes of anticipation work. *Abduction* has roots in pragmatist philosophy and is at the heart of doing grounded theory to which Leigh was committed for over three decades (Star 2007). In Leigh's (Star 1983) important early paper, *simplification* was initially framed as erasing the requisite work done to produce a scientific theory such that it could seem autonomous, freed of entanglements and politics, sleek, mobile, and capable of smooth traveling. In anticipation work, simplification efforts also erase the work and often expand to setting limits on what will be taken into account and the level of complexity engaged. Last, *hoping* is fuel, drive, process, product (Star 2009). My brief genealogy then documented the gradual entree of hope as affect into academic discourse post World War II, growing ever more explicit in the new millennium, especially in discourses about the promissory nature of biotechnology, biomedicine, and information technologies. Two exemplars illustrated the three processes of anticipation work in action.

In conclusion, I briefly turn to the shifting politics of responsibility for doing anticipatory labor. Anticipation work is very much like articulation work which Anselm defined as "the specifics of putting together tasks, task sequences, task clusters—even aligning larger units such as lines of work and subprojects—in the service of work flow. . . . The overall process of putting *all* the work elements together and keeping them together represents a more inclusive set of actions"—an articulation process (Strauss 1991, 100, emphasis in original). Articulation work is the glue that holds things together, generally invisible unless and until it fails when and if "things go wrong." And its doers are often invisible too.

Anticipation work includes but is not limited to gathering information, abducting, simplifying, guessing, deciding, planning, acting, and hoping against hope that the guesses made are good enough. It takes into ongoing account how guesses are always already "on probation" (Peirce in Fann 1970, 31) which makes the work of anticipation neverending, part of the housework of living. Inspired by Leigh's (Star 1991a,b) significant legacy and our feminist-inspired conversations, my main argument is that, like articulation work, much of the work of anticipation is routinely invisibled—both the tasks themselves and the people doing those tasks. And some of it is very costly labor in terms of sustained anxiety and hypervigilance.

Leigh argued (Star and Strauss 1998, 9): "No work is inherently either visible or invisible. We always 'see' work through a selection of indicators: straining muscles, finished artifacts, a changed state of affairs. The indicators change with context, and that context becomes a negotiation about the relationship between visible and invisible work." Moreover, making work visible is not intrinsically "a good thing." Lucy Suchman (1995) analyzes the complex trade-offs potentially involved. While recognition and visibility can mean legitimacy and rescue from obscurity or other aspects of exploitation, that same visibility can reify work, create new opportunities for surveillance, and/or increase burdens of group communication.[41] For Leigh Star, in the end, the point was that unrecognized anticipation work goes unappreciated.

Two recent developments linked to recognition of anticipation work would please Leigh immeasurably (e.g., Star 2007, 2009). First, an expanding "sociology of responsibility" offers analyses of the social mechanisms through which people come to care about the welfare of others. It attempts to render visible the moral and ethical commitments that undergird taking on anticipation work and articulation work (e.g., Heimer and Staffen 1998; Mattingly 2010). And extending earlier feminist articulations (Hochschild 1969, 1983), there is a fresh spate of sociological research on women's labor and care work—largely unacknowledged and unpaid feminized forms of economic productivity (e.g., Boris and Parreñas 2010; Landsman 2009; Lappe 2012; Puig de la Bellacasa 2011; Waldby and Cooper 2010). Significantly, given the "animal turn" across the social sciences and humanities (e.g., Ritvo 2007; Weil 2010), a considerable and growing subset of the literature on care work concerns transpecies efforts (e.g., Haraway 2003, 2008; Friese 2013; Casper 2014; Chen 2012).

Second, there is broadening recognition of the importance of affect and emotion across the disciplines, including "the affective turn" (Clough 2007, 2010; Puar 2009), which also legitimates recognition of many facets of anticipation work. In 2006, Leigh Star was among those who led this off when, as president of the Society for Social Studies of Science, she chose the quite radical theme of "Silence, Suffering and Survival" for the annual meetings to encourage exploration of overlooked dimensions, boundaries, actors, and artifacts of science and technology. Others have since pursued the importance of love in some research, among both participants and scientists (e.g., Lappe 2012; Singh 2010; Silverman 2011, 2012; Will 2010). For example, Silverman (2012) titled a paper "Desperate and Rational: Of Love, Biomedicine, and Experimental Community." Dimensions of political economy are no longer the only modes of rational explanation.

I will end on a classic scholarly note, with a call for further research. Empirical projects are needed that explicitly explore the varied kinds of work of anticipation, including the amassing of data or information that I have not discussed, along with the kinds of work I have: abduction, simplification, and hoping. How are they stratified? Under what conditions are they invisibled? Left visible and exalted? Who does which aspects of anticipation work? Under what conditions? With what kinds of recognition or invisibling? Why might it matter whether anticipatory labor is invisible? Anticipation work is trading in futures and worthy of study. After all, "there is always a chance for something else, unexpected, new" (Clough 2010, 224). And anticipation is one of "the practices which are at the heart of social transformation long before we are able to name it as such" (Papadopoulos, Stephenson, and Tsianos 2008, xii). Leigh would appreciate that.

Notes

1. In sorrow, this chapter is dedicated to the memory of Susan Leigh Star, my mentor, colleague, and friend of thirty years who taught me pragmatism, science and technology studies, and sibling. Very special thanks for generous comments go to Larry Busch, Monica Casper, Maria Puig de la Bellacasa, and Stefan Timmermans. Earlier versions were presented at the Workshop on Anticipation sponsored by the UC Humanities Research Institute and the Department of Social and Behavioral Sciences, UC San Francisco (March 2011), and at the memorial session for Leigh Star at the 2011 meetings of the Society for Social Studies of Science in Cleveland.

2. In describing her undergraduate research assistantship with Mary Daly at Radcliffe, Leigh wrote about being trained to resist "methodolatry," the thoughtless adherence to old methods, and instead "to find those things that 'haunt' forms of knowledge and representation . . . [for example] the deletion of female agency in talking about work done in and around the home" (Star 2009, 335). Her legacy demonstrates how she brilliantly extended those lessons.

3. On managing uncertainties, see, for example, Star 1985, 1988, 1989.

4. See Bowker and Star 1999 on the difficulties of coding nursing work for computer-based efficiency assessments. A goodly portion of nursing is articulation work—hidden management and/or execution of the nonroutine.

5. Star (1991, 275) wrote: "The counterpoint to articulation work is routine: that which is [can be] packaged up, taken for granted, black-boxed."

6. This was a bestseller, developed when *Fortune Magazine* sponsored him to do extensive interviews on the CEOs of corporations such as General Electric and Ford.

7. An example is Star's chapter in *Sorting Things Out* on the tragedies produced when the rules of apartheid in South Africa violated people's lived identities and they were no longer allowed to live with their families and/or communities (Bowker and Star 1999).

8. Thanks to Larry Busch for this clarification. It involves the construction of a logical model that is true by definition, arguing that empirical data are useless since the model can never be overturned since it is true by definition. In April 1945, *Reader's Digest* published a condensed version of Frederich von Hayek's book, *The Road to Serfdom*, and the Book-of-the-Month Club widely distributed this condensed version for five cents per copy, selling over a million; today Hayek is required reading in Texas high schools along with Adam Smith and Keynes (Schuessler 2010, 27).

9. On the history of grounded theory, see especially Bryant and Charmaz 2007 and Strübing 2007. On the transnational renaissance in qualitative approaches, see especially the website of the International Congress on Qualitative Inquiry founded by Norman Denzin, a major symbolic interactionist and former colleague of Leigh Star (http://www.icqi.org). See also http://ejournals. library.ualberta.ca/index.php/IJQM; http://www.qualitative-research.net/index.php/fqs/issue/ archive; http://www.qualitativesociologyreview.org/ENG/.

10. These lenses allowed Star and me to enter science, technology, and medicine studies already at full gallop in terms of analyzing the concrete practices of scientific work, a key focus of the specialty during the 1980s. Our premise, learned at Strauss's knee, was that doing science is just doing another kind of work (Clarke and Star 2008). From related perspectives have come analyses of calculation (e.g., Latour 1987; Callon and Muniesa 2003) and accountancy (e.g., Porter 1995; Strathern 2000). But by and large, focus there has emphasized that which was being assessed rather than the processes—work—per se.

11. On the concept of situation, see Clarke 2005 (21–23 and passim).

12. From the abstract of "Moved by Moving Images: Temporal Looping in Labs and 3D CAVES" by Joseph Dumit (UC Davis) and Natasha Myers (York University). Session 009: Looping Tempo-

ralities I at meetings of the Society for Social Studies of Science, Tokyo, 2010. See http://4sonline.org/files/print_program0903.pdf at page 71 of the program with abstracts pdf.

13. Reichertz (2007, 216) attributes it to Julius Pacius's effort to translate a concept of Aristotle's.

14. While Peirce used the terms "retroduction" and "abduction" for the same process (Fann 1970), I use only abduction.

15. When Peirce is quoted in Fann 1970, the exact lines from Peirce's numbered works are provided. I have not included them here as it seems too cumbersome for our largely nonphilosophical purposes.

16. There is also a wonderful association between abduction and Victorian detective fiction as both engaged with the nature and sufficiency of evidence. Peirce himself wrote about the detecting processes of Sherlock Holmes as abductive (Sebeok and Umiker-Sebeok 1979). Major philosophers have since written about both of them as well as the scientific investigations of Poe and Dickens (e.g., Eco [1983] 1988). *Conjectural thinking* is the term often applied to the efforts of doctors, historians, and detectives alike (ibid., 205), all of whom operate in uncertain venues where the sufficiency of evidence can usually be questioned.

17. Richardson and Kramer (2006) argue that abduction is part of all phases of research: exploration, specification, reduction, and integration. For excellent discussions of temporalities, see session on "Looping Temporalities" at the 2010 meetings of the Society for Social Studies of Science in Tokyo, http://4sonline.org/files/print_program0903.pdf; and sessions "Lag! I & II" at the 2011 meetings of the Society for Social Studies of Science in Cleveland, http://www.4sonline.org/files/Print_Program_Abstracts_final.pdf.

18. Such "hopeful suggestions" are what I see grounded theory codes to be (see, e.g., Charmaz 2006; Strauss and Corbin 1994).

19. Grounded theory sought to counter the haute positivism that reigned in the social sciences post World War II, discussed earlier (Reichertz 2007; Richardson and Kramer 2006; Strübing 2007; cf. Timmermans and Tavory 2012).

20. Timmermans and Tavory (2012) argue that Glaser and Strauss were remiss in not considering grounded theory as abductive rather than inductive. In contrast, along with Atkinson, Coffey, and Delamont (2003), Reichertz (2007, 215) asserts that "GT [grounded theory] was to a very small extent abductive from the very start and became more abductive in its later stage, at least in the work of Strauss. Thus the Glaser-Strauss controversy can be characterized, at least in part, as one between induction and abduction." See also Locke 2007.

21. Another pragmatist approach to inquiry is offered in Dewey's (1938) *Logic: The Theory of Inquiry*. He develops a circular model of problem solving that is an interesting comparison with Peircean abduction. Strübing (2007, 591) also offers a beautiful diagram of this process.

22. This parallels Ludvik Fleck's ([1935] 1979) concept of "thought collectives" which generate theories based in their shared assumptions.

23. I hope to generate what Schwalbe and colleagues (Schwalbe et al. 2000) call "generic processes" from Star's more empirically embedded analysis of work in neuroscience. Thus, generalizing here is not about the what/thing but rather about the "hows" and "how tos."

24. Star's (1983) explication is more complicated and nuanced than I present, using the computer and information science concepts of well- and ill-structured problems. I have simplified.

25. Such a focus on process rather than individuals is a hallmark of symbolic interactionist sociology and also of grounded theory analyses (e.g., Strauss 1987; Star 2007).

26. In grounded theory research, the analyst determines when to stop gathering further data by assessing whether she has reached a point of saturation, when new data have ceased to reveal further information, range of variation, or new patterns or processes deemed important to the overall analysis. Charmaz (2006) calls this the point of "interpretive sufficiency." This is extremely challenging to learn and trust as it requires the analyst to tolerate uncertainty. See Glazer and Strauss 1967, Strauss and Corbin 1994, Morse 1995, Charmaz 2006, and Bowen 2008.

27. See note 25.

28. A concern for the invisible, the silent and silences, was clear in the work of Anselm Strauss (e.g., Strauss 1988, 1993; Star and Strauss 1998), and such concerns still engage his students (e.g., Charmaz 2002; Clarke 2005; Casper and Moore 2009).

29. Whether mobile theories or even technologies travel intact—remain "the same"—or are transformed when taken up elsewhere was at the heart of a debate between Star (Star and Griesemer 1989) and Bruno Latour (1986). Latour famously argued for "immutable mobiles" while Star argued that boundary objects which dwelt in multiple social worlds were differently perceived, interpreted, and utilized according to local needs and agendas. Here again we see Star as a symbolic interactionist asserting a Meadian multiplicity of perspectives which Mol (2002) attempted to deny.

30. A vast and indeed ancient literature engages hope in health and medical domains, including nursing, from Hippocrates to the present; it often takes up divergent views of hope comparing patients, families, and clinicians (e.g., DelVecchio Good et al. 1990; Eliott 2005; Will 2010; Mattingly 2010; Roscigno et al. 2012).

31. *Century of the Self* is a superb BBC documentary series by Adam Curtis focused on how Sigmund Freud's nephew, Edward Bernays, applied Freudian theories of the unconscious and sexuality in the creation of the modern marketing disciplines of advertising and public relations. It explores the roots and methods of modern consumerism, representative democracy, commodification, and their implications. It also questions how we came to see ourselves as individuals. See Wikipedia and http://topdocumentaryfilms.com/the-century-of-the-self/.

32. See on the history of DARPA the Department of Defense's own site: http://www.darpa.mil/about/history/history.aspx.

33. As president of the American Sociological Association, Wright chose "Real Utopias: Emancipatory Projects, Institutional Designs, Possible Futures" as the theme of the annual meetings in 2012. http://www.asanet.org/am2012/am2012.cfm?CFID=1472370&CFTOKEN=55513720.

34. These included Michel Serres, Gayatri Spivak, Isabelle Stengers, Brian Massumi, and Michael Taussig. See Zournazi 2002a.

35. N. Brown (2005, 332) also challenged "claims to the dominance of the present and the now—the momentocentrism"—of much recent writing on temporality.

36. Interestingly, this shift from a more informational approach to a more promissory one parallels the change in advertising strategies early in the twentieth century as Edward Bernays drew on Freudian theory. (See note 26.) Advertising then shifted from an educational orientation focused on product merits to creating desire through emotions accomplished "through a heavy emphasis on illustration" linked to "visual clichés" and "abstract types" of people with whom consumers were to identify (E. Brown 2005, 168).

37. Brown (2007) himself later tried to "reconnect" them.

38. Thanks to Allan Regenstreif's discussions with colleagues at the Freudian School of Quebec.

39. To me this is reminiscent of Howie Becker's (1960) concept of side-bets in his paper on commitment. If the actual bet doesn't win, maybe the side-bet will.

40. Anselm Strauss developed a theoretical framework called "negotiated order" about the plethora of ways that individuals and collectivities negotiate (with themselves and others) what will be done, how, when, etc. See Strauss 1991, 71–232; see also Clarke (in prep.).

41. See also Star and Strauss 1998; Bowker and Star 1999; Timmermans, Bowker, and Star 1998.

References

Adams, Vincanne, Michelle Murphy, and Adele E. Clarke. 2009. "*Anticipation*: Technoscience, Life, Affect, Temporality." *Subjectivity* 28:246–265.

Almklov, P. G. 2008. "Standardized Data and Singular Situations." *Social Studies of Science* 38 (6): 873–897.

Atkinson, P., A. Coffey, and S. Delamont. 2003. *Key Themes in Qualitative Research: Continuities and Change*. Walnut Creek, CA: Alta Mira Press and Rowman and Littlefield.

Bateson, G. 2002. *Mind and Nature: A Necessary Unit*. Cresskill, NJ: Hampton Press.

Becker, Howard S. 1960. "Notes on the Concept of Commitment." *American Journal of Sociology* 66 (July): 32–40.

Becker, Howard S. 1970. *Sociological Work: Method and Substance*. New Brunswick, NJ: Transaction Books.

Becker, Howard S. 1986. *Doing Things Together*. Evanston, IL: Northwestern University Press.

Belfiore, Michael. 2009. *The Department of Mad Scientists: How DARPA Is Remaking Our World, from the Internet to Artificial Limbs*. New York: Smithsonian Books and Harper/HarperCollins Publishers.

Berlant, Laurent. 2006. "Cruel Optimism." *Differences: A Journal of Feminist Cultural Studies* 17 (3): 20–36.

Blumer, Herbert. 1969. *Symbolic Interactionism: Perspective and Method*. Englewood Cliffs, NJ: Prentice Hall.

Boris, Eileen, and Rhacel Salazar Parreñas. 2010. *Intimate Labors: Cultures, Technologies, and the Politics of Care*. Stanford, CA: Stanford University Press.

Bowen, Glen A. 2008. "Naturalistic Inquiry and the Saturation Concept: A Research Note." *Qualitative Research* 8 (1): 137–152.

Bowker, Geoffrey, and Susan Leigh Star. 1999. *Sorting Things Out: Classification and Its Consequences*. Cambridge, MA: MIT Press.

Brown, Elspeth H. 2005. *The Corporate Eye: Photography and the Rationalization of American Commercial Culture, 1884–1929*. Baltimore, MD: Johns Hopkins University Press.

Brown, Nik. 2003. "Hope against Hype: Accountability in Biopasts, Presents and Futures." *Science Studies* 16:3–21.

Brown, Nik. 2005. "Shifting Tenses: Reconnecting Regimes of Truth and Hope." *Configurations* 13 (3): 331–355.

Brown, Nik, and Mike Michael. 2003. "A Sociology of Expectations: Retrospecting Prospects and Prospecting Retrospects." *Technology Analysis and Strategic Management* 15:3–18.

Brown, Nik, Brian Rappert, and Andrew Webster, eds. 2000. *Contested Futures: A Sociology of Prospective Technoscience*. Aldershot, UK: Ashgate Pubs.

Borup, Mads, Nik Brown, Kornelia Konrad, and Harro van Lente. 2006. "The Sociology of Expectations in Science and Technology." *Technology Analysis and Strategic Management* 18 (3–4): 285–298.

Bryant, Antony, and Kathy Charmaz. 2007. "Grounded Theory in Historical Perspective: An Epistemological Account." In *Handbook of Grounded Theory*, ed. Antony Bryant and Kathy Charmaz, 31–57. London: Sage.

Burke, Brian J., and Boone Shear. 2013. "Beyond Critique: Anthropology of and for Non-Capitalism." *Anthropology News* (January/February): 17, 21.

Busch, Lawrence. 2012. *Standards: Recipes for Reality*. Cambridge, MA: MIT Press.

Callon, Michel, and F. Muniesa. 2003. "Les Marchés Économiques Comme Dispositifs Collectifs de Calcul." *Reseaux* 21 (122): 189–233.

Casper, Monica J. 2014. "A Ruin of Elephants: Trans-species Love, Labor, and Loss." *Oppositional Conversations: AfterCatastrophe* 1 (2), http://cargocollective.com/OppositionalConversations_Iii/A-Ruin-of-Elephants-Trans-Species-Love-Labor-and-Loss (accessed July 17, 2015).

Casper, Monica J., and Lisa Jean Moore. 2009. *Missing Bodies: The Politics of Visibility*. New York: NYU Press.

Charmaz, Kathy. 2002. "Stories and Silences: Disclosures and Self in Chronic Illness." *Qualitative Inquiry* 8 (3): 302–328.

Charmaz, Kathy. 2006. *Constructing Grounded Theory*. London: Sage.

Chen, Mel Y. 2012. *Animacies: Biopolitics, Racial Mattering, and Queer Affect*. Durham, NC: Duke University Press.

Clarke, Adele E. 2005. *Situational Analysis: Grounded Theory after the Postmodern Turn*. Thousand Oaks, CA: Sage.

Clarke, Adele E. 2010. "From the Rise of Medicine to Biomedicalization: U.S. Healthscapes and Iconography c. 1890–Present." In *Biomedicalization: Technoscientific Transformations of Health, Illness in the United States*, ed. Adele E. Clarke, Janet Shim, Jennifer Fishman, Jennifer Fosket, and Laura Mamo, 47–87. Durham, NC: Duke University Press.

Clarke, Adele E. 2012. "Feminisms, Grounded Theory and Situational Analysis." In *Handbook of Feminist Research: Theory and Praxis*, 2nd ed., ed. Sharlene Hesse-Biber, 388–412. Thousand Oaks, CA: Sage.

Clarke, Adele E. In prep. "Straussian Negotiated Order: A Fifty Year Retrospective." Forthcoming in Norman Denzin, ed., *Studies in Symbolic Interaction*.

Clarke, Adele E., and S. Leigh Star. 2008. "Social Worlds/Arenas as a Theory-Methods Package." In *Handbook of Science and Technology Studies*, ed. Ed Hackett, Michael Lynch, Olga Amsterdamska, and Judy Waczman, 113–137. Cambridge, MA: MIT Press.

Clarke, Adele E., Janet Shim, Jennifer Fishman, Jennifer Fosket, and Laura Mamo, eds. 2010. *Biomedicalization: Technoscientific Transformations of Health and Illness in the United States*. Durham, NC: Duke University Press.

Clough, Patricia Ticineto. 2007. "Introduction." In *The Affective Turn: Theorizing the Social*, ed. Patricia Clough and Jean Halley, 1–33. Durham, NC: Duke University Press.

Clough, Patricia Ticineto. 2010. "The Affective Turn: Political Economy, Biomedia, and Bodies." In *The Affect Theory Reader*, ed, Melissa Gregg and Gregory J. Seigworth, 206–228. Durham, NC: Duke University Press.

Cooper, Frederick, and Randall Packard. 2005. "The History and Politics of Development Knowledge." In *The Anthropology of Development and Globalization: From Classical Political Economy to Contemporary Neoliberalism*, ed. Marc Edelman and Angelique Haugerud, 126–139. Malden, MA: Blackwell.

Daly, Kerry. 1997. "Replacing Theory in Ethnography: A Postmodern View." *Qualitative Inquiry* 3 (3): 343–365.

DeGrazia, Victoria, with Ellen Furlough, eds. 1996. *The Sex of Things: Gender and Consumption in Historical Perspective*. Berkeley: University of California Press.

DelVecchio Good, Mary-Jo, Byron J. Good, Cynthia Schaffer and Stuart Lind. 1990. "American Oncology and the Discourse on Hope." *Culture, Medicine and Psychiatry* 14:59–79.

Dewey, John. 1938. *Logic: The Theory of Inquiry*. New York: Holt, Rinehart & Winston.

Duggan, Lisa, and José Esteban Muñoz. 2009. "Hope and Hopelessness: A Dialogue." *Women and Performance: A Journal of Feminist Theory* 19 (2): 275–283.

Dumit, Joseph. 2012. *Drugs for Life: How Pharmaceutical Companies Define Our Health*. Durham, NC: Duke University Press.

Eco, Umberto. [1983] 1988. "Horns, Hooves, Insteps: Some Hypotheses on Three Types of Abduction." In *The Sign of Three: Dupin, Holmes, Peirce*, ed. Umberto Eco and Thomas A. Sebeok, 198–220. Bloomington: Indiana University Press.

Eliott, Jaklin A. 2005. "What Have We Done with Hope? A Brief History." In *Interdisciplinary Perspectives on Hope*, ed. J. Elliott, 3–45. Hauppauge, NY: Nova Science.

Eriksson, Lina. 2011. *Rational Choice Theory: Potential and Limits*. London: Palgrave Macmillan.

Fann, K. T. 1970. *Peirce's Theory of Abduction*. The Hague: Martinus Nijhoff.

Fleck, Ludwik. [1935] 1979. *Genesis and Development of a Scientific Fact*. Chicago: University of Chicago Press.

Foucault, Michel. [1982] 1997. "Technologies of the Self." In *Ethics: Subjectivity and Truth, Vol. I*, ed. Paul Rabinow, 223–228. New York: The New Press.

Friedman, Milton. [1962] 2002. *Capitalism and Freedom: Fortieth Anniversary Edition*. Chicago: University of Chicago Press.

Friese, Carrie. 2013. "Realizing Potential in Translational Medicine: The Uncanny Emergence of Care as Science." *Current Anthropology* 54 (S7): S129–S138.

Fujimura, Joan H. 1987. "Constructing Doable Problems in Cancer Research: Articulating Alignment." *Social Studies of Science* 17:257–293.

Glaser, Barney G., and Anselm L. Strauss. 1967. *The Discovery of Grounded Theory: Strategies for Qualitative Research*. Chicago: Aldine; London: Weidenfeld and Nicolson.

Gordon, Avery. 1997. *Ghostly Matters: Haunting and the Sociological Imagination*. Minneapolis: Minnesota University Press.

Hansen, Bert. 2009. *Picturing Medical Progress from Pasteur to Polio: A History of Mass Media Images and Popular Attitudes in America*. New Brunswick, NJ: Rutgers University Press.

Haraway, Donna J. 1997. *Modest_Witness@Second_Millennium.FemaleMan©_Meets_Oncomouse™: Feminism and Technoscience*. New York: Routledge.

Haraway, Donna. 2003. *The Companion Species Manifesto: Dogs, People, and Significant Otherness*. Chicago: Prickly Paradigm Press.

Haraway, Donna. 2008. *When Species Meet*. Minneapolis: University of Minnesota Press.

Harrington, Ann. 1999. *The Placebo Effect: An Interdisciplinary Exploration*. Cambridge, MA: Harvard University Press.

Hayek, Freidrich A. [1944] 2007. *The Road to Serfdom: Text and Documents*, ed. Bruce Caldwell. Chicago: University of Chicago Press.

Heimer, Carole A., and Lisa R. Staffen. 1998. *For the Sake of the Children: The Social Organization of Responsibility in the Hospital and the Home*. Chicago: University of Chicago Press.

Hochschild, Arlie. 1969. "Emotion Work, Feeling Rules and Social Structure." *American Journal of Sociology* 85:551–575.

Hochschild, Arlie Russell. 1983. *The Managed Heart: Commercialization of Human Feeling*. Berkeley: University of California Press.

Hughes, Everett C. 1971. *The Sociological Eye. Vol. 2: Selected Papers on Work, Self and the Study of Society*. New Brunswick, NJ: Transaction Books.

Hughes, Thomas P. 1989. *American Genesis: A Century of Invention and Technological Enthusiasm, 1870–1970*. New York: Viking.

James, William. 1921. *Pragmatism: A New Name for Some Old Ways of Thinking: Popular Lectures On Philosophy*. New York: Longmans, Green.

Koopman, Colin. 2009. *Pragmatism as Transition: Historicity and Hope in James, Dewey, and Rorty*. New York: Columbia University Press.

Laclau, Ernesto, and Chantale Mouffe. 1985. *Hegemony and Socialist Strategy: Towards a Radical Democratic Politics*. London: Verso.

Lampland, Martha, and Susan Leigh Star, eds. 2009. *Standards and Their Stories: How Quantifying, Classifying and Formalizing Practices Shape Everyday Life*. Ithaca, NY: Cornell University Press.

Landsman, Gail Hiedi. 2009. *Reconstructing Motherhood and Disability in the Age of "Perfect" Babies*. New York: Routledge.

Lappe, Martine. 2012. "Anticipating Autism: Navigating Science, Uncertainty and Care in the Post-genomic Era." Ph.D. dissertation, Sociology, University of California, San Francisco.

Lappe, Martine. 2014. "Taking Care: Anticipation, Extraction and the Politics of Temporality in Autism Science." *BioSocieties* 9 (3): 304–328.

Lather, Patti. 1999. "Naked Methodology: Researching Lives of Women with HIV/AIDS." In *Revisioning Women, Health and Healing: Feminist, Cultural and Technoscience Perspectives*, ed. Adele Clarke and Virginia Olesen, 136–154. New York: Routledge.

Lather, Patti. 2001. "Postbook: Working the Ruins of Feminist Ethnography." *Signs* 27 (1): 199–227.

Lather, Patti. 2007. *Getting Lost: Feminist Efforts Toward a Double(d) Science*. Albany: State University of New York Press.

Latour, Bruno. 1986. "Visualization and Cognition: Thinking with Eyes and Hands." In *Knowledge and Society: Studies in the Sociology of Culture Past and Present6*, ed. Henrika Kuklick and Elizabeth Long, 1–40. Bingley, UK: JAI Press.

Latour, Bruno. 1987. *Science in Action.* Cambridge: Harvard University Press.

Law, John. 2007. "Making a Mess with Method." In *Handbook of Social Science Methodology*, ed. William Outhwaite and Stephen P. Turner, 595–606. Thousand Oaks, CA: Sage.

Lessor, R. 2000. "Using the Team Approach of Anselm Strauss in Action Research: Consulting on a Project on Global Education." *Sociological Perspectives* 43 (4): S133–S147.

Leys, Colin. 2005. "The Rise and Fall of Development Theory." In *The Anthropology of Development and Globalization: From Classical Political Economy to Contemporary Neoliberalism*, ed. Marc Edelman and Angelique Haugerud, 107–125. Malden, MA: Blackwell.

Locke, K. 2007. "Rational Control and Irrational Free-Play: Dual-Thinking Modes as Necessary Tension in Grounded Theorizing." In *Handbook of Grounded Theory*, ed. Antony Bryant and Kathy Charmaz, 565–579. London: Sage.

Lyon-Callo, Vincent. 2013. "Teaching for Hope?" *Anthropology News* 23 (January–February): 3–4.

Mackenzie, Adrian. 2013. "Programming Subjects in the Regime of Anticipation: Software Studies and Subjectivity." *Subjectivity* 6:391–405.

Maines, David. 1991. "Pragmatism." In *Encyclopedia of Sociology*, ed. Edgar Borgatta and Marie Borgatta, 1531–1536. New York: Macmillan.

Massumi, Brian. 2002. *Parables for the Virtual: Movement, Affect, Sensation.* Durham, NC: Duke University Press.

Mattingly, Sheryl. 2010. *The Paradox of Hope: Journeys through a Clinical Borderland.* Berkeley: University of California Press.

Miller, Peter, and Nikolas Rose. 1997. "Mobilizing the Consumer: Assembling the Subject of Consumption." *Theory, Culture & Society* 14 (1): 1–36.

Mol, Annemarie. 2002. *The Body Multiple: Ontology in Medical Practice.* Durham, NC: Duke University Press.

Moreira, Tiago, and Paolo Palladino. 2005. "Between Truth and Hope: On Parkinson's Disease, Neurotransplantation and the Production of the 'Self.'" *History of the Human Sciences* 18:55–82.

Morrison, Michael. 2012. "Promissory Futures and Possible Pasts: The Dynamics of Contemporary Expectations in Regenerative Medicine." *Biosocieties* 7 (1): 3–22.

Morse, Jan. 1995. "Editorial: The Significance of Saturation." *Qualitative Health Research* 5 (2): 147–149.

Munoz, Jose Esteban. 2009. *Cruising Utopia: The Then and There of Queer Futurity.* New York: New York University Press.

Murphy, Michelle. 2013. "Economization of Life: Calculative Infrastructures of Population and Economy." In *Relational Ecologies: Subjectivity, Sex, Nature and Architecture*, ed. Peg Rawes, 225–254. London: Routledge.

Nerlich, Brigitte, and Christopher Halliday. 2007. "Avian Flu: The Creation of Expectations in the Interplay Between Science and the Media." *Sociology of Health & Illness* 29 (1): 46–65.

Novas, Carlos. 2006. "The Political Economy of Hope: Patients' Organizations, Science, and Biovalue." *Biosocieties* 1 (3): 289–306.

Orr, Jackie. 2006. *Panic Diaries: A Genealogy of Panic Disorder*. Durham, NC: Duke University Press.

Papadopoulos, D., N. Stephenson, and V. Tsianos. 2008. *Escape Routes: Control and Subversion in the Twenty-First Century*. London: Pluto Press.

Peale, Norman Vincent. 1952. *The Power of Positive Thinking*. New York: Simon & Schuster.

Peirce, Charles Saunders. 1929. Guessing.*Hound and Horn* 2:267–282.

Polak, Frederick L. 1961. *The Image of the Future*. 2 vols. Leyden: A. W. Sythoff.

Pollock, Neil, and Robin Williams. 2010. "The Business of Expectations: How Promissory Organizations Shape Technology and Innovation." *Social Studies of Science* 40 (4): 525–548.

Porter, Theodore M. 1995. *Trust in Numbers: The Pursuit of Objectivity in Science and Public Life*. Princeton, NJ: Princeton University Press.

Prainsack, Barbara, Ingrid Geesink, and Sarah Franklin, eds. 2008. "Stem Cell Technologies 1998–2008: Controversies and Silences." *Science as Culture* 17 (4): 351–484.

Puar, Jasbir K. 2009. "Prognosis Time: Towards a Geopolitics of Affect, Debility and Capacity." *Women & Performance* 19 (2): 161–172.

Puig de la Bellacasa, Maria. 2011. "Matters of Care in Technoscience: Assembling Neglected Things." *Social Studies of Science* 41 (1): 85–106.

Reichertz, J. 2007. "Abduction: The Logic of Discovery of Grounded Theory." In *Handbook of Grounded Theory*, ed. A. Bryant and K. Charmaz, 214–228. London: Sage.

Rich, Adrienne. 1978. *On Lies, Secrets and Silences: Selected Prose, 1966–1978*. New York: W. W. Norton & Co.

Richardson, Rudy, and Eric Hans Kramer. 2006. "Abduction as the Type of Inference That Characterizes the Development of a Grounded Theory." *Qualitative Research* 6 (4): 497–513.

Ritvo, Harriet. 2007. "On the Animal Turn." *Daedalus* 136 (4): 118–122.

Roscigno, Cecelia I. 2012. "Divergent View of Hope Influencing Communications between Parents and Hospital Providers." *Qualitative Health Research* 22 (9): 1232–1236.

Rose, Nikolas. 2007. *The Politics of Life Itself: Biomedicine, Power, and Subjectivity in the Twenty-First Century*. Princeton, NJ: Princeton University Press.

Schatzki, Theodore R., Karin Knorr Cetina, and Eike von Savigny, eds. 2001. *The Practice Turn in Contemporary Theory*. London: Routledge.

Schuessler, Jennifer. 2010. "Hayek: The Back Story." *The New York Times Book Review*, July 11, 27.

Schwalbe, Michael, S. Goodwin, D. Holden, S. Schrock, S. Thompson, and Michelle Wolkomir. 2000. "Generic Processes in the Reproduction of Inequality: An Interactionist Analysis." *Social Forces* 79:419–452.

Sebeok, Thomas Albert, and Donna Jean Umiker-Sebeok. 1979. "You Know My Method": A Juxtaposition of Charles S. Peirce and Sherlock Holmes.*Theory and Methodology in Semiotics* 26 (3–4): 203–250.

Seigworth, Gregory J., and Melissa Gregg. 2010. "An Inventory of Shimmers." In *The Affect Theory Reader*, ed. Melissa Gregg and Gregory J. Seigworth, 1–28. Durham, NC: Duke University Press.

Shaffir, William, and Dorothy Pawluch. 2003. "Occupations and Professions." In *Handbook of Symbolic Interactionism*, ed. Larry T. Reynolds and Nancy J. Herman-Kinney, 893–914. Lanham, MD: Altamira and Rowman and Littlefield Pubs., Inc.

Silverman, Chloe. 2011. *Understanding Autism: Parents, Doctors, and the History of a Disorder*. Princeton, NJ: Princeton University Press.

Silverman, Chloe. 2012. "Desperate and Rational: Of Love, Biomedicine, and Experimental Community." In *Lively Capital: Biotechnologies, Ethics and Governance in Global Markets*, ed. Kaushik Sunder Rajan, 354–384. Durham, NC: Duke University Press.

Singh, Jennifer. 2010. "Autism Spectrum Disorders: Parents, Scientists, and the Interpretations of Genetic Knowledge." Ph.D. dissertation, Sociology, University of California, San Francisco.

Smith, Dorothy E. 1987. *The Everyday World as Problematic: Toward A Feminist Sociology*. Boston: Northeastern University Press.

Star, Susan Leigh. 1983. "Simplification in Scientific Work: An Example from Neuroscience Research." *Social Studies of Science* 13:208–226.

Star, Susan Leigh. 1985. "Scientific Work and Uncertainty." *Social Studies of Science* 15:391–427.

Star, Susan Leigh. 1986. "Triangulating Clinical and Basic Research: British Localizationists, 1870–1906." *History of Science* 24:29–48.

Star, Susan Leigh. 1988. "The Structure of Ill-Structured Solutions: Heterogeneous Problem-Solving, Boundary Objects and Distributed Artificial Intelligence." In *Proceedings of the 8th AAAI Workshop on Distributed Artificial Intelligence*, Technical Report, Department of Computer Science, University of Southern California. Reprinted in *Distributed Artificial Intelligence*, 2nd ed., ed. M. Huhns and Les Gasser, 37–54. Menlo Park: Morgan Kaufmann, 1989.

Star, Susan Leigh. 1989. *Regions of the Mind: Brain Research and the Quest for Scientific Certainty*. Stanford: Stanford University Press.

Star, Susan Leigh. 1991a. "The Sociology of the Invisible: The Primacy of Work in the Writings of Anselm Strauss." In *Social Organization and Social Process: Essays in Honor of Anselm Strauss*, ed. David R. Maines, 265–283. Hawthorne, NY: Aldine de Gruyter.

Star, Susan Leigh. 1991b. "Invisible Work and Silenced Dialogues in Representing Knowledge." In *Women, Work and Computerization: Understanding and Overcoming Bias in Work and Education*, ed. I. V. Eriksson, B. A. Kitchenham, and K. G. Tijdens, 81–92. Amsterdam: North Holland Press.

Star, Susan Leigh. 2007. "Living Grounded Theory: Cogitive and Emotional Forms of Pragmatism." In *Handbook of Grounded Theory*, ed. Antony Bryant and Kathy Charmaz, 75–94. London: Sage.

Star, Susan Leigh. 2009. "Susan's Piece: Weaving As Method in Feminist Science Studies: The Subjective Collective." In "Feminist Science and Technology Studies: A Patchwork of Moving Subjectivities," ed. Wenda K. Bauschspies and Maria Puig de la Bellacasa, special issue, *Subjectivity* 28:344–346.

Star, Susan Leigh, and James R. Griesemer. 1989. "Institutional Ecology, 'Translations' and Boundary Objects: Amateurs and Professionals in Berkeley's Museum of Vertebrate Zoology, 1907–39." *Social Studies of Science* 19:387–420. [See also chapter 7, this volume.]

Star, Susan Leigh, and Anselm Strauss. 1999. "Layers of Silence, Arenas of Voice: The Ecology of Visible and Invisible Work." *Computer Supported Cooperative Work: The Journal of Collaborative Computing* 8:9–30.

Strauss, Anselm L. [1971] 2001. *Professions, Work and Careers*. San Francisco: Sociology Press. 2nd ed. New Brunswick, NJ: Transaction Books.

Strauss, Anselm L. 1987. *Qualitative Analysis for Social Scientists*. Cambridge, UK: Cambridge University Press.

Strauss, Anselm L. 1988. "The Articulation of Project Work: An Organizational Process." *Sociological Quarterly* 29:163–178.

Strauss, Anselm L. 1991. *Creating Sociological Awareness: Collective Images and Symbolic Representation*. New Brunswick, NJ: Transaction Pubs.

Strauss, Anselm. 1993. *Continual Permutations of Action*. New York: Aldine de Gruyter.

Strauss, Anselm L., and Juliet Corbin. 1994. "Grounded Theory Methodology: An Overview." In *Handbook of Qualitative Research*, ed. Norman K. Denzin and Yvonna Lincoln, 273–285. Thousand Oaks, CA: Sage.

Strauss, Anselm, Shizuko Fagerhaugh, Barbara Suczek, and Carolyn Weiner. [1985] 1997. *The Social Organization of Medical Work*. 2nd ed. New Brunswick, NJ: Transaction Publishers.

Strübing, Jörg. 2007. "Research as Pragmatic Problem-Solving: The Pragmatist Roost of Empirically-Grounded Theorizing." In *Handbook of Grounded Theory*, ed. Antony Bryant and Kathy Charmaz, 580–602. London: Sage.

Suchman, Lucy. 1987. *Plans and Situated Actions: The Problem of Human-Machine Communication*. New York: Cambridge University Press.

Sunder Rajan, Kaushik. 2006. *Biocapital: The Constitution of Postgenomic Life*. Durham, NC: Duke University Press.

Sunder Rajan, Kaushik. 2012. *Lively Capital: Biotechnologies, Ethics, and Governance in Global Markets*. Durham, NC: Duke University Press.

Taylor, Peter. 2005. *Unruly Complexity: Ecology, Interpretation, Engagement*. Chicago: University of Chicago Press.

Thompson, Charis. 2005. *Making Parents: The Ontological Choreography of Reproductive Technologies*. Cambridge, MA: MIT Press.

Tillack, Allison. 2012. "Imaging Trust: Information Technologies and the Negotiation of Radiological Expertise in the Hospital." Ph.D. dissertation, Medical Anthropology, University of California, San Francisco.

Timmermans, Stefan, Geoffrey Bowker, and Leigh Star. 1998. "The Architecture of Difference: Visibility, Controllability, and Comparability in Building a Nursing Intervention Classification." In *Differences in Medicine: Unraveling Practices, Techniques and Bodies*, ed. Marc Berg and Annamarie Mol, 202–225. Durham, NC: Duke University Press.

Timmermans, Stefan, and Betina Freidin. 2007. "Caretaking as Articulation Work: The Effects of Taking up Responsibility for a Child with Asthma on Labor Force Participation." *Social Science & Medicine* 65 (7): 1351–1364.

Timmermans, Stefan, and Iddo Tavory. 2012. "Theory Construction in Qualitative Research: From Grounded Theory to Abductive Analysis." *Sociological Theory* 30 (3): 167–186.

Traweek, Sharon. 1999. "Warning Signs: Acting on Images." In *Revisioning Women, Health and Healing: Feminist, Cultural and Technoscience Perspectives*, ed. Adele E. Clarke and Virginia L. Olesen, 187–201. New York: Routledge.

Tutton, Richard. 2011. "Promising Pessimism: Reading the Futures to Be Avoided in Biotech." *Social Studies of Science* 41 (3): 411–429.

Valentine, David. 2007. *Transgender: Ethnography of a Category*. Durham, NC: Duke University Press.

van den Hoonaard, W. C. 1997. *Working with Sensitizing Concepts: Analytical Field Research*. London: Sage.

von Mises, Ludwig. [1962] 2006. *The Ultimate Foundation of Economic Science*. 2nd. ed. Princeton, NJ: Van Nostrand Company, Inc., and Indianapolis, IN: Liberty Fund.

Waldby, Catherine, and Melinda Cooper. 2010. "From Reproductive Work to Regenerative Labour: The Female Body and the Stem Cell Industries." *Feminist Theory* 11 (1): 3–22.

Weil, Kari. 2010. "A Report on the Animal Turn." *differences: A Journal of Feminist Cultural Studies* 21 (2): 1–23.

Whyte, William H. 1956. *The Organization Man*. New York: Simon and Schuster.

Wiener, Carolyn. 2007. "Making Teams Work in Conducting Grounded Theory." In *Handbook of Grounded Theory*, ed. Antony Bryant and Kathy Charmaz, 293–310. London: Sage.

Will, Catherine M. 2010. "The Management of Enthusiasm: Motives and Expectations in Cardio-vascular Medicine." *Health* 14:547–563.

Williams, Raymond. [1976] 1983. *Keywords: A Vocabulary of Culture and Society*. London: Fontana/Collins and New York: Oxford University Press.

Zournazi, Mary. 2002a. *Hope: New Philosophies for Change*. Annandale NSW, Australia: Pluto Press; London UK: Lawrence & Wishart; New York: Routledge, 2003. http://ro.uow.edu.au/cgi/viewcontent.cgi?article=1087&context=artspapers.

Zournazi, Mary. 2002b. "Hope, Passion, Politics with Chantal Mouffe and Ernesto Laclau." In *Hope: New Philosophies for Change*, 122–149. New York: Routledge.

Zournazi, Mary. 2002c. "Navigating Movements—with Brian Massumi." In *Hope: New Philosophies for Change*, 210–243. New York: Routledge.

Zukin, Sharon, and Jennifer Smith Maguire. 2004. "Consumers and Consumption." *Annual Review of Sociology* 30:173–197.

5 Living Grounded Theory: Cognitive and Emotional Forms of Pragmatism

Susan Leigh Star

Editors' Note: In this remarkable paper written late in her life, Leigh Star describes her often anguished struggles to find an intellectual home in which to pursue her PhD and a research method congruent with her established relationalism, feminism, antipositivism, and antiformalism. She speaks of both the love and rage involved in becoming a scientist, and of the need to learn to manage one's anxiety in order to remain open analytically during the often long process of pursuing a research project. For Star, research was always already a cognitive-emotional process.

Scientific writing often encodes powerful emotive narratives. Enshrouded, archived, hidden away in government white papers and documents, read and unread, lives every passion and drama common to all human activity. However, what scientists do (often including ourselves) is fundamentally inaccessible to most of the world. People may see a map of a genome or a syringe full of experimental medication. These are just the end products, however, of a web of relationships, what Lave and Wenger have called communities of practice. Lave and Wenger make the strong claim that membership in these communities *constitutes* learning and science (Adler and Obstfeld 2007; Bowker and Star 1999, chapter 10; Lave and Wenger 1991; Obstfeld 2005). These relationships are usually invisible to readers of science and technology (Star and Strauss 1999; Suchman 1987). Part of the reason for this is precisely that scientists rely on the relational world not the concrete reified world.[1] Relations between people, between different perceptions of objects, between nature and politics, between laboratories and administrations, to name but a few, *are part of the relational world.* Another aspect is that scientists are normatively discouraged to write directly about this invisible part, and untrained in its analysis. This includes the love, suffering, dedication, covering up, and forming selves in the scientific world (see Clarke 1998 for how this has appeared in the work of reproductive scientists and those studying sex). Most scientists write in a kind of encrypted voice, a language that relies on this invisible work, and its extremes of isolation and specialization (Latour 1987).

Popular notions of science support this quiet suppression of passion. This includes simplistic notions of "science as truth"; science or scientific medicine as concerning the heroic search for new discoveries and cures; and scientists as dispassionate judges of pure results. Although these norms are slowly changing in some fields, there are yet many barriers to overcome. The conventional way of writing even forbids the use of the first person (which, undeniably, is at best awkward with multiple authors, and "we" implies another valuation altogether). The standards and formal classifications that pervade science always represent treaties between conflicting passions and desires, yet what could look more innocuous or boring? (Bowker and Star 1999; Star 1999).

Thus forming a scientific self entails a peculiar kind of pain and of joy that remains almost unspeakable. It leaks out in so-called popular science articles: the rage of an ecologist at seeing habitats destroyed, for example. It leaks out in memoirs and biographies: for example, in Evelyn Fox Keller's biography of Barbara McClintock (Keller 1983). In our own social science, it leaks out in the form of reflexivity, personal narrative, poetry, visualization, and performance art (Star 1998). These genres are an accepted part of social sciences in a few places (although often at great cost, see Laurel Richardson's 1996 account of writing a poem instead of an article about her research on single mothers, and its reception at the American Sociological Association).

This chapter has a long genesis in my own learning, teaching, and living of grounded theory and Pragmatist philosophy. In all of the books and articles about grounded theory, I keep searching for a particular answer: how does it feel to do grounded theory? Am I alone in feeling intense emotion while doing analysis? Or in the feeling that I am, in some sense, always doing grounded theory? As a sociologist, I don't believe, of course, that anyone has pure or unique experiences, except as they combine to form a unique biography. C. Wright Mills's idea of "personal troubles and public problems" (2000) or the feminist notion that *the personal is political* is always the beginning heuristic for me, rather than the idea that I am alone or unique. This takes a few years to develop, I think. In other words, even to ask the question, Am I the only one? presupposes the answer: Of course not. So when I called my student, Olga Kuchinskaya, to talk about what I was writing, her voice filled with relief at my words, and she described her own feelings of doing analysis as quite similar to this. So encouraged, I took a deep breath and began.

As a graduate student, I searched for years for teachers who would not try to divorce me from my life experience, feelings, and feminist commitments. At the same time, I didn't want just a "touchy-feely" sort of graduate education; I also needed to satisfy the love for stringent analysis I had developed as an undergraduate. I wouldn't have known how to say it, exactly, then, but I was looking for a way simultaneously to incorporate formal and informal understandings of the world. I sought a methodological place that was faithful to human experience, and that would help me sift through the chaos of meanings and produce the eureka of new, powerful explanations. I also wanted a way of understanding the world that I could carry with dignity in my world as a feminist activist and as a working poet.

A tall order, to be sure.

A Pathway to Grounded Theory

After my undergraduate degree, I began my choice of graduate programs (as one often does at the age of twenty-one), by following a lover to Santa Cruz, California. I enrolled in a PhD program in philosophy of education at Stanford. This program was unusual in that it promised many of the things I looked for, including faculty interested in qualitative methods and in comparative, historical studies of learning. However, no doubt due to my own lack of knowledge of philosophy or of the world of education, I found myself mute in most of the community-building efforts there. As serendipity would have it, one of the key professors in the small program was on leave; and I was unable to make the translations between philosophy and social science. We read John Dewey and I was intrigued, but his writings were not yet animated for me. (Years later, with the addition of empirical data and a community of grounded theory scholars, I would fall in love with Dewey. I actually wept when I found out he was no longer alive. But back to my brief chronology.)

In the midst of my confusion and intellectual loneliness, I saw a poster in the Education Library, announcing the University of California at San Francisco's (UCSF) Program in Human Development. It said that UCSF offered an interdisciplinary program in adult development, which would include questions about how people choose career paths; ethnicity and how it intertwined with aging in different communities; and what developmental events in adulthood couldn't be predicted (even the concept of adult development seemed radical to me). I applied to the program and was accepted, and I began my studies at UCSF that autumn.

I had also noticed in passing, on my way to UCSF, that Anselm Strauss was an "adjunct" faculty in the program. I was happy about that, as Glaser and Strauss's *The Discovery of Grounded Theory* had been used successfully by a friend in her feminist, qualitative dissertation. I had even read *The Discovery of Grounded Theory* (Glaser and Strauss 1967), a manifesto for freedom from the sterile methods that permeated social sciences at the time. But I didn't know exactly how to use it. For my senior undergraduate thesis in psychology, I had instead used George Kelly's personal construct theory (1955, 1963), which I liked for its open, recombinant possibilities in eliciting people's priorities and categories. The personal construct method is an analytic tool for eliciting an individual's core repertoire of concepts, and how they are ordered for importance in use. The map that results is called a "repertory grid." In its spatial representations of concepts, it resembles aspects of Clarke's powerful situational analysis, which offers a methodological guide for mapping social worlds and larger-scale arena formed by the

intersection of many social worlds, as for example in the formation of a scholarly discipline. Her book is a seminal "second generation" example of Straussian grounded
theory (Clarke 2005).

I had used personal construct theory with the goal of finding out how women experienced (or did not experience) the paradigm shift to feminism that seemed to be going
on all around me (in the mid-1970s, in Boston). I added, in order to scale up the psychological approach, a pair of theoretical works that spoke to the formation of widespread shifts in consciousness (the term current at the time)—Thomas Kuhn's *The
Structure of Scientific Revolutions* (1970) and Mary Daly's *Gyn/Ecology* ([1978] 1990)—to
examine the nature of large-scale changes in thought. Unlike grounded theory, however, my methodological approach lacked much inveiglement in people's lives. That is,
without fieldwork, it was not possible to observe the personal constructs in the contexts of action, in the full spectrum of messy and formal acts in which humans participate. I had hopes of deepening this earlier work toward anthropological/qualitative
sociological methods when I arrived at UCSF.

There was much to recommend in the Program in Human Development. In classes,
I was introduced to phenomenological and dialectical psychology, in particular the
work of Klaus Riegel (1978) and of L. S. Vygotsky (1986). We read the emergent critiques of moral developmental psychologist Lawrence Kohlberg (1981). These included
Carol Gilligan's feminist alternatives to Kohlberg's developmental steps that she saw as
lacking context, uncertainty, and nuance in women's lives (Gilligan 1993). As well, I
had the opportunity to delve into the cross-cultural critiques of Kohlberg's U.S.-centric
moral developmental model. These many critics accused Kohlberg of taking American
individualism, cognitivism, and logic, and claiming that these are universal values. All
of these readings helped push me from the individual as a unit of analysis toward communities, organizations, and complex relations as foremost.

At the same time, however, I continued to lack a satisfying method, or a deeper
methodology that would allow me to move forward in my general intellectual project.
In retrospect, I would now say that I lacked a community of people who would help me
develop my earlier intuitions about eliciting dimensions important to respondents,
and how it might link with larger structural analyses. Most of the people in Human
Development found George Kelly "a bit old-fashioned," or just not comprehensible.
While accepting gender as a variable, as a group they were fairly uninterested in qualitative, empirical explorations of engendering in various social formations, different
experiences of becoming a woman or a man in various racial and ethnic groups, different cohorts, different sexual orientations, and so on. I found the statistical approaches
offered as methods completely hollow for answering my burning questions. Furthermore, the justifications for using statistics seemed rather scientist to me, for example,
"no one will believe you unless you use numbers" or "qualitative research is not generalizable" (without being able to say why not). As an activist, these blind and

authoritarian directives angered me; I had been well trained as an undergraduate by feminist science critic Ruth Hubbard and feminist philosopher Mary Daly to question authority in such academic statements. I began to remember Glaser and Strauss, and upon inquiry, realized that I was eligible to take classes in other departments at UCSF. However, when I went to my advisors in Human Development, I was discouraged from taking classes from Strauss, as he was, according to one advisor, "not a real sociologist." (And thus I was introduced to the politics of qualitative research.)

By the following year, I determined that I would take qualitative classes in spite of my advisors' resistance. I began the long sequence of courses in fieldwork and grounded theory analysis in the Social and Behavioral Science Department, working first with Leonard Schatzman and Virginia Olesen, and then with Barney Glaser and Anselm Strauss. I will not continue with this detailed personal chronology, beyond noting some of the lessons learned in my search for teachers and methods:

• Simply finding grounded theory was not self-evident. It meant walking a twisted path, full of contingency and accidental proximities.
• The formation of this path was/is not accidental, however full of contingencies it may be. Rather, it forms the basis for creating a critical map of my emerging intellectual commitments. In my first fieldwork class, a prerequisite for the grounded theory sequence, Leonard Schatzman's first lesson was to get us to write down "how we ended up here" (at UCSF). This itself became the data for beginning to analyze complex social processes, and how to name them and begin to seek dimensions in the search itself. 1 was astonished that such exploration could itself be analyzed as a sociological phenomenon. A posting on a bulletin board, a love relationship, a chance meeting or phrase, can be considered as data. In this, a reflexive move analyzing life decisions becomes a tool for deepening one's repertoire of concepts and commitments. Today, I would call it the building of an intellectual infrastructure.
• In bringing both contingencies and commitments to explicit, overt analysis, one creates the chance to reflect on a somewhat unconscious set of choices, and to include the heart of method as a part of lived experience.

Grounded theory is an excellent tool for understanding invisible things. It can be used to reveal the invisible work involved in many kinds of tasks, as I have written about elsewhere (Bowker and Star 1999; Star 1991a,b, 1998; Star and Ruhleder 1996; Star and Strauss 1999). This includes invisible work in the acquisition and practice of method. *The longer one practices grounded theory, the more deeply imbricated it becomes in daily life.* This, of course, results in examining the various forms of invisible work one does as an analyst. For example, among those I have found in my own life are:

• Carrying nineteenth-century heavy, dusty volumes of patient records from a consulting room to a small attic chamber; waiting to retrieve them until the consulting physicians were done with the room in which they were stored (Star 1989);

• Explaining bisexuality and various sexual practices to a young, eager, and naive respondent, in order to create a cordial space for conducting an interview;

• Having to ask elderly lesbians to sign a legal release form explaining that, since at the time sodomy and certain other sexual practices were illegal, they might be arrested if they spoke of having had fellatio or anal sex (the Human Subjects Committee at San Francisco State required us to do this, not distinguishing genders or life experiences: you can imagine how silly I felt explaining this to a seventy-five-year-old life-long lesbian);

• Learning local parking regulations and practices in some fifty different genetic laboratory locations, from Vancouver to Missouri and beyond, during multisited fieldwork to help develop a tool for communication across a scientific community.

Abstracting from this short list, one can see various forms of work that are not discussed in the final published reports. This invisible work includes managing my embarrassment at asking personal questions in an interview about sexual practices and identity/emotion work (Charmaz 1991; Corbin and Strauss 1988). It includes the manual labor of carrying the heavy records book from place to place. All fieldworkers have similar stories (never get an anthropologist talking about plumbing in faraway places, especially over dinner). In my work as a sociologist of science, I came up with the code, "deleting the work." Scientific journals are full of articles that delete the development, setting, communication practices, and "grunt work" involved in doing science. Of course, as a scientist, this insight applies to me as well.

Code: Getting to "Out of Bounds"

What is a code? When I have taught grounded theory, I have explained going through the data repeatedly, looking for several sorts of things. These include anomalies, distaste, liking one person more than another, a shock of recognition as a respondent uses a phrase in local jargon that captures something about the site or acts (an in vivo code). I've taken students through the classical teachings of fieldwork, including the especially helpful *Doing Fieldwork: Warnings and Advice* by Rosalie Wax (1971) and Sanjek's *Fieldnotes* (1990), an edited volume that begins with Jean Jackson's provocative essay, "I Am a Fieldnote." Why these two books in particular? I hadn't really thought about it in depth before, however, in writing this chapter I see that both are written in clear, deeply personal terms, and they do not take the object of analysis, or the methodological procedures for granted. They include emotions, especially joy, mourning, confusion, and anxiety. (Code: *When emotions break through.*)

What is a code? A code sets up a relationship with your data, and with your respondents. One of the core mandates of sociology is the ability to ask the question, Of what is this an example? For instance, when I studied nineteenth-century physiologists doing brain experiments on great apes, I was asking questions about, inter alia, the nature of experiments, how materials are obtained, the role of social movements in restricting science (antivivisection, for example), and how one simian was turned into part of "the brain," an abstract map of the human brain.

Abstracting includes this sort of dialogue with imaginary others: sociologists, or advisors, other writers, or clients. Abstracting means to drop away properties from the original object. (Code: *abstracting away intimacy.*) This does not require a full specification of properties in the tradition of analytic philosophy; rather, this occurs by comparison, in the pragmatic and grounded theory senses, as outlined earlier. In fact, one is simultaneously discovering new specific properties and then merging them or dropping them in the face of comparisons. This is an open-ended, sprawling type of research, indeterminate, and structured by one's own ability to manage "grounded abstractions" and "local emotions" while continuing to develop theoretical sensitivity.[2]

For example, a couple of months ago I saw a bobcat in my driveway; a beautiful wild animal, poised there. I wanted to pet it, to get to know it, to help it—all sorts of emotions welled up in me. (Code: *when emotions break through//wildness.*) Fear for my own smaller cats, wondering why it was in my driveway; after many years here, I had never seen one before. I begin to analyze the bobcat. (Codes: *Wild, pity, beauty, fear, out of bounds.*) Then, my thoughts were of myself and our small settlement here in the mountains. We are out of bounds, too. We have no gas or sewage, and coyotes, deer, and quail are much more common than people. But the bobcat and I are not the same. However, I keep thinking about out of bounds and what it means to each of us.

In a research project, I would have hundreds of codes and many sorts of comparisons to make. But in my daily life, I also think this way. As Everett Hughes put it, "What do a priest and a prostitute have in common?" It seems a shocking comparison, but then you begin to see the circumstances, the context, of their work. They both listen to people's confessions; they work with people one-on-one (usually) in a setting that invites these sorts of intimacies; they listen rather than reveal their own lives, and so forth (Hughes 1970). By comparing, yet going back again and again to the data, we preserve something of what we see in both of these lines of work; something also of the shock of the new, and the new way of seeing that is more abstract than before. However, as I dip back into the data, I refresh my image of the people over and over. (Coding, question of constant comparison: *abstracting away* and *the breaking through of emotions simultaneously* invoke *out of bounds. Can I take this comparison through more experiences and expand it?*)

Resources

I pause at this point for an incomplete list of resources for doing grounded theory in research, before discussing some of the affective and pragmatic approaches I take in my own work. Kathy Charmaz's recent textbook, *Constructing Grounded Theory: A Practical*

Guide through Qualitative Analysis (Charmaz 2006) provides an extremely clear, suggestion- and example-filled guide to beginning to use grounded theory. In particular, the chapter on coding makes much of my work in this chapter easier. It examines the *how* of coding, something that has been much discussed, but often with a lack of clarity, and confusing internal contradictions. An earlier attempt, in Barney Glaser's *Theoretical Sensitivity* is also full of good suggestions, but is somewhat cramped for some users by an idiosyncratic language of logic that was never broadly picked up in the sociological or professional communities (1978). The book is also a little difficult to obtain and read. (I speak here from teaching experience.) Thus, this welcome update from Charmaz represents and clarifies quite a bit of Glaser's work in *Theoretical Sensitivity*. Similarly, Strauss's *Qualitative Analysis for Social Scientists* (1987) has several compelling examples that help one intuit the nature of coding, and a Q&A section at the end that is helpful. For the simple mechanics of handling field notes as a beginning, Schatzman and Strauss's *Field Research* (1973) remains useful in dividing codes into types: methodological, observational, and substantive. However, its language does not connect well to later usage (that is, more advanced usage and later scholarship). Adele Clarke's excellent new book, *Situational Analysis* (2005), helps to fill in this gap. Her approach focuses on arenas and social worlds, and in particular the problems associated with analyzing disciplinary growth and change. This will be useful for those wishing to analyze changes in science, industry, politics, or social movements (or combinations of these), and in facing the initial "messiness" of the data and the nature of developments. It does not focus overly much on the epistemological processes of coding per se. However, combined with Charmaz's book and the insights provided in this volume, Clarke helps build an indispensable scaffold for moving between the cognitive, affective "close-in" aspects of grounded theory to the larger-scale changes in the current world. As with all grounded theory, all these works advocate moving from data to analysis, and back again, recursively.

Long ago, Herbert Blumer (Strauss's teacher and student of George Herbert Mead) called for the following of "hunches" through data, intrinsically a personal biographical approach at the beginning. Glaser spoke of "life cycle" sources for topics, although not for coding; for example, both he and Strauss lost a parent close to the time of the writing of several books on death and dying (Glaser and Strauss 1965, 1968; Strauss and Glaser 1970). This personal experience was always fascinating to me, but neither Glaser nor Strauss were very interested in exploring or discussing the nature of how their own emotions were used as sources for analysis (either in their methodological writing or in their classes)—which, of course, being obsessed as I am with invisible work makes me even more curious.

The preceding is a very incomplete sketch of work on grounded theory, meant primarily to point to new resources and to understand the dearth of material for answering the question, Where do codes come from?

A Recipe for the Cognitive-Emotional Generation of "A Code" in Grounded Theory

Object Relations

What is a code? What do a bobcat and a sociologist have in common? We are both a bit out of bounds, myself in what I am writing and thinking, and she in crossing the driveway and peering up at my house. Using the preceding codes of emotions breaking through and abstracting simultaneously, I enrich the notion of out of bounds. My first reaction comes from a sense of attachment (remembering that attachment can be positive or negative or both). At some emotional level, I want to know this bobcat. I feel kinship with her. She is beautiful. I feel somewhat attached to her, as if I would like to capture her wildness and beauty and make it part of me. But I know that to do so in real life would be to kill that beauty. So I begin to think with "grounded theory feelings" about this little relationship. As I think about this, and abstract the relationship, a code forms that both adds and subtracts from the experience. Out of bounds is not the total of my experience or relationship with the bobcat. However, the combination of empathy, or attachment, and abstraction, is even more powerful than either alone. In working with biologists, I have felt this dual vision: if they see a subject within their expertise, they are compelled to name it, often in Latin. Almost simultaneously, they will say something empathic, like "hey there, little fella," or "what a beauty," or, as frequently, sorrow about the condition of its habitat. In fact, each part of science has some version of this double vision (see Keller 1983; or Star and Ruhleder 1996, on "the worminess of the worm" in a genetics nematode laboratory). A code, then, is a matter of both attachment and separation. When I am able to hold both *simultaneously*, I experience the joy and grief of adulthood. To speak of "life cycle" reasons for topics: I grew up in a close-knit, rural family, one that had very specific ideas about how things are done. Yet my reading (I seemed to have been practically born reading) implied other worlds, where these assumptions were not taken for granted. Gradually I began to nurture the notion that *somewhere else* was a place to reconcile these things. This required that I step out of bounds with respect to my extended birth family. This was emotionally hazardous, and it took me many years to trust my own experience of the world, and to balance many forms of attachment and separation, or abstraction and intimacy. From this experience I have drawn many projects related to marginality, outsiders, membership, or lack thereof, and invisible work.

Within psychological theory, the simultaneous attachment-separation idea is found extensively in the object relations model of familial and social dynamics. Perhaps most notably, the work of D. W. Winnicott (1965) on separation and transitional objects captures the dynamics of this developmental process. Winnicott has been primarily remembered for his studies of infant and child development. He was also a vital, imaginative theorist, moving from his clinical work to theory (and then back again). Although he was a psychoanalyst, his links with the Freudian school do not appear (to me) to be overly dogmatic or even central. Concepts such as "separation anxiety," "pathological attachment," and "good-enough mothering" belong to him; he also was relentlessly both theoretical and material. His work on transitional objects has (not unlike the work of Gregory Bateson [1972]) impacted scholars from many disciplines, and has been used to revitalize much of the analysis of love and loss.

Winnicott believes the infant learns gradually (in the best case) to attach to the parent in many small ways, and to experience those separations that are bearable for it. Maturity means holding larger and multiple separations along with attachment, neither total separation (abstraction) nor total attachment (total intimacy), but a balance. In the words of Winnicott (1965):

> It is generally acknowledged that a statement of human nature is inadequate when given in terms of interpersonal relationships, even when the imaginative elaboration of function, the whole of fantasy both conscious and unconscious, including the repressed unconscious, is allowed for. There is another way of describing persons . . . of every individual who has reached to the stage of being a unit . . . it can be said that there is an *inner reality* to that individual, an inner world which can be rich or poor and can be at peace or in a state of war . . . if there is a need for this double statement, there is need for a triple one; there is the third part of the life of a human being, a part that we cannot ignore, an intermediate area of *experiencing*, to which inner reality and external life both contribute. It is an area which is not challenged, because no claim is made on its behalf except that it shall exist as a resting-place for the individual engaged in the perpetual human task of keeping inner and outer reality separate yet inter-related. (230; emphasis in original)

One way to achieve this maturity is to learn how to manage the anxiety produced by separation or the smothering of too much attachment, through the use of what Winnicott called a transitional object: with young children, this is often a blanket, a doll, or some other small prized possession. The transitional object belongs a little to each world: the old world of being attached, and the new world of growing up, leaving, going away, abstracting.

Codes in Grounded Theory Are Transitional Objects

Codes allow us to know more about the field we study, yet carry the abstraction of the new. When this process is repeated many times, and constantly compared across spaces and across data, it is also possible reflexively to grow. In grounded theory, this is known as theoretical sampling. Codes are part, also, of the third space of development, the "holding space" of experience. Theoretical sampling stretches the codes, forcing other sorts of knowledge of the object. The theory that develops repeats the attachment-separation cycle, but in this sense taking a code and moving it through the data. In so doing, it fractures both code and data. Again, it calls up some anxiety, and at the same time, perhaps causally, it calls for authority. There isn't any roadmap, and to make it worse, as one practices constant comparison across data sets and even outside "normal" ethnographic data (Strauss 1970), one constantly loses and gains, attaches and separates.

Let me give an example of this from my own work, and foreshadow how this appears in Pragmatist problem solving. Some years ago, I became interested in the gap between how people act, and how they are represented online and on paper. Part of this insight came from doing fieldwork in an artificial intelligence (AI) laboratory and a neurophysiology laboratory, another part of it a reflection on my training to self-censor my feelings in the narrative of my research. One of the first family of codes I developed concerned simplification in scientific work (Star 1983). I examined the different ways that scientists, in writing up their results, discarded "unruly" data such as that gathered from women, black people, and bald men. The ideal research object was a white, middle-class ten-year-old boy. Unruly data included several codes about the work itself: discarding anomalies, substituting reliability for validity (a code I would return to throughout my work), and all the sorts of formatting constraints that writing for scientific journals contain, and that implicitly inhibit the "unruly data" of the investigator as well as the subject.

As this study grew into various studies of computer use, creation of classification systems, and other ways scientifically of representing human behavior, it became increasingly important for me to immerse myself in pragmatism. Much of the modem current of computer critique, particularly in artificial intelligence and in classification, has been purely logical and cognitive, including the philosophy of mind and of language (but see the very important exception of Suchman 1987, and her colleagues). The inspirational philosophers included Heidegger, Wittgenstein, John Searle, and Quine. I wrote one paper, accepted for publication, that attacked much of this work by questioning how AI used humans as research material. Then I sat with the paper, and with myself, and realized that I had to choose between becoming a scientific gadfly, or getting on with the project of seeing *how* these philosophers and computer scientists worked together; on understanding the impact of different forms of computing, and what work was being done to create visions of the human mind.

I withdrew the paper, which had been accepted for *AI and Society*, as I made the choice to return to studying work. I felt ashamed of myself, as if I had come close to becoming a sort of muckraker. In any event, this decision stood me in good stead, as I began working more closely with computer scientists as colleagues, not themselves solely as ethnographic subjects. Following the precept of "analyze consequences, not antecedents," I took this aspect of Dewey's work and outlook to heart. I ended up writing my first book on the coordination of work from different lines of work in early (nineteenth-century) neurophysiology and brain research (Star 1989).

So the doing of grounded theory, at its most basic, is, among other things, an emotional challenge and a call to methodological maturity. With other sorts of analysis, such as focus groups or surveys, one primarily is seeking to extract data from respondents. A focus group is face-to-face, but not long-term; in some survey research, it is often delegated such that one never meets a respondent (Roth 1966). (Of course, this varies considerably, and I am *not* saying that the joining of experience/affect with abstraction is the unique purview of grounded theory.)[3]

The Long Haul

Over a lifetime of research, some people have a sense of a *life's work*. For Strauss's festschrift, I wrote:

Every passionate scientist has a mystery at the center of his or her research. In exactly the ancient senses of mystery and passion, there are questions or sets of questions that can never be solved, only wrestled with, embraced, and, one hopes, transformed. The primacy of work is just such a mystery in the work of Anselm Strauss, and his profoundly fertile transformations have given rise to a new way of framing an old sociological/philosophical question: the relationship between the empirical/material on the one hand, and the theoretical abstract on the other. Here I call this the relationship between the visible and the invisible. (Star 1991b, 265)

It may be a very useful exercise for one to question oneself, at any career stage, about the nature of one's own passions and mysteries. (For those interested, my own passions are most closely spelled out in Bowker and Star 1999; Star 1991b, 1998; Star and Griesemer 1989.) Of course, in setting out this comparison with object relations, I realize that there is an infinite regress in possibility. Where does attachment come from? If I say, as many would, that it arises from the body, the unconscious, and one's biography, this takes my point from a chapter to someone else's life-work. At the same time, I hope this recognition of affect, attachment, and deep feelings will legitimate aspects of doing grounded theory that are rarely written about, although frequently talked about in some circles, perhaps most often in a teacher's counsel during the dissertation process. I was lucky beyond measure in having Strauss as an advisor I trusted completely.

Pragmatism

Pragmatist philosophy challenges one to accept the invitation to adulthood offered by object relations as interpreted previously. It is an occupation of the third space, and through grounded theory, the implication of codes as transitional objects both emotionally and analytically. Learning grounded theory was not divided from learning about Pragmatist philosophy and also about early Pragmatist-informed sociology and social psychology. As graduate students, we trained simultaneously in method and theory, and were encouraged to use both philosophers and other sociologists as sources for comparison, coding, ontology, and epistemology. We nearly always read the originals, not textbooks.

We emphasized the early Chicago School and its relationship with Dewey, James, Mead, and later, Bentley. [4] We joined a community of practice with profound historical roots. I will take several Pragmatist tenets, and relate them to the attachment-separation balance I have presented.

Consequences, Not Antecedents

One of the simplest and most difficult tenets of pragmatism is that understanding is based on consequences, not antecedents. One does not build an a priori logic, philosophical analysis with preset categories or do so as "verificationist" social scientists, as Barney Glaser so mordantly termed them. Rather, the process is backward to most modes of analysis. In a sense, to follow the language of this chapter, one bares one's soul to the elements, and sees what happens. "What happens" is a matter of several things. An interruption to experience (or as I teach it, an anomaly that gets one's attention in the data) was actually consistent with Peirce's notion of abductive thinking. Both Dewey and Mead posited that a fact is the result of an interruption to experience. Earlier, William James's notion of how we form habits, and react to their disruption, presages this formulation.

A reflection, based in a community and dialogue, about the nature of the experience, results in a new object (a more formal code, or the result of theoretical sampling, in Strauss's terms). This idea of the chain *experience-interruption-reflection-object* dates back in complex ways to Dewey's first paper, "The Reflex Arc Concept in Psychology" ([1896] 1981). There, Dewey argued that perception is not continuous, but constantly interrupted and in a way that demands interpretation. A reflex is not a matter of a continuous stream of "information" hitting the black box of the brain, but rather a constant feat of interpretation, as routine acts and ways of seeing are interrupted, interpreted, and revised. Thus, from James's "blooming, buzzing confusion," we arrive at humanness through the constantly offered interpretations of our family, community, nature, media, art, animals, and all others.

Choosing among conflicting interpretations means constant struggle for selfhood, that is, how one shapes a body of interruptions and interpretations, and comes to incorporate integrity and authority in action. In this, Pragmatism is neither modem nor postmodern, but orthogonal to the terms of those debates, including positivism vs. interpretation (or realism vs. relativism). Choices are complexly mediated by close-in cultures, and also by cultures-at-a-distance, including media.

A glimpse here, then, of what underlies attachment and separation: the fluid whisperings (and sometimes demands) of others in childhood, mediated then and in adulthood by history and the growing self. It constantly challenges the input–output model with the attachment-separation-transitional object formation. Dewey's *The Quest for Certainty* (1929a) scales up from the "Reflex Arc" paper, and examines how philosophers and other scholars use preset models, concepts, and methods. Dewey deems this a quest for certainty, a shelter from the emotional storm in a sense, although he uses few affective words to describe this. I would say that this quest for certainty (for example, for a single model that pre-explains most events in the world) is a way of shielding ourselves from the powerful pain of the attachment-separation grief (in part, at least). In doing grounded research, this is a moment to return to the data, bringing the subject-object mediation back to coding and categorizing.[5]

The Objective Reality of Perspectives

The maturity implied by constantly practicing the object relations side of grounded theory/Pragmatism may actually be the scariest aspect of doing grounded theory. In so doing, one becomes a methodological maverick and, in the earlier stages of one's career, this can be costly. In addition, practicing this philosophy directly challenges the binary and well-guarded division between interpretation and reality. Interpretation is poetry. Reality is science. They are meant to be kept apart. Putting them together is asking for trouble.

I came to sociology of science as a Pragmatist, at the beginning of what later would be called "the science wars," a bitter, divisive brawl where "constructivist theorists" (or interpretive theorists) were pitted against "realist scientists." Positivist scientists saw the work of much of the new program of sociology of science (beginning in roughly 1976) as anti-science. This included their attempts to see science as just another kind of work, as something that changes historically and culturally, and as a process subject to developmental contingencies and politics. Constructivism is a complex argument, not merely a know-nothing attitude. Participants often had high stakes in their approach to science, much of it quite personal, as well as a social world (community of practice), shared ideology, and collective practice. Their examination of realist science was met with indignation by traditional scientists. For example, those exploring the cultural and historical aspects of physics were seen as mystical dreamers, or even fools: "Are you saying that if I jump off a building I won't fall to the earth?" Well, no. But constructivists were, for the most part, interested in exploring the meaning of, say, falling, as culturally constituted; of injury and the body as having different meanings in different times and places. The world, as scientists had explained it to date, appeared as both brutal and universalist (except perhaps for art, religion, and the like) and reactive in just the way that Dewey saw in the reflex arc model.[6]

Delicately dissecting, situating, and making the world ontologically and epistemologically open to revision was not of interest to the traditional science warriors. They were quite threatened by the prospect of cultural relativism as applied to science. Often citing "Nazi science" or "Lysenkoism" as examples of what may happen as a result of relativism or constructivism, this group of scientists ridiculed sociology of science as if we were know-nothing yahoos who sought to ridicule science.

From the Pragmatist view, the response to this quarrel is to examine questions of responsibility, location, consequences, and authorship. Pragmatists see "universalism" as agreements across a large number of communities practices and cultures, nothing more or less. It does not exist in some a priori analytic reality. People always interpret events from a situated and complexly principled point of view. For example, death means the end for some religions, a transition for others, a transformation for still others. These are radically varying views of an experience we all undergo. If one respects interpretation, they are not, however, universal. Nevertheless, those things with powerful scope and scale consensus are to be respected, most of the time as being just that; this does not, however, mean that one must sign on to the belief in ontological or epistemological universality.

The classic article of Pragmatist philosopher George Herbert Mead, "The Objective Reality of Perspectives," gives a kind of mandate for the *ontological primacy of interpretation* ([1927] 1964). Mead argues that a perspective is a way to stratify and order nature. That stratification, he asserts, comes from the development of a perspective, ancient and slowly accrued, or more novel, no matter. These stratifications are "the only form of nature that is there." This means that as an analyst, the confidence and authority accumulated by attachment-separation-transition (uncountable times, sometimes recursing, sometimes accruing, sometimes taking other shapes and textures) must come to the fore as an author, when one writes. This can be psychologically difficult.

I believe that this difficulty is akin to something Barney Glaser once called "flooding," a common experience in grounded theory. One has a plethora of codes, comparisons, and provisional ways of arranging the data. But so much seems worthy. How to weed? How to choose? In his book, *Memory Practices in the Sciences*, Bowker (2006) analyzes the continually changing nature of the past. In *The Mnemonic Deep*, he notes that even a name has multiple, tangled origins and aesthetics, for example in naming new species. The illusion of the completed and perfected is only that, a story we make up in order to give legitimacy to our own authority. But as we plumb the mnemonic deep, we find exactly the same challenge as with "contemporary" data. The gap between the romantic story of the past and the messy, attached, feeling-full past can be painful and misleading, just as can pluralistic ignorance ("I must be the only one") in social groupings. As Dewey (1929b) said in *Experience and Nature*: "Romanticism is an evangel in the garb of metaphysics. It sidesteps the painful, toilsome labor of understanding and of control which changes sets us, by glorifying it for its own sake. Flux is made something to revere, something profoundly akin to what is best within ourselves, will and creative energy. It is not, as it is in experience, a call to effort, a challenge to investigation, a potential doom of disaster and death" (51).

Here both Bowker and Dewey capture the visceral terror of authority in research (see also Becker 1986). Part of it is an attempt, as Mead would say, to nail down "the specious present" (1932). We live between the past and the present, poised and vulnerable, despite the romantic call to prior solutions. In this sense, pragmatism reverses our commonsense temporality, and challenges us to a profound heterochronicity.

The Human Skin: Philosophy's Last Line of Defense

I will conclude with some points from the Pragmatist political scientist Arthur Bentley (1935, 1954), who, like Dewey, had a long and productive career and produced several important books before joining forces with John Dewey. Much of his work on science and organizations, including that on relativity theory and on beliefs and fact (written in the 1920s and 1930s), resonates profoundly with more recent sociology of science. He is less well known than Dewey, except perhaps for their coauthored *Knowing and the Known* (Dewey and Bentley [1954] 1970), one of Dewey's last publications.

One essay of Bentley's, now available in a volume of his reprinted papers, has had a profound impact on my own work in science. The article "The Human Skin: Philosophy's Last Line of Defense" argues that philosophers use the skin (the edges of the body) as an epistemological border and barrier, in fact out of a kind of superstition (Bentley 1954). To the extent that the object of analysis in pragmatism and in grounded theory is nearly always a form of action, this idea crystallizes a central challenge for, in fact, all of sociology and philosophy of science. The idea of action as the central unit of analysis has been discussed in all of the grounded theory books cited here, and in Strauss's most complete theoretical statement, *Continual Permutations of Action* (Strauss 1993).

If action is the unit, some unknown interiority (perhaps the brain, perhaps the gene, perhaps memory or history) cannot form the antecedent basis for action. An action always ramifies and continues, at least those sorts of action of importance for sociological analysis. Actions traverse the skin. They do not originate in individuals, but rather as a result of relations, the "between-ness" of the world. Thus, Bentley's work, at least as much as Dewey's, is a radical refutation of individualism. It calls to Durkheim's notion of the *sui generis* aspect of social facts, that is, that relations between many people constitute a level and unit of analysis that are whole and infungible (1938). The continuous nature of the act, and how and/or when you take that as your basic unit of analysis, changes perception of the world. Perhaps we might call this continual permutations of analysis.

Conclusion: Living Grounded Theory

In conclusion, I would like to return to how grounded theory permeates my way of seeing the world, in connection with pragmatism. Embedded in my everyday action, it is a powerful tool, almost a spiritual tool to decenter my own assumptions and constantly remind me to try to take the role of the other, in Mead's words. This has developed over a long period of time (I have been doing this for about twenty-nine years, making me a baby, really, in terms of learning). However, I do have a sense of a life-work, as I have mentioned. So in this sense, both temporally and spatially, grounded theory helps to form my biography as well as my way of seeing and perceiving day to day. The risk of adding different studies together, of following various paths connected to science, technology, medicine, and information, is that I will fall into a confused compilation of theory, and measure it according to the conventional world of academia. The joy or goal, conversely, is to understand what I write as a wild, imaginative window on the world. I hope that I have the courage to allow one study to interrupt the experience of another.

How does one create a life-work and remain open; open to the data, open to being wrong, to redoing one's own work, to actively seek out new views and mistakes? For me, that has come through the privilege of teaching grounded theory, and of collaborating with people who like to work this way. That is, to embrace a continuous, embedded, imbricated, multiple, constantly compared way of making sense of myself. I hope this chapter is helpful in understanding the emotional depths and life-work of living grounded theory and pragmatism. The growing community of analysts, critics, and students is my ground of reflection, and we give each other the courage to go on.

Acknowledgments

Thanks to Olga Kuchinskaya for a good discussion of the emotional aspects of coding and choice; to Pat Lindquist for teaching me to write and to grow; to Geof Bowker, for anchoring my experience and helping me braid my projects together in my own idiosyncratic way; and to John Staudenmaier, for being my friend. Kathy Charmaz, Tony Bryant, David Obstfeld, and anonymous referees provided invaluable insights.

Notes

1. Another way of "seeing" the web of relationships in science is examining cases where people attempt to export the science in the absence of those relationships. For a brilliant example, see Wenda Bauchspies's ethnography of the acceptance of science as a "stranger" (1998), in the sense classically developed by Simmel ([1908] 1950).

2. My thanks to Kathy Charmaz for her very perceptive comment about this.

3. One *may* do grounded theory on any data, but qualitative analysis is really not very good at predicting elections or understanding large-scale demographic change. It may be very useful in companionship with other methods, but that is a statement more honored in the breach than the practice.

4. We did read Peirce as well, and some of the more minor philosophers, but at that time Peirce was not emphasized to us; he was presented by Strauss as being of a much older generation, and

few of us gravitated toward his work alone. Nor were we encouraged to do so. Kathy Charmaz notes that he was more emphasized in her cohort at UCSF than he was in mine.

5. My thanks again to Kathy Charmaz.

6. And reactive in just the way that Dewey saw in the reflex arc.

References

Adler, Paul S., and David Obstfeld. 2007. "The Role of Affect in Creative Projects and Exploratory Search." *Industrial and Corporate Change* 16 (1): 19–50.

Bateson, Gregory. 1972. *Steps to an Ecology of Mind*. San Francisco: Chandler.

Bauchspies, Wenda. 1998. "Science as Stranger and the Worship of the Word." *Knowledge and Society* 11 (2): 189–211.

Becker, Howard S. 1986. *Writing for Social Scientists: How to Start and Finish Your Thesis, Book, or Article*. Chicago: University of Chicago Press.

Bentley, Arthur. 1935. *Behavior, Knowledge, and Fact*. Bloomington, IN: Principia Press.

Bentley, Arthur. 1954. *Inquiry into Inquiries: Essays in Social Theory*. Ed. and with an introduction by Sidney Ratner. Boston: Beacon Press.

Bowker, Geoffrey. 2006. *Memory Practices in the Sciences*. Cambridge, MA: MIT Press.

Bowker, Geoffrey, and Susan Leigh Star. 1999. *Sorting Things Out: Classification and Its Consequences*. Cambridge, MA: MIT Press.

Charmaz, Kathy. 1991. *Good Days, Bad Days: The Self in Chronic Illness and Time*. New Brunswick, NJ: Rutgers University Press.

Charmaz, Kathy. 2006. *Constructing Grounded Theory: A Practical Guide through Qualitative Analysis*. Thousand Oaks, CA: Sage.

Clarke, Adele E. 1998. *Disciplining Reproduction: Modernity, American Life Sciences, and the Problems Of Sex*. Berkeley: University of California Press.

Clarke, Adele. 2005. *Situational Analysis: Grounded Theory after the Postmodern Turn*. Thousand Oaks, CA: Sage.

Corbin, Juliet, and Anselm Strauss. 1988. *Unending Work and Care: Managing Chronic Illness at Home*. San Francisco: Jossey-Bass Publishers.

Daly, Mary. [1978] 1990. *Gyn/ecoiogy: The Metaethics of Radical Feminism*. Boston: Beacon Press.

Dewey, John. 1929a. *The Quest for Certainty: A Study of the Relation of Knowledge and Action*. New York: Minton, Balch.

Dewey, John. 1929b. *Experience and Nature*. New York:George Allen and Unwin.

Dewey, John. [1896] 1981. "The Reflex Arc Concept in Psychology." In *The Philosophy of John Dewey*, ed. J. J. McDermott, 136–148. Chicago: University of Chicago Press.

Dewey, John, and Arthur Bentley. [1954] 1970. *Knowing and the Known*. Boston: Beacon Press.

Durkheim, Emile. 1938. *The Rules of Sociological Method*. Chicago: University of Chicago Press.

Gilligan, Carol. 1993. *In a Different Voice: Psychological Theory and Women's Development*. Cambridge, MA: Harvard University Press.

Glaser, Barney G. 1978. *Theoretical Sensitivity: Advances in the Methodology of Grounded Theory*. Mill Valley, CA: Sociology Press.

Glaser, Barney G., and Anselm L. Strauss. 1965. *Awareness of Dying*. Chicago: Aldine.

Glaser, Barney G., and Anselm L. Strauss. 1967. *The Discovery of Grounded Theory: Strategies for Qualitative Research*. Chicago: Aldine.

Glaser, Barney G., and Anselm L. Strauss. 1968. *Time for Dying*. Chicago: Aldine.

Hughes, Everett. 1970. *The Sociological Eye*. New York: Aldine.

Keller, Evelyn Fox. 1983. *A Feeling for the Organism: The Life and Work of Barbara McClintock*. New York: W. H. Freeman.

Kelly, George A. 1955. *The Psychology of Personal Constructs*. New York: W. W. Norton and Company.

Kelly, George A. 1963. *A Theory of Personality: The Psychology of Personal Constructs*. New York: W. W. Norton and Company.

Kohlberg, Lawrence. 1981. *The Philosophy of Moral Development: Moral Stages and the Idea of Justice*. San Francisco: Harper & Row.

Kuhn, Thomas. 1970. *The Structure of Scientific Revolutions*. Chicago: University of Chicago Press.

Latour, Bruno. 1987. *Science in Action*. Cambridge, MA: Harvard University Press.

Lave, Jean, and Etienne Wenger. 1991. *Situated Learning: Legitimate Peripheral Participation*. Cambridge: Cambridge University Press.

Mead, George Herbert. [1927] 1964. "The Objective Reality of Perspectives." In *Selected Writings*, ed. A. J. Reck, 306–319. Chicago: University of Chicago Press.

Mead, George Herbert. 1932. *The Philosophy of the Present*. La Salle, IL: Open Court Publishing Co.

Mills, C. Wright. 2000. *The Sociological Imagination*. Oxford: Oxford University Press.

Obstfeld, D.2005. "Social Networks, the *Tertius Iungens* Orientation, and Involvement in Innovation." *Administrative Science Quarterly* 50:100–130.

Richardson, Laurel. 1996. "Ethnographic Trouble." *Qualitative Inquiry* 2:27–30.

Riegel, Klaus F. 1978. *Psychology, Mon Amour: A Countertext*. Boston: Houghton Mifflin.

Roth, Julius A. 1966. "Hired Hand Research." *American Sociologist*, 190–196.

Sanjek, Roger. 1990. *Fieldnotes: The Makings of Anthropology*. Ithaca, NY: Cornell University Press.

Schatzman, Leonard, and Anselm L. Strauss. 1973. *Field Research: Strategies for a Natural Sociology*. Englewood Cliffs, NJ: Prentice Hall.

Simmel, Georg. [1908] 1950. "The Stranger." In *The Sociology of George Simmel*, ed. Kurt Wolff, 402–408. Glencoe, IL: Free Press.

Star, Susan Leigh. 1983. "Simplification in Scientific Work: An Example from Neuroscience Research." *Social Studies of Science* 13:205–228.

Star, Susan Leigh. 1989. *Regions of the Mind: Brain Research and the Quest for Scientific Certainty*. Stanford: Stanford University Press.

Star, Susan Leigh. 1991a. "Power, Technologies and the Phenomenology of Standards: On Being Allergic to Onions." In *Sociology of Monsters? Power, Technology and the Modern World*, ed. John Law, 27–57. Sociological Review Monograph 38. Oxford: Basil Blackwell.

Star, Susan Leigh. 1991b. "The Sociology of the Invisible: The Primacy of Work in the Writings of Anselm Strauss." In *Social Organization and Social Process: Essays in Honor of Anselm Strauss*, ed. David Maines, 265–283. Hawthorne, NY: Aldine de Gruyter.

Star, Susan Leigh. 1998. "Experience: The Link between Science, Sociology of Science and Science Education." In *Thinking Practices*, ed. Shelley Goldman and James Greeno, 127–146. Hillsdale, NJ: Lawrence Erlbaum.

Star, Susan Leigh. 1999. "The Ethnography of Infrastructure." *American Behavioral Scientist* 43:377–391.

Star, Susan Leigh, and James R. Griesemer. 1989. "Institutional Ecology, 'Translations' and Boundary Objects: Amateurs and Professionals in Berkeley's Museum of Vertebrate Zoology, 1907–39." *Social Studies of Science* 19:387–420. [See also chapter 7, this volume.]

Star, Susan Leigh, and Karen Ruhleder. 1996. "Steps toward an Ecology of Infrastructure: Design and Access for Large Information Spaces." *Information Systems Research* 7:111–134.

Star, Susan Leigh, and Anselm Strauss. 1999. "Layers of Silence, Arenas of Voice: The Ecology of Visible and Invisible Work." *Computer Supported Cooperative Work: The Journal of Collaborative Computing* 8:9–30.

Strauss, Anselm. 1970. "Discovering New Theory from Previous Theory." In *Human Nature and Collective Behavior: Papers in Honor of Herbert Blumer*, ed. Tomotsu Shibutani, 46–53. Englewood Cliffs, NJ: Prentice Hall.

Strauss, Anselm. 1987. *Qualitative Analysis for Social Scientists*. Cambridge: Cambridge University Press.

Strauss, Ansefm. 1993. *Continual Permutations of Action*. New York: Aldine de Gruyter.

Strauss, Anselm L., and Barney G. Glaser. 1970. *Anguish: A Case History of a Dying Trajectory.* Mill Valley, CA: Sociology Press.

Suchman, Lucy A. 1987. *Plans and Situated Actions: The Problem of Human-Machine Communication.* New York: Cambridge University Press.

Vygotsky, L. S. 1986. *Thought and Language.* Cambridge, MA: MIT Press.

Wax, Rosalie H. 1971. *Doing Fieldwork: Warnings and Advice.* Chicago: University of Chicago Press.

Winnicott, D. W. 1965. *The Maturational Processes and the Facilitating Environment; Studies in the Theory of Emotional Development.* London: Hogarth.

6   Misplaced Concretism and Concrete Situations: Feminism, Method, and Information Technology

Susan Leigh Star

Editors' Note: This working paper, written by Leigh for a Gender-Nature-Culture Feminist Network Conference at Arhus University in Denmark in 1994, has had an influence on feminist methodology far beyond the restricted distribution of the publication itself. This constitutes the first time it is brought to a wider audience. Nina Wakeford (chapter 3, this volume) makes extensive use of it. The concept of "misplaced concretism" was drawn from John Dewey. The close relationship both historically and conceptually between pragmatist philosophy and grounded theory was a core feature of Leigh's work. Her move to seeing objects not as unitary artifacts but as inherently ambiguous both develops this concept and situates it within feminist methodology.

As social theorists and activists, we often see multiplicity and heterogeneity as accidents or exceptions. The marginal person, who is for example of mixed race, is seen as the troubled outsider; just as the thing that doesn't fit into one category or another gets put into a "residual category." I don't know where this originated in Western scientific and political culture, but I do know that the habit perpetuates a cruel pluralistic ignorance. None of us are pure. None of us are even average. And all things inhabit someone's residual category in some category system. In recent years feminists and multicultural theorists have been working toward enriching our understanding of multiplicity and misfit, and decentering the idea of an unproblematic mainstream. During the same period, such issues have become increasingly of concern to some information scientists. As the information systems of the world expand and flow into each other, and more kinds of people use them for more different things, it becomes harder to hold to pure or universal ideas about representation or information. This chapter is an attempt to pull together these threads and define a common territory where feminism, race critical theory, multiculturalism, and information science meet, a space sometimes called "cyborg." Cyborg, as used for example by Donna Haraway (1991) and Adele Clarke (1993), means the intermingling of people, things (including information technologies), representations, and politics in a way that challenges both the romance of essentialism and the hype about what is possible technologically. It acknowledges the interdependence of people and things, and just how blurry the boundaries between them have become. My best hope for this cyborg essay is that it will produce in you a gestalt switch where multiple memberships and multiple meanings come to the fore- ground; and it will become purity and universality that routinely bear the burden of proof.

Introduction

"Just don't go all the way."

This was my mother's response upon my announcement, at the age of nineteen, that I had become a feminist and was studying feminist philosophy at university. Of course I knew exactly what she meant, just as I had known what she meant when she left me with the injunction "just don't kill us" as her last words to me before I left for university. "Not going all the way" meant:

- don't become a lesbian, and therefore in her eyes,
- don't become a man, therefore
- don't also become some sort of weird transsexual, therefore
- don't shame us with your monstrosity, and give up all hope for us having grandchildren.

"Not killing us" simply meant don't get pregnant if you're not married. The year was 1971 when I left for university; I became a feminist in 1973.

In a sense my mother's words at both times summed up both an immense shift in paradigm and a profound conservatism. Not going all the way marks the outer limits of what most people, inside and outside academia and science, have been able to understand about feminism. In its most extreme form, it is a rejection of traditional femininity, a sexualized crossing over that made the word "dyke"[1] intimately familiar to every feminist in those early years, whether or not she was intimately familiar in that exact way with other women. As Kate Millett's work and the *Woman-Identified Woman* manifesto of the early 1970s made so clear, "dyke" was the outer-limiting case of how far you could go with this feminism stuff.

An elegant, dense social movement, culture, and literature grew up on the ashes of homophobia and reductionism fueling those stereotypes. Over two decades, this new world order revisioned sexuality, put homophobia on trial, rejected the straight-gay dichotomy, and began to understand the complex relations between the sexualization of race and the racialization of sex, sexual choice, and gender (don't forget the work of distinguishing sex, sexuality, and gender in the first place, which hadn't yet been done).

Those early years were painful, heady explorations of all of these aspects jumbled up together. Were men mutants? Did they have testosterone poisoning? Or was it just culturalhistorical location? (Culturalhistorical became one word, and note the naive "just" thrown in to modify it.) Did women evolve from a different origin? Was lesbianism the natural state, and all heterosexuality really a form of enforced rape? Was collective, global withdrawal from men and separatism the only answer? Should we have babies in test tubes? Should we become celibate, should we all declare ourselves lesbians regardless of urges, should we have orgasms together in a circle?

In the mid-1970s I moved to the San Francisco area as had thousands of mavericks for many decades, including many radical lesbian feminists. We continued to argue and speculate about men and women, nature, and culture. An ironic event in the late 1970s and early 1980s was one of the many influences on the nature of those very lively discussions. We referred to it privately as the "turkey baster disaster." As this cohort of lesbians aged, many began to want children. Clearly, sleeping with men was out of the question, both for aesthetic reasons and for all the political reasons noted above. Besides, most of the men we knew were gay. So, a homegrown technology quickly evolved for conducting artificial insemination. You would find a smart, healthy gay man, convince him to make a donation to the cause, capture it in little plastic cup, put it in a turkey baster, and rush home to your lover. In a loving shared ritual, you would then inseminate the mother-to-be—whichever one of you had been chosen for this role. Hundreds of women became pregnant in this fashion over a period of a few years. "It's a turkey baster baby" become a funny and affectionate answer to the perennial question, "who is the father?" (see figure 6.1).

However, the "disaster" part was not to be discovered (collectively) for several months. It had some distinct material elements. Many lesbians couldn't afford to live right in the city of San Francisco, which was becoming increasingly expensive (partly due to the influx of gay men into areas such as the Castro district, driving up housing demand within this specialized and relatively affluent sector). So many lived in Oakland, an economically more modest city about fifteen miles away across the Bay Bridge. The story goes that driving back across the bridge from San Francisco with the little Donation, it would cool down. This cooling effect made it slightly more probable that the xy-carrying sperm, which were a bit lighter than the bigger xx-carrying sperm, would be shaken to the top of the baster, (the tip of it being held up vertically) and end up reaching the woman's body first. Thus, after a time it became apparent that almost all the lesbians that got pregnant this way were having boy children! It was an epidemic of boy children! (The estimates began to run to 85 percent.) None of the women had envisioned this outcome in the independent-woman scenario put forward at the time—most had dreamed of daughters, of expanding the woman-only, neomatriarchal community. Workshops for lesbian mothers of sons sprang up. Support groups. Signs

Rubber Bulb

Shaft

This is a picture of a turkey baster. After writing the article I learned that it is not a common cooking article in Scandanavia. Its purpose is to keep moist very large turkeys (often up to 10 kilos for the ceremonial roasted turkeys eaten by many Americans on Thanksgiving Day) during the long cooking process (up to 6 hours). The rubber bulb, when squeezed, forms a suction which draws the cooking juices into the shaft; the liquid is then squirted back over the bird. This basting process is done repeatedly during the cooking. With smaller dishes a spoon can be used, but with the larger birds you can burn yourself that way by bumping your hand against the side of the oven.

Figure 6.1
Diagram and description of a turkey baster

at meetings declaring "women-only space. No men or boys allowed" became more rare. Talk of men as mutants seemed to drop off sharply.

I don't pretend that this anecdote is demography or even very good physiology, and there were many other factors influencing the move away from separatism.[2] The whole event could have been coincidence or perceptual emphasis, and for all I know the actual ratio was 50–50. What was important about it for our purposes here, however, was a marked shift in the community from essentialist, "biologically" driven explanations to more complex, contradictory, heterogeneous ways of thinking about experience and situations. "Men" became modified by "our sons"; "mutants" were newly domesticated and intimate. And from this experience and many others, the vocabulary of essentialism was deeply scrutinized and abandoned (by many). Among other things, we took the misplaced concretism of sex and re-situated it within the concrete experience of gender and relationships.

I want to take us from low-tech turkey basters to high-tech computers via this example, because I think it says something important about feminism, method, and technology. One simplistic way of reading these events (and others of the late 1970s and early 1980s) is that the blunt reality of life experience interfered with an idealistic, ideological kind of talk—a form of realpolitik that was also a co-optation. Weren't we always making exceptions for "our men"? Wasn't that one of the first things to come under attack in early feminism? Had the issue stopped there, perhaps the criticism would have some weight. But it has not. Our sons of lesbian mothers was only one of dozens of contradictions and complexities that we have articulated and survived. So we are forced to a deeper reading of events about feminism and the philosophy of technology.

The kind of rethinking processes that lesbian feminists engaged in during this period are not extraordinary within feminism, but routine. The material conditions of our experiences, in a collective, forced us to go very deeply indeed into all sorts of metaphysical, ontological, and epistemological questions. What is a human? What is a category? Does challenging a category mean giving up a moral stand? There was a lot of agony about actions such as making exceptions, appearance vs. "reality," and the meaning of alliance. Thus we evolved the deconstruction of gender, the centering of gender/sexual ambiguity and multiplicity, the fight for erasure of gender differences under some circumstances, the interlocking nature of race, class, and gender oppression, and the honoring of historical and cultural traditions of masculinity and femininity in various ethnic cultures: *all at once.*

I hold that it is the all at once-ness that is at the core of feminist survival, and as a consequence at the core of our relationship to science and technology. The power to hold multiple, contradictory views in a moral collective is necessary in shaping the divergence between Big Brother and a positive, cyborg-inhabited multiverse. There is also the power to evolve and change through painful collective confrontation with experience. One way to think of this is in understanding feminism as a method and as a methodology. In this sense, only methods—not positions, systems, or artifacts—have a chance to go *all the way*.

Method

The nature of method is complex. You can have a method for many things: cleaning house, ordering shelves, organizing demonstrations, building technology, or doing science. A method is distinct from a recipe or formula, in exactly the sense that science is not embodied in a textbook and cooking is not a cookbook. It is a real-time, lived, and experiential form of ordering practice. In the words of Isabelle Stengers:

> Indeed, you do not follow a challenge, you do not obey it, it does not direct you. You have to invent the way to answer it, it proposes risks for your answers, but gives you no model. Thus it is consonant with my conception of science. It is consonant because our "social experience," the moral and political options which situate us cannot become self-conscious just by a process of honest self-examination. It must be created through an active process of learning. Learning how we are situated, inventing the situations from which we can learn more about our situation does not give power to emancipation over cognition. It associates both emancipation and cognition. (1993, 46)

In French *experience* is the word for *experiment*. The root in lived experience is important here. Recipes or lists may act as heuristics or aides-de-memoir, but they never substitute for the communication and community that order work historically. If your mother never taught you to cook, it is extremely difficult to learn it alone from a cookbook. If you never got chance to hang out in a scientific laboratory or a computer game arcade, you may similarly be at a disadvantage in learning aspects of science and technology, as Kiesler and Sproull have noted.

Method is a way of surviving experience. It is a word at once stronger than paradigm, in the sense that it often crosses, both historically and spatially, most uses of the Kuhnian term. It may be part of several paradigms; it may persist after other attributes of a paradigm have fallen away. Methods considered in this fashion may have many of the features of surviving experience, depending on the values of the community using them: they can become imperialistic or monolithic (if one only has a hammer, the world becomes a nail, etc.); they can become a means of enforcing fundamentalism (reducing the world to that which can be perceived using the method); or they can become ways of encompassing multiplicity, complexity, and ambiguity.

It is in this latter sense that feminism is important methodologically, I think, although we have sometimes used it in the monolithic or reductionist senses. Feminists have written some extremely powerful methodological pieces, not always recognized as such. This is Haraway's vision in "A Manifesto for Cyborgs": the geography of nature to be experienced by feminism is coyote, shifting, always partial. "How do I then act the bricoleur that we've all learned to be in various ways, without being a colonizer. . . . How do you keep foregrounded the ironic and iffy things you're doing and still do them seriously. . . . One reply that makes sense to me is, the subjects are cyborg, nature is coyote, and the geography is elsewhere" (Haraway 1991). But included in the cyborg image is the question: *how*—a fundamentally methodological question.

A similar methodological question is at the heart of bell hooks's *Yearning:* she speaks of the difficulty and complexity of crafting a counterhegemonic discourse, based in ordinary homely events, always shifting, always marginal, and drawing power from the double vision of that marginality (hooks 1990). It is in Adrienne Rich's (1977) *Women and Honor: Some Notes on Lying*, another methodological manifesto. How does one distinguish truth from lying? She says, "The truth is not one thing." It is a complex, woven carpet, with each strand a partial truth. The aggregate is a negotiated order, not an a priori set of structures or a formula to be applied. Methods again appear in Gloria Anzaldúa's "La conscientia de la mestiza": the doubleness and the ambiguity of the male/female, straight/gay, Mexican/American borderland can become a creative approach to surviving, a rejection of simplistic purity and of essentialist categories, while at the same time remembering physical and political suffering, racism, and sexism (1987).

Feminism as a method thus creates robust findings through the articulation of multiplicity, contradiction, and partiality, while standing in a politically situated, moral collective. Our scientific genius is found as much in the process of an impoverished old black woman surviving to speak her story as in the virtuoso performance of a Nobel laureate. Our validity and reliability are not the decontextualized terms of social science, where they refer to problems in measurement that have been stripped of feelings and experience. In Mary Daly's words, this means doing away with "methodolatry," or the fetishizing of particular techniques. In Shulamith Reinharz's terms, knowledge is inseparable from the process of strategic community building and understanding.

Considered formally, then, the attributes of feminist method that are particularly important are:

1. experiential and collective basis;
2. processual nature;
3. honoring contradiction and partialness;
4. situated historicity with great attention to detail and specificity; and
5. the simultaneous application of all of these points.

Patti Lather, a brilliant feminist methodologist, has written about this under the rubric *transgressive validity* (Lather 1993). Her map of attributes of a feminist method, which I read after I wrote the passage above, has some remarkable resonances. She asks the question, where, after poststructuralism, can we find validity? Her answer has four points that together equal a validity which rests reflexively on the contemporary crisis of representation: *Ironic validity* (which problematizes the single voice, realist representation of nature); *paralogical* validity (which emphasizes paradox and heterogeneity); *rhizomatic* validity (which undermines the taken-for-granted and keeps opening up new ways of situated seeing); and *voluptuous* validity (which precisely *goes too far*, and joins ethics and epistemology) (685–686).

Take for example a very large information system that keeps track of medical information about pregnancy and childbirth. An ironic validity would demand that the representation of pregnancy as an illness be challenged in the encoding, and include multiple views of what in fact constitutes a pregnancy. Paralogical validity might at the same time acknowledge that for some women an aborted fetus never constituted a pregnancy at all, while for others it did, and there is no resolving that question in any simple fashion. A rhizomatic validity would seek out the range of what we take for granted about pregnancy, childbirth, and sexuality, and keep challenging us to break boundaries of what we think we know about these processes—perhaps cross-culturally, but also in phenomenological explorations of daily routines connected with them, denying the taken-for-granted "every woman feels X": and being sure to listen for complex differences and similarities. A voluptuous validity in this sense might bring entirely other conceptions of birth and pregnancy, such as the kind of spiritual birth found in some religious cultures, or dreams and visions. From this base, feminism has some important things to say about technology, and especially information technology, in its capacity creatively and ethically to mingle people, things, and experience.

Feminism and Technology

At its most abstract, the design and use of information systems intimately involves linking experience gained in one time and place with that gained in another, via representations of some sort. Even mere replication and transmission of information from one place to another involves encoding and decoding as time and place shift. Thus the context of information shifts in spite of continuities; and this shift in context imparts multiplicity to the information itself. In order to be useful, the information must reside in more than one context. And then, in order to be meaningful, the contexts must be linked through some sort of judgment of equivalence or comparability on the part of the user. Consider the design of a computer-supported cooperative work (CSCW) system to support collaborative writing. Eevi Beck studied the evolution of one such system: "how two authors, who were in different places, wrote an academic publication together making use of computers," and she notes that "the work they were doing and the way in which they did it was inseparable from their immediate environment and the culture which it was part of" (1995, 53). To make the whole system work, they had to juggle time zones, spouses' schedules, and sensitivities about parts of work practice such as finishing each other's sentences as well as manipulating the technical aspects of the writing software and hardware. They had to build a shared context in which to make sense of the information. Beck is arguing against a long tradition of decontextualized design where only the technical, or narrowly construed, considerations about work hold sway.

None of this is new in theories of information and communication: we have long had models of signals and targets, background noise and filters, degradation of signals and quality controls. It becomes new, however, when people are added as active interpreters of information, who themselves inhabit multiple contexts of use and practice (Star 1991). What becomes problematic under these circumstances is the relationship between people and things, or objects, the relationship that creates representations and not just noise. Information is only information when there are *multiple* interpretations. One person's noise may be another's signal, or two people may agree to attend to something—but it is the tension between contexts that actually creates representation.

We lack good relational language here. There is a permanent tension between the formal and the empirical, the local/situated and attempts to represent information across localities. It is this tension itself which is underexplored and undertheorized; it is not just a set of interesting metaphysical observations, but also a pragmatic unit of analysis. How can something be simultaneously concrete and abstract? The same and yet different? We are not used to thinking in this fashion in science, although it is more common in art and literature, especially in surrealist art and Bakhtinian aspects of the novel—and in feminism. In Donna Haraway's words: "No layer of the onion of practice that is technoscience is outside the reach of technology of critical interpretation and critical inquiry about positioning and location; that is the condition of embodiment and mortality. The technical and the political are like the abstract and the concrete, the foreground and the background, the text and the context, the subject and the object" (Haraway 1993, 10).

As a sociologist, I see this tension itself as collective, historical, and partially institutionalized. The medium of an information system is not just wires and plugs, bits and bytes, but also conventions of representation, information both formal and empirical. A system becomes a system in design and use, not the one without the other. The medium is the message, certainly, and it is also the case that the medium is a political creation. Very large scale, global and national information systems (the Internet, the Digital Highway, National Information Infrastructure) as well as more restricted tools designed for long-distance collaboration are rapidly changing how we work, and to some degree how we play.

A fully developed method of multiplicity for information systems must draw on many sources and make many unexpected alliances (Star 1989a,b, chapter 1, c; Hewitt 1985; Goguen 1992). If both people and objects inhabit multiple contexts, and a central goal of information systems is to transmit information across contexts, then the pathway of representation must analytically include everything which populates those contexts. This includes people, things/objects, representations, and information about its own structure. The major requirements of such an understanding are thus:

1. how can objects inhabit multiple contexts at once, and have both local and shared meaning?
2. how can people living in one community, and drawing their meanings from other people and objects situated there, communicate in another?
3. what is the relationship between the two?
4. what range of solutions to the preceding three questions are possible, and what consequences attend each of them—*cui bono?*

Standardization has been one of the common solutions to this class of problems:[3] if the interfaces and formats are standard across contexts, then at least the first three questions become clear, and the fourth becomes moot. But we know from a long and gory history of attempts to standardize information systems that standards don't remain standard for very long, and that one person's standard is another's confusion and mess (Gasser 1986; Star 1991).

Why should computer scientists read African American poets? What does CSCW have to do with race critical theory and feminist methods and metaphysics? I propose that feminist, antiracist, and multicultural theory, and our collective experience in those domains is one of the richest places for which to understand these core problems in information systems design: how to preserve the integrity of information without *a priori* standardization and its attendant violence. In turn, if those lessons can be taken seriously within the emerging cyberworld, there may yet be a chance to base it in a democratic moral order.

Requirements for Feminist Method

Consider then the idea of a particular information context ecologically, as composed of people and things, themselves in ecological relation, with numbers of representations and signals, and ways of working. Lave and Wenger (1992) have called such contexts "communities of practice," a term which I like because it emphasizes the ways in which people work together and act together to form communities, not just traditional organizational forms and boundaries. It is an idea which has been welcomed by information scientists as well as a way of talking about a linked web of actions, people, and artifacts. The following two sections explore the idea of community of practice from the point of view of things, and of people.

Although we think we pretty much know what we mean by people and things, I'm not so confident. Let us begin with the question: "What is a thing?"

Objects and Communities of Practice

As Engestrom (1990) and other activity theorists note so well, activity is always mediated by tools and material arrangements. You never act in a vacuum or in a "pure" universe of doing, but it is always with respect to arrangements, tools, and material objects. Strauss (1993) has recently made a similar point, emphasizing the continuity and permeability of such arrangements—you are never starting from scratch. Both go to great lengths to demonstrate that an idea, or something you've learned, can also be considered as having material/objective force in its consequences and mediations.

Here I mean "object" to include all of these things: stuff and things, tools and techniques, and ideas, stories, and memories—those objects which are treated as things by community members (Clarke and Fujimura 1992a, 1992b). They are used in the service of an action, and mediate it in some way. Something actually becomes an object only in the context and action and use—it is then as well something that has force to mediate subsequent action.

A community of practice is defined in large part according to the co-use of such objects, since all practice is so mediated. The relationship of the newcomer to the community largely revolves around the nature of the relationship with the objects—and not, counterintuitively, directly with the people. Acceptance or legitimacy derives from the familiarity of action mediated by "member" objects.

But "familiarity" is a fairly sloppy word. I mean it here not instrumentally, as in proficiency, but relationally, as a measure of taken-for-grantedness. (An inept programmer can still be a member of the community of practice of computer specialists, albeit a low-status one in that she takes for granted the objects to be used.) A better way to describe the trajectory of an object in a community is as one of *naturalization*. By naturalization, I mean stripping away the contingencies of an object's creation and its situated nature. A naturalized object has lost its aura of anthropological strangeness, and is in a sense "de-situated" in that members have forgotten the local nature of the object's meaning.[4] We no longer think much about the miracle of plugging a light into a socket and obtaining illumination, and must make an effort of anthropological imagination to remind ourselves of contexts in which it is still unnaturalized.

Objects become natural in a particular community of practice over a long period of time (see Latour's 1987 arguments in *Science in Action* for a good discussion of this). Objects exist, with respect to a community of practice, along a trajectory of naturalization, which has elements of both ambiguity and duration. It is not predetermined whether an object will ever become naturalized or how long it will remain so—rather, practice/activity is required to make it so and keep it so. The more naturalized an object becomes, the more unquestioning the relationship of community to it; the more invisible the contingent and historical circumstances of its birth;[5] the more it sinks into the community's infrastructure. The turkey baster story is in part funny because it's so ordinary. Almost all Americans know about turkeys, even if it's not part of their regular diet, and a turkey baster is a naturalized object in many women's communities of cooking practice. We don't think twice about its nature, only about whether or not we can find it on Thanksgiving Day when we need it. Technologies are often a good example of this—in a sense they are a form of collective forgetting, or naturalization, of the contingent, messy work they replace. I am writing this paper on a Macintosh Power-Book; "cutting and pasting" with it are no longer phenomenologically novel operations, although I can remember when they once were.

So far we are only talking here about the relationship between an object and a single community of practice, so it is possible analytically to talk about a smooth trajectory of naturalization. But in fact this singular relationship doesn't exist in the real world, an important fact in carrying the methodology further.

People and Membership

We are all members of various groups that practice activities together. Membership in such groups is a complex process, varying in speed and ease, with how optional it is, and how permanent it may be. One is not born a violinist, as we all know, but gradually becomes a member of the violin-playing community of practice through a long period of lessons, shared conversations, technical exercises, and participation in a range of other related activities. Jean Lave and Etienne Wenger have investigated how this membership process unfolds, and how constitutive it is of learning.

People live, with respect to a community of practice, along a trajectory (or continuum) of membership, which like naturalization has elements of both ambiguity and duration. In Lave and Wenger's terms, they may move from *illegitimate peripheral participation* to full membership in a community of practice, and it is extremely useful in many ways to conceive of learning in this way. But to twist the concept around a little differently along the lines expressed above: if we include objects and their naturalization as the sine qua non of membership, then the trajectory of learning becomes a series of encounters with the objects of practice in the community.[6] Illegitimacy is seeing those objects as would a stranger—either as a naïf or by comparison with another frame of reference in which they exist. One does not have to be Isaac Stern to know fully and naturally what to do with a violin, where it belongs, and how to act around violins and violinists. But if you use a Stradivarius to swat a fly (not as part of an artistic counter-event!) you have clearly defined yourself as an outsider, in a way that a school child practicing scales has not.

Membership can thus be described individually as the *experience* of encountering objects, and increasingly being in a naturalized relationship with them. (Think of the experience of being "at home," and how one settles down and relaxes when surrounded by utterly familiar objects; think of how demented you feel in the process of moving house.) Someone's illegitimacy appears as a series of interruptions to experience (Dewey 1916, 1929), or a lack of smooth naturalization trajectory. In a way, then, individual membership processes are about the resolution of interruptions (anomalies) posed by the tension between the ambiguous (outsider, naive, strange) and the naturalized (at home, taken-for-granted) properties of objects. Collectively, membership can be described as the processes of managing the tension between naturalization (routines and conventions, cumulatively let us call this "transparency"), on the one hand, and the degree of openness to immigration on the other. Communities vary in their tastes for openness. Cults, for example, are one sort of collective which is low on the openness dimension and correspondingly high on the naturalization/positivism dimension—us vs. them. The paradoxes of multiplicity (paralogical validity) may be resolved by closing or opening the world. If one chooses an open model, then the combination of multiplicity, the links between partial perspectives becomes the problematic.

Borderlands and Boundary Objects

I have spent a number of years studying how people do in fact choose and craft linked partial perspectives in conducting scientific work. Science and technology are good places to study the rich mix of people and things brought to bear on complex problem-solving questions, but also good places to understand more about membership in communities. This has led me to try to understand people and things ecologically, both with respect to membership and with respect to the things they live with.

"Marginality" as a technical term in sociology refers to human membership in more than one community of practice. (Things, strictly speaking, do not have membership, in the sense of negotiated identity and legitimacy.) A good example of someone who is marginal is someone who belongs to more than one race, such as being half white and half Asian. Again, I am not using marginality here in the sense of center/margin or center/periphery (e.g., not "in the margins"), but rather in the old-fashioned sense of Robert Park's "marginal man," the one who has a double vision by virtue of having more than one identity to negotiate (Park 1952; Stonequist 1961; Simmel [1908] 1950; Schutz 1944). Strangers are those who "come and stay a while," long enough so that membership becomes a troublesome issue—they are not just nomads passing through, but people who sort of belong and sort of don't.

Marginality is an interesting paradoxical concept for people and things. On the one hand, we have membership defined in terms of naturalization of objects. On the other, everyone is a member of multiple communities of practice. Yet since different communities generally have differently naturalized objects in their ecology, how can someone maintain multiple memberships without becoming simply schizophrenic? How can they naturalize the same object differently, since naturalization by definition demands forgetting about other worlds?

There are as well some well-known processes in social psychology for managing these things: passing, or making one community the shadow for the other; splitting, or having some form of multiple personality/chameleon; fragmenting, or segmenting the self into compartments; becoming a nomad, intellectually and spiritually if not geographically.

One dissatisfaction I have with these descriptions is they all paint each community of practice as ethnocentric, as endlessly hungry and unwilling collectively to accommodate internal contradictions. There is also an implicit idea of a sort of imperialist uber-social world ("the mainstream") that is pressing processes of assimilation on the individual (e.g., "Americanization" processes in the early twentieth century). Communities vary along this dimension of open/closedness, and it is equally important to find successful examples of the nurturing of marginality (although it is possible that by definition they exist anarchically and not institutionally/bureaucratically). Just like brilliant nurturing teachers who do it in spite of the system. . . . Here again, feminism has some important lessons. An important theme in recent feminist theory is resistance to such imperializing rhetoric and the development of alternative visions of coherence without unconscious assumption of privilege. Much of it emphasizes the double vision of Anzaldúa, and the partiality and modesty of Haraway's cyborg.

Charlotte Linde's book on the processes of coherence in someone's life stories also provides some important clues; she especially emphasizes accidents and contingency in the weaving together of a coherent narrative (Linde 1993). The narratives she analyses are in one sense meant to reconcile the heterogeneity of multiply naturalized object relations in the person, where the objects in question are stories/depictions of life events. Linden (1993) and Strauss (1969) have made similar arguments about the uncertainty, plasticity, and collectivity of life narratives.

Boundary Objects

In studying scientific problem solving, I have been concerned for a number of years to understand how scientists could cooperate without agreeing about the nature of objects. Because scientific work is always composed of members of different communities of practice (1 know of no science that is not interdisciplinary in this way, especially if you include laboratory technicians and janitors, which I do), this is a pressing problem for modeling "truth," the putative job of scientists. In developing models for this work, I coined the term "boundary objects" to talk about how scientists do this. I do not think the term is exclusive to science, but I think science is an interesting place to study them because the push to make problem solving explicit gives one an unusually detailed amount of information about the arrangements.

Boundary objects are those scientific objects which both inhabit several communities of practice and satisfy the informational requirements of each of them. Boundary objects are thus objects which are both plastic enough to adapt to local needs and constraints of the several parties employing them, yet robust enough to maintain a common identity across sites. They are weakly structured in common use, and become strongly structured in individual-site use. These objects may be abstract or concrete. I first noticed the phenomenon in studying where the specimens of dead birds had very different meanings to amateur birdwatchers and professional biologists, but "the same" bird was used by each group. Such objects have different meanings in different social worlds but their structure is common enough to more than one world to make them recognizable, a means of translation. The creation and management of boundary objects is a key process in developing and maintaining coherence across intersecting scientific communities (Star and Griesemer 1989).

Boundary objects arise over time from durable cooperation between communities of practice, as working arrangements that resolve anomalies of naturalization without imposing naturalization from one community or from an outside. They are therefore most useful in analyzing cooperative and relatively equal situations; issues of imperialist imposition of standards, force, and deception have a somewhat different structure.

Given the above, we can see that the source of boundary objects comes from a combined willingness of the community of practice to accommodate marginality AND multiply naturalized objects. And this is as good a definition of cooperation as I can come up with, come to think of it.

It's worth thinking about here the difference between boundary objects which grow between two adult communities, and the attempt to create cooperation between two communities (or N communities) forced together, or where one is composed of children and the other of adults, or where one group is treated as powerless for whatever reason.

Borderlands and Monsters

In traditional sociology this model might have overtones of functionalism, in its emphasis on insiders/outsiders and their relations. But functionalists never considered the nature of objects or of multiple legitimate memberships. If we think in terms of a complex cluster of *multiple* trajectories *simultaneously* of both memberships and naturalizations, it is possible to think of a relational mapping: between borderlands[7] and monsters.[8] A monster occurs when an object refuses to be naturalized (Haraway 1992); a borderland occurs when two communities of practice coexist in one person (Anzaldúa 1987). And feminism has had a great deal to say about this, for borderlands are the naturalized home of those monsters known as cyborgs.

If we read monsters as persistent resisters of transparency/naturalization within some community of practice, then the experience of encountering an anomaly (such as that routinely encountered by a newcomer to science, for instance a woman, or man of color) may be keyed back into membership. You know that you don't belong when what seems like an anomaly to you is natural for everyone else—over time, collectively, it can become monstrous in the collective imagination (Hales1994). Here is what Haraway has called "the promise of monsters" and one of the reasons that for years they have captured the feminist imagination. Frankenstein peering in the warmly lit living room window; Godzilla captured and shaking the bars of his cage are intuitions of exile and madness, and symbols of how women's resistance and wildness has been imprisoned and reviled, kept "just outside."

In a practical sense, this is a way to talk about what happens to experience in the science classroom when someone comes in with no experience of formal science. It is not simply a matter of the strangeness, but of the politics of the *mapping* between the anomalies and the forms of strangeness/marginality.

In accepting and understanding the monsters and the borderlands there may be an intuition of healing and power, as Gloria Anzaldúa shows us in her brilliant and compassionate writing. It is not an easy healing and certainly not a magic bullet, but a complex and collective "twisted path," a challenge to easy categories and simple solutions. It is, in fact, a politics of ambiguity and multiplicity—this is the real possibility of the cyborg. For scholars, this is necessarily an exploration that exists in interdisciplinary borderlands and crosses the traditional divisions between people, things, and technologies of representation.

Engineered vs. Organic Boundary Objects

Most schools are terrible places to grow boundary objects,[9] because they both strip away the ambiguity of the objects of learning, and impose or ignore membership categories (except artificial hierarchically assigned ones). In mass schooling and standardized testing, an attempt is made to insist on an engineered community of practice, where the practices are dictated, and the naturalization process is monitored and regulated *while ignoring borderlands*. They are virtual factories for monsters. In the 1970s and 1980s many attempts were made to include other communities in the formula via affirmative action and multicultural initiatives. But where these lacked the relational base between borderlands and the naturalization of objects, they ran aground on the idea of measuring progress in learning. This is partly a political problem and partly a representational one. As we learned so painfully over the years as feminists, a politics of identity based on essences can only perpetuate vicious dualisms (remember the turkey baster disaster). That is, if I as a white male science teacher bring in an African American woman as a (Platonic) representative of African American-ness and/or woman-ness, then attempt to match her essential identity to the objects in the science class room (without attending much to how they are fully naturalized objects in *another* community of practice), costly and painful mismatches are inevitable. I risk causing serious damage to her self-articulation (especially where she is alone), and her ability to survive (a look at the dismal retention statistics of women and minority men in many sciences and branches of engineering will underscore this point). Any mismatch becomes her personal failure, since the measurement yardstick remains unchanged although the membership criteria seem to have been stretched. Again, both borderlands and anomalous objects have been deleted. Kal Alston, writing of her experience as an African American Jewish feminist, has referred to herself as a "unicorn"—a being at once mythical and unknowable, straddling multiple worlds.

But all people belong to multiple communities of practice—it's just that in the case of the African American woman in science, the visibility and pressure is higher, and her experience is especially dense in the skill of surviving multiplicity. Thus Patricia Hill Collins's title, "Learning from the Outsider Within" has many layers and many directions to be explored as we all struggle for rich ways of mapping that honors this experience and survival (Collins 1986). Karla Danette Scott (1995) has recently written about the interwoven languages of black women coming to university, and how language becomes resources for this lived complexity. They "talk black" and "talk white" in a seamless, context-driven web, articulating the tensions between those worlds as a collective identity. This is not just code switching, but braided identity, a borderland.

The Relational Nature of this Argument

The model proposed here takes the form of a many-to-many relational mapping, between multiple marginality of people (borderlands and monsters) and multiple naturalizations of objects (boundary objects and standards). Over time, the mapping is between the means by which individuals and collectives have managed the work of creating coherent selves in the border lands (e.g., Collins, Anzaldúa) on the one hand, and to create durable boundary objects on the other.

It is also not just many-to-many relational, but meta-relational. By this I mean that the map must point simultaneously to the articulation of selves and the naturalization of objects. One of the things that is important here is honoring (I won't say capturing) the work involved in borderlands and boundary objects. This work is almost necessarily invisible from the point of view of any single community of practice: as Collins points out, what white person really sees the work of self-articulation of the black person who is juggling multiple demands/audiences/contingencies? It is not just willful blindness (although it can be that), but much more akin to the blindness between different Kuhnian paradigms, a revolutionary difference. Yet the juggling is both tremendously costly and brilliantly artful. Every community of practice has its overhead:

paying your dues, being regular, hangin', being cool, being professional, people like us, conduct becoming, getting it, catching on. . . . And the more communities of practice one participates in, the higher the overhead—not just in a straightforwardly additive sense, but interactively. "Triple jeopardy" (i.e., being old, black, and female) is not just three demographic variables or conditions added together, but a tremendously challenging situation of marginality requiring genius for survival. The overheads interact.

Articulation Work/Invisible Work

What is the name for this work of managing the overheads and anomalies caused by multiple memberships on the one hand, and multiply naturalized objects on the other? Certainly, it is invisible. Most certainly, it is methodological, in the sense of reflecting on differences between methods and techniques. It is often invisible. Within both symbolic interactionism and the new field of computer-supported cooperative work, the term "articulation work" has been used to talk about some forms of this invisible "juggling" work (Schmidt and Bannon 1992; Gerson and Star 1986).

Canonically, articulation work is work done in real time to manage contingencies; work that gets things back "on track" in the face of the unexpected, that modifies action to accommodate unanticipated contingencies. It is richly found for instance in the work of head nurses, secretaries, homeless people, parents, and air traffic controllers, although of course all of us do articulation work in order to keep our work going. Modeling articulation work is one of the key challenges in the design of cooperative and complex computers and information systems. This is because real-time contingencies, or in Suchman's terms, situated actions (1987), always change the use of any technology (driving across the Bay Bridge with the turkey baster—what happens when there's a traffic jam? Designing an air traffic control system—what happens when an accident in Chicago backs up traffic into Paris on a holiday weekend, and extra help cannot be found?).

But articulation work is also about novelty and the ways in which one person's routine may be another's emergency or novelty—what happens when one clerk, Person A, entering data into a large database does not think of abortion as a medical matter, but as a crime, while another, Person B, thinks of it as a routine medical procedure? Person A's definition excludes abortion from the medical database, Person B's includes it. The resulting data will be, at the least, incomparable, but in ways that may be completely invisible to User C, compiling statistics for a court case arguing for the legalization of abortion based on prevalence. When articulation work is deleted, made invisible in this fashion, then voices are suppressed and we see for formation of "master narratives" and the myth of the mainstream universal.

Thus, we can see articulation work as partly about managing the mismatches between memberships and naturalization. One way to think about this is through the management of anomalies—as a tracer. Anomalies or interruptions, the cause of contingency, come when some person or object interrupts the flow of expectations. One reason that "glass box technology" or pure transparency is impossible is that anomalies always arise when multiple communities of practice come together, and useful technologies cannot be designed in all communities at once. Monsters arise when the legitimacy of that multiplicity is denied.

Transparency is in theory the endpoint of the trajectory of naturalization, as complete legitimacy or centrality is the end point of the trajectory of membership in a community of practice. However, due to the multiplicity of membership of all people, and the persistence of newcomers and strangers, a well as the multiplicity of naturalization of objects, this is inherently nonexistent in the real world. For those brief historical moments where it seems to be the case, it is unstable.

But what are the things that make objects and statuses seem given, durable, real? Several things coalesce.

Generalization

Some objects are naturalized in more than one world. They are not then boundary objects, but rather they become *standards* within and across the multiple worlds in which they are naturalized. A lot of mathematics, and in the West, much of medicine and physiology fits this bill. In the Middle Ages a lot of Christian doctrine fit this too. The hegemony of patriarchy arose from the naturalization of objects across a variety of communities of practice, with the exclusion of women from membership and the denial of our alternative interpretations of objects (Kramarae 1988; Merchant 1980; Croissant and Restivo 1995).

When an object becomes naturalized in more than one community of practice, its naturalization gains enormous power—to the extent that a basis is formed for dissent to be viewed as madness or heresy. It is also where ideas like "laws of nature" get their power, because we are always looking to other communities of practice as sources of validity, and if as far as we can look we find naturalization, the invisibility layers up and becomes doubly, triply invisible. Sherry Ortner's (1974) classic essay on man:culture/woman:nature shows that this has held for the subjugation of women even where specific cultural circumstances vary widely, and her model of the phenomenon rests on the persistent misunderstanding of borderlands and ambiguity in many cultures. Before her, Simone de Beauvoir wrote of the ethics of ambiguity, showing the powerful negative consequences of settling for one naturalized mode of interaction. We still need an ethics of ambiguity, still more urgently with the pressure to globalize, and the integration of systems of representation through information technologies across the world.

Casual vs. Committed Membership

Another dimension to take into account here is the degree to which membership demands articulation at the higher level. Being a woman and African American and disabled are three sorts of membership that are nonoptional, diffused throughout life, and embedded in almost every sort of practice and interaction.[10] So it's not equitable to talk about being a woman in the same breath as being a scuba diver—although there are ways in which both can be seen under the rubric "community of practice." But if we go to the framework presented above, there is a way to talk about it. To the extent that the joint objects are (a) multiply naturalized in conflicting ways, and (b) diffused through practices that belong to many communities, they will pose a "sticking point" that defies casual treatment. So for scuba diving it is primarily naturalized in a leisure world, and not especially central to any others. Its practice is restricted and membership is contained, neither contagious nor diffuse.

On the other hand, "learning math" is multiply naturalized across several powerful communities of practice (math and science teachers and practitioners); at the same time it is both strange AND central to others (central in the sense of a barrier to further progress). It is also diffused through many kinds of practices, in various classrooms, disciplines, and workplaces. (Hall 1990). Some communities of practice expect it to be fully naturalized—a background tool or a substrate/infrastructure—in order to "get on with" the business of being, for example, a scientist (Lave 1986). However, there is no map or sense of the strangeness of the object across other memberships. Here, too, information technologies are both diffused and strange, with rising expectations of "literacy" across worlds.

Conclusion

Things and people are always multiple, although that multiplicity may be obfuscated by standardized inscriptions. In this sense, with the right angle of vision, things can be seen as heralds of other worlds, and of a wildness that can offset our naturalizations in liberatory ways. Where we hold firmly to a relational vision of people-things-technologies, in an ethical political framework (which I name here as feminist), we have a chance to step off the infinite regress of measuring the consumption of an object naturalized in one centered world, such as the objects of Western science, against an infinitely expanded set of essentially defined members as consumers.

By relational here I mean to argue against the misplaced concretism of the title, and to affirm the importance of feminism as method. I also mean to take seriously the power of membership, its continual nature (i.e., we are never not members of some communities of practice), and the inherent ambiguity of things. Boundary objects, however, are not just about this perpetual ambiguity, not just temporary solutions to disagreements about anomalies, but rather durable arrangements between communities of practice. Forgetting this, as we routinely do, we empower the "objective voice" that creates the suffering of monsters in borderlands, and walk away from the gentle and generous vision of *mestiza* consciousness offered by Anzaldúa.

These relations define a space against which and into which information technologies of all sorts enter. These technologies of representation are entering into all sorts of communities of practice on a global scale, in design and in use. They are a medium of communication and broadcast, as well as of standardization. *Method* in the world of information systems design is increasingly coming to mean the challenge of cooperation across heterogeneous worlds, of modeling articulation work and multiplicity. But it is not all benign—we also face the risk of a franchised, dully standardized infrastructure ("500 channels and nothing on," in Mitch Kapur's words), or of an Orwellian nightmare of surveillance.

Feminism offers a tradition of reflective denaturalization, of a politics of simultaneity and contradiction intuited by the term "cyborg." Long ago we began with the maxim that "the personal is political," and that each woman's/community's experience has a primacy that we must all learn to afford. We went from turkey basters to cyborg politics in less than twenty years. We owe much of this to the hard work and suffering of communities of practice who had been made monstrous or invisible, especially women of color and their articulation of the layered politics of insider/outsider and borderlands. One part of the methodological lesson from feminism read in this way is that experience/experiment incorporates an ethics of ambiguity, with both modesty and anger. In Nell Morton's words, this means that how we hear each other is a matter of "listening forth" from silence—listening is active, not passive; it means stretching to affiliate with multiplicity.

Notes

I would like to thank Kal Alston, Marc Berg, Geof Bowker, Adele Clarke, David Edge, Penny Eckert, Jim Greeno, Rogers Hall, Simon Kaplan, John Law, Jason Lewis, Annemarie Mol, Kathy O'Connor, Randi Markussen, Susan Newman, Donna Haraway, Bruno Latour, Jean Lave, Charlotte Linde, Joseph Goguen, Jeremy Rochelle, Karla Danette Scott, Anselm Strauss, Karl Thoresen, and Etienne Wenger for their ideas and input into this chapter. Part of this work was supported by a fellowship at the Program in Cultural Values and Ethics, University of Illinois. Much of it was informed by lively conversations with colleagues at the Institute for Research on Learning and with the members of the WITS Colloquium, University of Illinois.

1. American slang for "lesbian."

2. Including the AIDS epidemic among gay men in San Francisco, which radically altered gay male-lesbian politics in tbe city, and gave rise to a number of cooperative projects and institu-

tions; the changing nature of the relationship between women of color and white women in the movement, and the rise of antiracism as an integral part of radical feminism.

3. The two other types are formal, or axiomatic approaches; and encyclopedic listings with flattened or standardized nomenclatures. Both present other sorts of equally interesting political problems (Star 1989b; Bowker and Star 1994; Timmermans, Bowker, and Star 1998).

4. The work of Schutz and subsequent ethnomethodologists investigates this naturalization process through language.

5. Deconstructing this invisibility is one of the major shared projects of ethnomethodology, symbolic interactionist studies of science and gender, and the Annalist school of historiography.

6. Clearly questions of languages are central here as well, and I do not mean to exclude them by emphasizing things. Language considered as a situated tool, in relationship with other tools and things, is part of this model.

7. Namely, in the sense of multiple memberships.

8. Namely, in the sense of interruptions to the experience of naturalization.

9. I borrow the phrase from Howard Becker's (1972) classic "A School Is a Lousy Place to Learn Anything In," an essay that covers related ground.

10. One of the intriguing features of electronic intersection is that it makes disclosure of these memberships voluntary, or at least problematic, where participants do not know each other in real life.

References

Anzaldúa, Gloria. 1987. *Borderlands/La Frontera: The New Mestiza*. San Francisco: Aunt Lute.

Beck, Eevi E.1995. "Changing Documents/Documenting Changes: Using Computers for Collaborative Writing over Distance." In *The Cultures of Computing*, ed. Susan Leigh Star, 53–68. Oxford: Basil Blackwell.

Becker, Howard S.1972. "A School Is a Lousy Place to Learn Anything In." *American Behavioral Scientist* 16 (12): 85–105.

Bowker, Geoffrey, and Susan Leigh Star. 1994. "Knowledge and Infrastructure in International Information Management: Problems of Classification and Coding." In *Information Acumen: the Understanding and Use of Knowledge in Modern Business*, ed. Lisa Bud, 187–213. London: Routledge.

Clarke, Adele E.1993. "Modernity, Postmodernity and Reproductive Processes, c1890–1990: or, 'Mommy, Where Do Cyborgs Come from Anyway?'" Paper presented at the Gender, Science and Technology Conference, Melbourne, Australia.

Clarke, Adele E., and Joan H. Fujimura, eds. 1992a. *The Right Tools for the Job: At Work in Twentieth-Century Life Sciences*. Princeton: Princeton University Press.

Clarke, Adele E., and Joan H. Fujimura. 1992b. "What Tools? Which Jobs? Why Right?" In *The Right Tools for the Job: At Work in Twentieth-Century Life Sciences*, ed. Adele E. Clarke and Joan H. Fujimura, 3–44. Princeton: Princeton University Press.

Collins, Patricia Hill. 1986. "Learning from the Outsider Within: The Sociological Significance of Black Feminist Thought." *Social Problems* 33:514–532.

Croissant, Jennifer, and Sal Restivo. 1995. "Science, Social Problems, and Progressive Thought: Essays on the Tyranny of Science." In *Ecologies of Knowledge: Work and Politics in Science and Technology*, ed. Susan Leigh Star, 39–87. Albany: SUNY Press.

Dewey, John. 1916. *Logic: The Theory of Inquiry*. New York: Holt, Rinehart and Winston.

Dewey, John. 1929. *The Quest for Certainty*. New York: Open Court.

Engestrom, Y. 1990. "When Is a Tool? Multiple Meanings of Artifacts in Human Activity." In *Learning, Working and Imagining*, ed. Y. Engestrom, 171–195. Helsinki: Orienta-Konsultit Oy.

Gasser, Les. 1986. "The Integration of Computing and Routine Work." *ACM Transactions on Office Communications Systems* 4: 205–225.

Gerson, E. M., and S. L. Star. 1986. "Analyzing Due Process in the Workplace." *ACM Transactions on Office Information Systems* 4:257–270.

Goguen, Joseph. 1992. "The Dry and the Wet." In *Information Systems Concepts*, ed. Eckhard Falkenberg, Colette Rolland, and El-Sayed Nasr-El-Dein El-Sayed, 1–17. Amsterdam: Elsevier North-Holland.

Hales, Mike. 1994. "Information Systems Strategy, a Cultural Borderland, Some Monstrous Behaviour." In *The Cultures of Computing*, ed. Susan Leigh Star, 103-117. Oxford: Blackwell.

Hall, Rogers P.1990. "Making Math on Paper: Constructing Representations of Stories about Related Linear Functions." Ph.D. dissertation, University of California, Irvine.

Haraway, Donna J.1991. *Simians, Cyborgs, and Women: The Reinvention of Nature*. New York: Routledge.

Haraway, Donna. 1992. "The Promises of Monsters: A Regenerative Politics for Inappropriate/d Others." In *Cultural Studies*, ed. Lawrence Grossberg, Cary Nelson, and Paula Treichler, 295–337. New York: Routledge.

Haraway, Donna J.1993. "Modest Witness@ Second Millenium. The FcmalcMan©Meets OncoMouse™." Paper given at the conference Located Knowledges: Intersections between Cultural, Gender, and Science Studies, April, University of California, Los Angeles.

Hewitt, Carol. 1985. "The Challenge of Open Systems." *BYTE* 10:223–242.

hooks, bell. 1990. "Choosing the Margin as a Space of Radical Openness." In *Yearning: Race, Gender and Cultural Politics*, 145–153. Boston: South End Press.

Kramarae, Cheris, ed. 1988. *Technology and Women's Voices: Keeping in Touch*. New York: Routledge.

Lather, Patti. 1993. "Fertile Obsession: Validity after Post-structuralism." *Sociological Quarterly* 34:673–693.

Latour, Bruno. 1987. *Science in Action*. Cambridge, MA: Harvard University Press.

Lave, Jean. 1986. "The Values of Quantification." In *Power, Action and Belief: A New Sociology of Knowledge?*, ed. John Law, 88–111. Boston: Routledge and Kegan Paul.

Lave, Jean, and Etienne Wenger. 1992. *Situated Learning: Legitimate Peripheral Participation*. Cambridge: Cambridge University Press.

Linde, Charlotte. 1993. *Life Stories*. Oxford: Oxford University Press.

Linden, R. Ruth. 1993. *Making Stories, Making Selves: Feminist Reflections on the Holocaust*. Columbus: Ohio State University Press.

Merchant, Carolyn. 1980. *The Death of Nature: Women, Ecology and the Scientific Revolution*. New York: Harper and Row.

Ortner, Sherry. 1974. "Is Female to Male as Nature Is to Culture?" In *Woman, Culture and Society*, ed. Michelle Rosaldo and Louise Lamphere, 67–87. Stanford: Stanford University Press.

Park, Robert Ezra. 1952. *Human Communities*. Glencoe, IL: Free Press.

Rich, Adrienne. 1977. *Women and Honor: Some Notes on Lying*.Pittsburgh, PA: Motherroot Publications.

Schmidt, K., and L. Bannon. 1992. "Taking CSCW Seriously: Supporting Articulation Work." *Computer Supported Cooperative Work: An International Journal* 1:7–40.

Schutz, Alfred. 1944. "The Stranger: An Essay in Social Psychology." *American Journal of Sociology* 69:499–507.

Scott, Karla Danette. 1995. "'When I'm with My Girls': Identity and Ideology in Black Women's Talk about Language and Cultural Borders." Ph.D. dissertation, Department of Speech Communication, University of Illinois, Champaign Urbana.

Simmel, Georg. [1908] 1950. "The Stranger." In *The Sociology of George Simmel*, ed. Kurt Wolff, 402–408. Glencoe, IL: Free Press.

Star, Susan Leigh. 1989a. "Layered Space, Formal Representations and Long-Distance Control: The Politics of Information." *Fundamenta Scientiae* 10:125–155.

Star, Susan Leigh. 1989b. *Regions of the Mind: Brain Research and the Quest for Scientific Certainty*. Stanford: Stanford University Press.

Star, Susan Leigh. 1989c. "The Structure of Ill-Structured Solutions: Heterogeneous Problem-Solving, Boundary Objects and Distributed Artificial Intelligence." In *Distributed Artificial Intelligence 2*, ed. M. Huhns and Les Gasser, 37–54. Menlo Park: Morgan Kaufmann.

Star, Susan Leigh. 1991. "Power, Technologies and the Phenomenology of Conventions: On Being Allergic to Onions." In *A Sociology of Monsters: Essays on Power, Technology and Domination*, ed. John Law, 26–56. London: Routledge.

Star, Susan Leigh, and James R. Griesemer. 1989. "Institutional Ecology, 'Translations' and Boundary Objects: Amateurs and Professionals in Berkeley's Museum of Vertebrate Zoology, 1907–39." *Social Studies of Science* 19:387–420. [See also chapter 7, this volume.]

Stengers, Isabelle. 1993. *The Invention of Modern Science*. Minneapolis: University of Minnesota Press.

Stonequist, Everett. 1961. *The Marginal Man*. New York: Russell and Russell.

Strauss, Anselm. 1969. *Mirrors and Masks: The Search for Identity*. San Francisco: The Sociology Press.

Strauss, Anselm. 1993. *Continual Permutations of Action*. New York: Aldine deGruyter.

Suchman, Lucy. 1987. *Plans and Situated Actions: The Problem of Machine-Human Communication*. Cambridge: Cambridge University Press.

Timmermans, Stefan, Geoffrey Bowker, and Susan Leigh Star. 1998. "The Architecture of Difference: Visibility, Discretion, and Comparability in Building a Nursing Intervention Classification." In *Differences in Medicine: Unraveling Practices, Techniques, and Bodies*, ed. Marc Berg and Annemarie Mol, 202–225. Durham, NC: Duke University Press.

II   Boundary Objects

7 Institutional Ecology, "Translations," and Boundary Objects: Amateurs and Professionals in Berkeley's Museum of Vertebrate Zoology, 1907–1939

Susan Leigh Star and James R. Griesemer

Editors' Note: The concept of boundary objects is likely the one for which Leigh Star is most widely known, not only in science and technology studies but in computer and information sciences, library sciences, sociology and beyond. The concept extends Anselm Strauss's social worlds/arenas theory in important directions. Notably, it challenges the concept of "immutable mobiles" from actor-network theory, arguing instead that when things travel across different communities of practice, they are constructed differently in different sites to meet the needs and goals of the local situation. Sustaining Leigh's pragmatist philosophical roots, boundary objects become robust through their usefulness across diverse situations.

Scientific work is heterogeneous, requiring many different actors and viewpoints. It also requires cooperation. The two create tension between divergent viewpoints and the need for generalizable findings. We present a model of how one group of actors managed this tension. It draws on the work of amateurs, professionals, administrators, and others connected to the Museum of Vertebrate Zoology at the University of California, Berkeley, during its early years. Extending the Latour-Callon model of *intéressement*, two major activities are central for translating between viewpoints: standardization of methods, and the development of "boundary objects." Boundary objects are both adaptable to different viewpoints and robust enough to maintain identity across them. We distinguish four types of boundary objects: repositories, ideal types, coincident boundaries, and standardized forms.

Most scientific work is conducted by extremely diverse groups of actors—researchers from different disciplines, amateurs and professionals, humans and animals, functionaries and visionaries. Simply put, scientific work is heterogeneous. At the same time, science requires cooperation—to create common understandings, to ensure reliability across domains and to gather information which retains its integrity across time, space, and local contingencies. This creates a "central tension" in science between divergent viewpoints and the need for generalizable findings. In this chapter we examine the development of a natural history research museum as a case in which both heterogeneity and cooperation are central issues for participants. We develop an analytical framework for interpreting our historical material, one which can be applied to studies similarly focused on scientific work in complex institutional settings.

The plan of the chapter is as follows. First we consider the ramifications of the heterogeneity of scientific work and the need for cooperation among participants for the nature of translation among social worlds. We suggest modifications of the *intéressement* model of Latour, Callon, and Law. We urge a more ecological approach and develop the concept of boundary objects to analyze a case study of a research natural history museum. We discuss the history of the Museum of Vertebrate Zoology at the University of California, Berkeley, and describe conceptions of it by participants from several distinct social worlds, including those of professional scientists, amateur naturalists, patrons, hired hands, and administrators. Our discussion is meant to be suggestive rather than conclusive at this stage, outlining an approach to case studies as well as providing a partial analysis of the case at hand. We conclude with further discussion of boundary objects and the allied issue of methods standardization.

The Problem of Common Representation in Diverse Intersecting Social Worlds

Common myths characterize scientific cooperation as deriving from a consensus imposed by nature. But if we examine the actual work organization of scientific enterprises, we find no such consensus. Instead, we find that scientific work neither loses its internal diversity nor is consequently retarded by lack of consensus. Consensus is not necessary for cooperation or for the successful conduct of work. This fundamental sociological finding holds in science no less than in any other kind of work.[1] However, scientific actors themselves face many problems in trying to ensure integrity of information in the presence of such diversity. One way of describing this process is to say that the actors trying to solve scientific problems come from different social worlds and establish a mutual modus operandi.[2] A university administrator in charge of grants and contracts, for example, answers to a different set of audiences and pursues a different set of tasks, than does an amateur field naturalist collecting specimens for a natural history museum.

When the worlds of these actors intersect a difficulty appears. The creation of new scientific knowledge depends on communication as well as on creating new findings. But because these new objects and methods mean different things in different worlds, actors are faced with the task of reconciling these meanings if they wish to cooperate. This reconciliation requires substantial labor on everyone's part. Scientists and other actors contributing to science translate, negotiate, debate, triangulate, and simplify in order to work together.

The problem of translation as described by Latour, Callon, and Law is central to the kind of reconciliation described in this chapter.[3] In order to create scientific authority, entrepreneurs gradually enlist participants (or in Latour's word, "allies") from a range of locations, reinterpret their concerns to fit their own programmatic goals, and then establish themselves as gatekeepers (in Law's terms, as "obligatory points of passage").[4] This authority may be either substantive or methodological. Latour and Callon have called this process *intéressement*, to indicate the translation of the concerns of the non-scientist into those of the scientist.

Yet, a central feature of this situation is that entrepreneurs from more than one social world are trying to conduct such translations simultaneously. It is not just a case of *intéressement* from nonscientist to scientist. Unless they use coercion, each translator must maintain the integrity of the interests of the other audiences in order to retain them as allies. Yet this must be done in such a way as to increase the centrality and importance of that entrepreneur's work. The *n*-way nature of the *intéressement* (or let us say, the challenge intersecting social worlds pose to the coherence of translations) cannot be understood from a single viewpoint. Rather, it requires an ecological analysis of the sort intended in Hughes's description of the ecology of institutions:

In some measure an institution chooses its environment. This is one of the functions of the institution as enterprise. Someone inside the institution acts as an entrepreneur . . . one of the things the enterprising element must do is choose within the possible limits of the environment to which the institution will react, that is, in many cases, the sources of its funds, the sources of its clientele (whether they be clients who will buy shoes, education or medicine), and the sources of its personnel of various grades and kinds. This is an ecology of institutions in the original sense of that term.[5]

An advantage of the ecological analysis is that it does not presuppose an epistemological primacy for any one viewpoint; the viewpoint of the amateurs is not inherently better or worse than that of the professionals, for instance. We are persuaded by Latour that the important questions concern the flow of objects and concepts through the network of participating allies and social worlds. The ecological viewpoint is antireductionist in that the unit of analysis is the whole enterprise, not simply the point of view of the university administration or of the professional scientist. It does, however, entail understanding the processes of management across worlds: crafting, diplomacy, the choice of clientele and personnel. Our approach thus differs from the Callon-Latour-Law model of translations and *intéressement* in several ways. First, their model can be seen as a kind of "funneling"—reframing or mediating the concerns of several actors into a narrower passage point (see figure 7.1). The story in this case is necessarily told from the point of view of one passage point—usually the manager, entrepreneur, or scientist. The analysis we propose here still contains a managerial bias, in that the stories of the museum director and sponsor are much more fully fleshed out than those of

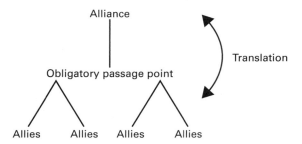

Figure 7.1
Translation in actor-network theory

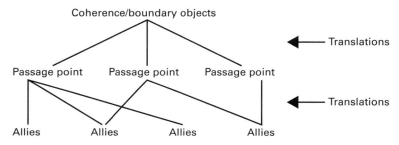

Figure 7.2
Translations and boundary objects

the amateur collectors or other players. But it is a many-to-many mapping, where several obligatory points of passage are negotiated with several kinds of allies, including manager-to-manager types (see figure 7.2).

The coherence of sets of translations depends on the extent to which entrepreneurial efforts from multiple worlds can coexist, whatever the nature of the processes which produce them. Translation here is indeterminate, in a way analogous to Quine's philosophical dictum about language.[6] That is, there is an indefinite number of ways entrepreneurs from each cooperating social world may make their own work an obligatory point of passage for the whole network of participants. There are, therefore, an indeterminate number of coherent sets of translations. The problem for all the actors in a network, including scientific entrepreneurs, is to (temporarily) reduce their local uncertainty without risking a loss of cooperation from allies. Once the process has established an obligatory point of passage, the job then becomes to defend it against other translations threatening to displace it.

Our interest in problems of coherence and cooperation in science has been shaped, in part, by trying to understand the historical development of a particular type of institution: natural history research museums. Museums of natural history originally arose when private collectors in the seventeenth century opened their cabinets of curiosities to public view. The display of wealth, polite learning, and emulation of the aristocracy, as well as development of reference collections for physicians and apothecaries, were common motives for making cabinets. Many such cabinets were arranged to display, and evoke wonder at, the variety and plenitude of nature or to represent the universe in microcosm. Such museums, in other words, developed as part of popular culture.[7] In the nineteenth century, many new museums were developed by amateur naturalists, rather than by members of the "general public," through their participation in societies for the amateur naturalist. These societies filled an important role in the development of the museum-based science to come.[8] The museum we studied, the Museum of Vertebrate Zoology (MVZ) at the University of California, Berkeley, is important as an example of a museum devoted to scientific research from its inception, aided by the alliance of an amateur naturalist/patron and an early West Coast professional scientist. The MVZ did not take on scientific research as an adjunct to public instruction or popular edification as had many eastern museums; if anything, the reverse is true. (A symbol of this tradition of research is the evident pride with which current museum staff members draw attention to an advertisement on the front door stating that there are "NO PUBLIC EXHIBITS.")

As such, the development of the natural history research museum represents an important stage in the professionalization of natural history work, as well as an example of the changing relationship between amateurs and professionals after the professionalization of biology in America had already begun. Unlike many well-documented cases of eastern institutions that looked to the European scientific community as a model and for legitimation, western biologists had to struggle to gain credibility in the eyes of the already professionalizing biological community in the eastern United States itself. Successful pursuit of the research problems through which the MVZ's scientists hoped to gain recognition depended on an evolving set of practices instituted to manage the particular sort of work occasioned by the intersection of the professional, amateur, lay, and academic worlds.[9] There, several groups of actors—amateurs, professionals, animals, bureaucrats, and "mercenaries"—succeeded in crafting a coherent problem-solving enterprise, surviving multiple translations.

Joseph Grinnell was the first director of the Museum of Vertebrate Zoology. He worked on problems of speciation, migration, and the role of the environment in Darwinian evolution. Grinnell's research required the labors of (among others) university administrators, professors, research scientists, curators, amateur collectors, private sponsors and patrons, occasional field hands, government officials, and members of scientific clubs.

Some objects of interest to all these social worlds included:

- species and subspecies of mammals and birds
- the terrain of the state of California

- physical factors in California 's environment (such as temperature, rainfall, and humidity)
- the habitats of collected animal species

Methods Standardization and Boundary Objects

It is normally the case that the objects of scientific inquiry inhabit multiple social worlds, since all science requires intersectional work. Varying degrees of coherence obtain both at different stages of the enterprise and from different points of view in the enterprise. However, one thing is clear. Because of the heterogeneous character of scientific work and its requirement for cooperation, the management of this diversity cannot be achieved via a simple pluralism or a laissez-faire solution. The fact that the objects originate in, and continue to inhabit, different worlds reflects the fundamental tension of science: how can findings which incorporate radically different meanings become coherent?

In analyzing our case study, we see two major factors contributing to the success of the museum: methods standardization and the development of boundary objects. Grinnell's managerial decisions about the best way to translate the interests of all these disparate worlds not only shaped the character of the institution he built, but also the content of his scientific claims.[10] His elaborate collection and curation guidelines established a management system in which diverse allies could participate concurrently in the heterogeneous work of building a research museum. It was a lasting legacy. Grinnell's methods are looked upon as quaint and overly fastidious by current generations of museum workers,[11] but they are still taught and practiced at the Museum of Vertebrate Zoology. (They were also adopted by several other museums around the United States during the first part of this century.)[12] For example, his course handouts for 1913[13] are similar to current field manuals for students in Zoology 107 at Berkeley.[14] There was an intimate connection between the management of scientific work as exemplified by these precise standards of collection, duration, and description, and the content of the scientific claims made by Grinnell and others at the museum.

The second important concept used to explain how museum workers managed both diversity and cooperation is that of boundary objects. This is an analytic concept of those scientific objects which both inhabit several intersecting social worlds (see the list of examples in the previous section) and satisfy the informational requirements of each of them.[15] Boundary objects are objects which are both plastic enough to adapt to local needs and the constraints of the several parties employing them, yet robust enough to maintain a common identity across sites. They are weakly structured in common use, and become strongly structured in individual site use. These objects may be abstract or concrete.[16] They have different meanings in different social worlds but their structure is common enough to more than one world to make them recognizable, a means of translation. The creation and management of boundary objects is a key process in developing and maintaining coherence across intersecting social worlds.

In the next section, we provide some background to the museum's evolution, then turn to a discussion of both methods standardization and boundary objects.

Grinnell and the Museum of Vertebrate Zoology, 1907–1939

The biological sciences in America were undergoing a number of transitions during this period. The educational and cultural functions of natural history were being subsumed under the research goals of scientists. Biological research was increasingly conducted in academic institutions such as universities and specialized research stations rather than in societies formed by amateurs. Professional biologists sought international credibility by distinguishing themselves from amateurs, establishing advanced degrees as credentials, establishing specialized journals for the dissemination of results, and increasingly eschewing the public's eclectic interests in science. For organism-based subdisciplines (for example, ornithology, mammalogy, herpetology), the central transition was a shift from studies of classification and morphology to studies of process and function. With this change of focus, methods and practices diversified. From mostly observational and comparative approaches, biological methods came to include experimental, manipulative, and quantitative techniques and natural history methods were refined so as to focus on increasingly specialized research problems.[17]

At the same time, a number of "inventorying" efforts of the federal government were coming to fruition in reports of collecting and surveying trips to the West. The Bureau of the Biological Survey, founded in 1905 as an arm of the U.S. Department of Agriculture, for example, made a massive effort to chart the flora and fauna of the states and territories. As the nineteenth century closed, these reports were used by their authors and others to go far beyond mere catalogs of materials. Their data were used, for example, to begin to develop general biogeographic principles of animal and plant distribution, most notably that of C. Hart Merriam, later an important influence on the Museum of Vertebrate Zoology workers.[18]

The participation of ecology in these changes meant both distinguishing itself from its basis in descriptive natural history and adopting new methods. On the one hand, ecologists adopted a set of problems originating in evolutionary theory (adaptation, natural selection), geography (distribution and abundance), and physiology (effects of physical factors such as heat, light, soil, and humidity on life-history). On the other, they learned new methods of quantification and analysis and the use of biological indicators.[19] Ecology emerged from the last century as a subdiscipline distinct from systematics, morphology, and genetics. Ecologists were concerned with (1) the bases for adaptation, (2) extending physiology to consider the dynamics of interacting groups of organisms, and (3) the quantification of the physical (physiographic) environment as it affects the life-histories of organisms. New theoretical work was beginning to emerge that distinguished ecology as well.[20]

Joseph Grinnell (1877–1939) extended his work in natural history to include ecological problems during this period of disciplinary shift. He studied at Stanford University under the tutelage of Charles H. Gilbert and David Starr Jordan.[21] At the tum of the century Stanford naturalists were bringing together problems of habitat and distribution along with evolutionary theory to form an emerging geographical conception of speciation. This merger was to become of central concern to evolutionists and ecologists alike later in this century.[22]

Alexander and Grinnell

The Museum of Vertebrate Zoology was founded at Berkeley in 1908 by Annie Montague Alexander (1867–1950). Alexander was heir to a Hawaiian shipping and sugar fortune and a dedicated amateur naturalist.[23] Inspired by paleontology courses she took at Berkeley and by safari experience with her father in Africa, Alexander decided to build a museum of natural history. As her first director, she chose Grinnell, at that time an instructor at Throop Polytechnic Institute in Pasadena (later to become CalTech).

Grinnell had been an enthusiastic and dedicated bird and mammal collector since his boyhood in Indian Territory. His father was a physician who worked with American Indians and he grew up with Oglala Sioux Indian playmates.[24] He was a founding Member of the Cooper Ornithological Club, a major western birdwatching and ornithological association. (Their bulletin would later become the journal *The Condor*, edited by Grinnell for many years.) When Alexander first met Grinnell in 1907, he had already made significant theoretical contributions and was an established scientist.[25] David Starr Jordan, for example, included him in a survey of zoologists supporting his "general law of distribution," discussed in his famous 1905 paper.[26] Grinnell became the founding director of the MVZ in 1908. In 1913, he finished his Stanford PhD and was appointed to the Berkeley Zoology Department.[27]

Beginning with Grinnell's and Alexander's own collecting efforts, the museum developed into an important repository of regional specimens of vertebrates. Alexander's contributions alone came to over 20,000 specimens.[28] As part of this work, Grinnell and his staff codified a precise set of procedures for collecting and curating specimens.[29] Many of Grinnell's descriptive monographs on the systematics, geography, and ecology of birds and mammals are still used today as important reference works. Grinnell also contributed important concepts to the literature on geographic distribution, ecology, and evolution. He extended C. Hart Merriam's life-zone concept to a hierarchical classification system for environments, and developed an important and influential concept of the "niche." He argued for trinominalism in systematics as essential to studies of speciation. He also worked toward a "two layer" theory of evolution that incorporated evolution of the environment as part of an explanation for natural selection.[30]

We have been talking so far about the goals and interests of only a few of the people essential for the MVZ's success as a going concern. The work at the museum, like that of scientific establishments everywhere, encompassed a range of very different visions stemming from the intersection of participating social worlds. Among these were amateur naturalists, professional biologists, the general public, philanthropists, conservationists, university administrators, preparators and taxidermists, and even the animals that became the research specimens.[31]

It is not possible to consider all of these visions equally in this chapter, so we are forced to consider most fully those of the entrepreneurs like Grinnell and Alexander. However, by considering the work of Grinnell and Alexander as a part of a network that spans a number of intersecting social worlds, we can begin the task of tracing the network into those other social worlds. An adequate account of *n*-way translation in this case awaits the results of tracing our way out and back again. It also requires conducting such tracings from a variety of starting points (that is, including some starting points which would be considered "peripheral" or "subsidiary" on a one-way translation model, such as the work of commercial specimen houses or taxidermists). Only with tracings from multiple starting points can we begin to test the robustness of the network.

The more limited work discussed in this chapter is, in part, conditioned by the historical record—for us as scholars, scientific publications are the boundary objects which are also obligatory passage points! Records concerning the entrepreneurs who served as administrators of the MVZ are kept in the central archives of the university that housed the museum. Records concerning the many other elements of the network of such amateur collectors who contributed specimens to the museum and articles to naturalist society newsletters are not equally centralized. Nevertheless, it is important not to mistake the search heuristic of starting with the centralized records for a theoretical model of the structure of the network itself. In the following section we adumbrate the central features of several visions of the MVZ and its work.

Grinnell's Vision

One of Grinnell's passions was the elaboration of Darwinian theory that was to be derived from the work of the MVZ. Darwin had argued that natural selection is the chief mechanism by which organisms adapt, but had said little about the precise nature of the environmental forces of change. Grinnell wanted to extend the Darwinian picture by developing a theory of the evolution of the environment as the driving force behind natural selection.[32]

Sadly, Grinnell died before he was able to express his views on evolution and speciation in a major theoretical monograph. (Some of his more important views are excerpted in a book posthumously produced by his students). While in the field, he did outline such a volume and his research program is perhaps best characterized by its title, *Geography and Evolution*. Grinnell's overarching theoretical concern was to bring the study of both physical and biotic environmental factors to bear on the problems of evolution. The chapter titles of his book outline serve to summarize the topics to which Grinnell had devoted his career. He felt that their synthesis would have fulfilled his theoretical programmatic:

1. The concept of distributional limitation; chronological versus spatial conditions.
2. The nature of barriers; examples of different sorts of barriers in mammals and birds.
3. Distributional areas defined: realms, life zones, faunal areas, association; the ecologic niche.
4. Bird migration as a phase of geographic distribution.
5. Kinds of isolation; degrees of isolation as influencing results; the significance of geographic variation.
6. "Plasticity" versus "conservatism" in different groups of birds and mammals.
7. The pocket gophers and the song sparrows of California.
8. Reconcilability of geographic concept with that of genetics; species and subspecies in nature defined.
9. "Orthogenesis" from the standpoint of geographic variation.
10. The bearing of geography and evolution upon human problems.[33]

From these titles, it is clear that Grinnell's approach to questions of evolution differed radically from, say, those of experimental genetics. His natural world was a large-scale, topographical one; his units of analysis and selection were subspecies and species, habitats and niches. This vision required vast amounts of highly detailed data about flora, fauna, and aspects of the environment. He needed a small army of assistants to collect these data.

Prior to the establishment of the MVZ, Grinnell and Alexander exchanged many letters in which they expressed their hopes and visions for its future. In one of these letters Grinnell stated his scientific and political goals:

First, as regards the working up of the Alaska mammals, it seems to me it should be done as far as possible by our own men. We want to establish a center of authority on this coast. I take it that was one purpose you had in mind founding the institution. I will grant that it would take our man, whoever he may be, longer to work up the paper, than the BS [Biological Survey] people. But in the former case we would be ever so much the stronger and better able to tackle the next problem . . .

I believe in buying desirable material where definitely in hand and subject to selection and in-spection. I have more faith, however, in the salaried field man who turns in everything he finds.[34]

Grinnell was clearly concerned to ensure that the materials collected by others met his scientific requirements. The odd specimen collected from here or there might serve as backup for work in taxonomy, but collection for ecological and evolutionary pur-poses required more thorough documentation. This included documenting the pres-ence of groups of animal species in a particular place at the same time of day and season of the year. It also required comparisons of samples over time—hence Grinnell's preference for the salaried field man. In other words, conducting scientific research on problems in ecology and evolution in a museum setting required more than just a change of interests and training on the part of the scientific staff; it also required changes in basic collecting and curating procedures. Moreover, Grinnell clearly had an institutional goal as well as a research goal and that was to build a center of authority. One means of doing this would be to build collections of scientific value which are not easily duplicated elsewhere, or which are tailored to particular research problems not well served by collections elsewhere. Grinnell focused his collecting efforts on the American West, a place distinguished from the East by its great geographical diversity, and asked scientific questions that could only be answered by careful consideration of such geographically based organic diversity on a finer scale than is available in muse-ums pursuing worldwide collections.

Grinnell needed accurate information in the form of carefully preserved animal specimens and documented native habitats over tens or hundreds of years. This placed constraints on the museum's physical organization.[35] In an essay titled "The Museum Conscience," Grinnell argued that the order and accuracy are the chief aims of the curator (once the specimens are safely preserved). On the subject of order, he wrote:

To secure a really practicable scheme of arrangement [of specimens, card indexes and data on specimen labels] takes the best thought and much experimentation on the part of the keenest museum curator. Once he has selected or devised his scheme, his work is not done, moreover, until this scheme is in operation through all the materials in his charge. Any fact, specimen, or record left out of order is lost. It had, perhaps, better not exist, for it is taking space somewhere; and space is the chief cost initially and currently in any museum.[36]

On the second aim, accuracy, Grinnell continued:

The second essential in the care of scientific materials is *accuracy*. Every item on the label of each specimen, every item of the general record in the accession catalog, must be precise as to fact. Many errors in published literature, now practically impossible to "head off," are traceable to mistakes on labels. Label-writing having to do with scientific materials is not a chore to be hand-ed over casually to "25-cent-an-hour" girl, or even to the ordinary clerk. To do this essential work correctly requires an exceptional genius plus training. By no means any person that happens to be around is capable of doing such work with reliable results.[37]

Grinnell's vision of environmental evolution reinforced his conception of collection and curation.[38] He designed the museum so that sampling from restricted locations over long periods of time would capture evolution in progress as environments changed.

Fulfillment of Grinnell's theoretical vision required that specimens and field notes collected over many years be painstakingly curated. In this fashion comparisons of materials could be made by scientists who would come to work for the museum after Grinnell himself was long gone. This concern was not unique to Grinnell or his museum,[39] but Grinnell was a master at articulating both the "museum conscience," as he called it, and his scientific goals.[40] That is, both preservation for posterity and hot new theoretical findings must be protected.

Grinnell, too, had a sense of urgency about "preserving California." Whereas the Smithsonian Institution in Washington, D.C., and the American Museum in New York had the entire world's natural species for their purview, Alexander, Grinnell, and their associates limited themselves to Californian birds and mammals and, later, reptiles and amphibians.[41] As Grinnell wrote to Alexander in the early years of the museum, "[T]here is nothing attractive about collecting in a settled-up, level country. But it ought to be done, and the longer we wait, the fewer 'waste lots' there will be in which to trap for native mammals."[42]

Again in May of the same year, Grinnell wrote: "It would surely be a fine thing if we could acquire a collection of fresh-water ducks, geese, waders, etc. All the species, with the possible exception of killdeer and herons, are decreasing in numbers rapidly, and it is at least certain that specimens will never be obtainable to better advantage than now. All thru [sic] San Joaquin Valley, many of the former marshy areas are not ditched or diked; and the great fields, where geese grazed, are being cut up into farms."[43]

However, to Grinnell, the important feature of preservation was recording information. The important preserved objects were ecological facts, not mere specimens used to educate the public about a vanishing wilderness.[44] Indeed, shortly after its founding, the MVZ decided not to pursue displays of its objects at all. Nevertheless, it was essential to Grinnell's success as a research scientist that he continue to attract Alexander's patronage. Grinnell shaped research problems that were suited to work in the region that Alexander wished to document in the form of collections. By seeking to establish a center of authority for problems well served by this regional focus, Grinnell simultaneously shaped his research goals and increased the value of Alexander 's continued support—not only would she be preserving a sample of California's native fauna for posterity, she would be contributing to the establishment of a research center.

Alexander's Vision

Annie Alexander, too, saw the flora and fauna of California disappearing under the advance of civilization. She felt that it should be meticulously preserved and recorded.[45]

As a passionate and single-minded patron of science, Alexander contributed funds and oversight sufficient virtually to control the MVZ as an autonomous organization on the Berkeley campus. She intended her museum to serve as a demonstration project to the public about what could be done in conservation and zoological research.[46]

As a rich, unmarried woman, Alexander had a degree of autonomy unusual for women during this period. Alexander's trips were primitive by comparison with the "ladylike" expeditions to Africa made by aristocratic women in a somewhat earlier time. Her scrapbooks and the museum archives contain pictures of her camping out, toting rifles, and scaling mountains. She was an indefatigable amateur collector. Along with her lifelong companion and partner, Louise Kellogg, she conducted many expeditions to gather specimens for the MVZ, the Museum of Paleontology, and the Herbarium.

In addition to collecting, Alexander served the museum in other capacities. She was its primary patron, funding the museum building, staff salaries, specimen and equipment purchases, and expeditions. She was as well a day-to-day administrator who approved expenditures in minute detail, including operating expenses and budget reports, hiring and firing personnel, reviewing productivity of the staff, and approving the nature and location of their expeditions (Grinnell, for example, reported to her and sought her advance approval of expeditions).

In none of these roles was Alexander a theoretical scientist. While she read some evolutionary theory, her primary "take" on the job of the museum came from her commitments to conservation and educational philanthropy. The MVZ was a way of preserving a vanishing nature, of making a record of that which was disappearing under the advance of civilization. For her, as for many social elites of the period, natural history was both a passionate hobby and a civic duty.

The Collector's Vision

In addition to its museums, California was imbued with a particularly vigorous conservation and nature-loving amateur constituency. John Muir and the Sierra Club, the Cooper Ornithological Club, the Society of Western Naturalists, and the Save the Redwoods League, among other organizations, all brought amateurs and academics together for purposes of collecting and conservation.

Amateur collectors wanted to play a role in the scholarly pursuit of knowledge by professional scientists. They sought legitimacy for their conservation efforts. They shared with both Alexander and Grinnell the sentiment that what was unique to California and the West should be described, preserved, and made available to the public.[47] The intrinsic beauty of nature should be shared and protected. The expeditions themselves were at once opportunities for peaceful observation and enjoyment of the natural world and a battle of wits between collectors on the one hand and recalcitrant animals and environments on the other. How does one persuade a reluctant and clever animal to participate in science? For the natural historian, there is a delicate balance between capturing an animal at all costs, and capturing an animal with the integrity of its valuable information unassailed. The animals must be brought in physically intact; their habitats must be detailed so that the specimen has scientific meaning. ("Without a label," says one zoologist friend, "a specimen is just dead meat.") The animals, as mentioned before, must be caught quickly, before the larger ecological balances change and they adapt to new conditions. In order to measure changes, Grinnell and other theorists also needed baseline data from which to proceed.

Animals within the museum present another kind of recalcitrance: they must be preserved against decay. The littlest allies, the dermestid beetles that clean the captured specimens so that skeletons can be used for research, are often the most difficult to discipline! They escape their bounds, eat specimens they should not and eat parts of specimens that are needed for other work. Such allies are coaxed and managed through containment and a certain amount of brute force.

A typical example of the struggles with recalcitrant animal allies may be found in Louise Kellogg's field notebook from an expedition in 1911:

March 20. We left the house at six and went to the Stop Thief traps first. Both had been robbed of their bait and the tracks of two animals, probably a civet and a coon were visible—in the one place the creature had reached through the trap and taken the bait without springing it and in the other had pushed aside a rock and got the bait out from above but in the scuffle the bait was caught in the trap and was found lying on one side partially eaten. I caught two microtus out of 21 traps set in the grass. The bait was eaten from two sets of dipodomys and the others were untouched.[48]

The Trappers' Vision

In fulfilling their interests in natural history and collecting, the amateur collectors were often on the front line, making contact with a host of other social worlds. These included farmers and townspeople on or near whose land the collectors searched for specimens, and trappers and traders who could provide them with specimens that were rare or difficult to capture. These people were often invaluable sources of information and other sorts of help (food, camping places)—sometimes for a price.

Many of the backwoods trappers being "interested" by the amateur collectors or the museum workers had little or no interest in either conservation or science as such. Their coin of exchange was money, information about hunting, or possibly the exchange of a less scientifically interesting but edible specimen for one valued by the collectors. Friction between viewpoints here was smoothed by such exchanges. For example, Alexander described a set of problems with a recalcitrant trapper who wanted to sell skins to the museum:

You will notice that two of the skulls are broken. It seems next to impossible to persuade a trapper to kill an animal without whacking him on the head. The bob cat is in rather a sorry plight that has a history. I am holding on to Knowles [a trapper) a little longer in the hope that he may get a panther and some coyotes. He has dogs with him. He set the no. 3 traps but the coyotes as he expressed it "did not throw them" although they walked all over them. He will have to set them finer. Knowles is about as good as the ordinary run of trappers who can't see anything in a skin except its commercial value—and the little extra care in skinning that we demand frets them.

The University Administration's Vision

Another important participant in the museum enterprise was the university administration. Their vision of natural history and California was different yet again from that of the staff of the museum and the amateur collectors. The University of California during this period was trying to become a legitimate, national-class university, and was also trying to begin seriously to compete with the eastern universities for scientific resources and prestige. It was at the same time clearly a local school, a pet charity for many of the San Francisco Bay Area elite, and a training ground for local doctors, lawyers, industrialists, and agriculturalists.

The university was willing to accommodate a natural history museum as long as Alexander was willing to fund it. The administration accepted Alexander's funding of the MVZ as part of this vision, measuring the museum's contribution to this goal by its own criteria: level of funding and prestige returned to the university as a whole. They had similar arrangements with local philanthropists Phoebe Hearst and Jane Sather for charitable research or library endeavors on campus. In turn, Alexander enjoyed an administrative power almost unheard of for single individuals at major universities today. She hired and fired museum staff, chose expedition sites, and managed administrative liaison with the regents of the university.

The different visions and economic values of participating worlds are clear from the sometimes-stormy correspondence between Alexander and the regents and university administration over autonomy for the museum. Here, Grinnell responds to Alexander's chagrin about the university president's vision of the museum in monetary terms:

I think the letter from President Wheeler is fine. You must consider his limitations (and those of the Regents) in forming any conception of the methods and aims of such an institution as the Museum. It seems nothing more than natural that these men should measure your work for the University in terms of the dollars involved. Money is the common standard, and, too, it is the money that makes the major part of our work possible. You deserve all the credit expressed, and more, on this score alone. It is nothing to be ashamed of, or to resent, if their appreciation seems to be prompted only by recognition of the money cost of the Museum. They don't know any better, and the intrinsic value of the Museum and your work for it remain the same.[49]

The MVZ was administratively separate from the department of zoology. It was publicly active in natural history circles and it was the home for meetings of local natural history clubs such as the Cooper Club and the Society of Western Naturalists. In this sense it helped meet the university's goal of being a local cultural center.

Analysis of Methods: Standardization and Boundary Objects

The worlds just described have both commonalities and differences. To meet the scientific goals of the museum, the trick of translation required two things: first, developing, teaching, and enforcing a clear set of methods to "discipline" the information obtained by collectors, trappers, and other nonscientists; and generating a series of boundary objects which would maximize both the autonomy and communication between worlds. Different social worlds maintained a good deal of autonomy in parallel work situations. Only those parts of the work essential to maintaining coherent information were pooled in the intersection of information; the others were left alone. Participants developed extremely flexible, heterogeneous economies of information and materials, in which needed objects could be bartered, traded, and bought or sold. Such economies maximized the autonomy of work considerations in intersecting worlds while ensuring "trade" across world boundaries.

From a purely logical point of view, problems posed by conflicting views could be managed in a variety of ways:

• via a "lowest common denominator" which satisfies the minimal demands of each world by capturing properties that fall within the minimum acceptable range of all concerned worlds; or
• via the use of versatile, plastic, reconfigurable (programmable) objects that each world can mold to its purposes locally; or
• via storing a complex of objects from which things necessary for each world can be physically extracted and configured for local purposes, as from a library; or
• each participating world can abstract or simplify the object to suit its demands; that is, "extraneous" properties can be deleted or ignored; or
• work in the worlds can proceed in parallel except for limited exchanges of standardized sorts; or
• work can be staged so that stages are relatively autonomous.

The strategies of the different participants in the museum world share several of these attributes; we now focus on two major varieties.

Methods and Collectors

What do you think of the system? It seems complex at first reading. But it means detailed, exact, and easily get-at-able records. And the better the records the more valuable the specimens.[50]

Specimens are preserved in a highly standardized way, so that specific information can be recovered later on when the specimen is stored in a museum. For example, it makes a great deal of difference to ease of measurement, handling, and storage whether the limbs are "frozen" at the sides of the body or outstretched, straight or bent. Color of pelage, scales, and so on, are usually not preservable, and color photographs or accurate notes may be the only feasible solution to this preservation problem. Whether soft parts (internal organs and fatty tissues) are preserved depends on the availability of techniques, the conventions for preserving external structures, and the parts commonly studied. If precise measurements of long bones are desired, for example, the animal must usually be taken apart to expose them.[51]

For geographical distribution work, and more especially for the ecological problem of inferring environmental factors limiting species' ranges from distributional data, the taxa to which specimens belong must be linked to a geographical location and to each other. The objects of interest are collections of taxa represented in a particular geographical location. Study of the factors responsible for presence or absence of particular taxa from a local area proceeds according to a method outlined by Grinnell and his colleagues: "In practice, the method used in this survey to get at the causes for differential occurrence as observed was, first, to consider the observed actual instances of restriction of individuals of each kind of animal; and second, to compare all the records of occurrence with what we know in various respects of the portion of the section inhabited, this is an attempt to detect parallels between the extent of presence of the animal and of some appreciable environmental feature or set of features."[52]

Thus, it is necessary to translate specimens into ecological units via a set of field notes. This creates a tension or potential incoherence between collectors and theorists. Let us examine the process of preservation of information. Once faunas are represented as lists of species (and subspecies) linked to a location, their distributional limits are established in terms of the overlapping ranges of their member taxa (or a subset of indicator species). These collections must in turn be linked to a distribution of potentially responsible environmental factors. Hence, in addition to the translation work of creating abstract objects (lists of species, lists of factors) from concrete, conventionalized ones (locations, specimens, field notes), a series of increasingly abstract maps must be created which link these objects together.

Reports of fieldwork begin with an itinerary and often a topographical map of the region explored. Taxa represented by specimens can be plotted on these maps, and if they also serve as indicators of life zones, faunas, or associations, an ecological map of these units can be constructed. In parallel, maps of environmental factor isoclines (quantitatively or qualitatively expressed) can be constructed from field notes and geographic maps. Then the environmental factors maps can be superimposed on the maps of ecological units, and the strongest concordances used to rank environmental factors as delimiters of species distributions.[53]

The specimens per se are not the primary objects of ecological study—the checklists of taxa represented in a local area are. These checklists are then mapped into ecological units (geographically identified groups of species and subspecies) by finding subsets of the checklist which are limited to a geographical subarea. A map is constructed in terms of ranges of the ecological units set by the species ranges (within the geographical area studied) of species taken to be indicators for the zones. Grinnell and Alexander were able to mobilize a network of collectors, cooperating scientists, and administrators to ensure the integrity of the information they collected for archiving and research purposes. The precise set of standardized methods for labeling and collecting played a critical part in their success. These methods were both stringent and simple—they could be learned by amateurs who might have little understanding of taxonomic, ecological, or evolutionary theory. They thus did not require an education in professional biology to understand or to execute. At the same time, they rendered the information collected by amateurs amenable to analysis by professionals. The professional biologists convinced the amateur collectors, for the most part, to adhere to these conventions—for example, to clearly specify the habitat and time of capture of a specimen in a standard format notebook. Grinnell's insistence on, and success with, standardized methods of collecting, preserving, labeling, and taking field notes is a testament to his skillful management of the complex multiple translations involved in natural history work. The methods protocols themselves, and the injunctions implied, are a record not only of the kinds of information Grinnell needed to capture for his theoretical developments, but also of the conflicts between the various participating worlds. In this sense, each protocol is a record of the process of reconciliation.

Propagating methods is not an easy task. In working with amateur collectors, a major problem is to ensure that the data coming back in from the field is of reliable quality; that it does not decay en route through sloppy collecting or preserving techniques; that the collectors give enough information about where they got the beasts from so that the locations can be precisely identified. However, directions for collectors cannot be made so complicated that they interfere with the already difficult job of camping out in the wilderness, capturing sneaky little animals, or bribing reluctant farmers to preserve intact their saleable specimens.

Another way of saying this is that the allies enrolled by the scientist must be disciplined, but cannot be overly disciplined. Each world is willing—for a price—to grant autonomy to the museum and to conform to Grinnell's information-gathering standards. It is only gradually that a scientist in Grinnell's position comes to be an authority. Part of this authority is exercised through the standardization of methods.

Standardizing methods is different from standardizing theory. By emphasizing how, and not what or why, methods standardization both makes information compatible and allows for a longer "reach" across divergent worlds. Grinnell was thus able to accomplish several things at once. First, and perhaps most important, methods standardization allowed both collectors and professional biologists to find a common ground in clear, precise manual tasks. Collectors do not need to learn theoretical biology in order to contribute to the enterprise. Potential differences in beliefs about evolution or higher-order questions tend to be displaced by a focus on "how," not "why." The methods thus provided a useful "lingua franca" between amateurs and professionals. They also allowed amateurs to make a substantial contribution both to science and to conservation. The standardized specimens, field notes, and techniques provided consistent information for future generations or for researchers at a distance.

Grinnell's methods emphasis thus translated the concerns of his allies in such a way that their pleasure was not impaired—the basic activities of going on camping trips, adding to personal hobby collections, and preserving California remained virtually untouched. With respect to the collectors, Grinnell created a mesh through which their products must pass if they want money or scientific recognition, but not so narrow a mesh that the products of their labor cannot be easily used.

One consequence of this strategy is that Grinnell created a large area of autonomy for himself from which he could move into more theoretical arenas. His carefully crafted relationship with Alexander involved them both in making a commitment to methods and preservation techniques. As a sponsor, Alexander was concerned with preserving a representative collection of Californiana, both for posterity and as a demonstration of good scientific practice. It is clear from their correspondence that Alexander had little concern for the contents of the scientific theory—but that she was quite concerned with curation and preservation methods. While necessary for the sort of sweeping ecological work undertaken by the MVZ, "methods control" alone was not sufficient. Other means were necessary to ensure cooperation across divergent social worlds. These were not engineered as such by any one individual or group, but rather emerged through the process of the work. As groups from different worlds work together, they create various sorts of boundary objects. The intersectional nature of the museum's shared work creates objects which inhabit multiple worlds simultaneously, and which must meet the demands of each one.

Boundary Objects

In natural history work, boundary objects are produced when sponsors, theorists, and amateurs collaborate to produce representations of nature. Among these objects are specimens, field notes, museums, and maps of particular territories. Their boundary nature is reflected by the fact that they are simultaneously concrete and abstract, specific and general, conventionalized and customized. They are often internally heterogeneous. We have, in the management strategies of the MVZ, a situation with the following characteristics:

1. Many participants share a common goal: preserve California's nature. Those that do not share this goal participate in the economy via a neutral medium—direct monetary exchange (note: this includes the university administration!).

2. All participants come to agree literally to preserve samples of California's flora and fauna, as intact and as well tagged as possible.
3. For some participants (amateur collectors, general public, trappers and farmers) this literal, concrete preservation of animals is sufficient for their purposes.
4. For others (Grinnell, university administration), literal concrete preservation is only the beginning of a long process of making arguments to professional audiences and establishing themselves as "experts" in some theoretical domain.

So, in the case of the museum, the different worlds share goals of conserving California and nature, and of making an orderly array out of natural variety. These shared goals are lined up in such a way that everybody has satisfying work to perform in each world. How does this happen?

In building the theories and in building the organization, Grinnell had to maintain the conventionality of the objects so that future collecting could go on. The concerns and technologies of the amateurs, farmers, and so on, needed to be preserved if they were to continue fully participating. At the same time, he had to overcome the conventionality in order to make his objects scientifically interesting. It would not be enough if all the worlds collected objects which were in some sense challenging old ways of thinking about nature, or arguing with other parts of science. How did Grinnell balance the need for argument with the need for building on the very conventional understandings about California that the amateur collectors and clubs had? How did he escape being limited by their concerns?

Grinnell and Alexander quite brilliantly began their enterprise by building on a goal they shared with several participants (the university presidents, nature lovers, sponsors, and local social elites): draw a line around the West (sometimes even around the state) and declare it a nature preserve. (As one current member of the museum staff has wryly stated: "When you get to the Nevada border, turn around and drive the other way!") For Grinnell, then, California became a delimitable "laboratory in the field" giving his research questions a regional, geographical focus. For the university administration, the regional focus supported its mandate to serve the people of the state. For the amateur naturalists concerned with the flora and fauna of their state, research conducted within its bounds also served their goals of preservation and conservation. This first constraint is a weak one with many advantages. It gives California itself the status of a boundary object, an object that lives in multiple social worlds and has different identities in each.

Grinnell then transformed this agreement into a resource for getting more money. He became one of the primary people in charge of preserving California. He made extensive alliances with conservation groups. This provided him with a definite but still weakly constrained and weakly structured base. Furthermore, the geographical concepts he wanted to advance were built on this kernel of support for California preservation. He needed a baseline for his geographical theories and comparisons, as the conservation movement needed and wanted information about the natural baseline threatened by development interests. At the core and beginning of his work, then, he placed a common goal and conventional understanding, with boundaries from several different worlds that coincide. These coincident boundaries, around a loosely structured, boundary object, provide an anchor for more widely ranging, riskier claims.[54]

From the standardized information Grinnell collected, he built an orderly repository. And from this library of specimens, he was able to build ecological theories different from those being developed in the rest of the country. His autonomy in this regard rested on solving the problems of boundary tensions posed by the multiple intersections of the worlds that met in the museum. Grinnell's work was highly abstract, with a strong empirical base and strikingly strong support from participating worlds.

In analyzing these translation tasks represented by the MVZ undertaking, we found four types of boundary objects. This is not an exhaustive list by any means. These are only analytic distinctions, in the sense that we are really dealing here with systems of boundary objects that are themselves heterogeneous.

1. *Repositories*. These are ordered "piles" of objects that are indexed in a standardized fashion. Repositories are built to deal with problems of heterogeneity caused by differences in unit of analysis. An example of a repository is a library or museum. It has the advantage of modularity. People from different worlds can use or borrow from the "pile" for their own purposes without having directly to negotiate differences in purpose.

2. *Ideal type*. This is an object such as a diagram, atlas or other description that in fact does not accurately describe the details of any one locality or thing. It is abstracted from all domains, and may be fairly vague. However, it is adaptable to a local site precisely because it is fairly vague; it serves as a means of communicating and cooperating symbolically—a "good enough" road map for all parties. An example of an ideal type is the species. This is a concept that in fact described no specimen, which incorporated both concrete and theoretical data and served as a means of communicating across both worlds. Ideal types arise with differences in degree of abstraction. They result in the deletion of local contingencies from the common object and have the advantage of adaptability.

3. *Coincident boundaries*. These are common objects that have the same boundaries but different internal contents. They arise in the presence of different means of aggregating data and when work is distributed over a large-scale geographic area. The result is that work in different sites and with different perspectives can be conducted autonomously while cooperating parties share a common referent. The advantage is the resolution of different goals. An example of coincident boundaries is the creation of the state of California itself as a boundary object for workers at the MVZ. The maps of

California created by the amateur collectors and the conservationists resembled traditional roadmaps familiar to us all, and emphasized campsites, trails, and places to collect. The maps created by the professional biologists, however, shared the same outline of the state (with the same geopolitical boundaries), but were filled in with a highly abstract, ecologically based series of shaded areas representing "life zones," an ecological concept.

4. *Standardized forms.* These are boundary objects devised as methods of common communication across dispersed work groups. Because the natural history work took place at highly distributed sites by a number of different people, standardized methods were essential, as discussed earlier. In the case of the amateur collectors, they were provided with a form to fill out when they obtained an animal, standardized in the information it collected. The results of this type of boundary object are standardized indexes and what Latour would call "immutable mobiles" (objects which can be transported over a long distance and convey unchanging information). The advantages of such objects are that local uncertainties (for instance, in the collecting of animal species) are deleted.

People who inhabit more than one social world—marginal people—face an analogous situation. Traditionally, the concept of marginality has referred to a person who has membership in more than one social world: for example, a person whose mother is white and father is black.[55] Park's classic work on the "marginal man" discusses the tensions imposed by such multiple membership, problems of identity and loyalty.[56] Marginality has been a critical concept for understanding the ways in which the boundaries of social worlds are constructed, and the kinds of navigation and articulation performed by those with multiple memberships. The strategies employed by marginal people to manage their identities—passing, trying to shift into a single world, oscillating—provide a provocative source of metaphors for understanding objects with multiple memberships. Can we find similar strategies among those creating or managing joint objects across social world boundaries?

A social world, such as the world of amateur natural history collectors, "stakes out" territory, either literal or conceptual. If a state of war does not prevail, then institutionalized negotiations manage ordinary affairs when different social worlds share the same territory (for instance, the United States government and the Mafia). Such negotiations include conflict and are constantly challenged and refined. Everett Hughes has talked about such overlaps and has described organizations which manage collisions in space sovereignty as "intertribal centers."[57] Gerson's analysis of resources and commitments provides a general model of sovereignties based on commitments of time, money, skill, and sentiment.[58] Gerson and Gerson, drawing on Hughes's earlier work, have discussed the complex management of such overlapping place perspectives.[59] In their analysis, the central cooperative task of social worlds which share the same space but different perspectives is the "translation" of each other's perspectives.

In this chapter, we are interested in that sort of n-way translation which includes scientific objects. In particular, we are interested in the kinds of translations scientists perform in order to craft objects containing elements which are different in different worlds—objects marginal to those worlds, or what we call boundary objects.[60] In conducting collective work, people coming together from different social worlds frequently have the experience of addressing an object that has a different meaning for each of them. Each social world has partial jurisdiction over the resources represented by that object, and mismatches caused by the overlap become problems for negotiation. Unlike the situation of marginal people who reflexively face problems of identity and membership, however, the objects with multiple memberships do not change themselves reflexively, or voluntarily manage membership problems. While these objects have some of the same properties as marginal people, there are crucial differences.

For people, managing multiple memberships can be volatile, elusive, or confusing; navigating in more than one world is a nontrivial mapping exercise. People resolve problems of marginality in a variety of ways: by passing on one side or another, denying one side, oscillating between worlds, or by forming a new social world composed of others like themselves. However, management of these scientific objects—including construction of them—is conducted by scientists, collectors, and administrators only when their work coincides. The objects thus come to form a common boundary between worlds by inhabiting them both simultaneously. Scientists manage boundary objects via a set of strategies only loosely comparable to those practiced by marginal people.

Intersections place particular demands on representations, and on the integrity of information arising from and being used in more than one world. Because more than one world or set of concerns is using and making the representation, it has to satisfy more than one set of concerns.

When participants in the intersecting worlds create representations together, their different commitments and perceptions are resolved into representations—in the sense that a fuzzy image is resolved by a microscope. This resolution does not mean consensus. Rather, representations, or inscriptions, contain at every stage the traces of multiple viewpoints, translations, and incomplete battles. Gerson and Star have discussed a similar collision in an office workplace,[61] and considered the problem of evaluating the standards which apply as reconciliation takes place—a problem which computer scientist Carl Hewitt has called "due process."[62] Gokalp has described some of the processes of collisions which arise when multiple fields come together; he calls these "borderland" disciplines.[63]

The production of boundary objects is one means of satisfying these potentially conflicting sets of concerns. Other means include imperialist imposition of representations, coercion, silencing, and fragmentation.[64]

Summary

The different commitments of the participants from different social worlds reflect a fascinating phenomenon—the functioning of mixed economies of information with different values and only partially overlapping coin. Andrews has a compelling example of this from a natural history expedition of the period to Mongolia: natives there used fossils for fang shui (geomancy), and were in the habit of dissolving them in liquid and drinking them![65] The sacred fossil beds were well protected against foraging paleontologists, who considered them equally valuable but for different reasons. The economy of the museum thus evolves as a mixture of barter, money, and complex negotiations: money in exchange for furs and animals from trappers; animals in exchange for other animals from other museums and collectors; scientific classification in exchange for specimens donated by amateur naturalists; prestige and legitimacy for economic support; food and bait in traps in exchange for animals' unwitting cooperation.

As the museum matures, and becomes more efficient, the scientists have made headway in standardizing the interfaces between different worlds. In the case of museum work, this comes from the standardization of collecting and preparation methods. By reaching agreements about methods, different participating worlds establish protocols that go beyond mere trading across unjoined world boundaries. They begin to devise a common coin that makes possible new kinds of joint endeavor.

But the protocols are not simply the imposition of one world's vision on the rest; if they are, they are sure to fail. Rather, boundary objects act as anchors or bridges, however temporary.

The central analytical question raised by this chapter is: how do heterogeneity and cooperation coexist, and with what consequences for managing information? The museum is in a sense a model of information processing. In the strategies used by its participants are several sophisticated answers to problems of complexity, preservation, and coordination. Our future work will examine these answers in different domains, including the history of evolutionary theory and the design of complex computer systems.

Notes

We would like to thank our colleague Elihu Gerson of Tremont Research Institute for many helpful conversations about the content of this chapter. We would also like to thank The Bancroft Library, University of California, for access to the papers of Joseph Grinnell, Annie Alexander and the Museum of Vertebrate Zoology. David Wake and Barbara Stein of the Museum of Vertebrate Zoology have graciously allowed us access to the museum's archives and assisted us in locating material; Howard Hutchinson of the Berkeley University Museum of Paleontology generously provided access to the Alexander-Merriam correspondence in the archives there. Annetta Carter, Frank Pitelka, Joseph Gregory, and Gene Crisman have provided valuable firsthand information about Alexander and Grinnell. We would also like to thank Michel Callon, Adele Clarke, Joan Fujimura, Carl Hewitt, Bruno Latour, John Law, and Anselm Strauss for their helpful comments and discussions of many of these ideas. Star's work on this chapter was supported in part by a generous grant from the Fondation Fyssen, Paris.

1. In general social science, this finding can most clearly be seen in the studies of workplaces by Chicago school sociologists. See, for example, Everett C. Hughes, *The Sociological Eye* (Chicago: Aldine, 1970). For evidence of this in science, see David Hull, *Science as a Process* (Chicago and London: University of Chicago Press, 1988); Bruno Latour and Steve Woolgar, *Laboratory Life* (Beverly Hills, CA: Sage Publications, 1979); Bruno Latour, *Science in Action* (Cambridge, MA: Harvard University Press, 1987); Martin Rudwick, *The Great Devonian Controversy* (Chicago and London: University of Chicago Press, 1985); Susan Leigh Star, "Triangulating Clinical and Basic Research: British Localizationists, 1870–1906," *History of Science* 24 (1986): 29–48.

2. Anselm Strauss, "A Social World Perspective," *Studies in Symbolic Interaction* 1 (1978): 119–128; Elihu M. Gerson, "Scientific Work and Social Worlds," *Knowledge* 4 (1983): 357–377; Adele Clarke, "A Social Worlds Research Adventure: The Case of Reproductive Science," in *Theories of Science in Society*, ed. T. Gieryn and S. Cozzens (Bloomington: Indiana University Press, 1990), 15–42.

3. Michel Callon, "Some Elements of a Sociology of Translation: Domestication of the Scallops and the Fishermen of St. Brieuc Bay," in *Power, Action and Belief, Sociological Review Monograph No. 32*, ed. John Law (London: Routledge & Kegan Paul, 1985), 196–230; Latour, *Science in Action*, note 1; Bruno Latour, *The Pasteurization of French Society* (Cambridge, MA: Harvard University Press, 1988).

4. John Law, "Technology, Closure and Heterogeneous Engineering: The Case of the Portuguese Expansion," in *The Social Construction of Technological Systems*, ed. Wiebe Bijker, Trevor Pinch, and Thomas P. Hughes (Cambridge, MA: MIT Press, 1987), 111–134; Michel Callon and John Law, "On Interests and Their Transformation: Enrollment and Counter Enrollment," *Social Studies of Science* 12 (1982): 615–625.

5. Everett C. Hughes, "Going Concerns: The Study of American Institutions," in *Sociological Eye*, note 1, 52–72, at 62.

6. W. V. O. Quine, *Word and Object* (Cambridge, MA: MIT Press, 1960).

7. See the excellent review by L. Daston, "The Factual Sensibility," *Isis* 79 (1988): 452–467.

8. See, for example, Sally G. Kohlstedt, "Curiosities and Cabinets: Natural History Museums and Education on the Antebellum Campus," *Isis* 79 (1988): 405–426. Although it has frequently been claimed that the rise of scientific biology coincided with the demise of natural history at the tum of the twentieth century, some have argued that natural history was "refined" rather than replaced. See, for example, K. Benson, "Concluding Remarks: American Natural History and Biology in the Nineteenth Century," *American Zoologist* 26 (1986): 381–384. On the distinction of the amateur naturalist from the public and from professional scientists, see S. Kohlstedt, "The

Nineteenth-Century Amateur Tradition: The Case of the Boston Society of Natural History," in *Science and Its Public*, ed. G. Holton and W. Blanpied (Dordrecht, Holland: D. Reidel, 1976), 173–190.

9. For an assessment of the effects on the structure of theoretical models produced, see also James R. Griesemer, "Modeling in the Museum: On the Role of Remnant Models in the Work of Joseph Grinnell," *Biology and Philosophy* 5 (1990): 3–36. Additional analysis of the role of amateur naturalists can be found in David Allen, *The Naturalist in Britain: A Social History* (London: Allen Lane, 1976).

10. Griesemer, ibid.

11. Frank Pitelka, personal communication to Griesemer.

12. E. R. Hall, *Collecting and Preparing Study Specimens of Vertebrates* (Lawrence: University of Kansas, 1962).

13. Joseph Grinnell Papers, The Bancroft Library, University of California, Berkeley.

14. See manuals of instruction by Grinnell's student E. R. Hall, Joseph Grinnell Papers, note 13; and S. Herman, *The Naturalist's Field Journal, A Manual of Instruction based on a System Established by Joseph Grinnell* (Vermillion, SD: Buteo Books, 1986).

15. Susan Leigh Star, "The Structure of Ill-Structured Solutions: Boundary Objects and Heterogeneous Distributed Problem Solving," in *Readings in Distributed Artificial Intelligence 3*, ed. M. Hubs and L. Gasser (Menlo Park, CA: Morgan Kaufmann, 1989), 37–54.

16. See Griesemer, *Modeling in the Museum,* note 9; Nancy Cartwright and H. Mendell, "What Makes Physics' Objects Abstract?" in *Science and Reality*, ed. J. Cushing, C. Delaney, and G. Gutting (Notre Dame, IN: University of Notre Dame Press, 1984), 134–152.

17. Garland Allen, *Life Science in the Twentieth Century* (Cambridge: Cambridge University Press, 1978); Kohlstedt, "Nineteenth-Century Amateur Tradition," note 8; Benson, "Concluding Remarks," note 8; Jane Maienschein, Ron Rainger, and Keith Benson, "Introduction: Were American Morphologists in Revolt?," *Journal of the History of Biology* 14 (1981): 83–87; Philip Pauly, *Controlling Life, Jacques Loeb & the Engineering Ideal in Biology* (New York: Oxford University, 1987); Ronald Rainger, "The Continuation of the Morphological Tradition: American Paleontology, 1880–1910," *Journal of the History of Biology* 14 (1981): 129–158; Garland Allen, "Morphology and Twentieth-Century Biology: A Response," ibid., 159–176.

18. William Goetzmann, *Exploration & Empire* (New York: W. W. Norton, 1966); Keir B. Sterling, *Last of the Naturalists: The Career of C. Hart Merriam* (New York: Amo Press, 1977, revised edition); M. Smith, *Pacific Visions, California Scientists and the Environment, 1850–1915* (New Haven, CT: Yale University Press, 1987); C. Hart Merriam, "Type Specimens in Natural History," *Science*, N. S. 5 (1897): 731–732; Merriam, "Criteria for the Recognition of Species and Genera," *Journal of Mammology* 1 (1919): 6–9; Merriam, "Laws of Temperature Control of the Geographic Distribution of Terrestrial Animals and Plants," *The National Geographic Magazine* 6 (1894): 229–241; Merriam, "Results of a Biological Survey of Mount Shasta, California," *Bureau of the Biological Survey,*

North American Fauna, vol. 16 (1899); J. Moore, "Zoology of the Pacific Railroad Surveys," *American Zoologist* 26 (1986): 331–341; L. Spencer, "Filling in the Gaps: A Survey of Nineteenth Century Institutions Associated with the Exploration and Natural History of the American West," ibid., 371–380; on the Bureau of the Biological Survey, see Donald Worster, *Nature's Economy* (New York: Cambridge University Press, 1988).

19. See W. C. Allee, A. E. Emerson, 0. Park, T. Park, and K. Schmidt, *Principles of Animal Ecology* (Philadelphia and London: W. B. Saunders, 1949); Sharon Kingsland, *Modeling Nature: Episodes in the History of Population Ecology* (Chicago: University of Chicago Press, 1985); R. Mcintosh, *The Background of Ecology* (Cambridge: Cambridge University Press, 1987).

20. E. Cittadino, "Ecology and the Professionalization of Botany in America, 1890–1905," *Studies in History of Biology* 4 (1980): 171–198; Joel Hagen, "Organism and Environment: Fredric Clements's Vision of a Unified Physiological Ecology," in *The American Development of Biology*, ed. R. Rainger, K. Benson, and J. Maienschein (Philadelphia: University of Pennsylvania Press, 1988), 257–280; William Kimler, "Mimicry: Views of Naturalists and Ecologists Before the Modem Synthesis," in *Dimensions of Darwinism*, ed. Marjorie Grene (Cambridge: Cambridge University Press, 1983), 97–128. See also Kingsland, *Modeling Nature*; Macintosh, *Background of Ecology*, note 19.

21. Hilda Grinnell, "Joseph Grinnell: 1877–1939," *The Condor* 42 (1940): 3–34.

22. Ernst Mayr, "Ecological Factors in Speciation," *Evolution* 1 (1947): 263–288; Mayr, "Speciation and Systematics," in *Genetics, Paleontology and Evolution*, ed. Glenn L. Jepsen, George Gaylord Simpson, and Ernst Mayr (Princeton, NJ: Princeton University Press, 1949), 281–298; David Lack, "The Significance of Ecological Isolation," ibid., 299–308.

23. See Hilda W. Grinnell, *Annie Montague Alexander* (Berkeley, CA: Grinnell Naturalists Society, 1958). That she was an amateur naturalist and not merely a financial backer is clear from Kohlstedt's (1976) discussion of the amateur tradition, "Nineteenth-Century Amateur Tradition," note 8. Amateurs were more interested in scientific investigation than the general public, which was largely interested in the exposition of ideas about nature as part of the general culture, but amateurs typically had a broad vision of the aims and nature of scientific research: ibid., 175. The dependence on amateurs/patrons declined as universities and governments took over the financial stewardship of science, and eventually amateurs such as Alexander virtually disappeared from scientific academia.

24. See Grinnell, ibid.; Alden Miller, "Joseph Grinnell," *Systematic Zoology* 3 (1964): 195–249.

25. J. Grinnell, "The Origin and Distribution of the Chestnut-Backed Chickadee," *The Auk* 21 (1904): 364–365, 368–378.

26. D. Jordan, "The Origin of Species through Isolation," *Science* 22 (1905): 545–562.

27. That Grinnell could become director of the MVZ without appointment in the Department of Zoology at Berkeley suggests that credentials for museum naturalists were not identical with those required for academic appointment. Grinnell's career marks a transition phase in the incorporation of research natural history into academic science. For histories of the Department of Zoology, including the MVZ, see Richard Eakin, "History of Zoology at the University of Cali-

fornia, Berkeley," *Bios* 27 (1956): 66–92. (Reprint available from the Department of Zoology at Berkeley.)

28. See Eakin, ibid.

29. These procedures included directives such as: to use a single serial set of identification numbers for all specimens collected during an expedition regardless of type, "roadkills," nests, eggs, wet preservations, and so on; to give precise data on the location of capture of a specimen including altitude and county; to "attend minutely to proper punctuation"; to observe the proper order for reporting data on both field tags and in field notebooks; to pack "miscellaneous material . . . with as great care as skins or skulls. Cheek pouch contents, feces, etc., should be placed in small envelopes or boxes, with labels inserted, and such containers packed in a stout box to prevent crushing"; and most importantly, to "write full notes, even at risk of entering much information of apparently little value. One cannot anticipate the needs of the future, when notes and collection are worked up. . . . Be alert for new ideas and new facts." These quotations were taken from a handout, "Suggestions as to Collecting," used in Grinnell's natural history course, Zoology 113, Grinnell Correspondence and Papers, Bancroft Library, University of Califomia, Berkeley. The handout was emended and used by a number of Grinnell's successors at Berkeley and elsewhere. See also Hall, *Collecting and Preparing Study Specimens*, note 12.

30. See Griesemer, "Modeling in the Museum," note 9.

31. On museums in general, see E. Alexander, *Museums in Motion: An Introduction to the History and Functions of Museums* (Nashville, TN: American Association for State and Local History, 1979); Laurence V. Coleman, *The Museum in America*, 3 vols. (Washington, DC: The American Association of Museums, 1939); George W. Stocking Jr., "Essays on Museums and Material Culture," in *Objects and Others: Essays on Museums and Material Culture*, ed. G. Stocking Jr. (Madison: University of Wisconsin Press, 1985), 3–14.

32. Joseph Grinnell, "Significance of Faunal Analysis for General Biology," *University of California Publications in Zoology* 32 (1928): 13–18.

33. On natural history museums in particular, see C. Adams, "Some of the Advantages of an Ecological Organization of a Natural History Museum," *Proceedings of the American Association of Museums*, vol. 1 (1907): 170–178; K. Benson, "From Museum Research to Laboratory Research: The Transformation of Natural History into Academic Biology," in Rainger, Benson, and Maienschein, *American Development of Biology*, note 20, 49–83; E. Colbert, "What Is a Museum?," *Curator* 4 (1961): 138–146; Joseph Grinnell, "The Methods and Uses of a Research Museum," *Popular Science Monthly* 77 (1910): 163–169; S. Kohlstedt, "Henry A. Ward: The Merchant Naturalist and American Museum Development," *Journal of the Society for the Bibliography of Natural History* 9 (1980): 647–661; Kohlstedt, "Natural History on Campus: From Informal Collecting to College Museums," paper delivered to the West Coast History of Science Association (Friday Harbor, WA: September 1986); Kohlstedt, "Curiosities and Cabinets," note 8; Kohlstedt, "Museums on Campus: A Tradition of Inquiry and Teaching," in Rainger, Benson, and Maienschein, *American Development of Biology*, note 20, 15–47; Ernst Mayr, "Alden Holmes Miller," National Academy of Sciences of the United States, *Biographical Memoirs*, vol. 33 (1973), 176–214; Ronald Rainger, "Just

before Simpson: William Diller Matthew's Understanding of Evolution," *Proceedings of the American Philosophical Society* 130 (1986): 453–474; Rainger, "Vertebrate Paleontology as Biology: Henry Fairfield Osborn and the American Museum of Natural History," in Rainger, Benson, and Maienschein, *American Development of Biology*, note 20, 219–256; Dillon Ripley, *The Sacred Grove: Essays on Museums* (New York: Simon & Schuster, 1969); A. Ruthven, *A Naturalist in a University Museum* (Ann Arbor: University of Michigan Alumni Press, 1963).

34. Joseph Grinnell to Annie Alexander, November 14, 1907, Joseph Grinnell Papers, The Bancroft Library, University of California, Berkeley.

35. Grinnell, "Methods and Uses," note 33. This essay, originally a director's report to the president of the University of California, was later published in *Popular Science Monthly* as an article outlining Grinnell's vision titled "The Methods and Uses of a Research Museum."

36. Grinnell, "The Museum Conscience" (1922), in op. cit. note 33, 107–109, at 108.

37. Ibid.

38. Joseph Grinnell, "Barriers to Distribution as Regards Birds and Mammals," *The American Naturalist* 48 (1914): 248–254; Grinnell, "An Account of the Mammals and Birds of the Lower Colorado Valley with Especial Reference to the Distributional Problems Presented," *University of California Publications in Zoology* 12 (1914): 51–294.

39. See Rainger, "Just before Simpson," note 33. See note 31 for similar considerations by W. D. Matthew in the American Museum.

40. For another early example, see C. Adams, "Some of the Advantages," note 33.

41. See E. M. Gerson, "Audiences and Allies: The Transformation of American Zoology, 1880–1930," paper presented to the conference on the History, Philosophy and Social Studies of Biology (Blacksburg, VA, June 1987); Eakin, "History of Zoology," note 27, reports the possibly apocryphal story that Jordan and Grinnell agreed that Stanford would get fishes and Berkeley would get birds and mammals.

42. Joseph Grinnell to Annie Alexander, February 13, 1911, Joseph Grinnell Papers, The Bancroft Library, University of California, Berkeley.

43. Joseph Grinnell to Annie Alexander, May 11, 1911, ibid.

44. See Grinnell, op. cit. note 31.

45. See Grinnell, *Annie Montague Alexander*, note 23, 7.

46. Annie Alexander to Joseph Grinnell, January 6, 1911, Annie M. Alexander Papers (Collection 67/121 c), The Bancroft Library, University of California, Berkeley.

47. See also Smith, *Pacific Visions*, note 18.

48. Louise Kellogg, 1911 field notebook, Fieldnote Room, Museum of Vertebrate Zoology, University of California, Berkeley. See also Annie Alexander, 1911 field notebook, for similar observations.

49. Joseph Grinnell to Annie Alexander, March 27, 1911, Joseph Grinnell Papers, The Bancroft Library, University of California, Berkeley.

50. Joseph Grinnell to Annie Alexander, November 14, 1907, Joseph Grinnell Papers, The Bancroft Library, University of California, Berkeley.

51. See Hall, *Collecting and Preparing Study Specimens*, note 12, for a full discussion of these preparation and preservation techniques.

52. Joseph Grinnell, J. Dixon, and Jean Linsdale, "Vertebrate Natural History of a Section of Northern California through the Lassen Peak Region," *University of California Publications in Zoology* 35 (1930): 1–594 and i–v.

53. See Griesemer, "Modeling in the Museum," note 9.

54. William C. Wimsatt, "Robustness, Reliability and Overdetermination," in *Scientific Inquiry and the Social Sciences*, ed. M. Brewer and B. Collins (San Francisco, CA: Jossey-Bass, 1981), 124–163.

55. Robert E. Park, "Human Migration and the Marginal Man," in his *Race and Culture* (New York: The Free Press, 1928, reprinted 1950), 345–356; E. C. Hughes, "Social Change and Status Protest: An Essay on the Marginal Man," in his *The Sociological Eye*, note 1, 220–228.

56. See also Everett V. Stonequist, *The Marginal Man: A Study in Personality and Culture Conflict* (New York: Russell & Russell, 1937, reprinted 1961).

57. E. C. Hughes, "The Ecological Aspect of Institutions', in *The Sociological Eye*, note 1, 5–13.

58. Elihu M. Gerson, "On 'Quality of Life,'" *American Sociological Review* 41 (1976): 793–806.

59. Elihu M. Gerson and M. Sue Gerson, "The Social Framework of Place Perspectives," in *Environmental Knowing: Theories, Research and Methods*, ed. G. T. Moore and R. Golledge (Stroudsberg, PA: Dowden, Hutchinson & Ross, 1976), 196–205.

60. See Star, "Structure of Ill-Structured Solutions," note 15.

61. Elihu M. Gerson and Susan Leigh Star, "Analyzing Due Process in the Workplace," *ACM Transactions on Office Information Systems* 4 (1986): 257–270.

62. Carl Hewitt, "Offices Are Open Systems," *ACM Transactions on Office Information Systems* 4 (1986): 271–287.

63. Iskander Gokalp, "Report on an Ongoing Research: Investigation on Turbulent Combustion as an Example of an Interfield Research Area," paper presented at the Society for the Social Studies of Science (Troy, NY, November 1985).

64. We are grateful to an anonymous referee for drawing our attention to the limits of the cooperation model, and the importance of conflict and authority in science making.

65. Roy Chapman Andrews, *Across Mongolian Plains* (New York: D. Appleton, 1921).

8　Sharing Spaces, Crossing Boundaries

James R. Griesemer

Often, boundary implies something like edge or periphery, as in the boundary of a state or a tumor. Here, however, it is used to mean a shared space, where exactly that sense of here and there are confounded.

(Star 2010, 602–603)

Marginal Man

Meeting Leigh Star meant sharing a boundary, in the sense used "here." That space we shared in the 1980s heightened in me an already growing sense of marginality according to the meaning I learned from her: Robert Park's notion of the marginal man, trying to figure out how to manage one's boundary crossings that join and separate social worlds, in which the marginal man lives but does not quite gain full acceptance. Should I assimilate, return, transcend? The space I was entering joined and separated primarily science studies specialties, though these were practiced mostly by Western, middle-class, white men and women of only a few ethnicities.

I was a not-quite philosopher of science, newly minted from the University of Chicago's Conceptual Foundations of Science Program. I was on the job market, which meant going to the "right" conferences in hopes of interviews and meeting people with jobs to offer but, having come from a science background (genetics) and having crossed into philosophy of biology at Chicago with the aid of Bill Wimsatt, David Hull, Bob Richards, and a few biologists, capital-P Philosophy didn't feel like a good fit. The Eastern APA (where philosophers go to seek jobs) was an awful experience. I met sociologists Susan Leigh Star and Elihu Gerson in 1982 at a four societies meeting—4S, HSS, PSA, SHOT—that were a better fit. I got a one-year temporary job at UC Davis in the Philosophy Department, not far from San Francisco. Maybe we could work together.

I arrived in Davis in September 1983 and began visiting Leigh Star, Elihu Gerson, Adele Clarke, Joan Fujimura, and Rachel Volberg in Gerson's Tremont Research Institute in San Francisco on a nearly weekly basis. I met Howie Becker, Anselm Strauss,

Bruno Latour, and Carl Hewitt when they visited the group as well. Somehow, they became "my people" while I was trying to figure out how to "be" in a mostly analytic philosophy department and start a history and philosophy of science program at Davis. I learned that what I was doing was working out strategies to "pass" as a philosopher.

The closest I could get to philosophical "tradition" was to read Bas van Fraassen, Nancy Cartwright, and Ian Hacking in 1983. I had already rejected consensus in favor of cooperation as a core practical aim of scientists due to the influence of David Hull while I was in graduate school and, in similarly contrarian spirit, I had begun to favor materialities such as specimens and diagrams and experimental practices over mathematical and conceptual abstractions in trying to understand scientific modeling practices of the sorts that van Fraassen and Cartwright were writing about. So my interests fit well with Leigh's interests in analyzing cooperative work in the absence of consensus (Star 2010).

What I learned early on from "my people," however, was that I wasn't in the mainstream of anything and I'd better make that work for me if I wanted to survive as an academic. I couldn't really "return" to philosophy, though I did get back in the car after each week's visit to San Francisco and return to UC Davis, where I felt the pull of my new colleagues in the Philosophy Department and as I worked with geneticist Francisco Ayala to coteach the philosophy of biology course Marjorie Grene had started. As Ayala was the recognized scientist of our shared teaching space, I began to feel like the philosopher by contrast, though his friendly competition for that title as well—he was elected to the American *Philosophical* Society as we taught together—positioned me more as *assistant* professor of philosophy of biology than as assistant professor of *philosophy*. The student definition of who is what according to which department hosts the professor's address worked well enough. And I wasn't too sure what it would mean to transcend the disciplinary boundaries whose shared and unshared spaces I was squatting in at UC Davis. Although my perspective on academia was atypical, coming as I did from the committee system at Chicago rather than from a department of philosophy, my conversations with Leigh led to a fairly broad "derangement" that she actively encouraged. San Francisco sociology of science was assimilating me as much as I was assimilating it into my UC Davis philosophy of science. The "tractor beam"-like pulls and repulsing magnet-like pushes in multiple directions that take hold after you step into a boundary space is part of what makes boundary crossings so dynamic but, as a result of their fine scale, the outcomes are largely unpredictable. Leigh always talked about tacking back and forth through the boundary space as an important mechanism by which daily use of a boundary object could be well structured on both sides of the boundary space yet ill structured within it. The whole San Francisco experience, beginning with my collaboration with Leigh, was valuable to me in the way it taught me to *link* my career experiences of the viscosity and density changes that came with my own boundary crossings to the scientific practices I was studying.

My job having become tenure-track, gaining tenure would have to mean acceptance as a philosopher in at least the sense that I would become "card carrying." Prudence seemed to dictate trying to "pass" as a philosopher. I had to figure out how to write for philosophy journals. The lure of working with sociologists of science rather than just writing about them was risky. My first papers were rejected. I tried more direct means of "passing": I mentioned Quine in the introduction to a philosophy paper and got a better hearing from reviewers (Griesemer 1984).[1]

I started working with Leigh as she was finishing her dissertation and I was looking for new, California projects. It was like working with a living koan. Her words seemed ordinary, yet they were astonishing. I tried to read her as I had learned in Chicago from David Hull to read logical arguments in science: a contradiction doesn't count if you don't try to draw any conclusions from it.[2] Working in the boundary space meant learning to tolerate the dissonances while you tack back and forth through it. It was like being back in college on the first day of a computer programming class: I put Hollerith punch cards in the machine, typed on a keyboard, and the machine punched little holes that represent my code, then I fed them into a card reader and waited awhile for my printed output results to be put into a bin by an assistant tending the mainframe's printer. Usually the first line of output said: "Job control error." At first, talking to Leigh was like that. I had a lot of job control errors trying to make my programmatic inputs return responses from her that I could interpret as successful communication, using what I knew from my well-ordered worlds of genetics, evolutionary biology, and Chicago-styled philosophy of science. We connected on topics like: heuristics, fieldwork, libraries, and museums.

I knew a fair bit about the Museum of Vertebrate Zoology (MVZ) at Berkeley before my collaborations with Leigh Star and later with Elihu Gerson, having been an undergraduate genetics major at Berkeley in the 1970s, having studied with Dave Wake (then MVZ director), and having friends who had done research in the MVZ. I don't recall to what extent my knowledge of the museum influenced Leigh's initial interest. My new experiences as an assistant professor of marginal-man-science-studies made working through Leigh's brand new, first-draft boundary objects idea a natural. I was working on the materiality of the museum research experience and how it could lead to theoretical abstractions: collecting animals in the field, killing them, processing them in a tent, taking notes about it, transporting them back to the museum, cleaning them further, cataloguing and documenting them, storing them in cabinets for systematic study. I was beginning to write about that extraction process as a material process of abstraction—abstraction as the literal subtraction of parts of things (with subtraction of properties as side effects) by making specimens out of dead animals and field notes, rather than the mental subtraction of properties to gain the essence of "triangle," for example, by mentally subtracting the matter out of which triangles are inscribed on blackboards, or on paper, or in sand. I was responding to Nancy Cartwright's early work on Aristotelian versus Platonic abstraction (Cartwright and Mendell 1984). It resonated in my work with Leigh that specimens were marginal objects—boundary objects, well structured in daily use, such as by hunters seeking food, pelts, money, or trophies and by scientists working with specimens curated for research purposes seeking evidence of migration, evolution, or climate change; but becoming ill structured in the shared uses between social worlds, where a dead animal was a specimen to the hunter if he could offer to sell it to the scientific collector, but no use to the collector if killed by knocking the animal on the head, thus smashing the skull. I suppose my "philosophical" ideas were similarly of some use to Leigh in trying to relate the world of academic philosophers to her experience, but the specimen philosophical concepts and methods I brought to her ("model," "abstraction,") tended to look to her, I suppose, like they had been knocked on their sociological heads. Our work together was as much an exploration of our attempting to exist and coexist in an STS (science and technology studies) boundary space as it was of taking our respective "catches" back to our home disciplines. Although in some ways Leigh was always inscrutable to me, I think we both became a bit deranged by the collaboration in the sense that we were both crafting modes of interdisciplinarity, of marginality in Park's sense, "where exactly that sense of here and there are confounded" to quote Leigh Star again.

Leigh set our nascent ideas to the music of her emerging conversations on ill-structured problems. We riffed off Bruno Latour's "immutable mobiles." I got an inkling of what it was like to have a "sociological eye" as Mead called it (though in my case the eye was a glass one I held up to the light to see the wonderful refraction sociologists could do to my thinking). It seemed more and more like the right thing to do. I would try to "transcend" rather than assimilate social studies of science or return to philosophy of science. I would do "biology studies." We wrote the paper (Star and Griesemer 1989). It was good. It was different. Leigh Star and Joan Fujimura and Adele Clarke finished their dissertations and got jobs. Rachel Volberg left the field. I kept coming to San Francisco to continue conversations with Adele Clarke and with Elihu Gerson. I know now that I will never quite see as a sociologist sees (and that is a *good* thing). It is a perpetual amusement, joy, and gift from Leigh that my most cited paper by far is in a sociology of science journal as junior author to a sociologist. We had planned that I would be first author of a follow-up paper we would write together and publish in a philosophy journal. Sadly, that never happened, though we talked about it from time to time when Leigh came through the Bay Area or we met at conferences. And now Leigh is gone.

The Concept of a Boundary Object

According to Star and Griesemer (1989), the concept of a boundary object is "an analytical concept of those scientific objects which both inhabit several intersecting social worlds *and* satisfy the informational requirements of each of them. . . . Boundary objects are objects which are both plastic enough to adapt to local needs and the constraints of the several parties employing them, yet robust enough to maintain a common identity across sites. They are weakly structured in common use, and become strongly structured in individual-site use. These objects may be abstract or concrete. They have different meanings in different social worlds but their structure is common enough to more than one world to make them recognizable, a means of translation. The creation and management of boundary objects is a key process in developing and maintaining coherence across intersecting social worlds" (393).

There was a third idea in that paper beyond interpretive flexibility and informational requirements which was equally important to our argument: standardization of methods (see Star 2010). We argued that "By emphasizing how, and not what or why, methods standardization both makes information compatible and allows for a longer "reach" across divergent worlds" (Star and Griesemer 1989, 405). This was key to the case we examined: the organization, development, and management of a research natural history museum—the Museum of Vertebrate Zoology at Berkeley. Methods standards allowed practitioners in intersecting worlds to not only find (ill-structured) common ground in their task work with boundary objects such as specimens and field notes, they also provided some hope that the information assembled and curated in the museum could be made available for future generations or researchers at a distance to use (ibid.).

It is gratifying that so many have found the concept useful to them. It has not gone untweaked or uncriticized by others, however. To some, our focus on objects was ancillary to their concerns or what they thought should be the main concern. Boundary work, boundary concepts (Gieryn 1999), boundary organizations (Bechky 2003; O'Mahony and Bechky 2008), boundary artifacts (Lee 2007), or trading zones (Galison 1997) might seem to present a better or more comprehensive ontological basis for STS work. And according to some of these proposals it appears that, in our focus on boundary objects and translation as the metaphor through which we discussed boundary crossing, we didn't leave enough room for these other categories. To others, it seemed that anything could be a boundary object, which would tend to make the concept vacuous.

My brief and somewhat generic response is to acknowledge these limitations on the concept of boundary object. But I think we had more mundane concerns in mind when we—or rather, Leigh Star (she was the brains of our operation)—worked up her idea. The concept of a boundary object was as much a tool *for us* to use in tracking our subjects as it was an assertion of a new ontological perspective on science in action. It is not the *only* category useful for the study of boundary crossings, for sure, but we found it useful because our scientist-subjects seemed to need some durable and yet plastic *objects* that afforded them flexibility in their dealings with people from/in other social worlds: the objects they co-crafted made boundary crossings among social worlds we wrote about. Yet they could serve the long-term goals of an archive of information for *scientists* of the future: they became strongly structured as specimens in the world of the museum scientists. Specimens and field notes, species and ecological structures were immutable mobiles at the heart of a material culture of scientific practices relevant to several social worlds.

And yes, I agree that anything *can* be a boundary object. But that doesn't rob the concept of meaning any more than "model" and "representation" are meaningless because anything *can* be a model or a representation of anything else. Current philosophical thinking on models and representations says that these are four- or five-place relations, not the simplistic two-place relations of words to world or abstractions to concrete phenomena (Giere 2006, van Fraassen 2008). A more nuanced view points to the dependence of the status of something as a model or representation on *use*: S uses X to represent Y to Z for purpose W. By the same token, any object *can* be a boundary object *if* users have an interest in tracking the movement of that object across pairs of interacting social worlds. Hunter H uses dead animal A as a specimen S for collector C for the purpose of economic exchange E. Collector C uses dead animal A as a specimen S for hunter H for the purpose of scientific research R. Two social worlds filled with different practices and purposes intersect at the dead animal/specimen. In the section on Tracking Commitments Reflect a Pragmatic Ethical-Epistemic-Ontology, I develop an example from studies of current research in the MVZ, where ecological localities are used as boundary objects in this five-place relational sense between scientific specialties I call "naturalists" and "ecologists": naturalists N use field note descriptions F to represent localities L to ecologists E for the purpose of reporting comprehensive collecting efforts CC; ecologists E use GPS coordinates G to represent localities L to naturalists N for the purpose of specifying where naturalists should conduct their representative collecting (RC) efforts. The ill-structuring of the concept of locality in the boundary spaces of these intersecting scientific worlds is key to understanding some intriguing problems for taking ecology online, in some of the same problematic ways that genomics has tried to go online, global and interoperable.

Whether an object will be a *good* boundary object—good to think with, good to work with, good to track with—depends on its suitability both *to the subjects* engaging the object as well as its suitability *for the STS researchers* studying those subjects. The object has to have "reach": it has to be durable enough to travel between worlds, but it also has to have "brackets" in place—sufficient standardization of methods, for example, that the object can be plastic as it moves among social worlds but can be custom-fit into local practice within each participating world (see Gerson 2008).

STS Research Tacks across Disciplinary Boundaries While Tracking Boundary Crossings of Its Subjects

If I were to rewrite the boundary objects paper today, I might emphasize methods standardization more than Leigh and I did in 1989, but more centrally, I would emphasize that the concept of a boundary object serves *dual* purposes. We always meant it as a tool for STS work, a heuristic methodological category to think with as much as an ontological category of objects to think about, and also as a subversive concept blurring distinctions between methodology and ontology.[3] The multiple utilities of the concept stemmed from its having served *us* as a meta-category for objects such as specimens, field notes, species, geographies, ecologies, and so forth that *we* needed to negotiate and translate in our STS work to cross *our* boundaries between Leigh's sociology, my philosophy, and our overlapping and shared but not yet common knowledge of the history of the museum. And because of the heterogeneity of our own work, we also needed a bookkeeping trick to keep track of those objects scientists and others made to serve *their* boundary crossings among *their* intersecting social worlds as they created and managed the museum. We called these "translations," following Latour, Callon, and Law, meaning movements from place to place or change of forms of concern in different social worlds. We didn't mean the merely linguist notion that Peter Galison (1997) took from our paper, though Galison was broadly sympathetic to our approach, and language is certainly not unimportant to how worlds may (or may not) intersect.

The 1980s and 1990s were rich with multifarious negotiations across disciplines about what STS should and would be and become: historians, sociologists, philosophers, anthropologists, educators, psychologists, theoretical computer scientists, business management specialists, and information technologists all had things to say. In navigating our own boundary crossings between disciplinary worlds, we had to make sense of many disparate sources of data (in the museum in jars, refrigerators, and cabinet shelves, in libraries, in personal archives, in interviews and oral histories, and in the published record), as well as many more possible sources from other data worlds we became aware of but couldn't do much about (e.g., anecdotal and diarist lives of trappers, railroad construction workers, trap manufacturers, gun and ammo retailers, paper and ink makers . . .). We were interested, too, in the theoretical implications for STS of taking the "network turn," yet we remained focused on centers of calculation or what Grinnell himself called a "center of authority" as the organizing foci of our STS work. *Following* boundary objects, even just in thought experiments of possible fieldwork, rather than looking only at where they "end up" in some center, allowed us to consider the network and gave us a means through which we might imagine traversing it, even when the archival job proved overwhelming. Twenty-two years of working on this project on and off, a multi-investigator National Science Foundation grant and two postdocs (Ayelet Shavit and Mary Sundeland) later, and we still face a daunting task to understand the now 105-year history of the Museum of Vertebrate Zoology. Using various objects of concern to our *subjects* that we discovered in our fieldwork had crossed boundaries among our subjects' social worlds seemed a good way to keep track of *our own* boundary crossings as well, between philosophical and sociological and historical concepts, methods, and materials of concern to us as we launched our own careers in intersecting social worlds. Boundary objects brought *Leigh and I* together and guided our searches through the archives. Sadly, boundary crossings can also be destructive: breaking some social bonds as they foster others. The letters we read in the archives told as much of frustrations of working with these objects as of joys: of animals that sprang traps and got away, of recalcitrant hired collectors and guides, of equipment malfunction, of getting lost, of wasted effort. It was no different for me and for Leigh. Working in a boundary space where nothing quite translates easily leads to the frustrations every interdisciplinarian knows firsthand: papers rejected as "not rigorous," as "not philosophy," as "not sociology," as, basically, someone else's terrain, expertise, problem.

The point is that Leigh brought the boundary object idea and I brought the material grounding of these object conceptions—specimens and species, environments and evolution, for example—to our joint work in a moment of methodological need. We needed to gain a handle on a complex historical-social dynamic, through the partial records available to us, which seemed at every turn to open onto wider and wider vistas of networked social-worlds interactions that we could not even map fully for the purposes of our study. So, today I emphasize that the concept of a boundary object can serve heuristically (see Wimsatt 2007), as part of one theoretical perspective among many to frame STS research activity as engaging processes of boundary crossings of *science-studies disciplinary worlds* at the same time it serves to track boundary crossings of STS research subjects.

The complementarity in my view of "boundary object" as methodological concept for us *and* our subjects *and also* as an ontological concept should not be surprising, since STS is after all engaged in some of the same kinds of work as our technoscientific subjects. It would be rather remarkable if at least some of the tools and methods that work for our subjects could not also work for us. Nevertheless, I agree with those critics and skeptics who would say that just because one has a hammer, one ought not to see *every* problem as a nail. As a multiperspectival pluralist (Griesemer 2000a), I do not believe there can be a single, best all-purpose model of science in action. And, further, I believe we need a variety of independent but complementary perspectives if we are ever to reach the kind of robust understanding of science that Leigh and I collaborated to search for. The STS toolbox, like the toolboxes of the sciences generally, needs a variety of tools.

Two points in the preceding discussion call for further comment to put our 1989 boundary object paper in the perspective of my work since then to show how deeply collaboration with Leigh has influenced me. In discussing the multiple utilities of the concept of a boundary object, I referred to some aspects of both STS research and MVZ work as "tracking work." I want to develop the idea of tracking work a little further because I think that the perspective that much scientific work is tracking work points to contemporary STS and philosophical themes that were not explicitly on my radar in 1989, in particular the need to integrate matters of ethical as well as epistemic and ontological concern. I have been much slower than Leigh was to trace implications of STS research to ethical concerns and I know her interest in those connections predated our collaboration. I am glad to acknowledge that she began to open that door for me and hope that the line of thought I am beginning to pursue now would please her.

Scientific Work Involves Tracking Work

A key challenge in any kind of empirical work is to *pay attention*. Engaging and attending while intra-acting (Barad 2007) to co-construct phenomena is required in order to observe, intervene, or understand phenomena on time- and size-scales relevant to one's projects and interests. The world around us does not come prepackaged in objects, properties, events, processes, or activities of shapes, sizes, or durations to which we can continuously engage and attend, to the extent often needed to gain understanding and control. By "control" here I mean control of our own bodies, not necessarily control of the "situation" or control of the subject or objects of research. Scientists, for example, frequently work by "following" a process to find out where it leads (or how to engage, evade or avoid it, or how to affect or control its direction), or to find out what happens when they interfere with its operation, or to find out how they might act in self-control to the benefit or cost of participants, patrons, or others with interests and concerns in "it." Phenomena are "packaged" in inquiry through the construction and use of research systems, often anchored by one or more "platforms" for research such as those built by MVZ workers (Griesemer 2014).

To follow a process of any extent *continuously*, however, is usually impossible. At the absolute limits of visual attention and stamina, you have to blink or your eyeballs will dry out. Instead, scientists (and many others) have learned to *mark* processes in order to *track* them with less than continuous, full engagement. Radioactive tracers, fluorescent stains, genetic markers, and embryonic transplants all facilitate tracking biological processes and determining how physiological, molecular, and genetic outcomes result from known inputs. Marks can be "noticings" of distinctive features that persist in subjects in stable relations to other features, to be noticed again at a later time or place, allowing an inference of continuity or connection between observed phases or stages without continuous engagement. Marks can also result from physical interventions that cause changes to features of subjects so as to be more "noticeable" or detectable to the intermittent observer or intermittently observant. In short, many of the devices, schemes, strategies, and technologies of science are designed for tracking work to facilitate empirical engagement without requiring continuous attention.

An effect of physical marking (or of using detectors) to track processes is that the collection of marks yields traces—series of marks—which can be used as data, but also serve as points of initiation of representations of the phenomena constructed through these marking interventions. Instruments (or practices) designed to collect these traces have been called "research enabling technologies" (see Rheinberger 2010, chap. 9).

Grinnell and his colleagues in the MVZ used various mapping techniques to make marks on maps that represented what they took to be distinctive features of the places they visited in their collecting expeditions on behalf of the museum (Griesemer 2011). "Three miles east of the intersection of county roads X and Y" is a place description that takes the road intersection as a noteworthy feature marking the place of interest three miles away. Grinnell and his colleagues also placed physical marks on the bodies of specimens: specimen tags with identification numbers, species names, collectors' names, place, and date of collection. All served to mark the specimen for subsequent tracking inside the museum (or to other places if the specimen was loaned out). More recently, ecologists, including those working in or with the MVZ, would trap, mark, release, and recapture animals in order to study their distributions and abundances without killing and making specimens of them. I don't think, therefore, that it is too much of a stretch to think about tracking MVZ work by noticing that certain objects, such as specimens, travel with our subjects among social worlds and that we in STS can use *them* (e.g., specimens, tags, and notes) as *our* marks to follow boundary crossing processes to see where the researchers and their interactants in other social worlds go.

Thus far I have been talking about tracking *epistemology*: the role of marking and tracking in generating data and representations together (in the form of traces) that play central roles in the production of scientific knowledge. I also want to comment on tracking ontology briefly. To track is to make an ontological commitment—a very *pragmatic* ontological commitment. Tracking is a subtle art, a bit like dancing with strangers (or like playing jazz as a Saturday night musician, see Becker 1986). You can't dance if all you do is literally follow your partner. You have to *anticipate* what your partner is *likely* to do in order to coordinate your movements with theirs and arrive at the same place at the same time or musicians to end at the same time. When scientists track phenomena they engage in a process of co-generation. Through their engagement with nature, they must anticipate, which is to say they must commit to, "what nature will be like" before they find out. Scientific ontology could be said to be about how things might well or will be, not about how things are. This is not exactly what is meant by saying that scientists don't engage in empirical work without an antecedent theory or hypothesis to test, but it is perhaps the element of truth in the old saw that "the scientific method" is "theory-laden." It's just that the antecedent should really be called the anticipant: something one anticipates rather than what came or comes previously. And the consequent is what results from the pragmatic commitment to act according to what is anticipated, thus bringing about something related to it. The idea that ontological commitment is *pragmatic* is that if you don't anticipate where your subject or partner is going, you often can't direct your attention, your body, or your instruments sufficiently quickly or deftly to keep up with the actions you deem salient to your project or purpose.

So, tracking requires commitment and commitments have implications for "the way things might," or will become, or will be, or indeed will have been, which is to say that they are ontological commitments. Only, these are ontological commitments that might be abandoned just seconds, minutes, or hours after they have been made if the tracking effort breaks down during the course of an experiment or observation. And they are of the most basic, practical kind. On the one hand, if you want to track chromosomes through cell division, you make an ontological commitment to properties and behaviors of cells and chromosomes that may surprise you and defeat your tracking procedures. For example, chromosomes are only compact, stainable bodies at some phases of their life cycles, it turns out, so trying to track them with staining alone as they dance their way through mitosis doesn't work. Other techniques are required. On the other hand, noticing the *recurrence* of similar banding patterns in similar chromosomes cycle after cycle permits the *inference* that genes on chromosomes can be tracked from cell generation to cell generation, provided you aren't so concerned about what happens to them during the time you *can't* attend to and track them with color-staining marks.

From the perspective of scientific tracking work, it appears that scientists wear their ontologies on their sleeves or hang them on their toolbelts. An ontology is something right at the surface of practical action that can be abandoned quickly if it doesn't serve a tracking use, but over time it can become deeply entrenched in scientific practice, as success at tracking and the corresponding representations provide feedback from theory to guide new tracking activity. On this view of science, ontology is superficial (until historically entrenched in shared practice); tracking work is fundamental.

Tracking Commitments Reflect a Pragmatic Ethical-Epistemic-Ontology

The link to ethical concerns is something I am exploring through the following train of thought. Ethics *begins* in attention—what matters to each of us is what we each attend to.[4] This does not mean what we attend to is right or good nor that what we do not attend to is wrong or bad, but attention is a core feature of any further ethical *consideration.*

Awareness, attention, and the embodied skills of attending in ways needed to do science have ontogenies—they *develop* in us; they are not inborn, instilled, or made by artifice. Thus, the roles values play in the attentional activities of science must also develop. There may be an important role in the ontogeny of scientific attentional skills for developmental "scaffolding"— through which humans develop knowledge, skills, and capacities to function as "autonomous" agents with the aid of other, usually more experienced members of socio-cultural groups (see Caporael, Griesemer, and Wimsatt, forthcoming). Our teachers (parents, other adults, peers, students, children) scaffold our behavioral development by bringing things to our attention that they think ought to matter to us. They create a (usually safe, but sometimes disastrously dangerous) training environment in which we can learn to make them matter to us (or not), protected from (but sometimes exposed to) the harms of valuing the wrong things or failing to value the right things, at the right or wrong times, or ways.

Attention—an aspect of what Ingold (2000) calls "the perception of the environment"—is the means through which values pervade human activities. Human activities include science. And at the heart of science is the activity of tracking. Values can guide tracking activity just as theoretical commitments to follow a certain model or representation can guide reasoning (see Griesemer 2000b), hence attention in tracking activity, hence the production of knowledge in boundary crossing work.

These two points—that scientific work involves tracking and that ethics begins in attention—merge in the practice of making marks in order to track processes. Karen Barad (2007) has argued that a conception of objective science cannot be merely "onto-epistemic"; it is a matter of "accountability to marks on bodies" and requires an ethics of practical action. As she puts it (390): "We are responsible for the world of which we are a part, not because it is an arbitrary construction of our choosing but because reality is sedimented out of particular practices that we have a role in shaping and through which we are shaped. . . . [T]he responsible practice of science requires a full genealogical accounting of the entangled apparatuses or practices that produce particular phenomena."

The animals marked in an ecological study, whether released to be recaptured or killed and tagged, did not consent to join the study. Scientists incur obligations to their subjects when they mark them and incur obligations to themselves and other humans affected by those marking and tracking activities. Accountability and the obligations incurred by marking are not yet ethics, but they are a beginning, a point—or rather a surface—of entry into matters of concern, a pragmatic ethical-onto-epistemic commitment.

My former postdoc, Ayelet Shavit, worked from 2005 to 2008 following a project of the MVZ to resurvey all of Grinnell's original field sites from the 1910s and 1920s. This resurvey, which began in 2003, was inspired by some of Grinnell's writings that Leigh and I quoted in 1989 on the methods and uses of a research museum that were posted on walls in the museum and which then-incoming MVZ director, Craig Moritz, had read in 2000 (pers. comm.). Ayelet became interested in the conduct of resurvey work in part because of the intersection of two specialist worlds inside the museum: field naturalists trained in the methods of comprehensive sampling who were descended from Grinnell, his students, or otherwise brought into the ecological tradition Grinnell pioneered in the early twentieth century, and postwar ecologists trained in mathematical and statistical methods of representative sampling for the sake of scientific hypothesis testing. The intersection took place in two theaters: the field sites themselves and in the museum through the computer databases they are building to archive the new resurvey data alongside the historical field note and catalog records of Grinnell's original surveys. Ayelet found that the concept of a specimen locality marked an important boundary between these two ecological sciences, which they discovered when they attempted to return to the "same places" Grinnell had collected (Shavit and Griesemer 2009). Localities are boundary objects that occupy the shared space between Grinnellian naturalists and postwar ecologists. The researchers tack between these social worlds as they try to construct databases that must manage the intersections of well-structured data objects from each world with a coherent set of "meta-data" (Shavit and Griesemer 2011a).

Modern ecological methods led to inferring localities from the historical field note information and using GPS to determine where to place a new trap as part of an effort to resurvey animals from Grinnell's sites. This can lead to tragic yet comical implementations of the concept of locality, for example when the GPS coordinates now point to a Pasadena parking lot or highway rather than an oak shrub and grassland habitat (Griesemer 2011). Do you place a trap in the middle of the road or in the nearest grassland? Does either choice represent "going back to the same locality" where an animal was caught in 1923? In the field, the conflict is between (1) the naturalists' perspective on placing traps or observers where the animals might now be found in order to collect comprehensively for all the species living in a *place*, and then to record GPS coordinates where the traps are placed versus, and (2) the ecologists' perspective on placing traps where GPS coordinates count as a translation of Grinnell's field note descriptions of collecting localities, and then to see whatever is trapped there in order to get a representative sampling of the fauna at those *coordinates* (Shavit and Griesemer 2011a).

Coupled to this boundary conflict between specialties is an ethical concern as well. Trap sampling effort should not result in traps placed where trapped animals would be exposed to the elements for too long because it would risk killing them—in the hot sun for example—before the scientists could return to mark and release them. Sampling effort, according to the statistical demands of modern post-Grinnellian ecology, however, requires equidistant trap placement for equivalent durations to insure *equality of effort* in each locality so as to guarantee *comparability* of data across trap lines and sites. So, a technical epistemological issue of how to collect information about animals properly is entangled with an ontological issue about what it means to "return" to the same place Grinnell visited—to step into the same river twice, as Heraclitus pondered—and moreover, because of the ways scientists intra-act with animals, resurveying is also entangled with the ethics of trapping animals.

Boundaries between these specialist worlds also become apparent within the museum as the naturalists and ecologists engage with computer scientists to develop interoperable, globally accessible databases (Shavit and Griesemer 2011b). We can see the boundaries literally displayed on the web pages, where the best the museum can do to manage the many conflicting kinds of "meta-data" about specimens is to juxtapose optical scans of historical field notes with electronically searchable data recorded according to the coding standards of a relational database. The "bridging" or "bracketing" techniques required to allow information to flow between the Grinnellian and post-Grinnellian worlds allow customization in each world, but create new challenges to facilitate and manage flows of information between them, the process Star called "tacking" (Star 2010).

Barad's notion of accountability brings ethical concerns inside the intimate practices of scientists engaging others "in nature" *within* the phenomena scientists co-generate together with their subjects. Science is social activity: doing things together (see Becker 1986). It is also heterogeneous work: the others may be other humans, but often they are nonhumans who work in different ways and for whom co-generation of phenomena with scientists results from force, violence, and intrusion. Science is a form of sociality in which a key goal of the scientists is to generate and understand phenomena, and in the process, commit to it ontologically and ethically (whether reflectively and reflexively or not).

Scientific practice can be understood broadly in terms of the activity of tracking processes by noticing or introducing marks on processes scientists track in order to "keep track" of (attend to) processes over the course of scientific engagement with subjects. The results of tracking work can be *subsequently described* as a mode of *inter*-action between scientists and their subjects, in which scientists keep track of the "intra-action" they enter into or join with their subjects and thereby *construct* a basis for representing phenomena as objective, in nature, and discovered by scientists. Entering and leaving intra-actions are ethical acts of world-*making* in the production of boundary *objects* just as surely as they are acts aimed at predicting, explaining, understanding, testing, or theorizing.

Conclusion

In revisiting our 1989 paper on boundary objects and methods standardization, on the sad and joyous occasion commemorating Leigh's death, I think the framing of scientific work in terms of tracking work is a pretty good fit. Boundary objects may serve as heuristic tools to guide empirical tracking work in ways that reveal tracking commitments *of STS researchers* as they negotiate the multiple disciplines needed to make sense of heterogeneous work in science, as well as to track commitments of their subjects. Or they may not. Tracking doesn't always or indeed usually doesn't work. That's the way it goes in science: most of the time it fails, and getting a "system" to "work" is what most scientific activity is about. But even the fact that it usually doesn't work is no argument against the utility of the concept.

Tracking commitments are pragmatic commitments that have ontological, epistemic, and ethical import: ontological commitments to track in particular ways, epistemic commitments to represent processes in terms of tracking results in the traces generated from marks, and ethical commitments to be accountable to bodies marked and configured, captured and released or killed, through empirical engagements. The merger of epistemic, ontological, and ethical considerations in a picture of pragmatic commitments, trackable through boundary crossings, reflects well, I think, the goals of the original Star and Griesemer paper, and I hope these ideas would please Leigh.

Acknowledgments

I thank Geof and Stefan for their encouragement, Joe Dumit and Ayelet Shavit for comments, and all the participants at the Chicago Knowledge/Values Workshop June 3–4, 2011, Marin Headlands STS camp June 17–21, 2011, and especially the Celebration of Susan Leigh Star in San Francisco, September 9–10, 2011, for their community and humanity. And of course I thank Leigh for cherished shared experiences that have had a deep impact on my career.

Notes

1. That backfired later when Star and I mentioned Quine in our discussion of "translation" in the boundary objects paper and we were then read by Peter Galison (1997, 47n48) as dealing only with language rather than in line with the full meaning of "translation" that Latour, Callon and Law had discussed: "translation" failed to translate fully between what Leigh and I were trying to doing and what Peter was doing. As Leigh and I assumed, and as I suggest below, perhaps it would be more illuminating to treat the concept of a boundary object as a heuristic and to focus on cases of failure or breakdown of boundary objectification than only to try to fit diverse social practices to it.

2. This is in contrast to logical arguments in logic and analytic philosophy, where contradictions are like code vulnerabilities in the Internet age: you know that as soon as a flaw or bug is found, it *will* be exploited in some pernicious way. Science, unlike logic or analytic philosophy, tends to keep a conceptual system going so long as the flaws can be mapped and avoided in practice.

3. The thing about subversion, though, is that subversive acts rarely tell you how to act once the subversion is achieved. If that's the critique of "boundary object," then I cheerfully accept the criticism.

4. Joe Dumit urges (pers. comm.) that this claim bears an important relation to Whitehead's concept of nature. Whitehead (1920, 28–29) writes:

In the philosophy of science we seek the general notions which apply to nature, namely, to what we are aware of in perception. It is the philosophy of the thing perceived, and it should not be confused with the metaphysics of reality of which the scope embraces both perceiver and perceived. No perplexity concerning the object of knowledge can be solved by saying that there is a mind knowing it. In other words, the ground taken is this: sense-awareness is an awareness of something. What then is the general character of that something of which we are aware? We do not ask about the percipient or about the process, but about the perceived. I emphasize this point because discussions on the philosophy of science are usually extremely metaphysical—in my opinion, to the great detriment of the subject. The recourse to metaphysics is like throwing a match into the powder magazine. It blows up the whole arena.

References

Barad, K. 2007. *Meeting the Universe Half Way: Quantum Physics and the Entanglement of Matter and Meaning*. Durham: Duke University Press.

Bechky, B. 2003. "Object Lessons: Workplace Artifacts as Representations of Occupational Jurisdiction." *American Journal of Sociology* 109 (3): 720–752.

Becker, H. 1986. *Doing Things Together*. Evanston, IL: Northwestern University Press.

Caporael, L., J. Griesemer, and W. Wimsatt, eds. Forthcoming. *Developing Scaffolds in Evolution, Cognition, and Culture*. Cambridge, MA: MIT Press.

Cartwright, N., and H. Mendell. 1984. "What Makes Physics' Objects Abstract?" In *Science and Reality: Recent Work in the Philosophy of Science*, ed. J. Cushing, C. Delaney, and G. Gutting, 134–152. Notre Dame: University of Notre Dame Press.

Galison, P. 1997. *Image and Logic*. Chicago: University of Chicago Press.

Gerson, Elihu M. 2008. "Reach, Bracket, and the Limits of Rationalized Coordination: Some Challenges for CSCW." In *Resources, Co- Evolution, and Artifacts: Theory in CSCW*, ed. Mark S. Ackerman, Christine Halverson, Tomas Erickson, and Wendy A. Kellogg, 193–220. New York: Springer-Verlag.

Giere, R. 2006. *Scientific Perspectivism*. Chicago: University of Chicago Press.

Gieryn, T. 1999. *Cultural Boundaries of Science*. Chicago: University of Chicago Press.

Griesemer, J. R. 1984. "Presentations and the Status of Theories." In *PSA 1984, Vol. 1*, ed. P. D. Asquith and P. Kitcher, 102–114. East Lansing, MI: Philosophy of Science Association.

Griesemer, J. 2000a. "Reproduction and the Reduction of Genetics." In *The Concept of the Gene in Development and Evolution, Historical and Epistemological Perspectives*, ed. P. Beurton, R. Falk, and H-J. Rheinberger, 240–285. Cambridge: Cambridge University Press.

Griesemer, J. 2000b. "Development, Culture and the Units of Inheritance." *Philosophy of Science* 67 (Proceedings): S348–S368.

Griesemer, J. 2011. "Baeolophus inornatus affabilis." In *Eine Naturgeschichte für Das 21. Jahrhundert: Hommage A zu Ehren von Hans-Jörg Rheinberger*, ed. S. Azzouni, C. Brandt, B. Gausemeir, J. Kursell, H. Schmidgen, and B. Wittmann, 138–140. Berlin: Max-Planck-Institut für Wissenschaftsgeschichte.

Griesemer, J. 2014 "What Salamander Biologists Have Taught Us about Evo-Devo." In *Conceptual Change in Biology: Scientific and Philosophical Perspectives on Evolution and Development*, ed. A. Love, 271–301. Boston Studies in the Philosophy of Science, vol. 307. Dordrecht: Springer Verlag.

Ingold, T. 2000. *The Perception of the Environment: Essays in Livelihood, Dwelling and Skill*. New York: Routledge.

Lee, C. 2007. "Boundary Negotiating Artifacts: Unbinding the Routine of Boundary Objects and Embracing Chaos in Collaborative Work." *Computer Supported Cooperative Work* 16: 307–339.

O'Mahony, S., and B. Bechky. 2008. "Boundary Organizations: Enabling Collaboration among Unexpected Allies." *Administrative Science Quarterly* 53:422–459.

Rheinberger, H.-J.2010. *The Epistemology of the Concrete: Twentieth-Century Histories of Science*. Durham: Duke University Press.

Shavit, A., and J. Griesemer. 2009. "There and Back Again, or, The Problem of Locality in Biodiversity Surveys." *Philosophy of Science* 76 (3): 273–294.

Shavit, A., and J. Griesemer. 2011a. "Transforming Objects into Data: How Minute Technicalities of Recording 'Species Location' Entrench a Basic Challenge for Biodiversity." In *Science in the Context of Application: Methodological Change, Conceptual Transformation, Cultural Reorientation*, ed. Martin Carrier and Alfred Nordmann, 169–193. Boston Studies in the Philosophy of Science, vol. 274. Dordrecht: Springer-Verlag.

Shavit, A., and J. Griesemer. 2011b. "Mind the Gaps: Why Are Niche Construction Processes So Rarely Used?" In *Transformations of Lamarckism: From Subtle Fluids to Molecular Biology*, ed. Snait Gissis and Eva Jablonka, 307–317. Cambridge, MA: MIT Press.

Star, S. L. 2010. "This Is Not a Boundary Object: Reflections on the Origin of a Concept." *Science, Technology & Human Values* 35 (5): 601–617.

Star, S. L., and J. R. Griesemer. 1989. "Institutional Ecology, 'Translations,' and Boundary Objects: Amateurs and Professionals in Berkeley's Museum of Vertebrate Zoology, 1907–39." *Social Studies of Science* 19:387–420. [See also chapter 7, this volume.]

van Fraassen, B. 2008. *Scientific Representation: Paradoxes of Perspectivism*. New York: Oxford University Press.

Whitehead, A. 1920. "The Concept of Nature." Project Gutenberg edition of 2006. http://www.gutenberg.org/files/18835/18835-h/18835-h.htm (accessed February 17, 2013).

Wimsatt, W. C. 2007. *Re-engineering Philosophy for Limited Beings: Piecewise Approximations to Reality*. Cambridge, MA: Harvard University Press.

9   So Boundary as Not to Be an Object at All

Brian Cantwell Smith

Dear Leigh,

It's been far too long since we talked. I've missed it—not just over the past year and a half, when it has been impossible, but for several years before that. Somehow I let those late night conversations with you and Geof slide, especially during those five decanal years, when the urgent preempted. My loss; my forever loss. I can't tell you how much I regret having let them slip away.

You've been on my mind, recently. There are so many things I'd like to talk to you about—questions I want to ask. Especially about *boundary objects*: that fabulous idea of yours that went platinum—that infectious term, that falls so trippingly off so many people's tongues these days, or at least off their keyboards. So I thought I'd try to write down some of the questions here.

Now I'm no sociologist, as you know. How often we laughed about that—about how, after my *Objects* book came out,[1] people said that I didn't have an ounce of social awareness in my body—that (such a layered irony) I was "missing the sociality gene." I'm not even much of an STS-er, though I've tried to be a flying buttress—a supporter from the outside. Peripheral, for sure—whether legitimate or not I'll let others decide. No matter. It never got in the way of our musings. That's one of the things I loved.

I have three questions—or maybe four. None are new. But what I have been wondering, recently, is whether they aren't all related, whether they don't all boil down to the same thing. So let me give them a shot.

The first question may seem minor. Some may even, uncharitably, think of it as carping. But I don't think it is. It even came up in your original paper with Jim.[2]

You say—as so many people have recited—that boundary objects are "objects which are both *plastic enough* to adapt to local needs and constraints of the several parties employing them, yet *robust enough* to maintain a common identity across sites."[3]

On the face of it, that seems innocent enough. Certainly inspiring. My question, though, has to do with the *voice* it is uttered in—with the perspective from which it surveys/surveils the situation. At a minimum, the statement is one that sees—characterizes, describes—the two (or more) sites in which boundary objects exist, in which they play a role, in which they are unsettlingly stabilized. Per se, that doesn't seem particularly challenging. Academic work, intellectual work, "registers," as I would put it (more on this in a moment), other peoples' and communities' practices. Perhaps especially sociological work (though how would I know?).

It's the phrase "robust enough" that is tricky. I've just never been able to get it out of my mind. It's like one of those songs that invades you, and won't let go. Robust enough *for whom*? Who gets to say so? And why?

If one were a naïve (or even sophisticated but still vanilla) realist, one might think that the object's robustness would be a *fact about it*—a property it has, relevant to its use in different sites—a fact or property that trumps or transcends or at least territorializes its employ by the parties implicated in the boundary practices.

But you're not naïve—and even if you are a realist, your flavor, if memory serves, is not vanilla. Moreover, Donna is lurking in the audience. And even if she isn't lurking in the audience, she is lurking somewhere. This is no God's eye mind-fuck view of robustness[4]—not from *you*. On the contrary, the stabilization of the object *as* an object—the norms and practices and pragmatic projects that identify and coalesce and harden the object and box it on the ears to make sure it behaves like an object—those things, according to you and your type (okay: my type too), are exactly specific to the kinds of "parties" that you refer to in the description.

So I get the boundary-negotiating parties. No problem there. What I'm wondering about is what party *you* are hanging out at. Who else was invited? Oh—and can I come?

One answer—not a deep answer, and not your answer, I'll hazard—is to back up and say that all that is going on, in these boundary object situations, is that the several parties to the negotiations *use the same term, fill out the same form, categorize under the same label, pile things up in the same repository*. But that's kind of vapid—and anyway, it doesn't answer the question. Who gets to say that they *are* the same term, form, label, repository? Homophones aren't thereby boundary objects, after all. There has to be some warrant for the "sameness," for the claim of commonality or contested identity across the boundary.

Another answer is that the robustness arises *from the durability and persistence of the common use*. But that can't be right, of course. Among other things, it's backward. What you say—and what you surely mean, since it is so sensible—is that the object is robust enough *for this durability and persistence to arise*. It's a *reason* for the durability; not simply a *name* of it.

Here's a funny example. I've always found the technical notion of a *variable* to be a fascinating case of a boundary object. It has played a role in catalyzing cooperation and collaboration between computer science and logic—in bringing them together, and also, interestingly, in keeping them apart. My sense is that the notion operates at the right scale, and that it structures work processes and normative practices appropriately. But variables—their values, their bindings, even what they are—are remarkably unsettled, as between the fields. Historically unsettled too. Torvalds and Russell and Quine precisify variables in intriguingly different ways. The differences lie fallow, allowing the fields to pursue their own concerns, and maintain their identity, yet feel—and genuinely be—collaborative.

But robust? Sure, in a way—I can see how one can say that. But I don't think that my saying so is innocuous. So anyway, that's my first question: what is it to say that a boundary object is "robust"—and who gets to say so, and how?

My second question has to do with *objects*. Not, I might say, with the qualifier. Boundaries are fine; I live there, pretty much, as so many of us do. ("Hey, you're a loner? I'm a loner, too! Let's get together!") *Objects*, though—I've never been sure about them. I guess that's why I had to write that book. I attribute it in part to coming from the North—or at least from outside. If you live in the city, hang out in offices, measure drinks in shots, you are certainly surrounded by objects: mutable mobiles, artifacts of human manufacture, chopped up and categorized and sold. But in the desolation of the tundra—in the crush of whitewater, in rocky canyons and flooded plains—I have no idea what or where the objects are.

Now I will say that you mean *less* by the term "object" than just about anyone I've ever read. And that's a good thing. Here are some of your words: an object is "a set of work arrangements that are at once material and processual";[5] an object is "something people . . . act toward and with";[6] and so on. So I am not accusing you of saying anything specific!

Nevertheless, I have two worries about objects—or maybe not worries, but at least questions. Put it this way: understanding the world in terms of objects is a *very particular* way to understand it—hugely useful but also fraught, consequential, and violent. Or to use my registration language: to register the world in terms of objects is a particular *kind* of registration—not universal, not ubiquitous, and certainly not innocent.

Over the last decade or so I have taught courses on what philosophers call "nonconceptual content," which for present purposes we can think of as ways to understand the world *not* in terms of objects. Think of the drenched smell of a wet forest in the spring as the snow recedes and the moldy mat of fall leaves begins to breathe. Or of the speed of your second tennis serve, just "that much slower" than the first. Or of the capitalist orientation of shopping malls. Or of what "it" refers to, in the sentence "it is raining." These aren't great examples, because I've referred to them—and reference objectifies. Over single malt I could do better. But you know, and I know, that you know what I mean.

For now, my question is this. Could the following be true:

1. That you know perfectly well that we don't just take the world in terms of objects—that is, don't take the world to consist of nothing but objects;[7] but
2. That it is exactly *object registration in particular* that you think plays the role of dynamic, nonconsensual, intercommunal sharing you were getting at?

There is some reason to suppose that this might have been your view. In their imposition of identity on some kind of background flux—in their abstraction or ignorance
of a wealth of fine details—objects may have "just the right stuff," may play just the
right kind of role in processes of negotiation and renegotiation, for you to feel that *they*,
in specific, are of the right ontological sort to play the boundary role. It is not a crazy
thought. Here is something I myself once wrote, along these very lines:

> I sometimes think of objects, properties, and relations[8] as the long-distance trucks and interstate
> highway systems of intentional, normative life. They are undeniably essential to the overall inte-
> gration of life's practices—critical, given finite resources, for us to integrate the vast and open-
> ended terrain of experience into a single, cohesive, objective world. But the cost of packaging up
> objects for portability and long-distance travel is that they are thereby insulated from the inex-
> pressibly fine-grained richness of particular, indigenous life—insulated from the ineffable rich-
> ness of the very lives they sustain.[9]

I don't know whether you would agree with any of that. But I will say this: you did
write, and not so long ago at that, that objects are "embodied, voiced, printed, danced,
and named."[10] So maybe your view isn't entirely different.

On the other hand—and this is why I've been making such heavy weather of all
this—for many years I have felt that I have been interested in what I take to be paradig-
matic examples of your platinum insight, or anyway in situations that *would be* para-
digmatic examples, except where this kind of objectual registration is *exactly not going
on.*

A couple of quick examples. Think of the alliance between right-wing Christians and
Zionist Jews, as regards their allegiance to American foreign policy toward Israel. Or
another example I remember talking to you about years ago, about cooperation between
the American Friends Service Committee (AFSC), Jewish and Communist social action
projects, and such Anabaptist groups as the Mennonites. It turns out, as I discovered
over several years, that these communities diverge, in fascinating ways, in their
understanding of the selfless approaches to social justice around which they all coalesce
(or to put it more carefully, in ways that hark back to my first question: that these
groups diverge in their attitude toward that which, given my background, I would call
"a selfless approach to social justice"). Or take another example from my own experience: back in the 1980s, a bunch of us organized a big multidisciplinary center, with
participation from Stanford and two Silicon Valley research groups at Stanford Research
Institute (SRI) and XeroxPARC.[11] By acclaim, the Center was dedicated to what we all
called *theoretical research*. A few years in, however, it emerged that to the Stanford folks,
"theoretical" more or less meant *nonexperimental*, whereas to researchers from the
industrial research labs, "theoretical" meant *nonapplied*. The difference created a vortex
of all three of the vital characteristics you take to surround the notion of a boundary
object.[12]

What strikes me about these examples is that it would be bizarre, or anyway so it seems to me, to call that on which they overlap an *object* of any sort.

My third question, to up the ante a bit, has to do with what has always struck me most about your characterization: your seeming presupposition that the contestation over boundary objects—struggle, negotiation, collaboration, discourse—is conducted solely between and// among *parties*. That pesky sociality again! (Frankly, I don't collaborate much. In fact the only C I ever got, in twenty-plus years of schooling, was in kindergarten, under the rubric "Plays well with others." With respect to others, in fact, I am far more interested in communion than in collaboration or cooperation. That's why I loved those late-night conversations with you and Geof.)

Anyway, what I find missing is that with which I myself primarily struggle and negotiate: not other sociologically salient souls, but *the stuff itself*. That is: what I find missing in your description—to put it contentiously—is *the world*.

Now I need to tread cautiously. Above, I tried to be careful to avoid calling you a closet naïve realist. Now I need to be sure that I don't come across as, or fall into the trap of, being one, too. And so, if pressed on what I mean by "world," I can scarcely answer without describing it in terms of ontological constructs (whether I speak in terms of objects, properties, and relations of the classical sort, or chronicle it in non-conceptual terms, or say anything at all—or even break out in song) that are particular to the contingencies, interests, biases, predilections, and so on, of my own projects. So it is a little dicey to know what to say, without running afoul of my first question.

The problem is not new, of course. Among other things, it has great religious pedigree. How can one say something about God, without limiting . . . him/her/it/they/whatever? One route is apophatic theology: the *via negativa*—talk only about what God is *not*. By analogy, here are some things that the world *isn't*. It isn't a great stew of objects, properties, relations, etc. Those are ontological or ontic structures, and inexorably plural. It isn't one—or two, or countable. It isn't what I think it is (the first step toward humility). But a *universa negativa* is not going to be enough to get at my fourth question—or to get at what this whole letter is about. So I need to do better.

In the *Objects* book, to deal with this problematic, I talk about *registration*, as I've said a couple of times already. The way I frame it, as you remember, and as I have already indicated here, is to say that one *registers the world in terms of objects, or features*, or whatever. More specifically, I use the world *ontology* ("what there is") for "the registered world"—in other words, for the world *as registered*. But now here's the critical move. *That which one registers*—that which founds or grounds our projects and processes of object registration, that which we and they are in and of and about—well, that is the world, *simpliciter*. So instead of calling it "God," or "The One," or "everything," I am going to call it *"that which."*

Okay, with this in mind, let's go back to the third question. You paint the process of nonconsensual collaboration—what I am characterizing as cooperation across registrational difference—in roughly paired terms, with two (or more) symmetrical parties. My own sense, to betray my own background again, is that one cannot understand these processes except as a trinitarian—except by including not just those two parties, but also what I will call "tw" (which you can read as "*the world*" or "*that which*," depending on your preference). It is not just that the parties use or overlap on the "same" object (which is to say, not the same object—that's your whole point). Nor is it just that they collaborate, or cooperate, or communicate. It is that, in doing so, they are *mutually engaged in the same tw.*

Just a few more points, and I am done. All of them have to do with how we should orient ourselves to the "that which"—to the inexorable tw.

To start with, I presume it is obvious that the tw—the "that which we register"—underlies all three of the issues I've brought forward.

With respect to the first question, about your voice or perspective, what I understand you (and anyone else, for that matter) to be doing, in saying of a boundary object that it is "robust," is to be *commenting on the adequacy of the registration of the "that which"* with which the boundary parties—and, crucially, you—are engaged. If one restricts oneself to the *ontic*—in other words, restricts one's attention to objects and properties, or situations, or actions and activities and practices, or processes, or features, or anything else *as registered*—then one deprives oneself of the resources needed in order to talk about the failure or success, or merits and demerits, or even, more to the point here, to the differential characteristics, to say nothing of overlap, of those objectual or other registration processes. If the "same" object differs, for example—as it does in the case at hand—then one lacks the resources necessary to say that the implicated parties are differentially registering the same or overlapping tw in and with which they are mutually engaged.

Robustness, I believe, to put the same point another way, is a *relation between an object and the "that which."* Sans ability to refer to the world, robustness must remain forever opaque. That is what I was never able to let go of in your initial characterization, these many years. And that, too, is why I have brought forward this talk of registration. In order to give voice to your robustness intuition, it seems to me, you have to get beyond or underneath talk that restricts itself to the registered (yes, I admit it: those are my words, "beyond" and "underneath"), in order to *disclose the registration activity itself*—a process that necessarily implicates the ineffable tw.

As regards the second question, about objects: if, as I don't recommend, one uncritically assumes that the world *is constituted of objects*—no matter how contested, contingent, multiply interpreted (actually that is odd: do we want to say that the *object* is differentially interpreted, or rather that different objects are the result of differential interpretations or registrations of the world, differential interpretations of the *that which*?)—anyway, if one takes the world to be constituted of objects, then one is liable to think that *objects* are the grounds of difference, or occlusion, or contestation. But I don't think the objects are the grounds of difference. I think *the world thereby registered* is the ground of difference. The divergences in individuated objects reflect differences that come out of differentially registering the shared "that which."

Third—but this is such a big issue that it is going to take a whole additional letter—if one takes the world to consist of objects, I don't think one can articulate an *ethics* worth a damn. For it is not just that one must be responsible *to the objects*, in my view, but, and perhaps even more seriously, one must be responsible *for the objects that one takes there to be*. That is, one must take responsibility *for one's (ontological) registration*. And without the "that which," that is not a recognition to which one can give voice. Or forget voice: it is not even a recognition over which one can share a conspiratorial smile.

This all leads directly to my fourth question—which is the last, and may be unfair, but I can't resist. It has to do with *you*, this time, more than with boundary objects— or perhaps with your personal relevance to these issues, rather than with their relevance to an idea of yours. What might seem unfair to some people, though I know it won't seem unfair to you, is that I want to talk, not about your intellectual work, but about your *life*—specifically, about your Jewish-Buddhist-Wiccan-mystical explorations, the paths you traveled and the insights and practices from those dimensions of your life.

Now as you know, I have no use at all for most of what people think the word "religion" names. I'm deeply anti-sectarian, and the rise of the religious right scares the bejeezus out of me. But I wasn't kidding—as you knew perfectly well—when I talked about the apophatic theology and the *via negativa*, and about the impossibility of saying anything about God that isn't restricting. I'm not interested in God, or in death, or in lots of other things one is supposed to be interested in—or, for that matter, in "being unto God," or "being unto death." I am interested in *the world*, and in *being unto the world*. In a nutshell, it is *being unto the world* that I have been going on about throughout this whole letter.

And here's the thing: I think it's ultimately the world in which you've been interested, too—that it is being unto the world to which your life has stood witness. Or anyway that's what I have—and maybe it has—loved about you. And I know you found partial expression (the only kind we get) in Buddhist practices, and in the coven, and in the Santa Cruz mountains, and perhaps more than anything in Geof, and in your friends. And I don't for a moment think that you *ever thought* that that toward which you were oriented, with and in those practices and people, can be captured in any kind of notion of an *object*, no matter how robust.

So my fourth question is this. Don't you think we can—don't you think it is time for us to—make good on what matters to us about these not-r-word practices and understandings and insights and communities, and infect or inflect our theoretical language, our academic deliberations, our intellectual inquiries, with what they know, with what we've come to know through them?

I'm not saying I know how to do that. But I will say this: It is that which I have been trying to do, in my own small way, in trying to get outside the object, in bringing forward reference to the "that which," in driving a wedge between ontology and metaphysics, in talking about registration. I am not saying that you will want to go this route. I would love to know where you foresee troubles, how you see it failing, what you feel it misses (besides sociality!). But mostly I know that you know that I have a ton of respect for these other dimensions of your life, as well as for your academic work—and also, and this is what matters, that I know full well, and have huge respect for the fact, that those other dimensions of your life were never *different*, or *other*, or *divorced* from any syllable you ever uttered, from any letter that you ever put to page. We all loved that: to hear you utter a single sentence was to be introduced to the whole full-blooded wondrous Leigh. So I thought it might help to put these few remarks into perspective—so that you could know, at least, what I have been trying to do, in making them.

Enough. I hope a bit of this makes sense—and who knows, maybe someday we will have a chance to talk with you about these things forever.

Meantime—well, I don't wish you rest. But I do wish you peace.

San Francisco
September 9, 2011

Notes

1. *On the Origin of Objects* (Cambridge, MA: MIT Press, 1996).

2. Susan Leigh Star and James R. Griesemer, "Institutional Ecology, 'Translations' and Boundary Objects: Amateurs and Professionals in Berkeley's Museum of Vertebrate Zoology, 1907–39," *Social Studies of Science* 19 (1989): 387–420. [See also chapter 7, this volume.]

3. Ibid., 393; emphasis added.

4. Donna Haraway, "Situated Knowledges," *Feminist Studies* 14, no. 3 (Autumn 1988): 575–599.

5. Susan Leigh Star, "This is Not a Boundary Object: Reflections on the Origin of a Concept," *Science, Technology, & Human Values* 35, no. 5 (2010): 601–617 at 604.

6. Ibid., 603.

7. Plus properties and relations; see note 8.

8. Like most philosophers, I take properties and relations to be object paraphernalia—entities inexorably associated with, tying together, distinguishing, etc., *objects* (an idea with etymological support: that properties are *proper* to objects).

9. "The Nonconceptual World," *Indiscrete Affairs*, vol. 2 (Cambridge, MA: Harvard University Press, forthcoming).

10. Star, see note 5, 603.

11. The Xerox Palo Alto Research Center (PARC).

12. Star, see note 5.

10   The Concept of Boundary Objects and the Reshaping of Research in Management and Organization Studies

Dick Boland

Walker Percy (1975) had an epiphany sitting at his desk in Covington, Louisiana, on an ordinary summer day in the 1950s, when he came to understand Helen Keller's magical experience of language on another summer day in Tuscumbia, Alabama, in 1887. In describing her experience, Helen Keller (1903, 23) wrote that she and her teacher, Anne Sullivan, had gone to a well house, where she observed the following:

Someone was drawing water and my teacher placed my hand under the spout. As the cool stream gushed over one hand, she spelled into the other the word *water,* first slowly, then rapidly. I stood still, my whole attention fixed upon the motion of her fingers. Suddenly I felt a misty consciousness as of something forgotten—a thrill of returning thought; and somehow the mystery of language was revealed to me. I knew then that "w-a-t-e-r" meant the wonderful cool something that was flowing over my hand. That living word awakened my soul, gave it light, hope, joy, set it free! . . .

I left the well house eager to learn. Everything had a name, and each name gave birth to a new thought. As we returned to the house, every object I touched seemed to quiver with life. That was because I saw everything with the strange new sight that had come to me.

I have read Percy's (1975) deeply engaging collection of nonfiction essays entitled *The Message in the Bottle* many times, but it wasn't until I began thinking carefully about Leigh Star's concept of the "boundary object" and reread his opening chapter on "the Delta Factor" that I felt I was finally beginning to "get" what Percy's triadic theory of language meant for the study of information and the work of social science. Percy (1975, 30) spoke of the effort required on his part to comprehend the significance of those "short paragraphs": "For a long time the conviction had been growing upon me that [those] short paragraphs in Helen Keller's *The Story of My Life* veiled a mystery, a profound secret, and that, if one could fathom it, one could also understand a great deal of what it meant to be Homo loquens, Homo symbolificus, man the speaking animal, man the symbol monger." Contemplating Star's naming of "boundary object" led to a similar type of epiphany for me, revealing something deeply true and mysterious.

Walker Percy was against the behaviorist approach to studying humans as an organism-in-an-environment, responding to signals from the environment like any other animal. He was against it because a stimulus-response theory could not account for what we do all the time—namely, talk to each other. The scientist, he would say, can use a behaviorist theory of stimulus-response to talk about anything in the world except about a scientist talking.

His insight from Helen Keller's experience at the well house was that we name things for each other; that naming things is something that requires two people, the person who names something and the person who makes the name meaningful. The stimulus-response model of human beings cannot account for the naming of a thing or event as a symbol, or for the necessity of two "organisms" in order for symbols to operate.

Percy was an MD, an accomplished southern author, an existential philosopher, and a scholar who dreamed of developing a semiotic theory that would help overcome the saturation of the social sciences with impoverished behaviorist models of man as a language user. He began with an adaptation of the semiotic triangle representing the relation among symbol, object, and meaning—or signifier, referent, and sense to Helen Keller as in figure 10.1.

The referent is the thing we wish to denote. The signifier is a word or image we associate with denoting the referent, and the sense for Helen Keller is how a human being gives meaning to the referent through use of the signifier. The relation of signifier to referent is arbitrary, adding to the multiplicity of how meaning can be made with symbols. The lesson he drew from the semiotic triangle was that the dyadic stimulus-response relations used in behaviorist models of human organisms might seem able to account for relations A-B or B-C, but could not account for A-B, B-C, and C-D together. That was a triadic relation that only occurred in human beings. No other organism has the experience of naming and making meaning with names that was experienced so forcefully by Helen Keller. We are the only ones who name and interpret names in this way, which Percy called the Delta Factor.

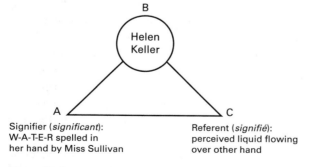

Signifier (*significant*):
W-A-T-E-R spelled in
her hand by Miss Sullivan

Referent (*signifié*):
perceived liquid flowing
over other hand

Figure 10.1
Walker Percy's triadic semiotic event

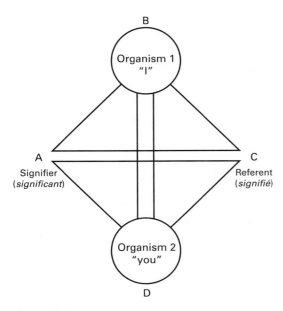

Figure 10.2
Percy's two-person semiotic model

Percy's unique contribution to semiotics was to assert the necessity of another self to whom the symbol is meaningful. His representation of this four-part, double semiotic triangle is shown in figure 10.2.

As Percy (1975) put it: "When a sign becomes a symbol, something profound happens." To paraphrase him, a symbolic meaning is different from a response to a stimulus. A stimulus triggers a response from an organism without anyone else being involved, or without anyone else having to exist. But for a symbol to be meaningful requires at least two people: "one who gives the name and one for whom the name is meaningful. The very essence of symbolization is an entry into a mutuality toward that which is symbolized" (256). He continued: "Here the terminology of object science falls short. One must use such words as mutuality or intersubjectivity, however unsatisfactory they may be. But, whatever we choose to call it, the fact remains that there has occurred a sudden cointending of the object under auspices of the symbol, a relation which of its very nature cannot be construed in causal language" (ibid., 257).

Through language and the process of naming for self and other, the signifier as a sign becomes a symbol. It moves from merely pointing to evoking interpretation. The movement from sign to symbol and back to sign is a dynamic of language at the heart of Star's naming of boundary object. Her naming asserts that there is a process of cointending in doing science work and that co-intending involves the use of symbols, in

which the names of things always carry multiple possible meanings and require interpretation. Like Percy, she saw that doing the work of science requires at least two persons, one person to name a thing or an action and another person for whom the
named thing or action is meaningful. I haven't seen reference to Wittgenstein in Percy's writing, but his ideas seem to parallel key ideas from Wittgenstein, such as the
impossibility of a "private language," and the unending multiplicity of naming and
making names meaningful in the language games of our forms of life.

As a tribute to Leigh, I propose that we take her naming of boundary object seriously. She did not name boundary object as an abstracted, universal concept-for-con-
sideration, but as a concrete, situated, concept-in-use. Boundary object does not
describe how language is built up from a combination of elements. Instead, boundary
object is a pointer to the magical way in which humans transcend the mechanisms of
language and experience its emergent property by successfully doing cooperative work
in the absence of a consensus on shared meanings. Boundary object denotes the miraculous, not the mechanical. In naming boundary object, Star was "calling it like it is,"
and so should we.

By naming boundary object, Star named a place in which actors with heterogeneous
knowledge can succeed in cooperating to do the work of science without having any
prior agreement on the nature of the objects, actions, measures, or goals that they were
working on. Before Star named boundary object for us, scientists collaborated across
disciplines by using the readily available, everyday names of the artifacts (things,
actions, practices) that they referred to in doing their cooperative work. But the use of
familiar, everyday names masks the diversity of meanings in play when actors from
multiple disciplines cooperate in working on a scientific project.

To draw from a prime example used in Star and Griesemer (1989), an artifact that
was being attended to by a group of scientists from different fields was referred to by its
everyday name of "map." Named as map, the artifact is still a symbol, recognized by all
as having multiple potential meanings, but that symbolic potency is masked. Named
as boundary object, the multiplicity of meanings and the evocative powers of the artifact as a symbol are brought to light. Leigh's shift from using the everyday names of the
objects that scientists focus attention on during their practices of inquiry, and toward
using the name boundary object, brings a new understanding of the practice of inquiry
not just in the cooperative work of science, but also in the cooperative work in organizations generally. The gift she gave us has a strong trajectory and a broad appeal, but is
still in its infancy with promise of further productive use across ever-broader segments
of the sciences, the professions, and the social world.

Star (1989) named boundary object to describe what she saw when she studied scientists working across disciplines. She saw their discourse and practices of cooperating
produce robust findings even though they did not have a consensus at the local level
of their work. They were able to "cooperate without having good models of each

other's work," "while employing different units of analysis, methods of aggregating data, and different abstractions of data," and "while having different goals, time horizons and audiences to satisfy" (Star 1989, 46). They were successfully doing science together without having consensus on the meanings of the basic elements drawn upon in their work.

Here is where things get sticky. Here is where we might begin relying on the idea that specialists share meanings and ways of working in a strong sense—that they have a form of consensus, an internal discipline-based consensus on all the things that we might expect to be unique and distinctly different across disciplines. And here is where I want to put words in Leigh's mouth—words she *almost* said in "This Is Not a Boundary Object." There, in response to the impertinent question, "Is there anything that isn't a boundary object?," I can hear her saying: We often assume that actors who belong to an established discipline use a set of basic terms that are defined in common as part of a shared meaning in their community. And sometimes they do—if they are doing what Kuhn would call "normal science" in their discipline. But "normal science" is an exceedingly rare phenomenon, and the vast majority of science work is in unique local settings, with nonstandard data about nonstandard materials, done under nonstandard constraints with nonstandard resources, even though every actor is a member in good standing of their home discipline. In these circumstances, the things and actions that they take into account are not an established part of their shared meaning. They are, in fact, closer to being boundary objects than infrastructures, even within a single disciplinary community.

I would like to use the following image in a thought experiment (see figure 10.3). The image is a representation and is thus composed of signs. Because the signs include animals and a human, they could be a type of icon, representing particular animals and a person that the artist knew. Or the image could rise above the level of a "portrait" and be taken as a symbol. Then, we would name it, and describe it as such by saying what it meant or could have meant.

• Granted, this image is complex, enigmatic and abstract. Nonetheless by naming it as a symbol, we first want to say what it is. We can then go on, step by step, adding layers of detail along the way and describing it in ways that seem to grow progressively more precise. For example—we could begin by saying it is a cave painting from Lascaux, created approximately 17,000 years ago, showing a bull that has been wounded by a spear. We could then add further details to that initial layer of description, and include that the bull's entrails are hanging out, that a man is laying on the ground beside the bull, and that there is a bird atop a stick to the lower left of them. We could continue adding layers of description, such as its colors, textures, line thickness, and decayed sections as well as details such as how the bull's head is turned, or the way his tail is crooked. But no matter how many descriptive terms we add, the meaning of it as a symbol escapes us and will continue to escape us until we move from describing it as a collection of signs to saying what the assembled signs as symbol mean. Here, following Star and Dewey, we hold that the image is

Figure 10.3
Boland's image for a thought experiment

meaningful to us if and to the extent that it helps us do work toward some end we want to accomplish.

• But what is the image symbolic of? We would have an interesting time speculating on how it could or should be interpreted. Each detail we used to name it as a representation is part of the raw material for our interpretation, and as we interpret it, we will identify more and more details to include, such as the shape of its entrails, the peculiar features of the human laying beside it, the type of bird atop the stick, the spiky hair on the bull's back, or perhaps even the expression on the bull's face.

• But how do we interpret it? What does it mean? I think the best answer is found in Gadamer's philosophical hermeneutics (1976). As he described it, we stand before the image as something alien to us—something by an unknown author, from an unknown culture, made with unknown intentions, used in unknown ways. At the same time, we recognize that we carry certain prejudice with us in "pre-judging" what it is. We have an initial sense of the whole that leads us to attend to some of the details and to use those details to read meaning into the image. Attending to certain details and interpreting their meaning as part of a larger whole creates a change in our initial prejudgment of the whole and its meaning. This changed sense of the whole leads us to attend to other details in new ways, which enables another, new sense of the whole and its meaning. The process of tacking back and forth from a sense of the whole to the details being attended to and the ways of attending to them, to a new sense of the whole and a new attention to details and ways of attending to them, and so on, is the dynamic structure of the hermeneutic theory of interpretation and is known as the hermeneutic circle.

• Leigh referred to this interpretive process as a tacking back and forth from the "shared" general form of the thing or object being considered to the uniquely contextualized and constrained local form of it in a particular instance of deliberation. As she describes it, the work of cooperating scientists is language games in forms of life. When scientists from different disciplines collaborate, actions and things are brought into focal attention to observe, analyze, and interpret. Interpreting an action or thing placed in focal attention instantiates it as a symbol with multiple potential interpretations and affirms that the action or thing is meaningful in many diverse ways.

A symbol such as a lion or a cross has multiple potential meanings and people can draw upon or reinvent them in remarkably diverse ways. Naming a symbol as a boundary object is different from naming it as a lion or a cross. When we name an action or a thing as a cross or a lion, the meanings that are made with it are those that emerge from considering that name as a symbol. Attending to lion or fire as symbols, we have a rich set of stereotypical meanings to draw on as we engage in an inquiry. When we name an action or thing as boundary object, the meanings do not come from a set of familiar uses, but are brought to it by the actors who are observing and interpreting it in their cooperative interaction.

A named symbol carries with it long traditions of multiple possible meanings, while a boundary object, which can be thought of as an unnamed symbol, or as meta-symbol, carries with it a blanker slate. In other words, a named action or event has meanings that are imported into the dialogue via its status as a symbol, whereas a boundary object has only the meanings that an individual in the dialogue attributes to it. Meaning making with a named symbol begins with the symbol and is modified by the individuals, but meaning making with a boundary object begins with the person who perceives it and must first name it, or rename it.

Because boundary objects are named as actions or things by those who perceive them, they represent each person's concerns and beliefs in a unique way. The lion as a named symbol can be interpreted as it is used in context, but a boundary object has to first be renamed by each inquirer, along with a context, to make it meaningful to them. As a result, studying people using named symbols in their inquiry dialogue is different from studying people who use boundary objects in their inquiry dialogue.

Dealing with actions and things as boundary objects instead of as named symbols results in a more open process of dialogue and a more open process of inquiry. This greater openness of inquiry and dialogue is a profound benefit of Star's boundary object concept. It is part of Leigh's legacy to us as social scientists, and will benefit us as long as we are able to continue using it in our inquiry into information, organization, and intelligence seeking social benefit.

My assertion is that in developing the concept of boundary object, Star gave us a tool for being more fully human in our social science research. When we use the name boundary object, it encourages us to expect that each individual who participates in an inquiry will exercise a distinctly original interpretive power. Each individual can be expected to bring their own unique naming of the boundary object as a symbol into the dialogue, one that is not shared fully with anyone else. Through their interaction and dialogue, the individuals will conduct inquiry and make decisions without having a common, shared understanding or a consensus. How do they accomplish that? This is the question that Leigh asked. The answer is a kind of mystery. Naming the unique symbolic use of boundary object does not solve the problem of how it is done.

Before the concept of boundary object was available to us, each symbol being attended to in an inquiry dialogue had a name. Each named symbol could be understood with multiple different meanings, but the range of those meanings is conditioned by our institutionalized understanding. So even though the meanings of a named symbol are theoretically multiple, the multiplicity was constrained by our expectations of what everyone knows—our socialized typifications. We seem to think that if we knew the context of use of a symbol, we would know the personal meaning for the individuals participating in the inquiry dialogue. When a named symbol of an action or thing is presented to us during an inquiry dialogue, we easily move from the presumption of a consensus on a shared understanding of the context to the presumption of a consensus-shared meaning that it has for all of us in that inquiry dialogue.

But if we are presented with a boundary object, instead of a named symbol, we are able to see the action or thing being considered as more open to each individual's own naming and less predictably a part of our supposed repository of consensus-shared understandings of its meaning. When presented with a boundary object during an inquiry dialogue, we are not led to believe we know what each person involved will name it or how they will make it meaningful, but with a named symbol, we believe we do.

As a final thought, I feel compelled to redraw a figure that Leigh Star (2010) used in her "This Is Not a Boundary Object" paper. There, she began to explore a longitudinal view of how boundary objects emerge from the residual categories generated when attempts at standardization inevitably prove inadequate for the variety and novelty encountered in doing science. This makes me think of how the dynamic of attempts at standardization, emergence of residual categories, and naming of new boundary objects is a never-ending cycling through the hermeneutic circle. Neither boundary objects nor infrastructures are ever stable, and are always becoming or dissolving.

I have taken the liberty of drawing this version of her transformational image of our struggle to grab onto the world with language (see figure 10.4). In the struggle, we are forever trying to close the heterogeneous, ill-structured phenomena encountered when doing work, into standardized objects, seeking to build infrastructure we can reliably work with. As residual categories start to leak out of what we thought had been stabilized and closed, we begin naming boundary objects that allow us an open space for working together without the comfort of reliably stable and singular language tools. As work proceeds with those boundary objects, we are confronted with a new multiplicity

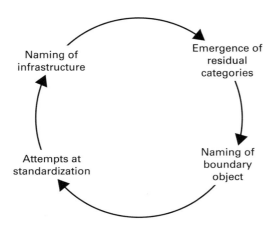

Figure 10.4
Boland's version of relationships between standards and residual categories

of concepts and understandings that threaten our ability to get work done. Then, we begin once more to seek closure through attempts at standardization. It is never so neat, though, and all four moments of these dialectical transformations are ongoing simultaneously.

I like the symmetry of this idea, and the way it takes us back to the distributed heterogeneous agents paper where Leigh explored the design of a distributed artificial intelligence system and proposed the concept of boundary object. I miss not hearing her objections.

References

Gadamer, Hans-Georg. 1976. *Philosophical Hermeneutics*. Berkeley: University of California Press.

Keller, Helen. 1903. *The Story of My Life*. New York: Doubleday.

Percy, Walker. 1975. *The Message in the Bottle: How Queer Man Is, How Queer Language Is, and What One Has to Do with the Other*. New York: Farrar, Straus and Giroux.

Star, Susan Leigh. 1989. "The Structure of Ill-Structured Solutions: Heterogeneous Problem-Solving, Boundary Objects and Distributed Artificial Intelligence." In *Distributed Artificial Intelligence 2*, ed. M. Huhns and Les Gasser, 37–54. Menlo Park: Morgan Kaufmann.

Star, Susan Leigh. 2010. "This Is Not a Boundary Object: Reflections on the Origin of a Concept." *Science, Technology & Human Values* 35:601–617.

Star, Susan Leigh, and James Griesemer. 1989. "Institutional Ecology, 'Translations' and Boundary Objects: Amateurs and Professionals in Berkeley's Museum of Vertebrate Zoology, 1907–39." *Social Studies of Science* 19:387–420. [See also chapter 7, this volume.]

11   Leigh Star and the Appearance of "The Structure of Ill-Structured Solutions"

Les Gasser

I met Susan Leigh Star (known as Leigh) in 1979 in San Francisco. We were both graduate students, and we met because of what people would now call our "social network"—a loosely knit collection of students, faculty, and researchers mainly circulating through Irvine; San Francisco; Cambridge, Massachusetts; and Paris. The activity of this network was in essence just "hanging out." The hangouts were kitchen tables, living rooms, cheap eating places, and an empty house claimed as a research institute. Travel was cheap—we could walk on to a $20 flight from Los Angeles up to our Bay Area friends for the weekend, so we did.

Rob Kling was first through the northern portal, and soon enough he dragged along his students Walt Scacchi and me. Leigh and the northern crowd accepted us as fellow travelers pretty quickly. In one form or another, we were all working across the boundaries of computer science and sociology, and thinking mostly about problems of distributed collective knowledge, distributed practices, and representation. Those terms seem pretty solid, accurate, and tame in 2013, but not so at the time. We struggled with notions, with words and with relevance. At one point we circulated a samizdat manifesto called "The Unnameable."[1]

An idea and a program bound us together: first, the vigorous belief that both social science and computer science had it wrong, and second a passionate program to change some foundational premises of those two spheres. We could talk with each other more precisely and more productively than we could with most of those in our own disciplines-of-origin. We wove together ideas, friendships, computer programs and applications, work products, papers, and travel. This weaving practice reflected the content of our ideas—that the computing was inherently social from the hardware on up through data, programming schemes, applications, "users," organizations, and institutions. And that social arrangements were inextricably bound to representational, technical, and mediated practices—all inseparable. But how to reflect on these things, how to represent them, how to defend and convince?

In the 1980s I got a job in a hard-core academic computer science department. Scacchi and I joked about how we couldn't use the "S" word ("social") around the department—the very idea of connecting hard technology to social settings and arrangements in foundational ways was an unthinkable thought, much less a pragmatic scientific enterprise on the tenure track.[2] Artificial intelligence (AI), though, was much in vogue—respectable and hard. It seemed clear that putting many AIs together was going to create a lot of the same issues that we'd been floating in the Unnameable discourse for years: dynamics of collective representation, social order, categories of knowledge, power relations of information, and so on. Even better, a small but influential community of AI and computer science (CS) researchers was just then embarking on studies of "distributed artificial intelligence" (DAI). I jumped on board.

In summer 1987 I was finishing up editing a collection on DAI; some AI researchers at SRI International, just thirty miles south of San Francisco, gave me a desk to work at for a couple of months. Leigh had a spare room where I could come and go, and the proximity got us talking once again more thoroughly about our work. Later that year Leigh moved to Computer Science at UC Irvine, and I organized the 1988 Workshop on Distributed AI in Arrowhead, California. My thoughts about the workshop were mostly about how to widen the community of participation and break open the intellectual landscape of DAI. For me, this meant building on the ideas and intellectual progress of the Unnameable group (though don't get the idea it was a "group" in any special or concrete sense, that's just a word I have now). Leigh being a newly certified computer science professor, I invited her, and she submitted what I believe is the initial boundary objects paper. The papers of the workshop were collected and edited by Mike Huhns and me, and came out in 1989 in a book on distributed AI.[3]

In the best tradition of Leigh's work the paper is reflexive. It crosses its own boundaries in form and in content—in form alone, notice that almost two-thirds of its references are CS papers. How many working sociologists then or now do that? In content, Leigh conceptualizes boundary objects from standpoints in classic AI (the frame problem; "blackboards" and multi-blackboard systems), biology (birds and bones), and work processes. As a work that spans and integrates, her hope was that it be interpretable across major intellectual divides that existed then, and that persist to this day.

Beyond context, I offer this historical sketch for another reason. Leigh's work exemplifies creativity and vision, but it also exemplifies an integration of the personal and the intellectual. She could not act without joining her personal experiences and sentiments to the content of her ideas. This connecting gives her work an unparalleled and enduring communicative power, as it radically couples the science and the humanity of research.

Notes

1. That manifesto is no longer accessible. For similar arguments, see Bendifalah and Scacchi 1989.

2. Fast-forward almost thirty years to today when the NSF Computer Science and Engineering Directorate has research programs with names like Social-Computational Systems.

3. In 2009 the book was cited as one of the "founding documents" of the field.

References

Bendifalah, Salah, and Walt Scacchi. 1989. "Work Structures and Shifts: An Empirical Analysis of Software Specification Teamwork." *ICSE '89: Proceedings of the 11th International Conference on Software Engineering*: 260–270.

12   The Structure of Ill-Structured Solutions: Boundary Objects and Heterogeneous Distributed Problem Solving

Susan Leigh Star

Editor's Note: Originally published in 1988, this chapter was the first in which Leigh Star articulated the concept of boundary objects, arguably one of her greatest contributions, which remained an intellectual interest throughout her career (see also Star and Griesemer 1989; chapter 7, this volume; and Star 2010). This chapter is also a lovely illustration of Star's ability to think simultaneously both on many levels and across disciplines. Here, she considers sociological theory (Durkheim) in the context of open systems development and artificial intelligence (computer science), which allows her to reveal some of the ways values become embedded in infrastructures. Star's keen ability to think across units of analysis and recognize both differences in discipline-based knowledge production and the mechanisms that allow people to work together despite such differences laid the groundwork for the boundary objects concept. It has since proven useful and important across numerous disciplines.

The chapter argues that the development of distributed artificial intelligence should be based on a social metaphor rather than a psychological one. The Turing test should be replaced by the "Durkheim test," that is, systems should be tested with respect to their ability to meet community goals. Understanding community goals means analyzing the problem of due process in open systems. Due process means incorporating different viewpoints for decision-making in a fair and flexible manner. It is the analog of the frame problem in artificial intelligence. From analyses of organizational problem solving in scientific communities, the paper derives the concept of boundary objects, and suggests that this concept would be an appropriate data structure for distributed artificial intelligence. Boundary objects are those objects that are plastic enough to be adaptable across multiple viewpoints, yet maintain continuity of identity. Four types of boundary objects are identified: depositories, ideal types, terrains with coincident boundaries, and forms.

Introduction: Larger than Life and Twice as Natural

Artificial intelligence has long relied on natural and social metaphors in a variety of ways, ranging from a source of inspiration for design to attempts at modeling natural information processing.[1] Why?

The reasons have fallen roughly into two categories: attempts at intelligence and attempts at intelligibility. *Attempts at intelligence* have had as their long-term goal the creation of a human or biological simulacrum, however that is defined—something that will pass the Turing test. Metaphors have long been a way of bridging the enormous gap between the current capabilities of machines and the state of the art in computer science, and the complexity and sophistication of natural information processing systems. *Attempts at intelligibility* have had as their long-term goal the production of something that will be usable and understandable by human intelligence. Metaphors used for these purposes point to the embeddedness of systems, user-friendliness, situated action, and so forth.

Yet in the metaphoric use of natural information processing, some important considerations become implicit. This especially includes understanding the relationship between the original source of metaphors and the final artifact. Some of the methodological debates in artificial intelligence reflect a deep uncertainty about the status of natural metaphors. Would a completely formal system allow them at all? If one is committed to a formal system, wherein does the fidelity to nature lie? Or do natural and artificial systems share formal properties to be discovered? (Hall and Kibler [1985] review these issues.) Many of these concerns are being brought to light by research in distributed artificial intelligence. This is first because the original Turing goal could not be met by distributed work, and second because the social, not the psychological or the biological, appears to many researchers in the field both as an important metaphor and as part of the system.

From the Turing Test to the Durkheim Test

The original Turing test (Turing [1950] 1963) involved a computer being able to mimic a woman well enough so that a human observer could not distinguish between a human male and a "female" computer. The test was predicated on a closed-universe model using "discrete state digital computers":

The prediction which we are considering [of all future states] is, however, rather nearer to practicability that that considered by Laplace. The system of the "universe as a whole" is such that quite small errors in the initial conditions can have an overwhelming effect at a later time. . . . Even when we consider the actual physical machines instead of the idealized machines, reasonably accurate knowledge of the state at one moment yields reasonably accurate knowledge any number of steps later. . . . Provided it could be carried out sufficiently quickly the digital computer could mimic the behavior of any discrete-state machine. The imitation game could then be played with the machine in question and the mimicking digital computer and the interrogator would be unable to distinguish them. Of course the digital computer must have an adequate storage capacity as well as working sufficiently fast. Moreover, it must be programmed afresh for each new machine which it is desired to mimic. . . . This special property of digital computers . . . is described by saying that they are *universal* machines.

Later in the article Turing reiterates that these computers can meet any new situation, as long as they have enough storage capacity.

Turing's model is more than a quaint, outdated vision of what computers can do. By going back to the original source, some fundamental values (and value conflicts) in the field of artificial intelligence are revealed, and therein some of the reasons for the ambivalence and confusion about metaphors. Turing's test world is closed, as already pointed out. But it also has the following properties that are being hotly contested by distributed artificial intelligence at this time:

- testing is done by individuals, not communities. There is no doubt in the tester's mind about what constitutes a valid result (in this case, stereotyped female behavior);
- computers, because they are programmable, are universal. Once a situation can be formally analyzed, it becomes amenable to understanding through this universal language;
- the only restriction on intelligence is lack of storage capacity (or processing power).

Critiques of these propositions have been coming from distributed artificial intelligence for some time. For example, Hewitt's open-systems model posits that all non-trivial real-world systems are open. They include properties of the real world, including distributed information processing, asynchronous updates, arms' length relationships between components, negotiation, and continual evolution (Hewitt and DeJong 1983; Hewitt 1986, 1988). These systems are open in several senses: there is no global temporal or spatial closure, and there is an absence of a central authority. Thus, rigid a priori protocols that will homogenize data and decision making both beg the question of openness and limit the problem-solving capacity of the system in the real world. Flexibility and evolution are the central concerns.

No amount of increased storage capacity can bypass the problems posed by open systems. The structure of the original Turing test, relying solely on a fixed repertoire of rules in order to mimic a range of behaviors, cannot accommodate this type of distributed system. The reasons are the same as for Hewitt's original critique: it could not analyze conflicting viewpoints within the system, and the fundamentally open nature of real world systems inevitably gives rise to such conflicts.

The conceptual struggle in distributed artificial intelligence has been with the tensions implied by the idea of a universal formal language and the inconsistency which arises from the distributed, open nature of the system itself. For example, Durfee and

Lesser (1987) propose the idea of partial global plans that dynamically model and incorporate the findings from distributed nodes of a system, maintaining the openness of the system but achieving coherence across nodes. Cammarata, McArthur, and Steeb (1983) state: "A main challenge to distributed problem solving is that the solutions which a distributed agent produces must not only be locally acceptable, achieving the assigned tasks, but also they must be interfaced correctly with the actions of other agents solving dependent tasks. The solutions must not only be reasonable with respect to the local task, they must be *globally coherent* and this global coherence must be achieved by *local computation alone.*"

In response to this challenge, the metaphors in this line of work have gone from single humans or human psychology[2] to organizations, interactions, negotiation, blackboards, networks, and communities. For example, Fox (1981) discusses the "technology transfer" possible between human organizations and artificial intelligence systems; Gasser (1987) calls for cooperation between distributed artificial intelligence and other fields of study concerned with coordinated action and distributed problem solving. I propose that this change in metaphoric base be recognized by replacing the vision of the Turing test with a test adequate to meet the challenges of distributed open systems: the Durkheim test.

Emile Durkheim (1858–1917) was a French sociologist who attempted to demonstrate the irreducible nature of what he called "social facts." For example, you could not understand differential suicide rates in different locations by simply saying that each case was pathological; something was happening at the "system level" that did not reduce to the terms of lower levels. Social facts, he said, are thus sui generis (or irreducible). He proposed the following law: "The determining case of a social fact should be sought among the social facts preceding it and not among the states of the individual consciousness," followed by the codicil: "The function of a social fact ought always to be sought in its relation to some social end" (Durkheim 1938).

The test of intelligence of a distributed open system is necessarily an ecological one. This means that it is sui generis at the social/system level, incorporating all parts of the system. Testing one node will not give reliable results; testing the whole open system is never possible (see e.g., Lesser and Corkill 1981). In the words of Davis and Smith (1983), "When control is decentralized, no one node has a global view of all activities in the system; each node has a local view that includes information about only a subset of the tasks." Thus, the very concept of a test must change in order to deal with such systems. Following Durkheim, we can say that it would be communal, irreducible, distributed, and dynamic. It is also important to note that it cannot be applied solely after a design is complete. In order to understand the acceptance and use of a machine in and by a community, that community must be actively present as it evolves.

So the Durkheim test would be a real-time design, acceptance, use, and modification of a system by a community. Its intelligence would be the direct measure of usefulness applied to the work of the community—in other words, its ability to change and adapt, and to encompass multiple points of view while increasing communication across viewpoints or parts of an organization. Such a test also changes the position of metaphors with respect to design and use considerations. In an open, evolving system, the boundaries between design and use, between technology and user, between laboratory and workplace, necessarily blur. Neither is the organization of work something that can be added after the design process (Kling and Scacchi 1982). Chang (1987) develops a model of this he calls participant systems. Thus, social metaphors may remain sources of inspiration, or guidelines for human–computer interface. But if we are stringently to apply the principles of open systems to design, and account for differing viewpoints and evaluation criteria at every step of the way, social systems become deeply implicated at all times.

The futility of the Turing test comes not from lack of storage capacity or processing power, but from a fundamental misunderstanding of the nature of computers and society as closed, centralized, and asocial. As that misunderstanding gets replaced by an open system, ecological, and political model of organizations, workplaces, and situations (which include both machines and human organization), the Turing test will be replaced by different forms of evaluation. (For a discussion of this from both sociological and computer science perspectives, see Bendifallah et al. 1988.)

Due Process, the Frame Problem, and Scientific Communities

As noted, the distributed and open nature of real systems gives rise to the existence of different viewpoints within the system. A viewpoint in this sense can occur at any level of organizational scale, from hardware to human organization. It can arise from, for example, asynchronous updates to a knowledge base, resulting in different ways of processing information at different nodes based on differences in a knowledge base. At higher levels, it can result from differences in the structure of tasks performed, different commitments, or different long- or short-term goals.

The simultaneous existence of multiple viewpoints and the need for solutions that are coherent across divergent viewpoints is a driving consideration in distributed artificial intelligence. Hewitt (1986) and Gerson (1987) have discussed aspects of this as the problem of *due process:* a legal phrase that refers to collecting evidence and following fair trial procedures. The due process problem in either a computer or human organization is this: in combining or collecting evidence from different viewpoints (or heterogeneous nodes), how do you decide that sufficient, reliable, and fair amounts of evidence have been collected? Who, or what, does the reconciling, according to what set of rules?

Davis (1980) notes that cooperation is necessary in order to resolve this class of problems, but that many researchers who came to distributed processing via attempts to synthesize networked machines see cooperation as a form of compromise "between potentially conflicting views and desires at the level of system design and configuration." The two motivations he suggests for cooperation are insolubility by a single node and compatibility (joining of forces).

The interdependence suggested by these motivations would seem to work against pluralism of viewpoints. How can two entities (or objects or nodes) with two different and irreconcilable epistemologies cooperate? If understanding is necessary for cooperation, as is widely stated in the distributed artificial intelligence literature, what is the nature of an understanding that can cooperate across viewpoints?

There is a fundamental similarity between these concerns about cooperation, that is, the due process problem, and the frame problem in artificial intelligence. The frame problem, as Hayes (1987) notes, "arises when the reasoner is thinking about a changing, dynamic world, one with actions and events in it. . . . [I]t only becomes an annoyance when one tries to describe a world of the sort that people, animals and robots inhabit." It is a problem, he states, not in computation, but in representation; it occurs in the presence of spatial or temporal change.

Spatial or temporal change is significant in this regard because of the epistemological incompatibilities that such change may bring about. As an actor moves through time and space, new information or new axiomatic requirements evolve (or devolve, depending on viewpoint), thus shifting the assumptive frame. Which axioms to retain or change, depending on which things can be taken for granted (or not) is at the heart of the frame problem (Pylyshyn 1987).

From the viewpoint of open systems, the problems of due process and the frame problem are figure-ground to one another. In the problem of due process, viewpoints evolve and change with new information and new situational constraints. The concept of due process means evaluating and synthesizing potentially incompatible viewpoints in the decision-making process: adducing evidence. The problem is one of drawing on different evidentiary bases. It is the differences in situation and viewpoint that make for epistemological incompatibility. In open systems, the lack of a sovereign arbiter means that questions of due process must be solved by negotiation, rules and procedures, case precedents, and so on (see Hewitt 1988).

The frame problem arose in the context of dealing with moving actors, absorbing information in a fashion that threatens the stability of their axiomatic structure. A robot, moving through novel open space, must find a robust way to deal with that novelty without having to add so many new axioms that it becomes bogged down in a "combinatorial implosion." But the problem is *not* really one of moving *through* neutral territory: in fact, it is an interactional problem. *Environment* really means a series of interactions with other objects: actors, events, and new kinds of ordered actions. In other words, the moving robot is forced to evaluate a series of interactions by picking and choosing from the heterogeneous, evolving, potentially incompatible viewpoints of other actors outside its original closed world.

The reconciliation between multiple viewpoints in the frame problem has thus been mischaracterized as a single actor problem. In fact, viewed temporally, and taking the actual content of the changing environments into account, the frame problem can be seen as a reconciliation between old and new experience in the same actor through a series of actions in open, distributed space.[3] The content of this experience is interactional because environments are a set of new actors and events. Solving the frame problem means adjudicating decisions about which evidence is important for which circumstances, and which can be taken for granted. The continuity of the robot's actions relies on a set of metarules that are structurally identical to the due process problem: What data does it take from which viewpoint? What is kept and what is discarded (thus the many discussions of relevance and inertia in the frame problem literature)? How can a decision be reached that incorporates both novelty and sufficient closure for action?

Human actors routinely solve both the frame problem and the due process problem. They do so in a variety of ways, as noted both in the social science literature and in the frame problem literature, and in variably democratic ways. In the remainder of this chapter I present one class of strategies employed by two scientific communities I have studied in some detail.

The studies began as an exploration of the scientific community metaphor in a long collaboration with Carl Hewitt. We analyzed issues that arose in the context of artificial intelligence research by looking at how human communities resolved them. These included issues such as due process (Gerson 1987), the resolution of conflict in a distributed community (Star 1989a), triangulation of evidence from domains with incompatible goals (Star 1986), resolution of local uncertainty into global certainty (Star 1985), local constraints on representing complex information (Star 1983), and the management of anomalous information (Star and Gerson 1987).

After some years, with the development of the open-systems model and the evolution of our own social science work, the "metaphor gap" seems to be closing.[4] The status of the social/community metaphor in the face of real-world systems embedded in organizations has shifted as the boundaries of "computer," "system," and "actors" are perceived as being larger and wider than Turing's closed-world model. Because advances in both artificial intelligence and social science call for the development of new *ecological units of analysis, methods, and concepts*, both the *content and role of metaphors have shifted.*

The concept of boundary objects to follow thus is simultaneously metaphor, model, and high-level requirement for a distributed artificial intelligence system. The more seriously one takes the ecological unit of analysis in such studies, the more central human problem-solving organization becomes to design—not simply at the traditional level of human–computer interface, but at the level of understanding the limits and possibilities of a form of artificial intelligence (Star 1989b).

The Scientific Community and Open Systems

Kornfeld and Hewitt (1981) proposed that the scientific community be taken as a good source of metaphors for open-systems work. Because real-world information systems are distributed and decentralized, they evolve continuously, embody different viewpoints, and have arms-length relationships between actors requiring negotiation. The internal consistency of an open system cannot be assured, due to its very character as open and evolving. The information in an open system is thus heterogeneous, that is, different locales have different knowledge sources, viewpoints, and means of accomplishing tasks based on local contingencies and constraints.

Scientific workplaces are open systems in Hewitt's sense of the term. New information is continually being added asynchronously to the situation. There is no central "broadcasting" station giving out information simultaneously to scientists. Rather, information is carried piecemeal from site to site (when it is carried at all), with lags of days, months, or even years.

Scientific work is distributed in this way. Thus, there is no guarantee that the same information reaches participants at any time, or that people are working in the same way toward common goals. People's definitions of their situations are fluid and differ sharply by location; the boundaries of a locality or workplace are simultaneously permeable and fluid (Latour 1988). Scientific theory building is deeply heterogeneous: different viewpoints are constantly being adduced and reconciled.

Yet within what may sound like near chaos, scientists manage to produce robust findings. They are able to create smooth-working procedures and descriptions of nature that hold up well enough in various situations. Their ability to do so was what originally fascinated Kornfeld and Hewitt. In the absence of a central authority or standardized protocol, how is robustness of findings (and decision making) achieved? The answer from the scientific community is complex and twofold: they create objects that are both plastic and coherent through a collective course of action.

Any scientific workplace can thus be described in two ways: by the set of actions that meets those local contingencies that constantly buffet investigators, or the set of actions that preserves continuity of information in spite of local contingencies (due process and the frame problem simultaneously). Understanding this requires a different appreciation of scientific theories than that traditionally put forward by philosophers. Scientific truth *as it is actually created* is not a point-by-point closed logical creation. Rather, in the words of ecologist Richard Levins, "our truth is the intersection of independent lies" (in Wimsatt 1980). Each actor, site, or node of a scientific community has a viewpoint, a partial truth consisting of beliefs, local practices, local constraints, and resources—none of which are fully verifiable across all sites. *The aggregation of those viewpoints is the source of the robustness of science.*

Heterogeneous Problem Solving and Boundary Objects

In the face of the heterogeneity produced by local constraints and divergent viewpoints, how do communities of scientists reconcile evidence from different sources? The problem is an old one in social science; indeed, one could say it reflects the core problem of sociology. One major concern of early sociologists, such as Robert Park and Georg Simmel, was to describe interaction between participants from groups (or worlds) with very different "definitions of the situation." This concern gave rise to a series of case studies of ethnicities, work groups, and subcultures now grouped loosely under the rubric "Chicago school sociology." Everett Hughes, a leader of this group, argued for an ecological approach to understanding the participation of heterogeneous groups within a workplace, neighborhood, or region. By this he meant that the different perspectives, or viewpoints, of the participants need to be understood in a sui generis fashion, not simply as a compilation of individual instances, and as situated action.

Some findings from our studies of scientists of potential interest to distributed artificial intelligence are that scientists

1. cooperate without having good models of each other's work;
2. successfully work together while employing different units of analysis, methods of aggregating data, and different abstractions of data;
3. cooperate while having different goals, time horizons, and audiences to satisfy.

They do so by creating objects that serve much the same function as a blackboard in a distributed artificial intelligence system. I call these *boundary objects*, and they are a major method of solving heterogeneous problems. Boundary objects are objects that are both plastic enough to adapt to local needs and constraints of the several parties employing them, yet robust enough to maintain a common identity across sites. They are weakly structured in common use, and become strongly structured in individual-site use.

Like the blackboard, a boundary object "sits in the middle" of a group of actors with divergent viewpoints. Crucially, however, *there are different types of boundary objects depending on the characteristics of the heterogeneous information being joined to create them.* The combination of different time horizons produces one kind of boundary object; joining concrete and abstract representations of the same data produces another. Thus, this chapter presents not just one blackboard, but a system of blackboards structured according to the dynamic, open-systems requirements of a community (including both machines and humans).

Types of Boundary Objects

In studying scientists, I identified heterogeneous subgroups within the scientific work-place. The analysis of boundary objects presented here draws on two case studies that incorporated radically different viewpoints in the conduct of work. First, I conducted a study of a community of neurophysiologists at the end of the nineteenth century in England. This group included both clinical and basic researchers, as well as hospital administrators, attendants, experimental animals, journalists, and patients (Star 1989a). Second, my colleagues and I conducted a study of a zoological museum from 1900 to 1940 at Berkeley (Star and Griesmer 1989; Gerson 1987). This group included professional biologists, amateur collectors, university administrators, animals, and local trappers, farmers, and conservationists.

What is interesting about these studies from the point of view of distributed artificial intelligence is that the structure and attributes of the information brought in from the different participants were distributed and heterogeneous, yet were successfully reconciled. Space prohibits a detailed discussion of all the differences in viewpoint, but two salient ones are summarized below.

First, in comparing clinical and basic research evidence, the following differences obtain: clinical research operates with a much shorter time horizon (cure the patient, not find the theoretical generalization) than basic research; the case is the unit of analysis for clinicians (an instance-based form of explanation), whereas for basic researchers it is analytic generalizations about classes of events. In clinical research, attention is directed toward concrete events such as symptoms, treatments, and patient trajectories. Diagnosis draws on medical theory to validate concrete observations of this nature. In basic research, attention is directed toward analytic generalizations such as refinements to others' theories, statements about the applicability of an experiment to a larger body of knowledge. Work proceeds from the experimental situation and is directed outward toward a body of knowledge. Finally, for the clinician, interruptions to work come in the form of complications, which are side effects to be dealt with locally and discarded from the evidentiary body (they never make their way into publication of the cases). Interruptions to work for the basic researcher come in the form of anomalies that must be accounted for in the body of evidence, either by controlling them or introducing them into the findings.

Second, in the world of the natural history museum, one primary source of comparison is between amateur and professional biologists. There are some similar differences as between clinicians and basic researchers. For the amateur collector of specimens, the specimen itself is the unit of analysis—a dead bird or a bone found in a specific location. Collecting, like clinical work, is the art of dealing on an instance-by-instance basis with examples and local contingencies. For the professional biologist, on the other hand, the specimens collected by amateurs form a part of an abstract generalization about ecology, evolution, or the distribution of species. The particular bug or beetle is not as important as what it represents. Furthermore, the work organization is highly distributed, ranging from the museum in Berkeley to various collecting expeditions throughout the state of California.

In analyzing these types of heterogeneity, I found four types of boundary objects created by the participants. The following is not an exhaustive list by any means. These are only analytic distinctions, in the sense that we are really dealing here with *systems* of boundary objects that are themselves heterogeneous.

Repositories

These are ordered piles of objects that are indexed in a standardized fashion. Repositories are built to deal with problems of heterogeneity caused by differences in unit of analysis. An example of a repository is a library or museum. They have the advantage of modularity (see figure 12.1).

Ideal Type or Platonic Object

This is an object such as a map or atlas that in fact does not accurately describe the details of any one locality. It is abstracted from all domains, and may be fairly vague. However, it is adaptable to a local site precisely because it is fairly vague; it serves as a means of communicating and cooperating symbolically—a sufficient road map for all parties. Examples of platonic objects are the early atlases of the brain, which in fact described no brain, which incorporated both clinical and basic data, and which served as a means of communicating across both worlds. Platonic objects arise with differences in degree of abstraction such as those that obtain in the clinical/basic distinction.

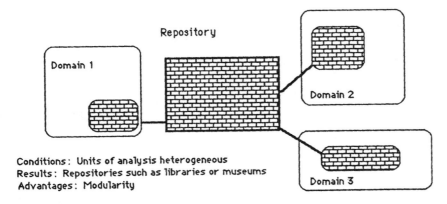

Conditions: Units of analysis heterogeneous
Results: Repositories such as libraries or museums
Advantages: Modularity

Figure 12.1
Boundary object: repositories

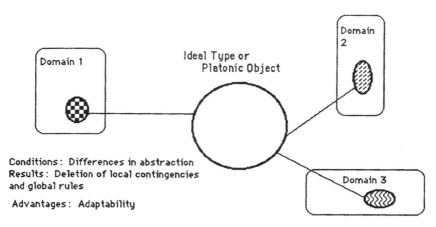

Figure 12.2
Boundary object: platonic object

They result in the deletion of local contingencies from the common object, and have the advantage of adaptability (see figure 12.2).

Terrain with Coincident Boundaries
These are common objects that have the same boundaries but different internal contents.[5] They arise in the presence of different means of aggregating data and when work is distributed over a large-scale geographic area. The result of such an object is that work in each site can be conducted autonomously, but cooperating parties can work on the same area with the same referent. The advantage is the resolution of different goals. An example of coincident boundaries is the creation of the state of California itself as a boundary object for workers at the museum. The maps of California created by the amateur collectors and the conservationists resembled traditional roadmaps familiar to us all, and emphasized campsites, trails, and places to collect. The maps created by the professional biologists, however, shared the same outline of the state (with the same geopolitical boundaries), but were filled in with a highly abstract, ecologically based series of shaded areas representing "life zones," an ecological concept (see figure 12.3).

Forms and Labels
These are boundary objects devised as methods of common communication across dispersed work groups. Both in neurophysiology and in biology, work took place at highly distributed sites, conducted by a number of different people. When amateur collectors obtained an animal, they were provided with a standardized form to fill out. Similarly, in the hospital, night attendants were given forms on which to record data

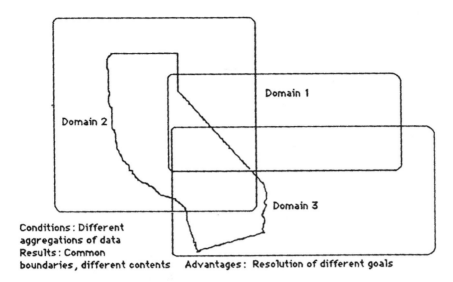

Conditions: Different
aggregations of data
Results: Common
boundaries, different contents   Advantages: Resolution of different goals

Figure 12.3
Boundary object: terrain with coincident boundaries

about patients' symptoms of epileptic fits in a standardized fashion; this information was later transmitted to a larger database compiled by the clinical researchers attempting to create theories of brain and nervous system function. The results of this type of boundary object are standardized indexes and what Latour would call "immutable mobiles" (objects that can be transported over long distance and convey unchanging information). The advantages of such objects are that local uncertainties (for instance, in the collecting of animals or in the observation of epileptic seizures) are deleted. Labels and forms may or may not come to be part of repositories (see figure 12.4).

Summary and Conclusions

What are the implications for distributed artificial intelligence of understanding the creation of boundary objects by scientists? First, boundary objects provide a powerful abstraction of the sort called for by Chandrasekaran 1981 to organize blackboards. They are, to use his terminology, neither committee nor hierarchy. They bypass the sort of problems of combinatorial implosion feared by Kornfeld and also bypass hierarchical delegation and representation. Unlike Turing's universal computer, the creation of boundary objects both respects local contingencies and allows for cross-site translation. Instead of a search for a logical Esperanto, already proved impossible in a distributed open-systems context, we should search for an analysis of such objects. Problem solving in the contexts described previously produces workable solutions that are not, in

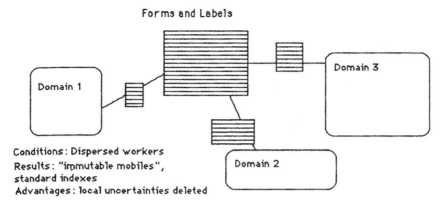

Figure 12.4
Boundary object: forms/labels

Simon's terms, well structured. Rather, they are ill structured: they are inconsistent, ambiguous, and often illogical. Yet, they are functional and serve to solve many tough problems in distributed artificial intelligence.

The problems of instantiating descriptions in distributed systems (Pattison, Corkill, and Lesser 1987) require a device similar to the creation of boundary objects for accounting for shifting constraints and organizational structures. Durfee, Lesser, and Corkill 1987 suggest a system that relies on cooperation and plan-based nodes that arrive at locally complete solutions for distributed problem solving. Again, the notion that systems of actors create common objects that inhabit different nodes in different fashions, and are thus locally complete but still common, should be useful here.

Future directions for research on these questions would include the following:

1. expanding the taxonomy of boundary objects and refining the conceptions of the types of information used in their construction;
2. examining the impact of combinations of boundary objects, and beginning to develop a notion of systems of such objects;
3. examining the problem of scaling up or applying an ecological, human/machine analysis to what is called multigrained systems in Gasser, Braganza, and Herman 1986.

The Durkheim test referred to in the beginning of this chapter is important in evaluating the construction and use of boundary objects. That is, the construction of such objects is a community phenomenon, requiring at least two sets of actors with different viewpoints. Analysis of the use of such an object at only one point in the system, or apart from its relationship to other nodes, will produce a systematic reductionist bias of the sort described by Wimsatt (1980). Heuristics used in such a fashion will reflect the neglect of the sui generis nature of the system. Furthermore, if the ecological unit of analysis recommended here and elsewhere in artificial intelligence is adopted, it should be noted that human designers, users, and modifiers of the computer systems involved will *make* boundary objects out of the information systems at every stage of the information-processing trajectory.

Acknowledgments

Conversations with Geof Bowker, Lee Erman, Les Gasser, James Griesemer, Carl Hewitt, Rob Kling, Steve Saunders, Randy Trigg, and Karen Wieckert were very helpful in formulating the ideas expressed here.

Notes

1. There are numerous descriptions of attempts to use such models; see, e.g., Ericsson and Simon 1979 for a review of sources of evidence on cognition.

2. I include network models of cognition here; I mean that the metaphors have moved away from individualist, black-boxed models of single actors.

3. Sociologists discuss this as the problem of continuity of identity (see Strauss 1969). The problem of inertia is structurally similar to the track-record heuristic described by Hewitt (1986) in his discussion of open systems.

4. Another factor may contribute to closing the gap. The metaphor, as a source of inspiration, models, or design specifications, works both ways: artificial intelligence is also a metaphor for sociological research (see Star 1989a for a discussion of this process)!

5. See Wimsatt 1980 for a fuller discussion of these issues.

References

Bendifallah, S., F. Blanchard, A. Cambrosio, J. Fujimura, L. Gasser, E. M. Gerson, A. Henderson, 1988. "The Unnamable: A White Paper on Socio-Computational 'Systems.'" Unpublished draft manuscript available from Les Gasser, Department of Computer Science, University of Southern California, Los Angeles, CA 90024.

Cammarata, S., D. McArthur, and R. Steeb. 1983. "Strategies of Cooperation in Distributed Problem Solving." In *IJCAI-83: Proceedings of the 8th International Joint Conference on Artificial Intelligence*, ed. Alan Bundy, 767–770. Los Altos, CA: Morgan Kaufmann.

Chandrasekaran, B. 1981. "Natural and Social System Metaphors for Distributed Problem Solving: Introduction to the Issue." *IEEE Transactions on Systems, Man, and Cybernetics* SMC-11 (1): 1–5.

Chang, E. 1987. "Participant Systems." In *Distributed Artificial Intelligence*, ed. M. N. Huhns, 311–339. Los Altos, CA: Morgan Kaufmann.

Davis, R. 1980. "Report on the Workshop on Distributed Artificial Intelligence." *SIGART Newsletter* 73:42–52.

Davis, R., and R. G. Smith. 1983. "Negotiation as a Metaphor for Distributed Problem Solving."*Artificial Intelligence* 20:63–109.

Durfee, E., and V. Lesser. 1987. "Using Partial Global Plans to Coordinate Distributed Problem Solvers." In *Proceedings IJCAI-87: Proceedings of the 10th International Joint Conference on Artifical Intelligence*, ed. John P. McDermott, 875–883. Los Altos, CA: Morgan Kaufmann.

Durfee, E., V. Lesser, and D. Corkill. 1987. "Cooperation through Communication in a Distributed Problem Solving Network." In *Distributed Artificial Intelligence*, ed. M. N. Huhns, 29–58. Los Altos, CA: Morgan Kaufmann.

Durkheim, Emile. 1938. *The Rules of Sociological Method*. New York: Free Press.

Ericsson, K. A., and H. A. Simon. 1979. "Sources of Evidence on Cognition: An Historical Overview." C.I.P. Working Paper No. 406. Carnegie Mellon Dept. of Psychology, October.

Fox, M. 1981. "An Organizational View of Distributed Systems." *IEEE Transactions on Systems, Man, and Cybernetics* SMC-11 (1): 70–80.

Gasser, L. 1987. "Distribution and Coordination of Tasks among Intelligent Agents." In *First Scandinavian Conference on Artificial Intelligence*, Tromsö, Norway, March.

Gasser, L., C. Braganza, and N. Herman. 1986. "MACE: A Flexible Testbed for Distributed AI Research." Technical Report CRI 87–01, Computer Research Institute, University of Southern California.

Gerson, E. M. 1987. "Audiences and Allies: The Transformation of American Zoology, 1880–1930." Paper presented at the Conference on the History, Philosophy, and Social Studies of Biology, Blacksburg, VA, June.

Hall, R. P., and D. F. Kibler. 1985. "Differing Methodological Perspectives in Artificial Intelligence Research." *AI Magazine* (Fall): 166–178.

Hayes, P. J. 1987. "What the Frame Problem Is and Isn't." In *The Robot's Dilemma: The Frame Problem in Artificial Intelligence*, ed. Z. W. Pylyshyn, 123–137. New York: Ablex.

Hewitt, C. 1986. "Offices Are Open Systems." *ACM Transactions on Office Information Systems* 4: 271–287.

Hewitt, C. 1988. "Organizational Knowledge Processing." Paper presented at the *8th AAAI Conference on Distributed Artificial Intelligence*, Lake Arrowhead, California, May.

Hewitt, C., and P. DeJong. 1983. "Analyzing the Roles of Descriptions and Actions in Open Systems." In *Proceedings of the National Conference on Artificial Intelligence (AAAI 1983)*, ed. Michael R. Genesereth, 162–167. Washington, DC: AAAI.

Kling, R., and W. Scacchi. 1982. "The Web of Computing: Computer Technology as Social Organization." *Advances in Computers* 21:1–90.

Kornfeld, W., and C. Hewitt. 1981. "The Scientific Community Metaphor." *IEEE Transactions on Systems, Man, and Cybernetics* SMC-11 (1): 24–33.

Latour, B. 1988. *Science in Action*. Cambridge, MA: Harvard University Press.

Lesser, V., and D. Corkill. 1981. "Functionally Accurate, Cooperative Distributed Systems." *IEEE Transactions on Systems, Man, and Cybernetics* SMC-11 (1): 81–96.

Pattison, H. E., D. Corkill, and V. Lesser. 1987. "Instantiating Descriptions of Organizational Structures." In *Distributed Artificial Intelligence*, ed. M. N. Huhns, 59–96. Los Altos, CA: Morgan Kaufmann.

Pylyshyn, Z. W., ed. 1987. *The Robot's Dilemma: The Frame Problem in Artificial Intelligence*. Norwood, NJ: Ablex.

Star, S. L. 1983. "Simplification in Scientific Work: An Example from Neuroscience Research." *Social Studies of Science* 13:205–228.

Star, S. L. 1985. "Scientific Work and Uncertainty." *Social Studies of Science* 15:391–427.

Star, S. L. 1986. "Triangulating Clinical and Basic Research: British Localizationists, 1870–1906." *History of Science* 24:29–48.

Star, S. L. 1989a. *Regions of the Mind: Brain Research and the Quest for Scientific Certainty*. Stanford, CA: Stanford University Press.

Star, S. L. 1989b. "Human Beings as Material for Artificial Intelligence: Or, What Computer Science Can't Do." Paper presented to the American Philosophical Association, Berkeley, CA, March.

Star, S. L., and E. Gerson. 1987. "The Management and Dynamics of Anomalies in Scientific Work." *Sociological Quarterly* 28:147–169.

Star, S. L., and J. R. Griesemer. 1989. "Institutional Ecology, 'Translations' and Coherence: Amateurs and Professionals in Berkeley's Museum of Vertebrate Zoology, 1907–39." *Social Studies of Science* 19 (3): 387–420. [See also chapter 7, this volume.]

Star, S. L. 2010. "This Is Not a Boundary Object: Reflections on the Origin of a Concept." *Science, Technology & Human Values* 35:601–617.

Strauss, A. 1969. *Mirrors and Masks: The Search for Identity*. San Francisco: Sociology Press.

Turing, A. [1950] 1963. "Computing Machinery and Intelligence," In *Computers and Thought*, ed. E. Feigenbaum and J. Feldman, 11–35. New York: McGraw-Hill. Originally published in *Mind* 59 (236): 433–460. doi:10.1093/mind/LIX.236.433.

Wimsatt, W. C. 1980. "Reductionist Research Strategies and Their Biases in the Units of Selection Controversy." In *Scientific Discoveries: Case Studies*, ed. T. Nickles, 213–259. Dordrecht: D. Reidel.

III   Marginalities and Suffering

13   Power, Technology, and the Phenomenology of Conventions: On Being Allergic to Onions

Susan Leigh Star

Editors' Note: Known familiarly as Leigh's "onions paper," this chapter first appeared in a book edited by John Law (1991) titled *A Sociology of Monsters: Essays on Power, Technology and Domination.* It explores precisely those themes along with the processes and consequences of being marginalized—of simply "not fitting," a life-long concern of Leigh's both personally and intellectually. Here she offers feminist, interactionist, and antiracist alternative frameworks for considering multiplicity—multiple simultaneous selves and commitments. Her concept of "torque" as the wrenching and twisting of marginalized lives developed in *Sorting Things Out: Classification and Its Consequences* (Bowker and Star 1999) further extends her powerful analysis initiated here.

On the one hand, recent studies in sociology of science and technology have been concerned to address the issue of heterogeneity: how different elements, and different perspectives are joined in the creation of sociotechnical networks. At the same time, there is concern to understand the nature of stabilization of large scale networks, by means that include processes of standardization. This chapter examines the model of heterogeneity put forth in the actor network model of Latour and Callon, particularly as a managerial or entrepreneurial model of actor networks. It explores alternative models of heterogeneity and multivocality, including splitting selves in the face of violence, and multiple membership/marginality, as for example experienced by women of color. The alternative explanations draw on feminist theory and symbolic interactionism. A theory of multiple membership is developed, which examines the interaction between standardizing technologies and human beings qua members of multiple social worlds, as well as qua "cyborgs"—humans-with-machines.

Introduction

Today I was reading about Marie Curie:
she must have known she suffered   from radiation sickness

her body bombarded for years   by the element
she had purified
It seems she denied to the end
the source of the cataracts on her eyes . . .
She died   a famous woman   denying
her wounds
denying
her wounds   came   from the same source as her power
(Rich 1978)

I guess what I am saying is that in the university and in science the boundary between insider and outsider for me is permeable. In most respects, I am not one or the other. Almost always I am both and can use both to develop material, intellectual, and political resources and construct insider enclaves in which I can live, love, work, and be as responsible as I know how to be. So, once more I am back to the dynamic between insider and outsider and the strengths that we can gain from their simultaneous coexistence and that surprises and interests me a lot (Hubbard, in Hubbard and Randall 1988, 127).

Is it not peculiar that the very thing being deconstructed—creation—does not in its intact form have a moral claim on us that is as high as the others' [war, torture] is low, that the action of creating is not, for example, held to be bound up with justice in the way those other events are bound up with injustice, that it (the mental, verbal, or material process of making the world) is not held to be centrally entailed in the elimination of pain as the unmaking of the world is held to be entailed in pain's infliction? (Scarry 1985, 22)

This is an essay about power.

Contrast the following three images of multiple selves or "split personalities":

1. *An executive of a major company presents different faces.* The executive is a middle-aged man, personable, educated, successful. To tour the manufacturing division of the plant, he dons a hard hat and walks the floor, speaking the lingo of the people who work there. In a board meeting he employs metaphors and statistics, projects a vision of the future of the company. On weekends he rolls up his sleeves and strips old furniture, plays lovingly with his children that he has not seen all week.

2. *A self splits under torture.* The adolescent girl sits on the therapist's couch, dressed as a prostitute would dress, acting coyly. Last week she wore the clothes of a matronly, rather somber secretary, and called herself by a different name. Her diagnosis is multiple personality disorder. Most cases of this once-thought-rare disorder arise from severe abuse, sexual or physical torture.

3. *A Chicana lesbian writes of her white father.* The words are painful, halting, since they are written for an audience finding its identities in being brown, or lesbian, or feminist. As in all political movements, it is easier to seek purity than

impurity. Cherríe Moraga (1983) writes of the betrayal that paradoxically leads to integration of the self, La Chingara, the Mexican Indian woman who sleeps with the white man, betrays her people, mothers her people. Which self is the "real" self here?

Bruno Latour's powerful aphorism, "science is politics by other means," coined in the context of his discussion of Pasteur's empire-building and fact-creating enterprises has been taken up by most of the research in the new sociology of science, in one form or another (Latour 1987). The central image of Pasteur is that of the executive with many faces: to farmers, he brings healing, to statisticians, a way of accounting for data, to public health workers, a theory of disease and pollution that joins them with medical research. He is stage manager, public relations person, behind-the-scenes planner. It is through a series of translations that Pasteur is able to link very heterogeneous interests into a mini-empire, thus, in Latour's words, "raising the world" (Latour 1983).

The multiplicity of selves that Pasteur is able to unite is an exercise of power of great importance. And from Latour's work, and that exploring related themes, we also understand that the enrollment does not just involve armies of people, but also of nature and technologies. Explanations and explorations, *intéressement*, extends to the nonhuman world of microbes, cows, and machines. A new frontier of sociological explanation is found through links between traditional interests and politics, and those usually ignored by such analyses, of nature and technique.

The multiplicity of Pasteur's identities or selves is critical to the kind of power of the network of which he is so central a part. Yet this is only one kind of multiplicity, and one kind of power, and one kind of network. Its power rests, as Latour, Callon, and others who have written about this sort of power in networks themselves attest, upon processes of delegation and discipline (Callon 1986). This may be delegation to machines, or to other allies—often humans from allied worlds who will join forces with the actor and attribute the fruits of their action back to him, her, or them. And the discipline means convincing or forcing those delegated to conform to patterns of action and representation. This has important political consequences, as Fujimura has written:

While Callon and Latour might be philosophically correct about the constructed nature of the science-society dichotomy (who represents nonhumans versus who represents humans), the consequences of that construction are important. . . . I want to examine the practices, activities, concerns and trajectories of *all the different participants*—including nonhumans—in scientific work. In contrast to Latour, I am still sociologically interested in understanding why and how some human perspectives win over others in the construction of technologies and truths, why and how some human actors will go along with the will of other actors, and why and how some human actors resist being enrolled. . . . I want to take sides, to take stands. (Fujimura 1991, 222)

The two other kinds of multiplicity I mention above—multiple personality and marginality—are the point of departure for feminist and interactionist analyses of power and technology. We become multiple for many reasons. These include the multiple personalities that arise as a response to extreme violence and torture and extend to the multiplicity of participating in many social worlds—the experience of being marginal. By experience and by affinity, some of us begin not with Pasteur, but with the monster, the outcast.[1] Our multiplicity has not been the multiple personality of the executive, but that of the abused child, the half-breed. We are the ones who have done the invisible work of creating a unity of action in the face of a multiplicity of selves, *as well as*, and at the same time, the *invisible work* of lending unity to the face of the torturer or of the executive. We have usually been the delegated *to*, the disciplin*ed*.[2] Our selves are thus in two senses monstrous selves, cyborgs, impure, first in the sense of uniting split selves and second in the sense of being that which goes unrepresented in encounters with technology. This experience is about multivocality or heterogeneity, but not only that. We are at once heterogeneous, split apart, multiple—and through living in multiple worlds *without* delegation, we have experience of a self unified only through action, work, and the patchwork of collective biography (see Fujimura 1991 and Strauss 1969 for discussions of this latter point).

We gain access to these selves in several ways:

1. By refusing those images of the executive in the network which screen out the work that is delegated. That is, in the case of Pasteur or any executive, much of the work is attributed back to the central figure, erasing the work of secretaries, wives, laboratory technicians, and all sorts of associates. When this invisible work (Star 1991; Shapin 1989; Daniels 1988) is recovered, a very different network is discovered as well;
2. By refusing to discard any of our selves in an ontological sense—refusing to "pass" or to become pure, and this means in turn;
3. By acknowledging the primacy of *multiple membership* in many worlds at once for each actor in a network. This *multiple marginality* is a source not only of monstrosity and impurity, but also of a power that at once resists violence and encompasses heterogeneity. This is at its most powerful a collective resistance, based on the premise that the personal is political.

All of these ways of gaining access imply listening, rather than talking on behalf of. This often means *refusing* translation—resting uncomfortably but content with that which is wild to us.

The Background in Science Studies

A number of recent conversations in the sociology of technology concern the nature of this relationship between people and machines, human and nonhuman (see e.g.,

Latour 1988; Callon 1986). Some focus on the divide between them: where should it be placed? There is a fierce battle, for instance, among several British and French sociologists of science on precisely this question. The British sociologists involved, on the one hand, argue that there is, and should be, a moral divide between people and machines, and attempts to subvert it are dehumanizing ones. They return us to a primitive realism of the sort we had before science studies. The French, on the other hand, focus against "great divides," and seek a heuristic flattening of the differences between people and machines in order to understand the way things work together. These often break conventional boundaries. A third strand, which I shall loosely call American feminist, argues that people and machines are coextensive, but in a densely stratified space, and that the voices of those suffering from abuses of technological power are among the most powerful analytically. A fourth strand, European and American phenomenology or ethnomethodology, argues that technology is an occasion to understand the way understanding itself—social order, meaning, routines—is constituted and reconstituted dynamically, and that reflexive analysis of technology is thus paramount. (Several of these essays appear in Pickering 1991.)

In this midst of these conversations, I have found myself asking, "what *is* technology?" or sometimes, "what *is* a human being?" As a result of the discussions I mentioned above, we walk in a very interesting landscape these days in science and technology studies. There are cyborgs, near-animate doors, bicycles, and computers, "conversations" with animals and objects, talk that sounds quite ecological and Green, if not downright pagan, about the continuum of life and knowledge; talk that opens doors on topics like subjectivity, reflexivity, multivocality, nonrational ways of knowing. In the policy field, things are scarcely less lively. On the one hand, critics of technology (Kling, Dreyfus) are labeled Luddites and scathingly attacked by those developing state-of-the-art technology. On the other, utopian advocates of new systems envision global peace through information technology, genetic maps, or cyberspace simulations. A third side invokes visions of techno-ecological disaster, accidents out of control, a world of increasingly alienated work where computers are servants of a management class. At the same time, people from all sides of the fray are blurring genres (fiction and science, for example), disciplines, or familiar boundaries.

Sociologists of science have helped[3] create this landscape through a heretical challenging of the biggest sacred cow of our times: the truthfulness of science as given from nature, the inevitability of scientific findings, their monolithic voices. Even in severely criticizing science for biases of gender, race, or militarism, science critics had not previously ventured far into this territory. Although often implicit, an early message from science criticism had been that science done right would not be biased. The message from sociology of science has consistently been: the "doing right" part *is* the contested territory. There are a few people asking the question about whether doing science at *all* can constitute doing right, or whether the entire enterprise is not necessarily flawed, but these are relatively rare: Restivo (1988) and Merchant (1980) are among them.

There is much disagreement in science studies about the nature of the politics by other means in science, both descriptively and prescriptively. We are recognizing that in talking of the central modern institutions of science and technology, we are talking of moral and political order (see Clarke 1990a). But do we have a fundamentally *new* analysis of that order (or those orders)? Are science and technology different? Or are they just new, interesting targets for social science?

Since few of us are interested in merely adding a variable to an extant analysis, most sociologists of science would hold that there *is* something unique about science and technology (but see Woolgar 1991 for a critique of this notion in the recent "turn to technology" in science studies). These include the ideas that

• science is the most naturalized of phenomena, helping form our deepest assumptions about the taken-for-granted;
• technology freezes inscriptions, knowledge, information, alliances, and actions inside black boxes, where they become invisible, transportable, and powerful in hitherto unknown ways as part of socio-technical networks;
• most previous social science has focused exclusively on humans, thus ignoring the powerful presence, effects, and heuristic value of technologies in problem solving and the moral order;
• science as an ideology legitimates many other activities in a meta-sense, thus becoming a complex, embedded authority for rationalization, sexism, racism, economic competitiveness, classification, and quantification;
• technology is a kind of social glue, a repository for memory, communication, inscription, actants, and thus has a special position in the net of actions constituting social order.

There is as well a persistent sense in science studies that technology in particular is terra incognita for social scientists, perhaps because of the myth of "two cultures" of those who work on machines vs. those who study or work with people.

Power in the Current Problems of Sociology of Technology

This sense of a new territory and a unique set of problems have prompted a number of historical reconstructions, where the participation of scientists, technologies, and various devices and instruments are included in the narrative. Many sociologists of science claim that taking these new actors into account gives a new, more complete analysis of action. "Politics by other means" is underscored by looking at how traditional power tactics, such as entrepreneurship or recruitment, are supported by new activities, such as building black boxes, or translating the terms of a problem from scientific language to some other language or set of concerns.

In the terms of Latour and Callon, this latter is the power of *intéressement*—the process of translating the images and concerns of one world into that of another, and then disciplining or maintaining that translation in order to stabilize a powerful network. The networks include people, the built environment, animals and plants, signs and symbols, inscriptions, and all manner of other things. They purposely eschew divides such as human/nonhuman and technology/society.

Another discourse about "politics by other means" concerns groups traditionally dispossessed or oppressed in some fashion: ethnic minorities, women of all colors, the old, the physically disabled, the poor. Here the discourse has traditionally been about access to technology, or the effects of technology (often differential) upon a particular group. Some examples include the sexist design and impact of reproductive technologies; the lack of access to advanced information technologies by the poor, further deepening class differences; the racist and sexist employment practices of computer chip manufacturers; and issues of deskilling and automation relating to labor.

Some writers in the science studies area have begun to bring these two concerns together, although others have begun to drive them apart in acrimonious battle (see e.g., Scott 1991). From one point of view, discussions of racism and sexism use reified concepts to manipulate tired old social theory to no good ends except guilt and boredom. From another, the political order described in actor network theory, or in descriptions of the creation of scientific facts, they describe an order that is warlike, competitive, and biased toward the point of view of the victors (or the management). Yet both agree that there are important joint issues in opening the black boxes of science and technology, in examining previously invisible work, and, especially, in attempting to represent more than one point of view within a network. We know how to discuss the process of translation from the point of view of the scientist, but much less from that of the laboratory technician, still less from that of the lab's janitor, much as we agree in principle that all points of view are important. There is a suspicion from one side that such omissions are not accidental, from the other, that they reflect the adequacy of the available material, but are not in principle analytic barriers.

The purpose of this chapter is to attempt to provide some tools hopefully useful for several of these discourses, and perhaps as well as show some ways in which technology reilluminates some of the oldest problems in social science. I can see two leverage points for doing this. These are 1) the problem of standards, and their relationship with invisible work; and 2) the problem of identity, and its relationship to marginality.

There are many challenges associated with adopting the stance that each perspective is important in a network analysis. One is simply to find the resources to do more work on traditionally underrepresented perspectives (see e.g., Shapin 1989; Star 1991; Clarke and Fujimura 1992). Another is using multiplicity as the point of departure for *all* analysis, instead of adding perspectives to an essentially monolithic model. Yet another is methodological: how to model (never mind translate or try to find a universal language for) the deep heterogeneities that occur in any juxtaposition, any network (Star and Griesemer 1989; Star 1988, 1989; Callon 1986, 1991). This methodological issue is a state-of-the-art one in many disciplines, including science studies, but also including organization studies, computer science (especially distributed artificial intelligence and federated databases), and literary theory.

This chapter speaks to the second point: how to make multiplicity primary for some of the concerns about power appearing now in science studies. The following example illustrates some common aspects of the problems of standards and invisible work.

On Being Allergic to Onions

I am allergic to onions that are raw or partially cooked. When I eat even a small amount, I suffer stomach pain and nausea that can last for several hours. In the grand scheme of things this is a very minor disability. However, precisely because it is so minor and yet so pervasive in my life, it is a good vehicle for understanding some of the small, distributed costs and overheads associated with the ways in which individuals, organizations, and standardized technologies meet.

The Case of McDonald's

Participation in McDonald's rituals involves temporary subordination of individual differences in a social and cultural collectivity. By eating at McDonald's, not only do we communicate that we are hungry, enjoy hamburgers, and have inexpensive tastes but also that we are willing to adhere to a value system and a series of behaviors dictated by an exterior entity. In a land of tremendous ethnic, social, economic, and religious diversity, we proclaim that we share something with millions of other Americans. (Kottak 1978, 82)

One afternoon several years ago I was very late to a meeting. Spying a McDonald's hamburger stand near the meeting, I dashed in and ordered a hamburger, remembering at the last minute to add, "with no onions." (I hadn't eaten at McDonald's since developing the onion allergy.) Forty-five minutes later I walked out with my meal, while all around me people were being served at lightning speed. Desperately late now and fuming, I didn't think about the situation, but merely felt annoyed. Some months later, I was again with a group, and we decided to stop to get some hamburgers at another McDonald's. I had forgotten about my former experience there. They all ordered their various combinations of things, and when it came to my turn, I repeated my usual, "hamburger with no onions." Again, half an hour later, my companions had finished their lunches, and mine was being delivered up by a very apologetic counter server. This time the situation became clear to me.

"Oh," I said to myself, "I get it. They simply can't deal with anything out of the ordinary." And indeed, that was the case. The next time I went to a fast-food restaurant I ordered along with everyone else, omitted the codicil about onions, took an extra plastic knife from the counter, and scraped off the offending onions. This greatly expedited the whole process.

The Curious Robustness of Disbelief on the Part of Waiters

I travel a lot. I also eat out at restaurants a lot. I can state with some certainty that one of the more robust cross-cultural, indeed cross-class, cross-national phenomena I have ever encountered is a curious reluctance by waiters to believe that I am allergic to onions. Unless I go to the extreme of stating firmly that "I don't want an onion on the plate, near the plate, in the plate, or even hovering *around* the food," I will get an onion where I have requested none (approximately four times out of five), at restaurants of all types, and all levels of quality, all over the world.

The Cost of Surveillance

In my case, the cost of surveillance about onions is borne entirely by me (or occasionally by an understanding dinner partner or host). Unlike people on salt-free, kosher, or vegetarian regimes, there exists no recognizable consumer demand for people allergic to onions. So I often spend half my meal picking little slivers out of the food or closely examining the plate—a state of affairs that would probably be embarrassing if I were not so used to doing it by now.

Anyone with an invisible, uncommon, or stigmatized disorder requiring special attention will hopefully recognize themselves in these anecdotes. If half the population were allergic to onions, no doubt some institutionalized processes would have developed to signal, make optional, or eliminate them from public eating-places. As things stand, of course, such measures would be silly. But the visible presence of coronary patients, elders, vegetarians, orthodox Jews, and so on, has led many restaurants, airlines, and institutional food suppliers to label, regulate, and serve food based on the needs of these important constituencies.

When an artifact or event moves from being presumed neutral to being a marked object—whether in the form of a gradual market shift or a stronger one such as barrier-free architecture for those in wheelchairs or deaf-signing for the evening news—the nature of human encounters with the technologies embedded in them may be changed. This is one form where politics arise in connection with technology and technological networks. These are politics that come to bear a label: "handicapped access," "reproductive technologies," "special education," even "participant-centered design."

But the signs that bear labels are deceptive. They make it seem as if the matter of technology were a matter of expanding the exhaustive search for "special needs" until they are all tailored or customized; the chimera of infinite flexibility, especially in knowledge-based technologies, is a powerful one.

There are two ways in which this illusion can be dangerous. The first is in the case of things like onions: there are always misfits between *standardized* or *conventional* technological systems and the needs of individuals (Star 1990 discusses this with respect to high-technology development). In the case of McDonald's, a highly standardized and franchised company, changes can be made only when market niches or consumer groups arise that are large enough to affect the vast economies of scale practiced by the firm. Thus, when dieters and Californians appear to command sufficient market share to make a difference, salad bars appear in McDonald's; non-onion entrees are far less likely. Even where there are no highly standardized production technologies (in most restaurants, for instance), a similar phenomenon may appear in the case of highly conventionalized activities—thus chefs and waiters automatically add onions to the plate, because most people eat them. It is easier to negotiate individually with nonstandardized producers, but not guaranteed. The lure of flexibility becomes dangerous when claims of universality are made about any phenomenon. McDonald's appears to be an ordinary, universal, ubiquitous restaurant chain. *Unless* you are: vegetarian, on a salt-free diet, keep kosher, eat organic foods, have diverticulosis (where the sesame seeds on the buns may be dangerous for your digestion), housebound, too poor to eat out at all—or allergic to onions.

The second illusion about perfect flexibility is a bit more abstract, and concerns not so much exclusion from a standardized form, but the ways in which membership in multiple social worlds can interact with standard forms. Let's say for the sake of argument that McDonald's develops a technology which includes vegetarian offerings, makes salt optional, has a kosher kitchen attached to every franchise, runs its own organic farms for supplies, includes a meals-on-wheels program and free lunches for the poor, and all sorts of modular choices about what condiments to add or subtract. But that morning I have joined the League to Protect Small Family-Owned Businesses, and, immune to their blandishments, walk down the street and bypass all their efforts. I have added a self to which they are blind, but which affects my interaction with them.

We have some choices in the sociology of technology about how to conceptualize these phenomena, which are obviously exemplary of many forms of technological change. First is a choice about what is to be explained. It is true that McDonald's franchises appear in an astonishing number of places; they are even more successful than Pasteur at politics by other means, if extension and visible presence are good measures. Is that the phenomenon to be explained—the enrollment and *intéressement* of eating patterns, franchise marketing, labor pool politics, standardization and its economics? It is also true that McDonald's screens out a number of clients in the act of standardizing its empire, as we have just discussed. Should *that* be the phenomenon we examine—the experience of being a McDonald's nonuser, a McDonald's resister, or even a McDonald's castaway? In the words of John Law, sociologist of technology and of McDonald's:

In particular, the McDonald's marketing operation surveys its customers in order to obtain their reaction to the adequacy of their experience in the restaurant on a number of criteria: convenience, value, quality, cleanliness and service . . . these criteria are in no way "natural" or inevitable. Rather they must be seen as cultural constructs. The idea that food should be fast, cheap, or convenient would be anathema, for instance, to certain sections of the French middle class. . . . These reasons for eating at McDonald's might equally well be reasons for *not* eating there in another culture. (1984, 184)

There are two kinds of phenomena going on here, and both miss another aspect of the transformation of the sort captured very well by semioticians in discussions of rhizomatic metaphors, or that which is outside of both the marked and unmarked categories, which resists analysis from inside *or* outside. In this case, this means living with the *fact* of McDonald's no matter where you fall on the scale of participation, since you live in a landscape with its presence, in a city altered by it, or out in the country, where you, at least, drive by it and see its red and the gold against the green of the trees, hear the radio advertising it, or have children who can hum its jingle.

The power of feminist analysis is to move from the experience of being a nonuser, an outcast, or a castaway, to the analysis of the fact of McDonald's (and by extension, many other technologies)—and implicitly to the fact that "it might have been otherwise"[4]—there is nothing necessary or inevitable about the presence of such franchises. We can bring a stranger's eye to such experiences. Similarly, the power of actor-network theory is to move from the experience of the building of the empire of McDonald's (and by extension, many other technologies) and from the enormous amount of enrollment, translation, and *intéressement* involved—to the fact that "it might have been otherwise"—there is nothing necessary or inevitable about any such science or technology, all constructions are historically contingent, no matter how stabilized.

One powerful way these two approaches may be joined is in linking the "nonuser" point of departure with the translation model, returning to the point of view of that which cannot be translated: the monstrous, the Other, the wild. Returning again to John Law's observation about the way McDonald's enrolls customers:

It creates classes of consumers, theorizes that they have certain interests, and builds upon or slightly diverts these interests in order to enlist members of that group for a few minutes each day or each week. It does this, group by group and interest by interest, in very particular ways. . . . Action is accordingly induced not by the abstract power of words and images in advertising, but rather in the way that these words and images are put into practice by the corporation, and then *interpreted* in the light of the (presumed) interests of the hearer. Advertising and enrolment work if the advertiser's theory of (practical) interests is workable. (Law 1984, 189)

He goes on to discuss the ways in which McDonald's shares sovereignty with other enterprises that seek to order lives, and of coexisting principles of order that in fact stratify human life.

But let our point of departure be not that which McDonald's stratifies, nor even the temporally brief but geographically extensive scope it enjoys and shares with other institutions, nor the market niches which it does not (yet?) occupy. Let it be the work of scraping off the onions, the self which has just joined the small business preservation group, the as-yet unlabeled. This is not the disenfranchised, which may at some point be "targeted"; not the residual category not covered in present marketing taxonomies. This is that which is permanently escaping, subverting, but nevertheless in relationship with the standardized. It is not nonconformity, but heterogeneity. In the words of Donna Haraway, this is the cyborg self: "The cyborg is resolutely committed to partiality, irony, intimacy, and perversity. It is oppositional, utopian, and completely without innocence. No longer structured by the polarity of public and private, the cyborg defines a technological polis based partly on a revolution of social relations in the oikos, the household. Nature and culture are reworked; the one can no longer be the resource for appropriation or incorporation by the other" (Haraway 1991, 151).

In a sense, a cyborg *is* the relationship between standardized technologies and local experience; that which is between the categories, yet in relationship to them.

Standards/Conventions and Their Relationship with Invisible Work: Heterogeneous "Externalities"

To speak for others is to first silence those in whose name we speak.
(Callon 1986, 216)

One problem in network theory is that of trying to understand how networks come to be stabilized over a long period of time. Michel Callon has tackled this problem in his essay, "Techno-Economic Networks and Irreversibility" (1991). There are some changes that occur in large networks which are irreversible, no matter what their ontological status. The initial choice of red as a color in traffic lights that means, "stop," for example, is now a widespread convention that would be functionally impossible to change, yet it was initially arbitrary. The level of diffuse investment, the links with other networks and symbol systems, and the sheer degree of interpenetration of red as "stop" renders it irreversible. We are surrounded by these networks: of telephones, computer links, road systems, subways, the post, all sorts of integrated bureaucratic record-keeping devices.

Irreversibility is clearly important for an analysis of power and of robustness in networks in science studies. A fact is born in a laboratory, becomes stripped of its contingency and the process of its production to appear in its facticity as Truth. Some Truths and technologies, joined in networks of translation, become enormously stable features of our landscape, shaping action and inhibiting certain kinds of change. Economically, those who invest with the winners in this stabilization process may themselves win big as standard setters. Later, others sign on to the standardized technologies in order to gain from the already-established structures, and benefit from these *network externalities*. Just as city dwellers benefit from the ongoing positive externalities of theaters, transportation systems, and a density of retail stores, network dwellers benefit from externalities of structure, density of communications populations, and already-established maintenance. Any growing network evidences this, such as the community of electronic mail users in academia. One can now sign on and (more or less) reliably communicate with friends, benefiting from a network externality that didn't exist just a few years ago.

Understanding how, and when, and whether one can benefit from network externalities is an essentially sociological art: how does the individual join with the aggregate, and to whose benefit? Once arrangements become standard in a community, creating alternative standards may be expensive or impossible, unless an alternative community develops for some reason. Sometimes the expense is possible and warranted, and may in fact lead to the development of another community, as in Becker's analysis of maverick artists (1982).

Becker raises the question of the connection among work, communities, and conventions in creating aesthetics and schools of thought. He begins with a series of simple, pragmatic questions: why are concerts two hours long? Why are paintings the size that they are (in general)? By examining the worlds that intersect to create a piece of art, *and valuing each one in his analysis*, he restores some of the normally hidden aspects of network externalities. There are contingencies for musicians' unions in prescribing hours of work, but also for those parking the cars of symphony goers, those cleaning the buildings after hours, and these contingencies, as much as considerations of more publicly acknowledged traditions, are equally important in forming aesthetic traditions.

So most composers write for concerts that are about two hours long, most playwrights write plays of similar length; most sculptures fit in museums and the backs of transport vans, and so forth. Those artists who are mavericks play with these conventions, opposing one or more. Occasionally, a naïve artist—with little knowledge of any of the conventions—will be picked up and accepted into the art world—and for that reason is especially sociologically interesting for illuminating the usually taken-for-granted.

The phenomenon Becker is pointing to in art is equally true in science and technology, if not more so, because there are so few instances of solitary or naive scientists (inventors are possibly a counterexample). Scientists and technologists move in communities of practice (Wenger 1990; Lave and Wenger 1991) or social worlds (Clarke 1990b) that have conventions of use about materials, goods, standards, measurements, and so forth. It is expensive to work within a world and practice outside this set of standards; for many disciplines (high-energy physics, advanced electronics research, nuclear medicine), nearly impossible.

Yet these sets of conventions are not always stable. At the beginning of a technological regime; when two or more worlds first come together; when a regime is crumbling—these are all periods of change and upheaval in worlds of science. As well, *the sets of conventions are never stable for nonmembers*. McDonald's may provide sameness and stability for many people—in John Law's words, it may order five minutes of their world each day—but for me and for others excluded from their world, it is distinctly not ordered. Rather, it is a source of chaos and trouble.

Network or Networks: That Is the Question

There is thus a critical difference between stabilization within a network or community of practice, and stabilization between networks, and again critical differences between those for whom networks are stable and those for whom they are not, where those are putatively the "same" network. Again we have a choice for a point of departure: does McDonald's represent a stable network, a source of chaos, or a third thing altogether?

Politics by Other Means or by the Same Old Means?

Bruno Latour explicates some of the features of actor-network theory, and the mix between humans and nonhumans involved in socio-technical systems, in his article on "The Sociology of a Door." He advocates an ecological analysis of people and objects, looking at the links between them, the shifts with respect to action, and the ways that duties, morality, and actions are shifted between humans and nonhumans: "The label 'inhuman' applied to techniques simply overlooks translation mechanisms and the many choices that exist for figuring or de-figuring, personifying or abstracting, embodying or disembodying actors" (Latour 1988, 303).

The analytic freedom accorded by this heuristic is considerable; in fact Latour and Callon's work has opened up a whole new way of analyzing technology. However, the problem remains with respect to humans and the question of power that such mixes may seem to sidestep traditional questions of distribution and access: "As a technologist, I could claim that, provided you put aside maintenance and the few sectors of population that are discriminated against, the groom does its job well, closing the door behind you constantly, firmly and slowly" (Latour 1988, 302).

There is no analytic reason to put aside maintenance and the few sectors of population that are discriminated against, in fact, every reason not to. As Latour himself notes in response to criticism of the actor-network theory for the political implications of its "leveling" of human/nonhuman differences, heuristic flattening does not mean the same thing as empirical ignoring of differences in access or experience. Rather, it is a way of breaking down reified boundaries that prevent us from seeing the ways in which humans and machines are intermingled.

However, one of the features of the intermingling that occurs may be that of exclusion (technology as barrier) or violence, as well as of extension and empowerment. I think it is both more analytically interesting and more politically just to begin with the question, *cui bono*? than to begin with a celebration of the fact of human/nonhuman mingling.

Network Externalities and Barriers to Entry: Physical and Cultural

One of the interesting analytic features of such networks is the question of the *distribution of the conventional*. How many people can get in and out of doors, and how many cannot? What is the phenomenology of encounters with conventions and standardized forms, as well as with new technologies? And here an opportunity for new ground in science studies arises: given that we are multiply marginal, given that we may interweave several selves with our technologies, both in design and use, where and what is the meeting place between "externalities" and "internalities"? I say this not to invoke another "great divide," but to close one. A stabilized network is only stable for some, and that is for those who are members of the community of practice who form/use/maintain it. And part of the public stability of a standardized network often involves the private suffering of those who are not standard—who must use the standard network, but who are also nonmembers of the community of practice.

One example of this is the standardized use of the pseudogeneric "he" and "him" in English to refer to all human beings, a practice now changing in many places due to feminist influence. Social psychologists found that women who heard this language form understood its meaning, but were unable to project a concrete example, and unable to place themselves within the example, whereas men could hear themselves in the example (Martyna 1978). Women thus both used and did not use the technology of this expression, and, with the advent of feminist analyses of language, were able to bring that experience to public scrutiny.

When standards change, it is easier to see the invisible work and the invisible memberships that have anchored them in place. But until then it may be difficult, at least from the managerial perspective. A recent article by Paul David (1989), an economist of standards, looks at a familiar problem for economists of information technology, called "the productivity paradox." For many firms, and even at the level of national economies, the introduction of (often very expensive) information technology has resulted in a decline in productivity, contrary to the perceived productivity benefits promised by the technology. David makes a comparison with the introduction of the general-purpose electric dynamo engine at the beginning of the century, which saw a similar decline in productivity. He refers to the work of several economists on the "transition regime hypothesis"—basically, that large-scale technological change means a change in economic regime, which carries its own—often invisible to standard analyses—costs.

The Transition Regime Hypothesis: Whose Regime? Whose Transition?

From the viewpoint of the analysis put forth here, the productivity paradox is no paradox at all. If much work, practice, and membership goes unrepresented in analyses of technology and socio-technical networks, then the invisible work that keeps many of them stabilized will go unaccounted for, but appear as a decline in productivity. Just as feminist theory has tried to valorize housework and domestic labor as intrinsic to large-scale economics, the invisible work of practice, balancing membership and the politics of identity is critical for the economics of networks.

Who carries the cost of distribution, and what is the nature of the personal in network theory? I believe that the answers to these questions begin with a sense of the multiplicity of human beings and of objects, and of a commitment to understanding all the work that keeps a network standardized for some. No networks are stabilized or standardized for everyone. Not even McDonald's.

Cyborgs and Multiple Marginalities: Power and the Zero Point

In torture, it is in part the obsessive display of agency that permits one person's body to be translated into another person's voice, that allows real human pain to be converted into a regime's fiction of power.
(Scarry 1985, 18)

It is through the use of standardized packages that scientists constrain work practices and define, describe, and contain representations of nature and reality. The same tool that constrains representations of nature can simultaneously be a flexible dynamic construction with different faces in other research and clinical/applied worlds. Standardized packages are used as a dynamic interface to translate interests between social worlds.
(Fujimura 1992, 205)

To translate is to displace. . . . But to translate is also to express in one's own language what others say and want, why they act in the way they do and how they associate with each other; it is to establish oneself as a spokesman. At the end of the process, if it is successful, only voices speaking in unison will be heard.
(Callon 1986, 223)

Several years ago I taught a graduate class in feminist theory at a large university in California. The first day of class eight women and one other person showed up. I couldn't tell whether the ninth person was male or female. S/he gave his/her name as "Jan," an ambiguous name. In the course of our class discussions, it turned out that Jan was considering transsexual surgery. S/he'd taken some hormone shots, and thus begun to grow breasts, and was dressing in a gender-neutral way, in plain slacks and short-sleeved shirt. S/he said that s/he wasn't sure if s/he wanted to go ahead with the surgery; that s/he was enjoying the experience of being ambiguous gender-wise. "It's like being in a very high tension zone, as if something's about to explode," she said one day. "People can't handle me this way—they want me to be one thing or another. But it's also really great, I'm learning so much about what it means to be neither one nor the other. When I pass as a woman, I begin to understand what feminism is all about. But this is different somehow."

I was deeply moved by Jan's description of the "high tension zone," though I didn't really know what to make of it at the time. A few weeks into the class we became friends, and she told me more about the process she was going through. She worked for one of the high technology firms in Silicon Valley, one that offered very good health insurance. But the health insurance company, Blue Cross, was unsure about paying for the extremely expensive process of transsexual surgery. Furthermore, the "gender identity clinic" where Jan was receiving psychotherapy and the hormone shots was demanding that s/he dress more like a conventionally feminine woman to "prove" that s/he was serious in her desire for the surgery. She told me that they required you to live for two years passing as a woman.

Around the Christmas holidays we fell out of touch. I was amazed to receive a phone call from Jan in February. "Well, congratulate me. I've done it," she exclaimed into the phone. "What?" I said, puzzled. "I've had the surgery, I'm at home right this minute," she said. I asked her how she was feeling, and also how it had happened. "Did (the company) decide to pay for it?" I questioned. "No," she replied. "Blue Cross decided to pay for the whole thing. And then the doctor just said, 'better do it now before they change their minds.' So I did!"

In the years that followed I saw Jan's (now Janice) name once in a while in local feminist club announcements; she became an active leader in the women in business groups in the area. I never saw her again after that February, but continued to be haunted by the juxtaposition of the delicate "high tension zone," the greed and hypocrisy of the insurance companies and physicians involved, and her own desperation.

Another friend has told me of a similar phenomenon within the gender clinics which require candidates for transsexual surgery to dress and act *as stereotyped females*, and deny them surgery if they do not: "They go from being unambiguous men, albeit unhappy men, to unambiguous women" (Stone 1989, 286). She goes on to recommend that the transsexual experience become an icon for the twin experiences of the high tension zone and the gender stereotype/violence:

Here on the gender borders at the close of the twentieth century . . . we find the epistemologies of white male medical practice, the rage of radical feminist theories and the chaos of lived gendered experience meeting on the battlefield of cultural inscription that is the transsexual body: a meaning machine for the production of ideal type. . . . Given this circumstance a counterdiscourse is critical, but it is difficult to generate a discourse if one is programmed to disappear. The highest purpose of the transsexual is to erase his/herself, to fade into the "normal" population as soon as possible. What is lost is the ability to authentically represent personal experience. (Stone 1989, 294)

Here is a socio-technical network, an exercise of power—and a certain kind of loss. What would it have taken to preserve the "high tension" of Jan's nonmembership, the impurity of being neither male nor female? This high tension zone is a kind of zero point between dichotomies (see Latour 1987; in *Irreductions*, in Pickering 1991) or between great divides: male/female, society/technology, either/or.

Elaine Scarry's extraordinary *The Body in Pain: The Making and Unmaking of the World* (1985) is a book about torture and war. Her argument is that during torture (and in similar ways during war) the world is created and uncreated. The torturer shrinks the world of the tortured, by taking the uncertainty of experienced pain and focusing it on material objects and on the verbal interchange between them. Old identities are erased, made immaterial.[5] We never really know about the pain someone else experiences, argues Scarry, and this uncertainty has certain political attributes that are explored during torture and war as the private becomes made public and monovocal. The visible signs of violence are transported to the public, and through a series of testaments, modifications, and translations become belief.

There are striking similarities between the making of the world Scarry describes and the making of the world by Pasteur described by Latour, or the successful process of translation Callon analyzes, although there seems to be no violence in these latter. A set of uncertainties is translated into certainties: old identities are discarded, and the focus of the world is narrowed into a set of facts.

The unity and closedness of the world of the torturer/tortured are seen as aberrant and outside the normal world by most people—far outside our normal realm. But Scarry argues that it is precisely this distancing that is one of the factors that makes torture possible, *because* it makes invisible to us what are in fact the pedestrian ingredients of making the world outside the extreme of torture. Simone de Beauvoir (1948) and Hannah Arendt ([1965] 1977) have made similar arguments about anesthetization to violence and the banality of evil. We always have elements of uncertainty about the personal world of another, especially about pain and suffering; we often leave one world for another, or narrow our experience without betrayal or permanent change—for example, in the dentist's chair, when we can think only of the immanent pain.

If we shift our gaze from the extremes: torture, or the enormous success of Pasteur, to something as simple and almost silly as an allergy to onions, it becomes clear that similarly quotidian events form part of a pattern. Stabilized networks seem to insist on annihilating our personal experience, and there is suffering. One source of the suffering is denial of the co-causality of multiple selves and standards, when claims are made that the standardized network is the only reality that there is. The uncertainties of our selves and our biographies fall to the monovocal exercise of power, of making the world. My small pains with onions are on a continuum with the much more serious and total suffering of someone in a wheelchair barred from activity, or those whose bodies in other ways are "nonstandard." And the work I do: of surveillance, of scraping off the onions, if not of organizing non-onion eaters, is all prior to giving voice to the experience of the encounters. How much more difficult for those encounters which carry heavier moral freight?

Networks, which encompass both standards and multiple selves, are difficult to see or understand except in terms of deviance or "other" as long as they are seen in terms of the executive mode of power relations. Then we will have doors that let in some people, and not others, and our analysis of the "not others" can't be very important, certainly not central. The torture elicited by technology, especially, because it is distributed over time and space, because it is often very small in scope (five minutes of each day), or because it is out of sight, is difficult to see as world making. Instead it is the executive functions, having enrolled others, which are said to raise the world.

The vision of the cyborg, who has membership in multiple worlds, is a different way of viewing the relationship between standards and multiple selves. And this involves weaving in a conception of multiple membership, of a cyborg vision of nature, along with the radical epistemological democracy between humans and nonhumans. In the words of Donna Haraway:

There's also the problem, of course, of having inherited a particular set of descriptive technologies as a Eurocentric and Euro-American person. How do I then act the bricoleur that we've all learned to be in various ways, without being a colonizer. . . . How do you keep foregrounded the ironic and iffy things you're doing and still do them seriously. Folks get mad because you can't be pinned down, folks get mad at me for not finally saying what the bottom line is on these things: they say, well do you or don't you believe that non-human actors are in some sense social agents? One reply that makes sense to me is, the subjects are cyborg, nature is coyote, and the geography is elsewhere. (in Penley and Ross 1990/1991, 10)

But there is a problem with this conception, and that has to do with the simultaneous poverty of our analyses of human/nonhuman, and of multiple membership for humans between human groups:

You can't work without a conception of splitting and deferring and substituting. But I'm suspicious of the fact that in our account of both race and sex, each has to proceed one at a time . . . there is no compelling account of race and sex *at the same time*. There is no account of any set of differences that work other than by twos simultaneously. Our images of splitting are too impoverished . . . we don't actually have the analytical technologies for making the connections. (in Penley and Ross 1990/1991, 15–16)

What would a richer theory of splitting involve, bringing together the following elements:

- multiple membership;
- maintaining the high tension zone while acknowledging the cost of maintaining it;
- the cost of membership in multiple arenas; and
- multivocality and translation?

Multiple Memberships, Multiple Marginalities

Every enrollment entails both a failure to enroll and a destruction of the world of the nonenrolled. Pasteur's success meant simultaneously failure for those working in similar areas, and a loss and world-destruction for those outside the germ theory altogether. We are only now beginning to recover the elements of that knowledge: immunology, herbal wisdom, acupuncture, the relationship between ecology and health. This had not to do with Pasteur vs. Pochet, but the ecological effects of Pasteurism and its enrollment.

One of Haraway's suggestions is that the destruction of the world of the nonenrolled is rarely total. While torture, or the total institution, is one end of a continuum, the responses to enrollment are far more varied along a much richer continuum. The basic responses, outside of signing on, have to do with a multiplicity of selves, partial signings-on, partial commitments. Ruth Linden's (1989) courageous and moving study of survivors of the Nazi Holocaust, interwoven with her own biography as an American Jew, testifies to this rich complexity. Adele Clarke's study (1990a, 1991) of the different communities of practice which joined together in creating modern reproductive science shows how multiple memberships, partial commitments, and meetings across concerns in fact constitute science.

Becker's analysis of commitments and "side bets" is apposite here. In his decoupling of commitment from consistency, there is a metaphor for decoupling translation and enrollment. How can we explain consistent human behavior? he asks. Ruling out mentalist explanations, functionalist explanations of social control, or purely behaviorist explanations, he instead offers that commitments are a complex of *side bets* woven by the individual, ways of involving his or her action in a stream of "valuable actions" taken up by others. Following Dewey's theory of action, he notes that we involve ourselves in many potential actions; these become meaningful in light of collective consequences, jointly negotiated (Becker 1960).

Similarly, our experiences of enrollment and our encounters with standards are complexly woven and indeterminate. We grow and negotiate new selves, some labeled and some not. Some are unproblematic in their multiplicity; some cause great anguish and the felt need for unification, especially those that claim sovereignty over the entire self.

One of the great lessons of feminism has been about the power of collective multiplicity. We began with the experience of being *simultaneously* outsiders and insiders (Hubbard and Randall 1988). In the end, it is the simultaneity that has emerged as the most powerful aspect of feminism, rather than the outsiderness. The civil liberties/ equal rights part of feminism would not have fundamentally extended political theory; but the double vision, and its combination of intimacy, ubiquity, and collectivity, has done so (Smith 1987). It's not so much that women have been left out, but that we were both in and out at the same time.

Sociology and anthropology have long traditions of studying the marginal person— the one who both belongs and does not belong, either by being a stranger (this is especially strong in the work of Simmel and Schutz) or by being simultaneously a member of more than one community. The person who is half black and half white, androgynous, of unknown parentage, the clairvoyant (who has access to another, unknown world)—all are either venerated or reviled in many cultures. The concept of the stranger, or strangeness to our own culture, as a window into understanding culture, is fundamental to many branches of anthropology and to ethnomethodology and its fruitful investigations into the taken-for-granted (see e.g., Garfinkel 1967 and its many references to Schutz).

Sociologist Everett Hughes extended Simmel's concern with the stranger, drawing on the work of his teacher Robert Park. He considered the anthropological strangeness of encounters between members of different ethnic groups who worked and lived together, and developed an analysis of some of the ways in which multiple membership plays itself out in the ecology of human relations. In "Dilemmas and Contradictions of Status," for example, he explores what happens when a person working in an organization belongs to two worlds simultaneously, and the prescriptions for action and membership are different (Hughes [1945] 1970, 141–150). He used the example of a female physician, or a Black chemist. Later sociologists used a related concept, "role strain," but that is one which fails to convey the sense of "high tension zone" or the complexity of the relationships involved in simultaneous multiple membership.

Another student of Park's, Everett Stonequist, reviewed various forms of marginality in his monograph, *The Marginal Man: A Study in Personality and Culture Conflict* ([1937] 1961). He discussed the stories of various racial and cultural hybrids: in Hawaii, in Brazil, in the United States and South Africa, as well as the phenomenon of cultural hybridism, as among immigrants and denationalized peoples, and the Jews. What is interesting about his work is that he places marginality at the center of *all* sociology: "It is the fact of cultural duality which is the determining influence in the life of the marginal man.

His is not a clash between inborn temperament and social expectation, between congenital personality tendency and the patterns of a given culture. His is not a problem of adjusting a single looking-glass self, but two or more such selves. And his adjustment pattern seldom secures complete cultural guidance and support, for his problem arises out of the shifting social order itself" (217).

But we are all implicated in this changing social order, Stonequist goes on to say—through technology, through shifts in the meaning of race and nationality, and through the diffusion of peoples across lands.

Because, in analyzing power and technology, we are involved in understanding precisely such shifts and precisely such shifting social orders, we could take a similar mandate. We know that the objects we are now including in the sociology of science and technology belong to many worlds at once. One person's scrap paper can be another's priceless formula; one person's career-building technological breakthrough can be another's means of destruction. Elsewhere I have analyzed the ways different social worlds construe the objects which inhabit more than one shared domain between scientists and others involved in the science-making enterprise, such as amateur collectors (Star and Griesemer 1989; Star 1988, 1989). People inhabit many different domains at once, as well, and the negotiation of identities, within and across groups, is an extraordinarily complex and delicate task. It's important not to presume either unity or single membership, either in the mingling of humans and nonhumans or among humans. Marginality is a powerful experience. And we are all marginal in some regard, as members of more than one community of practice (social world).

Conclusion: Metaphors and Heterogeneity

Because we are all members of more than one community of practice and thus of many networks, at the moment of action we draw together repertoires mixed from different worlds. Among other things, we create metaphors—bridges between those different worlds.

Power is about *whose* metaphor brings worlds together, and holds them there. It may be a power of the zero point or a power of discipline; of enrollment or affinity; it may be the collective power of not-splitting. Metaphors may heal or create, erase or violate, impose a voice or embody more than one voice. Figure 13.1 sketches some of the possible configurations of this sort of power.

This chapter is about a point of departure for the analysis of power. I do not recommend enfranchising or creating a market niche for those suffering from onion allergy; or a special needs assessment that would try to find infinitely flexible technologies for all such cases. Nor am I trying to say that conventions or standards are useless, or can be done without. But there is a question about where to begin and where to be based in our analyses of standards and technologies. If we begin with the zero point, like my

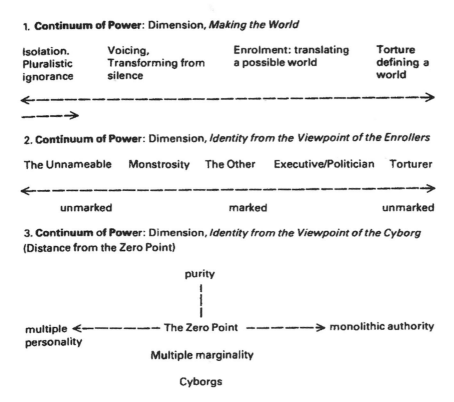

1. **Continuum of Power**: Dimension, *Making the World*

| Isolation. Pluralistic ignorance | Voicing, Transforming from silence | Enrolment: translating a possible world | Torture defining a world |

2. **Continuum of Power**: Dimension, *Identity from the Viewpoint of the Enrollers*

The Unnameable   Monstrosity   The Other   Executive/Politician   Torturer

unmarked   marked   unmarked

3. **Continuum of Power**: Dimension, *Identity from the Viewpoint of the Cyborg* (Distance from the Zero Point)

purity

multiple personality ← — — — — — The Zero Point — — — —→ monolithic authority

Multiple marginality

Cyborgs

Figure 13.1
Three power continuums

friend Jan, we enter a high tension zone which may illuminate the properties of the more conventionalized, standardized aspects of those networks which are stabilized for many. Those who have no doors, or who resist delegation—those in wheelchairs, as well as door makers and keepers, are good points of departure for our analysis, because they remind us that, indeed, it might have been otherwise.[6]

Acknowledgments

Geof Bowker and John Law made many helpful comments on this manuscript. A conversation with Bruno Latour illuminated the importance of the executive metaphor in understanding multiple personality. Conversations with Allan Regenstreif about the relationship between severe child abuse and multiple personality were extremely helpful. Their work and friendship, and that of Adele Clarke, Joan Fujimura, and Anselm Strauss are gratefully acknowledged.

Notes

1. Monsters are the embodiment of that which is exiled from the self. Some feminist writers have argued that monsters often represent the wildness that is exiled from women under patriarchal nomination, perhaps the lesbian self, and that apparently dichotomous pairs such as Beauty and the Beast, Godzilla and Fay Wray are actually intuitions of a healthy female self.

2. There are many courses for managers whose specialty is teaching executives how to delegate things to their secretaries and others below them in the formal hierarchy. Traditionally, of course, and still for the most part, this is male-to-female delegation.

3. Along with antiracist theorists, third world writers on de-centering, deconstructionists, literary theorists, feminist activists and theorists, and critical anthropologists, among others.

4. A methodological dictum of Everett Hughes ([1945] 1970).

5. This has striking resonances with the creation of the world in the "total institution" described by Goffman (1961) in his classic book *Asylums*. Sagerhaugh and Strauss (1979) as well describe a similar shrinkage of identity and of the world in their *Politics of Pain Management*.

6. This is one place where ethnomethodology and symbolic interactionism richly complement each other in exploring the taken-for-granted. See Becker 1967.

References

Arendt, Hannah. [1965] 1977. *Eichmann in Jerusalem: A Report on the Banality of Evil*. 2nd ed. Harmondsworth: Penguin.

Becker, Howard. 1960. "Notes on the Concept of Commitment." *American Journal of Sociology* 66:32–40.

Becker, Howard. 1967. "Whose Side Are We On?" *Social Problems* 14:239–247.

Becker, Howard. 1982. *Art Worlds*. Berkeley: University of California Press.

Callon, Michel. 1986. "Some Elements of a Sociology of Translation: Domestication of the Scallops and the Fishermen of St. Brieuc Bay." In *Power, Action and Belief*, ed. John Law, 196–233. London: Routledge & Kegan Paul.

Callon, Michel. 1991. "Techno-Economic Networks and Irreversibility." In *A Sociology of Monsters: Essays on Power, Technology and Domination*, ed. John Law, 132–164. London and New York: Routledge.

Clarke, Adele. 1990a. "A Social Worlds Research Adventure: The Case of Reproductive Science." In *Theories of Science in Society*, ed. Susan Cozzens and Thomas Gieryn, 15–42. Bloomington: Indiana University Press.

Clarke, Adele. 1990b. "Controversy and the Development of Reproductive Sciences." *Social Problems* 37:18–37.

Clarke, Adele. 1991. "Social Worlds/Arenas Theory as Organizational Theory." In *Social Organization and Social Processes: Essays in Honor of Anselm L. Strauss*, ed. David Maines, 119–158. Hawthorne, NY: Aldine de Gruyter.

Clarke, Adele, and Joan Fujimura, eds. 1992. *The Right Tools for the Job in Twentieth Century Life Sciences: Materials, Techniques, Instruments, Models and Work Organization*. Princeton, NJ: Princeton University Press.

Daniels, Arlene Kaplan. 1988. *Invisible Careers: Women Civic Leaders from the Volunteer World*. Chicago: University of Chicago Press.

David, Paul. 1989. "Computer and Dynamo: The Modern Productivity Paradox in a Not-Too-Distant Mirror." Center for Economic Policy Research Publication 172. Stanford University.

de Beauvoir, Simone. 1948. *The Ethics of Ambiguity*. New York: Philosophical Library.

Fujimura, Joan. 1991. "On Methods, Ontologies and Representation in the Sociology of Science: Where Do We Stand?" In *Social Organization and Social Processes: Essays in Honor of Anselm L. Strauss*, ed. David Maines, 207–248. Hawthorne, NY: Aldine de Gruyter.

Fujimura, Joan. 1992. "Crafting Science: Standardized Packages, Boundary Objects, and 'Translation.'" In *Science as Practice and Culture*, ed. Andrew Pickering, 168–214. Chicago: University of Chicago Press.

Garfinkel, Harold. 1967. *Studies in Ethnomethodology*. Englewood Cliffs, NJ: Prentice-Hall.

Goffman, Erving. 1961. *Asylums: Essays on the Social Situation of Mental Patients and Other Inmates*. New York: Anchor Books.

Haraway, Donna. 1991. *Simians, Cyborgs and Women: The Reinvention of Nature*. New York: Routledge.

Hubbard, Ruth, and Margaret Randall. 1988. *The Shape of Red: Insider/Outsider Reflections*. San Francisco: Cleis Press.

Hughes, Everett. [1945] 1970. *The Sociological Eye*. Chicago: Aldine.

Kottak, C.1978. "Rituals at McDonald's." *Natural History* 87: 75–82.

Latour, Bruno. 1983. "Give Me a Laboratory and I Will Raise the World." In *Science Observed: Perspectives on the Social Study of Science*, ed. K. Knorr-Cetina and M. Mulkay, 141–170. Beverly Hills, CA: Sage.

Latour, Bruno. 1987. *The Pasteurization of French Society, with Irreductions*. Cambridge, MA: Harvard University Press.

Latour, Bruno. 1988. "Mixing Humans and Non-Humans Together: The Sociology of a Door-Closer." *Social Problems* 35:298–310.

Lave, Jean, and Etienne Wenger. 1991. *Situated Learning: Legitimate Peripheral Participation*. Cambridge: Cambridge University Press.

Law, John. 1984. "How Much of Society Can the Sociologist Digest at One Sitting? The 'Macro' and the 'Micro' Revisited for the Case of Fast Food." *Studies in Symbolic Interaction* 5:171–196.

Law, John, ed. 1991. *A Sociology of Monsters: Essays on Power, Technology and Domination*. London and New York: Routledge.

Linden, Ruth. 1989. "Making Stories, Making Selves: The Holocaust, Identity and Memory." PhD dissertation, Department of Sociology, Brandeis University.

Martyna, Wendy. 1978. "What Does 'He' Mean? Use of the Generic Masculine." *Journal of Communication* 28:131–138.

Merchant, Carolyn. 1980. *The Death of Nature: Women, Ecology and the Scientific Revolution*. San Francisco: Harper & Row.

Moraga, Cherríe. 1983. *Loving in the War Zone: Lo Que Nunca Pasó por sus Labios*. Boston: South End Press.

Penley, Constance, and Ross, Andrew. 1990/1991. "Cyborgs at Large: Interview with Donna Haraway." *Theory/Culture/Ideology* 25/26: 8–23.

Pickering, Andrew, ed. 1991. *Science as Practice and Culture*. Chicago: University of Chicago Press.

Restivo, Sal. 1988. "Modern Science as a Social Problem." *Social Problems* 33:206–225.

Rich, Adrienne. 1978. "Power." In *The Dream of a Common Language, Poems 1974–1977*. New York: W. W. Norton and Company.

Scarry, Elaine. 1985. *The Body in Pain: The Making and Unmaking of the World*. Oxford: Oxford University Press.

Scott, Pam. 1991. "Levers and Counterweights: A Laboratory That Failed to Raise the World." *Social Studies of Science* 21:7–35.

Shapin, Steven. 1989. "The Invisible Technician." *American Scientist* 77:554–563.

Smith, Dorothy E. 1987. *The Everyday World as Problematic: A Feminist Sociology*. Boston: Northeastern University Press.

Star, Susan Leigh. 1988. "The Structure of Ill-Structured Solutions: Heterogeneous Problem-Solving, Boundary Objects and Distributed Artificial Intelligence." In *Distributed Artificial Intelligence, Vol. 3*, ed. M. Huhns and L. Gasser, 37–54. Menlo Park, CA: Morgan Kaufmann.

Star, Susan Leigh. 1989. *Regions of the Mind: Brain Research and the Quest for Scientific Certainty*. Stanford, CA: Stanford University Press.

Star, Susan Leigh. 1990. "Layered Space, Formal Representations and Long-Distance Control: The Politics of Information." *Fundamenta Scientiae* 10:125–155.

Star, Susan Leigh. 1991. "The Sociology of the Invisible: The Primacy of Work in the Writings of Anselm Strauss." In *Social Organization and Social Processes: Essays in Honor of Anselm L. Strauss*, ed. David Maines, 265–284. Hawthorne, NY: Aldine de Gruyter.

Star, Susan Leigh, and James R. Griesemer. 1989. "Institutional Ecology, 'Translations' and Coherence: Amateurs and Professionals in Berkeley's Museum of Vertebrate Zoology, 1907–39." *Social Studies of Science* 19:387–420. [See also chapter 7, this volume.]

Stone, Allucquere R. 1989. "The *Empire* Strikes Back: A Posttranssexual Manifesto." In *Bodyguards: The Cultural Politics of Sexual Ambiguity*, ed. Kristina Stroub and Julia Epstein, 221–235. New York: Routledge.

Sagerhaugh, S., and A. Strauss. 1977. *Politics of Pain Management; Staff-Patient Interaction*. Menlow Park, CA: Addison-Wesley.

Stonequist, Everett V. [1937] 1961. *The Marginal Man: A Study in Personality and Culture Conflict*. New York: Russell & Russell.

Strauss, Anselm. 1969. *Mirrors and Masks: The Search for Identity*. San Francisco: The Sociology Press.

Wenger, Etienne. 1990. "Toward a Theory of Cultural Transparency: Elements of a Social Discourse of the Visible and the Invisible." Ph.D. dissertation, Department of Information and Computer Science, University of California, Irvine.

Woolgar, Steve. 1991. "The Turn to Technology in Science Studies." *Science, Technology & Human Values* 16:20–50.

14 Anatomy Is Frozen Physiology, Or How I Learned to See the Process That Is Everywhere

Gail A. Hornstein

On January 18, 1985, Susan Leigh Star inscribed and sent to me a copy of her just-published poetry chapbook, *The Zone of the Free Radicals* (Star 1984). Here is the first page:

the zone of the free radicals
is a metaphor borrowed from nuclear/theoretical physics.
For every world of matter, there is a corresponding world of
anti-matter—a world which appears exactly the same, but whose
atoms have the opposite charge. In the anti-world to ours, for
example, electrons would be positively charged and protons
negatively charged. When even tiny particles from worlds of
matter and anti-matter collide, tremendous explosions occur.
At the interface between the worlds, then, there is a zone which
is neither matter nor anti-matter, tremendously hot and explosive,
against which both worlds "bounce." Every so often a particle
will escape into this zone from one side or another . . . becoming a
free radical. Free radicals have very short lives, brilliant colors,
and move with great speed and intensity.

By that point, almost thirty years ago, I already knew Leigh to be a person deeply immersed in boundaries—in studying them, in dissolving them, in trying to live outside them. It's part of what helped her to appreciate so fully the process that underlies everything. If there are no boundaries, or only diffuse and shifting ones, categorical distinctions start to dissolve and the dynamic processes that create the sense of structure stand out more sharply. What at first appear to be fixed categories are thus revealed as accomplishments—the result of actions, decisions, and choices, not simply inherent features of whatever system is being analyzed.

Psychiatry, for example, appears at first glance to be deeply static, a field primarily characterized by the sorting of symptoms into diagnostic categories, either those prescribed by the American Psychiatric Association in its *Diagnostic and Statistical Manual of Mental Disorders* (*DSM*), or, in certain parts of the world, into the alternative framework of the World Health Organization in its International Classification of Diseases. The constant attention given to diagnosis makes psychiatry seem quintessentially like anatomy, the study of the component parts of some larger pattern ("nature at its joints," in the classic phrase), a form of work which consists mainly of cutting through surface layers to expose the already-constituted organs or illnesses that lay beneath. Even the basic distinction between mental health and mental illness is assumed (by contemporary biological psychiatrists) to be like this—two clearly defined categories that can be teased apart, with little overlap between them.

For the two hundred years that psychiatry has existed, defining and redefining the classification of symptoms and sharpening the distinction between health and illness has been the field's consuming interest. (We see this vividly illustrated at present, in the massive amount of time, energy, and money being devoted to revising the DSM for the fourth time.) But as soon as we apply Leigh's wonderful insight that "anatomy is frozen physiology" (a framing I have loved since she first told it to me in the 1980s), we see immediately that psychiatry is actually just the opposite of static—a field where process is everywhere, but kept out of sight by particular kinds of invisible work, some of which I will outline here. (An especially apt topic for this chapter, since it was the focus of the only published piece Leigh and I ever wrote together, on the creation of universalities, a particularly robust kind of category, requiring constant maintenance by many forms of invisible work reinforcing one another simultaneously.)

Before starting in, I have to highlight two key features of psychiatry that distinguish it sharply from other medical specialties. First, there is astonishingly little consensus about core questions of the field. Psychiatrists today are still struggling with quandaries that first took shape in the eighteenth century: What causes mental illness? What forms does it take? How common are these? Do certain treatments work? How should psychiatrists be trained? Other fields of medicine are based on widely shared assumptions and standardized procedures; their debates center on narrow technical questions, not central issues like these.

Second, psychiatrists have always had a very different kind of relationship to their patients, as compared to their colleagues in other medical specialties. The insularity of asylums, where the overwhelming majority of psychiatrists have been trained and spent their whole careers until recently, meant that patients and their doctors often literally lived in the same buildings, on the same isolated grounds—thrown together, yet cut off from other people. (This is still true in the "therapeutic communities" that exist in rural locations in the United States and Europe today.) Psychiatrists and patients shared a deep interest in understanding the enigma of mental illness, and they both took to writing about causes, symptoms, and methods of treatment. At least one thousand accounts of mental illness by patients have now been published, taking every conceivable form (memoir, novel, diary, theoretical treatise, historical analysis, etc.), starting in the fifteenth century and rapidly increasing in number in recent years. (I am counting here only works in English, and published as books; no one knows how many other first-person accounts of madness have appeared in other languages or in the form of oral histories, testimonies, articles, blogs, etc.)

These two distinguishing features of psychiatry—little consensus on core questions, and an outspoken and often highly critical patient population with its own ideas about mental illness and treatment—have imbued psychiatric categories with a degree of instability and volatility that is simply not present in other medical fields. As a consequence, the processes holding this system together are far more apparent than is the case in more settled, less contentious areas of inquiry.

In her crucial paper "The Sociology of the Invisible" (Star 1991), Leigh wrote: "Work is the link between the visible and the invisible" (265). This was one of her analytic insights that I've always found particularly stimulating. So I will focus here on three kinds of invisible work in the world of psychiatry: first, the work of making diagnostic categories and methods of treatment seem robust enough to be widely used, despite their stunning lack of empirical support; second, the work that patients do to understand and cure themselves when the ideas and methods prescribed by their doctors do not work; and third, a kind of invisible *non*-work, the work that all of us *should* be doing but typically are not, of questioning the science, politics, and morality of how mental illness continues to be treated and understood.

I will focus especially on the second of these, the invisible work done by patients themselves, the vast majority of which is, even now, completely unknown to most mental health professionals. The reason it is even possible for me to talk about something this invisible is that over the past decade, I have been a participant observer in the alternative social worlds and political movements that psychiatric patients have created for themselves. I have analyzed countless first-person narratives and oral histories, participated in hundreds of peer-led support groups, conferences, and strategy sessions, and met with activist patients from all over the world. Their energetic efforts—largely outside the mental health system—to reframe pathologized identities, develop strategies to cope with distressing states and feelings, and sabotage the system or manipulate it into providing benefits they want are part of an international movement of current and former patients known generally as the psychiatric survivor movement. I have written in detail about a range of its ideas and accomplishments in *Agnes's Jacket: A Psychologist's Search for the Meanings of Madness* (Hornstein 2009); in this chapter, I will focus on one of the most influential parts of this movement, the Hearing Voices Network (HVN). Key to HVN's success is the work that gets done in the hundreds of peer support groups that people who hear voices have created for themselves in locations all over the world as an alternative to the medical model of "hallucination" espoused by most professionals.

Learning about mental health from crazy people is not as mad as it might seem. When a crucial resource is threatened, you learn to preserve what is still there. It's like coming to understand water conservation from people who live in dry climates and simply have to figure out how to survive there. Traumatic circumstances focus the mind, the way a pathogen stimulates the development of antibodies in the immune system. And you don't actually have to *be* one of the people who is personally afflicted to benefit, just as you don't have to live in a desert to learn about water conservation. Preserving mental health isn't actually that different, and the first step is learning to notice the feelings and ways of thinking and acting that deplete the resource versus those that help to restore it. This work was largely invisible until psychiatric patients began to write and talk about it.

We're all familiar with people who've had heart attacks and then learned to pay attention to their diet and stress levels to lessen their risk of a relapse. In similar fashion, people who hear voices or have other distressing mental experiences can learn what triggers their episodes or helps to lessen or prevent them. Mental health professionals typically start from the assumption that psychiatric symptoms don't vary much across contexts or over time, so they don't encourage patients to notice triggers or develop coping strategies to deal with distress *before* it becomes overwhelming. Patients have to learn—often only after years of failed treatment by professionals—what to do to help themselves. These insights aren't relevant only to them; most of the rest of us are pretty unclear about how our own minds work. We can all benefit from what people who've had to learn the hard way about the invisible work of maintaining mental health have to say about how it's done.

Let me offer an analogy to clarify what I mean here. While visiting the recent *Abstract Expressionism* exhibit at the Museum of Modern Art in New York, I spent a long time in front of Ad Reinhardt's 1963 piece, elliptically titled *Abstract Painting*. It's a large, entirely black canvas. According to Reinhardt, three distinct shades of black are present: "a black which is old and a black which is fresh. Lustrous black and dull black, black in the sunlight and black in shadow." People find the piece intriguing and are interested in how it works, but they aren't used to having to look at a painting for a long time in order to make sense of it. As I sat in front of it, the man standing next to me asked his wife, "Que pasa?" ("What's happening?") He looked for just a second, and then gave up and wandered off.

But a group of ten-year-old girls, on a tour of the exhibit, loved Reinhardt's work. They had a museum docent taking them around, and she encouraged them to look more carefully at the black and hinted at the shimmering faint cross that most people glimpse after extended viewing. Now I haven't tested this hypothesis, but I'd bet that people who have been deeply depressed can see the variety in Reinhardt's shades of black more readily than others can. Having spent so much time inside darkness, they've learned to navigate even in deep shadow.

My main claim in this chapter is that marginality and suffering in the realm of mental health are fundamentally shaped by the presence or absence of three kinds of invisible work:

First is the work of pushing processes—the "physiology" of the system—out of sight, so as to make the fixed (anatomical) categories appear to be more robust than they actually are. For example, to those who don't study them closely (which is most of us), DSM diagnoses seem discrete, reasonably reliable, and grounded in the data of neuroscience. This is certainly what the APA wants us to think about its work. But since the most minimal scrutiny of the data reveals the opposite to be true, the perceived robustness of DSM categories has to be seen as an extraordinary sociological achievement, which needs to be analyzed to be appreciated. Especially puzzling is how these categories manage to retain their influence and power even after acrimonious political battles among psychiatrists drafting the next revision of the system spill into the media, and more and more of the sham science that the APA uses to clothe its efforts are revealed. Serious work is being done here to make the edifice of biological psychiatry resemble a building instead of a house of cards. How, exactly, does this occur?

Some of the key processes include those that Leigh and I analyzed in our paper on universality biases (Hornstein and Star 1990). One strategy involves what we called "invisible referents, the invoking of forces or explanatory units that cannot be directly measured with the technology at hand" (428). Much of the justification for the constant revising of the DSM is cast in terms of the "progress of neuroscience," with such data the presumed basis of diagnostic categories. But since there is no agreement within psychiatry about how the disorders in the manual actually arise (with one exception, which I will get to a bit later), the precise ways that mental illnesses are "biologically determined" remain unspecified. The manual itself is a (very) long list of symptoms with no reference whatsoever to etiology. But by talking (vaguely) about "chemical imbalances in depression" and "genes for schizophrenia," psychiatrists make it seem as if these "invisible referents" that cannot be measured at present will ultimately reveal the bases for psychiatric symptoms.

The other strategy we mentioned in that paper, drawing our example from psychiatry, is the "jettisoning [of] uncertainties and empirical tests into the future" (Hornstein and Star 1990, 430). By consistently maintaining that all forms of psychopathology will someday be shown to have a biological basis, psychiatrists have been able—for more than a century—to successfully rebuff challenges to their claims. "Stressing the technical complexities of such research and insisting that it will only be 'a matter of time' until the necessary evidence comes in" (430), they put off into the indefinite future any test of their assumptions. The huge profusion of neuroscience studies and the media hype that surrounds them also contribute to the illusion that a firm body of evidence underlies diagnostic categories.

But a far more disturbing way that psychiatrists have maintained the robustness of DSM categories, one that goes beyond the strategies that Leigh and I discussed in that 1990 paper, is the punishing of patients who do not conform to its classifications. Consider this example:

In the 1970s, psychiatrists started to notice an increasing number of patients who were not, in general, "out of touch with reality," yet at times seemed to be much more disturbed than typical neurotics were. These patients had episodes so distressing they seemed close to psychotic, except for the fact that they would then resume lives that seemed more "normal." Since the stability and persistence of symptoms have always been considered key parts of any mental disorder—that is, if you're psychotic, that's how you are, it's not a state that comes and goes over short periods—these inconsistencies were troubling. These patients seemed essentially to be on the borderline between neurosis and psychosis, with much more variability than psychiatric classifications allowed for.

The discovery of a new type of patient who did not fit the existing system could, of course, have led to a questioning of the categories themselves. This is not, in fact, what happened. Instead, these hard-to-categorize patients were given their own special category. The conceptual problem of their seeming to be on the borderline between two categories in *psychiatrists'* system of classification was resolved by transforming this problem into the core pathology of the *patients*. In other words, rather than being seen as a challenge to the classification system, these "borderline symptoms" became the defining characteristic of a new category of mental illness. Then, in a final step that further shored up the diagnostic categories already in place, this new "borderline" pathology was classified as a "personality disorder." There it joined "schizoid," "narcissistic," "antisocial," and "paranoid" personality disorders as a fixed pattern of behavior assumed by most psychiatrists not to be amenable to treatment of any kind. In other words, the existence of these "borderlines" (as they soon derisively came to be known) not only failed to challenge the existing diagnostic system, in addition these patients were actually ejected from the daily practice of most physicians. Since there weren't really any treatments for personality disorders because they have no "symptoms" per se, just a more generalized "pathological identity," as soon as the patient was assigned a diagnosis of borderline personality disorder (BPD), she did not have to be seen again. (And indeed "she" is the correct pronoun here; according to recent estimates, women outnumber men by a ratio of 3:1 in the diagnosis of BPD [Bjorklund 2006].)

In their classic book *Making Us Crazy* (about the creation of the DSM, one of Leigh's favorite texts), Kutchins and Kirk (1997) quote a psychiatrist who says: "Borderline is a wastebasket diagnosis; the diagnosis is given to patients who therapists don't like, or are troublesome or are hard to diagnose and treat . . . [many of whom] have histories of sexual abuse and incest" (199). In *Trauma and Recovery*, psychiatrist Judith Herman

(1992) says borderline personality disorder has become a term "frequently used within the mental health professions as little more than a sophisticated insult" (123).

One of my favorite parts of the research for *Agnes's Jacket,* a book that describes the many alternative frameworks that patients have created for themselves, took place at an event in Exeter, England, in 2004 called "Extra Ordinary People." It was a gathering of patients who had been given personality disorder diagnoses and were now rebelling against them, creating their own approaches to understanding and dealing with their difficulties. They ridiculed the absurdities of the DSM—"Why don't we add 'Possible BPD' and then *everyone* could be diagnosed as borderline," was one suggestion—and they got back at the stigmatizing diagnoses of their doctors by wearing T-shirts emblazoned with their own choice of diagnostic category. (I purchased one to wear to academic meetings; in big black letters across the chest, it says "Antisocial Personality.") These actions by current and former patients highlight the punishing absurdities of diagnoses like "borderline," and undo some of the damage that these labels can cause to people's identities and relational lives.

Turning now to the more general forms of work that psychiatric patients do for themselves, we see a quintessential example of invisible work, despite its having been performed at every point in psychiatry's history. Psychiatric survivors call this "the expertise of experience" (in contrast to professionals' "expertise of training"), to capture their redefinition of what constitutes knowledge, skill, and effectiveness. This concept resonated deeply with Leigh's own intellectual work and her approach to training students. The 1999 paper "Layers of Silence, Arenas of Voice: The Ecology of Visible and Invisible Work" that she wrote with Anselm Strauss refers to "the expertise often hidden from view" (Star and Strauss 1999, 11), which is exactly the same thing as patients are talking about. In challenging the pejorative labels they are often assigned—like "treatment resistant"—psychiatric patients are moved to create their own alternative interventions (e.g., peer support groups) which offer possibilities for healing denied by those treating them.

The means by which patients acquire their expertise are as invisible to professionals as the skills themselves are. Consider the typical psychiatric ward, either in a mental institution or, more commonly today, in a general hospital. Patients sit around watching TV, or listening to their voices, or spacing out from the many medications they are prescribed. They appear to be doing nothing at all, paying little attention to each other or to subtleties of either their inner or outer experience. They are also demeaned, reduced to what Star and Strauss call "nonpersons" in all the ways they outline in that 1999 paper—given the food that no one else wants, locked in their rooms, forbidden to make autonomous life choices, and surrounded by people talking about them as if they were not there (Star and Strauss 1999, 18).

As a result, the sabotage and silent resistance that otherwise powerless groups always rely on in these settings are commonly seen, enabling the re-creation of a "person" from that "nonperson," via processes invisible to those in charge. This is crucial to creating a sense of positive identity to take the place of the one that was stigmatized (e.g., "I am a voice hearer" rather than "I am a schizophrenic patient").

A key element in the invisible work that patients do for themselves is articulation. In her wonderful book about the theater, *And Then, You Act: Making Art in an Unpredictable World*, Anne Bogart (2007) makes clear what this entails: "Articulation is a stroke on the canvas, an eloquent gesture, a harmonic chord, a lucid description. Articulation is born from the attempt to create bridges from the realm of private suffering to the outside world. From the heat of experience, you signal to others. Fueled by thought and feeling, its objective is clarity" (18–19).

An especially important kind of articulation work that psychiatric patients perform is the creation of narrative out of fragmentary memories of traumatic experience. Since biological models in psychiatry render trauma and, most especially, its connection to present distress (i.e., "symptoms") invisible, patients have to do this crucial work on their own, or in support groups (or in psychotherapy if they are able to access it). Putting words to traumatic experience allows it to be understood and dealt with. Connecting un-integrated bits of memory, somatic states, perceptions, and feelings makes them meaningful, the first step in creating a sense of agency. This is why Vietnam veterans, furious when their flashbacks, panic attacks, and other symptoms got diagnosed as "schizophrenia," lobbied so hard for the APA to add post-traumatic stress disorder (PTSD) as a diagnostic category (which finally occurred in the 1980 revision of the DSM). PTSD remains anomalous in the manual even today, the only category of disorder with an etiology explicitly acknowledged in its formulation.

Trauma is fundamentally a break in the narration of a life, between the self before and the self after. It is this narration process by patients that is most erased in psychiatrists' accounts of mental illness. So it is no wonder that hundreds of patients have written/spoken/published narratives of their experience, making the links that would otherwise remain hidden or obliterated. (For a list of the one thousand titles that have been published as books, see my *Bibliography of First-Person Narratives of Madness in English*, www.gailhornstein.com). Leigh's own intellectual work was deeply enmeshed in articulating these sorts of processes, a legacy of her training with Anselm Strauss; she always said that "one of Anselm's great gifts was in listening forth stories" (Clarke and Star 1998, 342).

There are many discussions—especially in the feminist literature—of "writing as healing" (e.g., DeSalvo 2000). But Leigh's insistence was always to move beyond these broad characterizations to see the precise work that was being done—the creation of links through chronology or episode, the commonality of suffering that comes from constructing shared accounts (especially of bizarre or horrific perceptions or feelings), the situating of events within broader patterns of meaning.

Narratives need scaffolding if they are to hold together and perform these ordering functions. Erecting this scaffolding requires a series of actions: deciding what types of events to try to recall, what specific information about these events is most relevant, how to contextualize each one, and so on. One crucial function of peer support groups (like the hundreds of HVN groups I've attended in Massachusetts and London over the past ten years) is that they explicitly help people to create accounts of their own experiences. Through empathic listening and uncritical noting of inconsistencies, gaps, or areas in need of further exploration, the group fosters an attentiveness to triggers, patterns, and explanations that a person might not notice or be able to formulate on his or her own.

Another key type of invisible work that constantly gets done in these hearing-voices peer support groups (and in many other kinds of psychiatric survivor groups) is the creation of new identities capable of challenging the stigmatizing labels assigned by psychiatrists. For example, every textbook in psychiatry and "abnormal psychology" claims that psychotic patients—especially those diagnosed schizophrenic—are "narcissistic," "egocentric," "lacking in empathy" and "unable to take the role of the other." This pejorative dismissal of patients' human qualities has blinded many mental health professionals to the many ways that people experiencing extreme states can help one another. In peer support groups, empathy and role taking are constant presences, as members help each other to become aware of talents and strengths that directly contradict doctors' dire prognostications.

What especially surprised me when I first started to participate in these groups was seeing people in states of extreme distress performing striking acts of empathy toward one another. And those not currently in those states were even more attuned to the needs of others, remembering all too well what it was like to have their "finer feelings," as one called it, denied by others. In patient-led support groups like the ones I write about in *Agnes's Jacket*, people do everything possible to make each other feel heard, understood, and validated as fellow sufferers. They listen acutely, ask one another subtle questions, take and offer advice, and laugh together in bitter recognition of the ironies of their shared circumstances. Beyond the concrete help and validation this provides, people leave these meetings having seen evidence of parts of themselves that others (especially psychiatrists) either ignore or claim not to exist. This helps to build a person's capacity for resistance ("you may think I lack insight, but I and others don't agree with that"), and their ability to think for themselves (which is hard to do when you are repeatedly told that you are "seeing things" or "hallucinating" or that your beliefs are "delusional"). Creating "possible selves" that challenge the pessimism of doctors is fundamental to the work of all psychiatric survivor groups, and is often astonishing in its effectiveness.

I want to note briefly a third kind of invisible work operating in psychiatry. This is really a kind of invisible nonwork, a work that *should* be occurring but *isn't*, a profound kind of silencing. (Here Leigh's refusal ever to exclude the moral dimension of any question is especially relevant.) This is the work of asking questions to expose the Emperor's embarrassing lack of evidence in which to clothe his science. Here are just a few examples of the questions every one of us should be asking psychiatrists, but typically don't: How can you say that there is a chemical imbalance in the brains of depressives if no test exists to measure these substances, and thus to tell us whether they are "balanced" or not? Why aren't people who are raped or sexually abused as children routinely diagnosed with PTSD? Why do psychiatrists continue to ignore the notion that most of the rest of us take for granted, that when bad things happen to people, it can drive them crazy?

Leigh never lost sight of the enduring insight of the feminist movement, as true today as when it was first formulated in the 1970s, that "the personal is the political." Indeed all her work on infrastructure can be read as an attempt to show precisely *how* the political becomes personal and vice versa. In this sense, for all her love of magic, Leigh was deeply an empiricist, wanting to see all the steps in the process, with nothing left unstated.

I will end with this quote from her paper "Experience: The Link between Science, Sociology of Science, and Science Education" (Star 1998) that precisely summarizes my argument here:

> There is no such thing as stress or malingering. There is only the loneliness of pain which has no categories or no allies; there is only suffering which falls mute because it is displaced from the known world. It is immoral to presume that someone in pain does not have the best knowledge of that pain. What is moral is to find the translation key, to listen, to recognize, and never to confuse muteness with lack of experience. (136)

References

Bjorklund, Pamela. 2006. "No Man's Land: Gender Bias and Social Constructivism in the Diagnosis of Borderline Personality Disorder." *Issues in Mental Health Nursing* 27:3–23.

Bogart, Anne. 2007. *And Then, You Act: Making Art in an Unpredictable World.* New York: Routledge.

Clarke, Adele E., and Susan Leigh Star. 1998. "On Coming Home and Intellectual Generosity." *Symbolic Interaction* 21:341–352.

DeSalvo, Louise. 2000. *Writing as a Way of Healing: How Telling Our Stories Transforms Our Lives.* Boston: Beacon Press.

Herman, Judith Lewis. 1992. *Trauma and Recovery.* New York: Basic Books.

Hornstein, Gail A. 2009. *Agnes's Jacket: A Psychologist's Search for the Meanings of Madness.* New York: Rodale Books.

Hornstein, Gail A., and Susan Leigh Star. 1990. "Universality Biases: How Theories about Human Nature Succeed." *Philosophy of the Social Sciences* 20:421–436.

Kutchins, Herb, and Stuart A. Kirk. 1997. *Making Us Crazy: DSM, the Psychiatric Bible and the Creation of Mental Disorders*. New York: Free Press.

Star, Susan Leigh. 1984. *The Zone of the Free Radicals*. Berkeley, CA: Running Deer Press.

Star, Susan Leigh. 1991. "The Sociology of the Invisible: The Primacy of Work in the Writing of Anselm Strauss." In *Social Organization and Social Process: Essays in Honor of Anselm Strauss*, ed. David Maines, 265–283. Hawthorne, NY: Aldine de Gruyter.

Star, Susan Leigh. 1998. "Experience: The Link between Science, Sociology of Science and Science Education." In *Thinking Practices in Mathematics and Science Learning*, ed. James G. Greeno and Shelley V. Goldman, 127–146. New York: Routledge.

Star, Susan Leigh, and Anselm Strauss. 1999. "Layers of Silence, Arenas of Voice: The Ecology of Visible and Invisible Work." *Computer Supported Cooperative Work* 8:9–30.

15   Categorizing Life and Death: The Denial of Civilians in U.S. Robot Wars

Jutta Weber and Cheris Kramarae

Our discussion below, part of our homage to Leigh presented at the 2011 Workshop Celebrating Contributions of Leigh Star, is based primarily on categories and numbers in use at that time to document the U.S. drone wars. The wars continue, of course, although the types, targets, and the classification systems change somewhat as goals and policies change. Susan Leigh Star's work encourages us not only to count casualties, in the material we are looking at, but also to be aware of causalities. In this chapter, we consider some of the casualties and causalities of the categories used in the earlier days of the drone wars.

"What is a computer?" That's the strikingly simple and endlessly complex question offered and encouraged by Susan Leigh Star (1995, 1), who pointed out that computers are both products and producers of media, at once objects of analysis and means of analysis. In this chapter we focus on issues involving computers and war—with regard to the so-called Revolution in Military Affairs, recent robot wars with unmanned combat systems, and imprecise killings with precision weaponry. We investigate the classifications—often unspoken—used to differentiate between combatants and civilians. Throughout we focus on vocabulary and categories.

Our study of computers and war, with a focus on robots and drones, owes much to Susan Leigh Star's work. Leigh continually pursued the links between lived experience, technologies, and silences (Star 2007, 227), and through her writing and conversations taught us much about how to make those links.

Always, she encouraged us to study the silences, the things we think are missing, the invisible, the people and the topics in the corner. Always, she provided us with stories to carve out the links between her research questions, methodology, and her own life. She made visible to us her own creative and powerful abilities to see and analyze what had often been overlooked by others, at the same time making connections to some of the circumstances in her own life which prompted her questions.

Always, she thought and wrote about language. Her work has taught us that we need to pay attention to the naming of things, how they are defined and redefined over time, and what is ignored, overlooked, or denied. Language is a prison in which we are wrapped, but it can also be an escape from the prison.

Technologies in Leigh's work are not just objects but also systems of relationships and cultural practices. Once when asked how the history of ideas would differ if technological issues were emphasized, Leigh replied that it would include "mess, suffering, exclusion, manual labor, and invisible work" (e.g., Star and Bowker 2007). We want to include some of that in this chapter on computers, drone wars, and the categorizing of death.

Bug Splat

The U.S. forces as well as the CIA are increasingly relying on armed pilotless aircraft, operated by remote control, to attack enemies in several countries of the South such as Iraq, Afghanistan, Pakistan, Yemen, Somalia, and Libya. Before Hellfire missiles are launched from the U.S. drones, the control team asks for the computer simulation of the "bug splat" of the attack, a read-out of the potential impact the missile will have on its ground target (see figure 15.1). "Bug splat" is also the term used by U.S. authorities

Figure 15.1
Photo 2009.12.18 House Destroyed, Ob50/Noor Behram (Bureau of Investigative Journalism 2011)

when human beings are successfully killed with drone missiles. For the U.S. controllers "bug splat" is the label for "militants" who have just been "neutralized"; for those on the ground, in many cases, the mess may represent a family that has been shattered, or a home destroyed, which are described in quite different words (Pincus 2011; Robinson 2011).

Bug Splat is also the name of a kid's video game in which you double-click bugs to kill them (http://www.cmyarcade.com/play-20981-Bug_Splat.html). Does this language, referring to the simulation of the impact of a drone on its potential target as a bug splat, reflect a "PlayStation mentality" (Alston 2010, 25) introduced by high-tech warfare, which leads to a growing devaluation of (non-high-tech affiliated, non-Western) human life?

The reference to killed Pakistani as "bugs" reminds us of some of the terminology used during the Vietnam war when GIs called their Vietnamese enemies "gooks" and "slopies," derisive labels to distance "others" during wars so as to trivialize their injuries and deaths—judging them to be "bugs" rather than humans, even if the people are not directly engaged in the fighting (see figure 15.2). (To counter the dehumanizing idea of the bug splat, in 2014 an artist group project gave the victims in North Waziristan a face; see http://edition.cnn.com/2014/04/09/world/asia/pakistan-drones -not-a-bug-splat/).

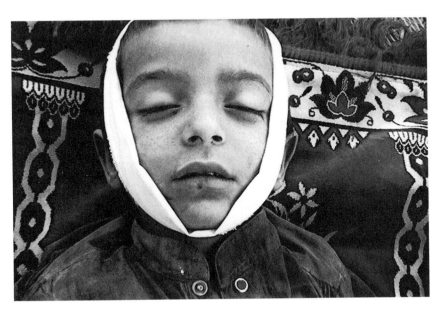

Figure 15.2
Photo 2009.08.21 Syed Wali Shah, aged 7 Ob32/Noor Behram (Bureau of Investigative Journalism 2011)

According to international humanitarian law (IHL) and law of war (LOW), in declared wars, opponents are legally permitted to capture or kill combatants—but not civilians. Here, we want to ask questions: What are considered to be the "legitimate military targets" for remote-controlled weapons in today's "war on terror."? What are the categories and classifications that differentiate between combatants and civilians with regard to the killing missions of unoccupied combat aerial vehicles in Afghanistan, Yemen, or Pakistan by the U.S. forces or even the CIA—missions that are mostly monitored from control centers in Nevada (U.S. Forces) or Virginia (CIA)—thousands of kilometers away from the battleground? As Geoffrey Bowker and Leigh Star have made clear: "Each standard and each category valorizes some point of view and silences another. This is not inherently a bad thing—indeed it is inescapable. But it is an ethical choice, and as such it is dangerous—not bad, but dangerous. . . . We are used to viewing moral choices as individual, as dilemma, and as rational choices. We have an impoverished vocabulary for collective moral passages" (Bowker and Star 1999, 5). They have also pointed out that "a category is something like a treaty or a cover of some sort that hides the messier version of what is inside" (Star and Bowker 2007, 273).

What are the major categories that are involved in organizing lives and deaths in the drone attacks—based on whose point of view and whose exclusion? What are the labels and definitions that support the categorizing, while hiding some of "the mess"?

Leigh (Star 1995) wrote about "dirty work" that may be physically clean but is spoiling or stigmatizing—such as weapons deployment. The required use of dehumanizing words and practices for one's enemies might also make for "dirty work"—as well as a pseudohumanizing rhetoric when recent weaponry is labeled as smart and precision weaponry. For example, Eyal Weizman (2006) argues that so-called precision weaponry (like robot drones for targeted killing) paradoxically produces more civilian casualties than does "traditional" bombing, because the rhetoric of precision "gives the military-political complex the necessary justification to use explosives in civilian environments where they cannot be used without endangering, injuring or killing civilians" (71) (see figure 15.3). Remote-controlled war can also be seen as an attempt to portray high-tech war as clean, easy, quick (as well as economic): "Violence can thus be projected as tolerable, and the public encouraged to support it" (72). Obviously, labels and definitions can be a matter of death—as well as of new technologies.

Computers and Remote-Controlled Wars

The long history of computers and the military is one we won't deal with here except to briefly note that the use of current drones, packed with sensors and sophisticated video technology that can view through clouds and in the dark, was preceded by uses of computers that helped crews drop bombs more accurately during World War II and the wars in Korea and Vietnam.[1] Because of government intelligence classification, the

Figure 15.3
Photo Shoes (photo: Noor Behram)

history of the military use of computers is still partly obscured; however, Paul Edwards (1996) notes that an automated defense with centralized control that could be directed far from the battlefield became a goal of the U.S. government during the Vietnam War. The resulting programs have, he suggests, resulted in the desire by the government to take human error and lag time out of the equation, and make the systems as fully automated as possible. Theorist Jordan Crandall suggests that from the 1950s on in fields such as operation research, game theory, and cybernetics the "development of computing became allied with the communication, command, simulation, and control imperatives of the Cold War" (Crandall 2005, no pagination given). This took on two primary forms: real-time tracking interfaces and distributed interactive simulation.

The Revolution in Military Affairs

In the last fifteen years, there has been ongoing reorganization and restructuring of the Western military called "Revolution in Military Affairs" (RMA) (Tilford 1995; Cohen 1996; Nye and Owens 1996). In brief, RMA is the plan of a highly connected, satellite-based and network-centric warfare. Many politicians and militaries are advocates for a remote-controlled war—a war with a Joint Battle Space Environment relying on light technology, precision weaponry, smart machines, and the Internet. In this futuristic picture, robots would operate autonomously in water, air, and on land, deploying central tasks of warfare; invulnerable cyborg soldiers with exoskeletons would use intelligent weapons and munitions. All entities would be connected via the Global Information Grid (GIG) based on the military Internet, which will enable a panoptic (God's) perspective within which the four-dimensional battle space would be totally controlled. The goal is to disable the enemy's information and communication structures and thereby to achieve "globespanning dominance based on a near-monopoly of space and air power" (Graham 2005, 175; see also Dillon 2002; Dandeker 2006), all this to handle new warfare scenarios of unconventional conflict.

According to this "philosophy" war is fought not between front lines but in a wide, simultaneous four-dimensional battle arena which leads to a totalizing and speeding up of war where former civilian spaces are becoming ubiquitously militarized—a development that started already in the Vietnam War with the invention of new technologies such as the helicopter and of so-called free fire zones in which "all humans present in the zone, no matter their age, gender, or occupation, were to be considered legitimate military targets. People without weapons, including children and senior villagers, were simply unarmed soldiers" (Blackmore 2005, 38).

The Persian Gulf War (1990) is regarded as the first network-centric warfare in which the U.S.-American military worked with "precision weaponry," extensive use of communication via satellite and sophisticated surveillance, reconnaissance, and intelligence-gathering systems. During the Gulf War and in the first months of the Afghanistan war this approach was celebrated as the magic bullet to win every future war and to finally overcome the widespread perception that any U.S. involvement in war activities might lead to another Vietnam—the so-called Vietnam syndrome (Bacevich 2009). In 2001, George W. Bush praised the newly invented and armed drone named "Predator" as a very good example of the RMA and its "high-tech weaponry [that] can shape and then dominate an unconventional conflict. . . .This unmanned aerial vehicle is able to circle over enemy forces, gather intelligence, transmit information instantly back to commanders, then fire on targets with extreme accuracy" (Bush 2001).

This futuristic scenario of warfare is accompanied by the "[old] fantasy of a better, safer tomorrow in which bad people alone are killed [immediately] and the innocent spared" (Blackmore 2005, 9). So RMA as techno-sophisticated warfare comes with a rhetoric that attempts to avoid international embarrassment over the ongoing killing by referring to "precision weaponry" or "smart weapons." Stephen Graham (2009) has pointed out that the U.S.-American dream of the invincibility of one's troops is also a longing for reinstating one's "waning military, economic and political supremacy" (xii).

On the techno-imaginary level, the vision of RMA has been to scare the enemies so thoroughly that they will "throw down their arms and fall on their knees in recognition of their wrongdoing" (Blackmore 2005, 8) and worship the technoscientific and intellectual superiority and high-tech, masculine capability of the force (Weber 2009).

However, techno-euphoria about RMA diminished as the wars in Iraq and Afghanistan continued and the number of dead Western soldiers increased. Many critics started questioning the idea that technology alone can win over terrorism and many draw parallels between the Vietnam and Afghanistan wars (i.e., Barry and Thomas 2009; Etheridge 2009).

It seems that RMA has not been able to determine the outcome of a war, and the concept of victory seems to be wiped from the discourses of the military(-industry-media) complex (Der Derian 2009; Lenoir and Lowood 2003; Kaplan 2013) and replaced with the concept of (endless) counterinsurgency (Bacevich 2009; Gregory 2011).[2] The vision of war as a "giant computer game" obviously didn't work out. But in the face of the growing financial weakness of the United States, the military has continued to rely on cost-efficient uninhabited aircraft in its seemingly endless counterinsurgence in the "war on terror" and in its efforts to save soldiers from ground "action" in areas such as North or South Waziristan. There are also rumors that drone technology supports the U.S. government's preferences for killing over capturing because the legal and political situation made "aggressive interrogation a questionable activity anyway" (Kenneth Anderson quoted in McKelvey 2011). At the same time, remote killing also has the "favor" that it avoids "potentially messy questions of surrender" (ibid.).

In the endless war on terror, RMA and especially drone technology is used today for remote-controlled targeted killings of so-called militants, without too many (known) civilian deaths, with low financial expense (relative to the costs of what other military technology such as jet bombers incurs), and minimum Western alliance casualties—all in the name of protecting the nation (and, often unnamed, the nation's industries). But what does a robot, or specifically drone, war mean for low-tech nations and especially for their civilian populations?

Categorizing Civilians, Counting Victims—From Drone Attacks in Pakistan

What are "legitimate" military targets for remote-controlled weapons in today's "war on terror"? What are the categories and classifications that differentiate between combatants and civilians with regard to the killing missions of unoccupied combat aerial vehicles in Afghanistan, Yemen, or Pakistan by the U.S. forces or other U.S. agencies such as the CIA—missions which are mostly monitored from control centers such as Nevada (U.S. forces) or Virgina (CIA)—both thousands of kilometers away from the battleground?

In the following, we want to focus on the categories used for killings via drone attacks, especially in Pakistan where the number of drone attacks is steadily increasing. During George W. Bush's administration, about 60 drone strikes were carried out in Pakistan between 2004 and January 2009. After the beginning of Barack Obama's presidency in January 2009, drones were more massively deployed—about 250 times. (The number of drone attacks and deaths in Pakistan decreased from 2011 on; see Bureau of Investigative Journalism 2014a). The conflicting reports of the number of "militant" and "civilian" deaths illuminate the political nature of the category system, the problematic civilian/militant binary division, the obscuring of the stories behind the labels, and the "residual" suffering (e.g., psychological trauma) that is ignored by the category system.

According to the Bureau of Investigative Journalism (Woods 2011b) more than 2,300 were killed by drone attacks in Pakistan—at least 392 civilians; many more civilians were reported injured. (Today the Bureau estimates 2,427–3,929 deaths and 416–959 civilians killed; Bureau of Investigative Journalism 2015). The liberal U.S. think-tank New America Foundation estimates 1,435–2,283 deaths by drone attacks of whom 1,145–1,822 were described as militants in reliable press accounts" (Bergen and Tiedemann 2011). Though it is still unclear whether the category "militant" stands for the sum of combatants and so-called unlawful combatants (while the relevant category of IHL is that of combatant), this way of categorizing would result in the number of about 300–500 killed civilians. These "reliable press accounts" of the number of "militants" killed are mostly from Western media such as the "Associated Press, Reuters, Agence France-Presse, CNN, and the BBC—and reports in the leading English-language newspapers in Pakistan—such as the *Daily Times*, *Dawn*, the *Express Tribune*, and the *News*— as well as those from Geo TV, the largest independent Pakistani television network" (New America Foundation 2011; italics added). Peter Bergen and Katherine Tiedemann suggest that statistically only "one out of every seven U.S. drone attacks in Pakistan kills a militant leader. The majority of those killed in such strikes are not important insurgent commanders but rather low-level fighters, together with a small number of civilians. In total, according to our analysis, *less than two percent of those killed by U.S. drone strikes in Pakistan have been described in reliable press accounts as leaders of al Qaeda or allied groups*. Not a single drone strike had targeted Osama bin Laden before he was killed by U.S. Navy SEALs on May 2" (Bergen and Tiedemann 2011; emphasis added).

The differences in counting and categorizing become more visible if we look at the estimation of specific years. For example, the *Long War Journal*, a project of the U.S. neoconservative think tank "Foundation for Defense of Democracies," counts only 43 civilian deaths by drone attacks in 2009 in Pakistan. The New America Foundation estimate approximately 120 civilian deaths by drone strikes, while the Pakistani newspaper *Dawn* counted more than 700 civilian deaths killed in 44 drone strikes in 2009 ("Over 700 killed" 2010).

In the context of IHL and the LOW the central question remains: How many civilians have been killed during the drone attacks? How does it happen that organizations or media end up with very different numbers of killed civilians that differ between 43 and 700?

Let's look on the methodology of the classification and the resulting numbering of dead civilians: The two U.S. think tanks made their counts relying on (mostly) Western media reports; nevertheless they still arrived at different results though the overall numbers are much lower than that of the Pakistani media report. The latter used the government records of killed persons and subtracted the number of Al-Qaeda/Taliban high-value targets killed. While it is unclear how Western media classify militants (the most often used category) as well as civilians and on which testimony these reports rely, the Pakistani newspaper *Dawn* relied on official Pakistani government records—records that seem to be irrelevant to most Western media. Unlike Western media, the reports for *Dawn* only counted high-value Al-Qaeda and Taliban fighters as combatants. These differences in counting and methodology are mostly not discussed, and projects such as the *Long War Journal* or the New America Foundation seem to take their own statistics for granted.

According to this logic, the *Long War Journal* project praises the drones as precision weapons that kill only the (bad) Taliban/Al-Qaeda while sparing more and more civilians every year: "Despite the sharp increase in both the frequency and total number of casualties resulting from Predator strikes since mid-2008, civilian casualties have remained relatively low. Naturally, it is difficult to determine the exact number of civilians killed in Predator strikes for many reasons—including intentional exaggeration by Taliban spokesmen, and vague accounts by Pakistani media sources which frequently report that a certain number of 'people' were killed in a strike, but rarely offer a follow-up report identifying which victims were civilians and which were militants. However, it is possible to get a rough estimate of civilian casualties by adding up the number of civilians reported killed from the media accounts of each attack. According to this method, a total of 94 civilians were reported killed as a result of all strikes between 2006 and 2009" (Roggio and Mayer 2010).

According to the *Long War Journal*, the problem of reliable numbers of civilian casualties lies in the vague reports of the Pakistani media. Roggio and Mayer accuse the Taliban spokesmen (and a question is whether they identify tribesmen automatically with Taliban spokesmen) of propaganda by exaggerating the numbers of civilian casualties.

The ACLU—the American Civilian Liberties Union—has a different idea about and different standards of information policy. The ACLU long urged the U.S. government to reveal the number of civilians killed in CIA drone strikes in Pakistan. But the government neither was admitting nor denying that there were any records on civilian casualties on drone strikes (ACLU 2011). In June 2011, President Obama's counter-terrorism advisor, John Brennan, claimed that the United States had been—at least since August 2010—"exceptionally precise and surgical in terms of addressing the terrorist threat. And by that I mean, if there are terrorists who are within an area where there are women and children or others, you know, we do not take such action that might put those innocent men, women and children in danger." He went on: "*there hasn't been a single collateral death because of the exceptional proficiency, precision of the capabilities that we've been able to develop*" (Dilanian 2011; emphasis added). Again we experience the rhetoric of clean, sterile warfare according to which only the evil ones get killed and the good ones are safe. (Current director of the CIA John Brennan still claims that CIA drone strikes are "a last resort to save lives" but that "the CIA should not be doing traditional military activities and operations" (Bureau of Investigative Journalism 2014b).

In July 2011, the Bureau of Investigative Journalism published a report on at least forty-five civilian deaths from CIA drone strikes in Pakistan between August 2010 and June 2011 (Woods 2011a). Some months earlier, at the end of 2010, the "Campaign for Innocent Victims in Conflict" (CIVIC) released a report written by Christopher Rogers, financed inter alia by the Harvard Law School Frederick Sheldon Traveling Fellowship. This report pointed to the high numbers of civilian casualties in North Pakistan. Different from the studies of the *Long War Journal* and the New America Foundation, CIVIC undertook not only a media analysis but also interviews with Pakistani and U.S. policy makers, with officials from NGOs and "over 160 Pakistani civilians with direct losses from the conflict, including the death of a family member, injury or damaged or destroyed homes or essential property." These interviews include nine with the relatives or friends of drone victims. The investigator(s) came to the conclusion that thirty civilians were killed in just the nine cases of drone attacks they were able to analyze (CIVIC/Rogers 2010, 15). The detailed methodology of the CIVIC report sheds light on the difficulties of gaining a reliable overview of the situation in conflict-affected areas in Pakistan, stressing that access to conflict-affected areas is very difficult, for example because the Pakistani government closed the areas of ongoing military operations. Foreigners do not get access to areas close to conflict zones in which many displaced people now live in shelters. In militant-controlled areas such as North Waziristan access is very risky and difficult for everybody because of the threat of getting kidnaped or killed by militant groups (see CIVIC/Rogers 2010, 6f). The areas where most drone strikes are conducted are South and North Waziristan. (Today the drone strikes shift toward Afghanistan, Iraq, Yemen, and Gaza; for the latter, see Gregory 2014.) While the former area is closed for everybody not in the military because of ongoing military operations, the latter is mainly controlled by assumed militants. As there are no U.S. forces on the ground other than some small Special Forces units, they can hardly evaluate civilian casualties or damage of civilian properties directly.

But it seems also the case that the United States does not gather information from the Pakistani officials or Pakistani intelligence. Without this information it is not possible to plausibly minimize collateral damage (CIVIC/Rogers 2010, 20). Very recently, the Independent think tank Oxford Research Group (ORG) claimed in a report (Breau, Aronsson, and Joyce 2011) that "Hit and Run attacks" might break humanitarian law. In ORG's interpretation, the Geneva Convention requires that parties responsible for the drone attacks search for and identify all persons killed in strikes. Susan Breau, professor of international law at Flinders University in Australia and lead author of the report, said to the media: "It is high time to implement a global casualty recording mechanism which includes civilians so that finally every casualty of every conflict is identified" ("'Hit and run' drone strikes" 2011). She stressed that this is "not asking for the impossible. The killing of Osama bin Laden suggests the lengths to which states will go to confirm their targets when they believe this to be in their own interest. Had the political stakes in avoiding mistaken or disputed identity not been so high, Osama bin Laden and whoever else was in his home would almost certainly have been typical candidates for a drone attack" (Oxford Research Group 2011). It becomes obvious that the life and death of some people are worthwhile to be investigated while the dying of the civilians is made mostly invisible by either denying their deaths and suffering or blaming them for getting involved in the targeted killings (Butler 2009; Weber 2010).

Indications are that the accounting of the drone war in North Waziristan might change. Recently there was a peace rally organized by the Pakistani opposition politician Imran Khan and some U.S. peace activists (including members of Code Pink and others) who led thousands of supporters to Tank, a town at the edge of the tribal districts; they were blocked by officials from entering South Waziristan. The rally received a great deal of media coverage.

Around the same time a report "Living under Drones: Death, Injury, and Trauma to Civilians from US Drone Practices in Pakistan" and a video on the same topic were produced by researchers at the International Human Rights and Conflict Resolution Clinic at Stanford Law School and the Global Justice Clinic at NYU School of Law and attracted a great deal of media interest (International Human Rights 2012). The purpose of the study was to conduct an independent investigation into whether and to what extent the drone strikes conformed to international law and caused harm and injury to civilians. The researchers interviewed 130 victims, witnesses, and experts. They examined extensively the categories of civilian and "militant." They noted that the term "militant" is never defined and criticized the fact that most major media outlets in Europe, the United States, or Pakistan seem to believe that any militant is a legitimate target of lethal force. They pointed to the problems of how targeted killings and drone strike practices undermine respect for the rule of law and international legal protections" (Cavallaro, Sonnenberg, and Knuckey 2012, viii), and stated: "Journalists and media outlets should cease the common practice of referring simply to 'militant' deaths, without further explanation. All reporting of government accounts of 'militant' deaths should include acknowledgement that the US government counts all adult males killed by strikes as 'militants,' absent exonerating evidence" (Cavallaro, Sonnenberg, and Knuckey 2012, x).

So the question remains how civilians and militants are defined. The quite different estimations of killed civilians are due not only to the difficulties of assessment but also to the very different ways of categorizing civilians—leading to varying counts of victims, for example in the war on terror in Pakistan. What is obvious is the fact that the numbers differ widely depending upon the political standpoint and moral choices of those who classify and count—from the U.S. government, mainstream Western media, non-mainstream media, non-Western, Pakistani media, national think tanks as well as human rights organizations such as CIVIC or the ACLU.

The U.S. government does not offer a binding definition of "civilians" in its ongoing wars "on terrorism." The International Committee of the Red Cross has given the following definition: "For the purposes of the principle of distinction in international armed conflict *all persons who are neither members of the armed forces of a party to the conflict* nor participants in a *levée en masse* [mass conscription or mass uprising] *are civilians and, therefore entitled to protection against direct attack unless and for such time as they take a direct part in hostilities*" (ICRC and Melzer 2009, 20; emphasis added).

But what does it mean to take part in hostilities? We know that many militant leaders live with their often-large families. Do these women and children lose their status as civilians because they share roof and table of a militant and are often dependent on the head of the family? Is a tribal elder who provides political support to militants an acceptable target? Christopher Rogers writes that the situation in Pakistan is such that people are very often afraid of the Taliban/Al-Qaeda fighters, so they give food or shelter to them without being militants themselves—not even being sympathetic with the militants. But in several cases houses were destroyed by a drone attack after people sheltered Taliban or Al-Qaeda people. When Pakistani government officials were asked about this problem, a high-level official replied: "'Don't give shelter or protection to state's enemies . . . if they have an agent of al-Qaeda or whomever in their house, then that it is the cost that they pay' (interview with anonymous government official)" (CIVIC/Rogers 2010, 15). Punishing those who live close to and are accused of cooperating with combatants is a well-known phenomenon in warfare but it is at the same time made invisible by the rhetoric of precision strikes and smart weapons. After 9/11, George W. Bush determined "that none of the provisions of Geneva apply to our conflict with al-Qaida in Afghanistan or elsewhere through the world" (Bush 2002). There is still a lively discussion going on in U.S.-American law and international relations about the categories relevant for the law of war—whether it is only combatants and civilians or whether the category of the unlawful combatant has to be included. Nevertheless, the category of the civilian officially was never at stake, although making those dependent on combatants to combatants themselves is a more than questionable policy.

But the question of categorizing and classifying is even more complicated: Can a drone controller working in Nevada and walking in the streets be regarded as a legitimate target for remote controlled killing? The UK Approach to Unmanned Aircraft Systems (2011) asks: "Would an attack by a Taliban sympathizer [on a drone controller walking in the streets of his U.S. home town] be an act of war under international law or murder under the statutes of the home state?" (Ministry of Defense 2011, 5–8) While U.S. officials are taking their right to kill combatants and militants on Pakistani ground for granted in their global armed conflict against Al-Qaeda, Taliban, and so-called and vaguely defined "associated forces," the question would be whether Taliban combatants could use targeted killing of U.S. drone controllers in any place according to IHL.

Human Rights Watch asked in their letter to the president, in November 2010, how the U.S. administration defines the global battlefield and whether lethal force can be used "in accordance with the laws of war, against a suspected terrorist in an apartment in Paris, a shopping mall in London, or a bus station in Iowa City? Do the rules governing targeted killing vary from one place to another?" (Human Rights Watch 2010).

Capabilities of Reconnaissance: On Differentiating Combatants and Civilians

Does it seem that the controllers of the drones cannot distinguish among categories of people on the ground? The background information is that every drone carries several high-resolution cameras which would—at least in theory—make it easy to have a close look on the terrain before one deploys drones to make sure that only combatants or "militants" and not civilians are nearby.

Intelligence, surveillance, and reconnaissance (ISR) is a central task in network-centric warfare to ensure permanent air surveillance over wide areas, to detect activity, and to monitor key places for hours and days. Before drone attacks, the uninhabited aerial vehicles (UAVs) are harvesting (additional) intelligence material. The reconnaissance capabilities of the drones deployed in Pakistan, Iraq, and Afghanistan, such as the Reaper or Predator drones, are enormous. In 2010 Jack Wade and Pauline Shulman from Z Microsystems wrote in the NASA journal technology briefs: "The reconnaissance capabilities of the forward camera on the Predator and Reaper, known as the "MTS Ball" or MQ-16 electro-optical/infra-red camera, can read a driver's license from 20,000 plus feet" (Wade and Shulman 2010). The drones are also equipped with highly sensitive sensor systems—for example, infrared sensors as well as Synthetic Aperture Radar (SAR). The latter permits day as well as night operations in most weather conditions. With the help of full-motion video (FMV), an on-demand, close-up view of the combat zone becomes possible. The FMV via the onboard cameras is the basis for remote controlling the drones over thousands of kilometers: "Full-motion video adds a fourth dimension to imagery: the ability to track activity over time. FMV provides outstanding event fidelity, seamless event progression, and a full context regarding the nature of the location and activities being viewed." Jack Wade and Pauline Shulman (ibid.) also discuss the technical problems of FMV; because it has very high bandwidth requirements, it needs huge digital storage space, and often only delivers low-quality pictures. They also hint to the high demands these new vision capabilities put on the humans controlling the drones.

In general, drone pilots are capable of assessing the battlefield in a close-up way, while completely out of danger like nobody else on the battlefield. So we wonder how it happens that so many civilians are killed by drone strikes. If it is possible to follow the targets in question for hours and days, it seems highly likely that drone controllers know whether civilians are close by when they decide to deploy the strike (although the question is still open about who counts as civilian at all). Perhaps the drone controllers are unable to handle the technical capabilities in an appropriate way. However, since the CIA as well as the U.S. military forces can access the video footage of every drone, it should be easy to evaluate the actions in the forefront of the strike as well as its result. Evidently, these efforts are not regarded as worthwhile or we would expect them to be made public in view of the ongoing discussions of the lawfulness of U.S. targeted killings in the tribal areas of Pakistan.

We have focused on the labels and definitions of militants and civilians. But this binary category system used to describe the effects of drone strikes ignores many other kinds of suffering caused by the strikes. The "Living under Drones" report points out that the government accounts fail to record many other injuries of the strikes, including not only serious and long-lasting physical injuries, but also the reluctance of neighbors and first responders to quickly rescue and provide medical assistance (because of fear of strikes on the rescue teams), the mental health problems (caused by the constant fear of being attacked), the reduced educational opportunities (because the schools have been struck), and the damage to the economic, social, and cultural life (because many people believe paid informants help the CIA identify potential targets, destroying trust within the communities). These injuries are some of the "mess" that is hidden by the category system used by the government.

"To Classify Is Human": On Moral Choices

As Bowker and Star (1999) make clear, classifying is a ubiquitous human activity that shapes our lives, ordering human interaction and producing advantage or suffering. They have discussed the significance of classifications as part of the infrastructure of people's lives, and the importance of being self-critical about the creation and use of classifications. We have focused on some of the classifications used in drone attacks to raise the visibility of people who are caught in classification schemes, sometimes with deadly results.

We have followed the dehumanizing as well as pseudohumanizing use of categories in the U.S. war on terror and the so-called Revolution in Military Affairs. The latter builds on a techno-imaginary of globe-spanning dominance, invincibility, and cost effectiveness. In this logic, the latest high-tech weapons are labeled as "smart" and (surgically) "precise"—which is a strategy of pseudohumanizing weapons and their killings and a means to legitimize violence in civilian environments. In this context, high-tech war is portrayed as clean, easy, and bloodless, seemingly as a way to get public support for it. At the same time (supposed) militants killed in a drone attack are categorized as "bug splat"—a well-known strategy of dehumanizing the enemy.

We have discussed several problems of assessing the killing of civilians by U.S. drones in Pakistan. For example, there are no U.S. government or military agencies that systematically and publicly investigate or collect data on civilian casualities or injuries, although there would be improved possibilities to do so with drone technology. In addition, we have found that agencies use quite different categories and classifications to differentiate among combatants, militants, and civilians, and thus statements about the number of deaths by drone attacks vary greatly, even among "reliable" (Western) agencies.

The definitions are not just a matter of statistical quibbling. If the standard as applied by the U.S. government to determine who is a combatant is very broad (including, for example, a person interacting or living with people believed to be militants), then many of the people killed may be innocent of militant intent or actions against the United States since economic and political realities may make engagement with militants necessary. (Under international law, guilt by association is not justified.) Distinguishing militants from civilians may be difficult in some cases, but without clear definitions the actual numbers of civilians and militants killed will also not be clear.

We are very aware that we can be criticized as overlooking military considerations, and thus being ignorant of what's "really going on." A criticism of this sort would certainly be a way of trying to exclude our concerns, and analyses. Our intent is not to give a military account but to enhance the discussion of some of the complex ways that computers in war are being used by people in various relationships, and of the ways that classifications can construct people's lives—and deaths.

We are thankful to Leigh for all the ways she taught us to inspect the language, the labels, and the relationships, and to see that many areas that do not receive a lot of attention from philosophers and sociologists, such as weapons development, are also forms of social organization. She and Adele Clark (Clarke and Star 1998) wrote that their teacher Anselm Strauss continually suggested that they study the unstudied, always listening for and being respectful of untold stories and the corners of social life where suffering is compounded by silence.

In this chapter, we sought to follow Leigh in exploring the unstudied, the neglected, and the silenced.

Notes

1. See, for example, information on the Norden bombsight at the National WWII Museum. See http://www.nationalww2museum.org/see-hear/collections/artifacts/norden-bombsight.html(accessed April 11, 2015).

2. We originally relied on a paper presented by Caren Kaplan in 2008 at the HASTAC Conference,"Electronic Techtonics: Thinking at the Interface," titled "'Everything Is Connected': Aerial Perspectives, the 'Revolution in Military Affairs,' and Digital Culture." Sadly, it is no longer accessible. See Kaplan 2013 for related discussions.

References

ACLU (American Civilian Liberties Union). 2011. "Civilian Deaths from CIA Drone Strikes: Zero or Dozens?," July 19. http://www.aclu.org/blog/national-security/civilian-deaths-cia-drone-strikes-zero-or-dozens (accessed November 4, 2011).

Alston, Philip. 2010. *Report of the Special Rapporteur on Extrajudicial, Summary or Arbitrary Executions*. www2.ohchr.org/english/bodies/hrcouncil/docs/14session/A.HRC.14.24.Add6.pdf (accessed November 4, 2011).

Bacevich, Andrew J. 2009. "Social Work with Guns." *London Review of Books* 31 (24) (December 17): 7–8.

Barry, John, and Evan Thomas. 2009. "Could Afghanistan Be Obama's Vietnam?" *Newsweek*, January 30. http://www.newsweek.com/could-afghanistan-be-obamas-vietnam-77749 (accessed November 4, 2011).

Bergen, Peter, and Katherine Tiedemann. 2011. "Washington's Phantom War: The Effect of the U.S. Drone Program in Pakistan," July/August. http://www.foreignaffairs.com/articles/67939/peter-bergen-and-katherine-tiedemann/washingtons-phantom-war (accessed January 18, 2015).

Blackmore, Tim. 2005. *War X*. Toronto: University of Toronto Press.

Bowker, Geoffrey, and Susan Leigh Star. 1999. *Sorting Things Out: Classifications and Its Consequences*. Cambridge, MA: MIT Press.

Breau, Susan, Marie Aronsson, and Rachel Joyce. 2011. "Drone Attacks, International Law, and the Recording of Civilian Casualties of Armed Conflict," June. http://www.oxfordresearchgroup.org.uk/publications/briefing_papers_and_reports/discussion_paper_2 (accessed November 4, 2011).

Bureau of Investigative Journalism. 2011. "Obama 2009 Strikes," August 10. http://www.the-bureauinvestigates.com/2011/08/10/obama-2009-strikes (accessed November 4, 2011).

Bureau of Investigative Journalism. 2014a. "All Estimated Casualities in Pakistan," July 17. http://www.thebureauinvestigates.com/wp-content/uploads/2012/07/All-Totals-Dash54.jpg (accessed January 18, 2015).

Bureau of Investigative Journalism. 2014b. "John Brennan Marks First Year at CIA amid No Confirmed Civilian Drone Deaths in Pakistan," March 11. http://www.thebureauinvestigates.com/2014/03/11/john-brennan-marks-first-year-at-cia-amid-no-confirmed-civilian-drone-deaths-in-pakistan/ (accessed January 18, 2015).

The Bureau of Investigative Journalism. 2015. "Get the Data: Drone Wars," January 18. http://www.thebureauinvestigates.com/category/projects/drones/drones-graphs/ (accessed January 18, 2015).

Bush, George W. 2001. "U.S. President George W. Bush Addresses the Corps of Cadets," December 12. http://www.citadel.edu/root/presbush01 (accessed July 27, 2015).

Bush, George W. 2002. "Text of Order Outlining Treatment of al-Qaida and Taliban Detainees," February 7. http://lawofwar.org/Bush_torture_memo.htm (accessed November 4, 2011).

Butler, Judith. 2009. *Frames of War: When Is Life Grievable?* London: Verso.

Cavallaro, James, Stephan Sonnenberg, and Sarah Knuckey. 2012. *Living under Drones: Death, Injury and Trauma to Civilians from US Drone Practices in Pakistan*. Stanford, CA: International Human Rights and Conflict Resolution Clinic, Stanford Law School; New York: NYU School of Law, Global Justice Clinic. https://www.law.stanford.edu/publications/living-under-drones-death-injury-and-trauma-to-civilians-from-us-drone-practices-in-pakistan (accessed April 11, 2015).

CIVIC (Campaign for Innocent Victims in Conflict)/Rogers, Christopher. 2010. *Civilians in Armed Conflict. Civilian Harm and Conflict in Northwest Pakistan*. http://civiliansinconflict.org/our-work/countries/pakistan/pakistan-publications (accessed April 11, 2015).

Clarke, Adele, and Susan Leigh Star. 1998. "On Coming Home and Intellectual Generosity. An Adele E. Clark and Susan Leigh Star Guest Edition." Special Issue: Legacies of Research from Anselm Strauss. *Symbolic Interaction* 21:341–351.

Cohen, Eliot A. 1996. "A Revolution in Warfare." *Foreign Affairs* 95:37–54.

Crandall, Jordan. 2005. "Operational Media." *ctheory*. http://www.ctheory.net/articles.aspx?id=441 (accessed January 7, 2012).

Dandeker, Christopher. 2006. "Surveillance and Military Transformation: Organizational Trends in Twenty-First Century Armed Services. In *The New Politics of Surveillance and Visibility*, ed. K. Haggerty and R. Ericson, 225–249. Toronto: University of Toronto Press.

Der Derian, James. 2009. *Virtuous War*. Rev. ed. New York; London: Routledge.

Dilanian, Ken. 2011. "U.S. Counter-terrorism Strategy to Rely on Surgical Strikes, Unmanned Drones." *Los Angeles Times*, June 29. http://articles.latimes.com/2011/jun/29/news/la-pn-al-qaeda-strategy-20110629 (accessed November 4, 2011).

Dillon, Michael. 2002. "Network Society, Network-centric Warfare and the State of Emergency." *Theory, Culture & Society* 19 (4): 71–79.

Edwards, Paul N. 1996. *The Closed World. Computer and the Politics of Discourse in Cold War America*. Cambridge, MA: MIT Press.

Etheridge, Eric. 2009. "The Vietnam War Guide to Afghanistan." *The New York Times*, October 12. http://opinionator.blogs.nytimes.com/2009/10/12/the-vietnam-war-guide-to-afghanistan/?_r=0 (accessed April 11, 2015).

Graham, Stephen. 2005. 'Switching Cities Off: Urban Infrastructure and US Air Power." *City* 9 (2): 169–194.

Graham, Stephen. 2009. *Cities under Siege. The New Military Urbanism*. London: Verso.

Gregory, Derek. 2011. "The Everywhere War." *Geographical Journal* 177 (3): 238–250.

Gregory, Derek. 2014. "Keeping Up with the Drones." http://geographicalimaginations.com/2014/11/20/keeping-up-with-the-drones/ (accessed November 18, 2015).

"'Hit and Run' Drone Strikes Are 'Breaking Laws of War.'"2011. *Channel 4*, June 23. http://www.channel4.com/news/hit-and-run-drone-strikes-are-breaking-laws-of-war (accessed November 4, 2011).

Human RightsWatch. 2010. "Letter to Obama on Targeted Killings and Drones." December 7. http://www.hrw.org/en/news/2010/12/07/letter-obama-targeted-killings?print (accessed November 4, 2011).

International Committee of the Red Cross (ICRC) and Nils Melzer. 2009. "Interpretive Guidance on the Notion of Direct Participation in Hostilities under International Humanitarian Law." http://www.icrc.org/eng/assets/files/other/icrc-002-0990.pdf (accessed November 4, 2011).

Kaplan, Caren. 2013. "Precision Targets: GPS and the Militarization of Everyday Life." *Canadian Journal of Communications* 38:397–420.

Lenoir, Tim, and Henry Lowood. 2003. "Theaters of War: The Military-Entertainment Complex." In *Kunstkammer, Laboratorium, Bühne. Schauplätze des Wissens im 17. Jahrhundert*, ed. J. Lazardzig, H. Schramm, and L. Schwarte, 432–464. Berlin: Walter de Gruyter Publishers.

McKelvey, Tara. 2011. "Inside the Killing Machine." *Newsweek*, February 13. http://mag.newsweek.com/2011/02/13/inside-the-killing-machine.htm (accessed April 11, 2015).

Ministry of Defence. 2011. "The UK Approach to Unmanned Aircraft Systems," March 30. https://www.gov.uk/government/publications/jdn-2-11-the-uk-approach-to-unmanned-aircraft-systems (accessed April 11, 2015).

New America Foundation. 2011. *The Year of the Drone. An Analysis of U.S. Drone Strikes in Pakistan, 2004–2011.* http://www.defense-unmanned.com/article/113/an-analysis-of-u.s.-drone-strikes-in-pakistan%2C-2004_2011.html (accessed April 11, 2015).

Nye, Joseph S., Jr., and William A. Owens. 1996. "America's Information Edge." *Foreign Affairs*, March/April, 20–36. http://www.unz.org/Pub/ForeignAffairs-1996mar-00020 (accessed April 11, 2015).

"Over 700 killed in 44 drone strikes in 2009." 2010. *Dawn.com*, January 2. http://www.dawn .com/news/958386/over-700-killed-in-44-drone-strikes-in-2009 (accessed April 11, 2015).

Oxford Research Group. 2011. "Discussion Paper: Drone Attacks, International Law, and the Recording of Civilian Casualties of Armed Conflict." *Reliefweb*, June 23. http://reliefweb.int/ node/421916 (accessed November 4, 2011).

Pincus, Walter. 2011. "Are Drones a Technological Tipping Point in Warfare?" *The Washington Post*, April 25. http://www.washingtonpost.com/world/are-predator-drones-a-technological-tipping -point-in-warfare/2011/04/19/AFmC6PdE_story.html (accessed November 4, 2011).

Robinson, Jennifer. 2011. "'Bugsplat': The Ugly US Drone War in Pakistan." *Al Jazeera*, November 29. http://www.aljazeera.com/indepth/opinion/2011/11/201111278839153400.html (accessed January 22, 2012).

Roggio, Bill, and Alexander Mayer. 2010. "Analysis: US Air Campaign in Pakistan Heats Up." *The Long War Journal*, January 5. http://www.longwarjournal.org/archives/2010/01/analysis_us_air _camp.php (accessed November 4, 2011).

Star, Susan Leigh. 1995. "Introduction." In *The Cultures of Computing*, ed. S. L. Star, 1–28. Oxford: Blackwell.

Star, Susan Leigh. 2007. "5 Answers." In *Philosophy of Technology: 5 Questions*, ed. J.-K. B. Olsen and E. Selinger, 223–231. Copenhagen: Automatic Press/VIP.

Star, Susan Leigh, and Geoffrey Bowker. 2007. "Enacting Silence: Residual Categories as a Challenge for Ethics, Information Systems, and Communication. *Ethics and Information Technology* 9:273–280.

Tilford, Earl H., Jr. 1995. "The Revolution in Military Affairs: Prospects and Cautions." Monograph. http://www.strategicstudiesinstitute.army.mil/pubs/display.cfm?pubid=238 (accessed November 4, 2011).

Wade, Jack, and Pauline Shulman. 2010. "Advancements in Full-Motion Video for Military UAS Surveillance Applications," September 1. http://www.techbriefs.com/component/content/ article/8504 (accessed November 4, 2011).

Weber, Jutta. 2009. "High-Tech Masculinities, Oriental Outlaws & Armchair Warfare." Paper presented at the conference Men and Masculinities, Moving On! Embodiments, Virtualities, Transnationalisations, April. Centre for Gender Excellence, Linköping University.

Weber, Jutta. 2010. "Armchair Warfare 'on Terrorism.' On Robots, Targeted Assassinations and Strategic Violations of International Law." In *Thinking Machines and the Philosophy of Computer Science: Concepts and Principles*, ed. Jordi Vallverdú, 206–222. Hershey, PA: Information Science Reference.

Weizman, Eyal. 2006. *Lethal Theory*, May, 53–78. http://www.jstor.org/discover/10.2307/ 41765087?uid=3739560&uid=2&uid=4&uid=3739256&sid=21105965453531 (accessed April 11, 2015).

Woods, Chris. 2011a. "US Claims of 'No Civilian Deaths' Are Untrue." *The Bureau of Investigative Journalism*, July 18. http://www.thebureauinvestigates.com/2011/07/18/washingtons-untrue-claims-no-civilian-deaths-in-pakistan-drone-strikes/ (accessed November 4, 2011).

Woods, Chris. 2011b. "Covert Drone War." *The Bureau of Investigative Journalism*, August 10. http://www.thebureauinvestigates.com/2011/08/10/most-complete-picture-yet-of-cia-drone-strikes/ (accessed December 23, 2011).

16 Infrastructures for Remembering

Janet Ceja Alcalá, Susan Leigh Star, and Geoffrey C. Bowker

Although archives have sometimes stabilized the heritage of communities through systems of organization that value preservation, broad representation, and access, they cannot be completely grounded. The contextual understanding incorporating the knowledge of the objectified has historically gone missing. Sometimes their "whispers" remain in the archives as Jeannette A. Bastian (2004) referred to the enduring traces of African slaves in records created by the Danish during their rule in the West Indies. These whispers reveal valuable information about the colonized, and as one Virgin Islander pointed out, "This history [in the records] is linked to us, it's our history!" (32). If the linkages between the records created by the colonizers of the colonized are evident and valued by the inhabitants who live and breathe this history, how are these ties made apparent in archives? Are we listening to the whispers?

No one was listening in 1917 when the United States purchased the West Indies from the Danish and took the records back to Denmark. Nor in the 1930s, when other colonial records of the now U.S. Virgin Islands were taken to the U.S. National Archives and filed away under a new bureaucratic recordkeeping infrastructure. After all, the United States had inherited them. The records twice displaced in settings other than the locale in which they were originally created left the people on the territory without much access to an important part of their recorded history.

This case is not unique and plenty others exist in which the colonized are not the recorders, but the recorded. Today, the whispers are being heard by archivists who now ask what the responsibilities of their profession are in restoring knowledge of postcolonial societies historically denied in archives (see, for example, Pugh 2011). This query falls into a space of uncertainty, especially as archival documents become digital objects, born-digital materials continue to increase in size, and people represented transgress individualities that have traditionally identified with the nation-state, or one homogenous culture. Additionally, digital materials travel in a way that facilitates their trade and distribution within infrastructures that often lack the indigenous guidance in the handling of these materials. That is, many collections are distributed among previous colonizer nations and are now being contested by formerly colonized groups who seek to forge their own histories.

In this chapter, we look at one of many borderlands of the information age: those spaces between offline and online representations. Like other borderlands, they bleed into one another, face each other, deny each other, and form chasms into which those on both sides—the many sides—must face themselves. The borderland is inhabited by classification systems, standards, many sorts of forms, and, our case in point, archives. All become, in Adrienne Rich's words, "a book of myths/in which our faces do not appear" (Rich [1973] 1994).

Gloria Anzaldúa (1987) posited a formal model of the borderland: we belong to two or more life-shaping worlds; we learn first to speak with a twisted tongue, imposters, crossers of life worlds, seers. Archival spaces are very interesting borderlands, populated with data (objects, languages, images, memories) that belong to multiple worlds. The objects are often gathered offline and then migrated online, over a chasm fraught with risks of erasure and simplification (Star 1983). As Ellen Balka (2009) has shown in her study of accidents in remote areas, these sorts of handoff from one context to another are never simple. There are silences, conversions, ad hoc fill-ins. In her examples of movement among paper, radios, medical classifications of different sorts, and computers, even after handoff, all forms continue to cohabit an uneasy and unclear ecology.

The "computerization" literature (from the 1980s through the late 1990s, see, e.g., Kling 1996) often discussed the "impact" of computers on previously uncomputerized organizations. Today the term is being superseded by a more constructivist approach that also includes the impact of humans on technologies. Yet those early questions remain: How has work changed? What new cognitive tasks have been imposed on users? Today, with the new forms of computerization, the ubiquity of applications, and a range of more interactive Web 2.0 possibilities, these concerns have morphed into an exploration of all aspects of life. Instead of impacting, they now take the form of coexisting knowledge modes, inertias, and pastiches.

For this reason, in this chapter we are focused on precisely that bricolaged ecology between traditional Western archival guidelines, and those proposed by minority and indigenous groups, specifically, Native Americans. This is but one of many emerging cases as people of color, indigenous peoples, and their allies face overwhelmingly Western approaches to preservation, representation, and access measures. We overflow the comfortable bounds of checklists and standard canons; we want our communities and our mestizo/a heritage to be part of the story. We present a case that illustrates the existing archival heritage of the United States with the growing concerns of Native American communities who now seek to integrate their work practices into the professional culture of archives. This inheritance is manifest in the "Protocols for Native American Archival Materials" (hereinafter referred to as Protocols), a set of guidelines that represent the values and concerns of Native American communities. However, before the case study, we give a little of each of our histories, so that you may better hear our concerns. This is a part of the infrastructure of remembering in indigenous worlds.

What We Keep and What We Throw Away: Why This Means So Much to Us

We are all affected by the ravages of memory and forgetting. We begin here with experiences, as we remember them—experiences that forced us to look into the abyss of nothingness.

Ceja Alcalá

I used to sit for hours at a desk in a museum rewinding through audiotape recordings trying to make out the Spanish language accounts of a Chumash elder. Back then, I was a volunteer and had been transcribing and translating oral histories because of my interest in linguistics and language preservation. I was interested in listening to the Chumash words that in theory had gone extinct, but actually survived by morphing with the dominant language—Spanish. I happily took on a task I valued for its familiarity in my own home with translations I did for my parents from English to Spanish and vice versa. I wanted to know more about the cultural encounters of linguistic traditions that had created an association for me through the inheritance of a colonial lingua.

In listening to the stories of the Chumash elder, I not only heard two languages morph into one, but I also recalled the expressions of my grandfather. Both talked slowly and vividly about the land and animals they cared for out in the hills somewhere . . . in different parts of the Americas. The Chumash elder's words mediated by magnetic media had frozen them in time, while my memories had been retrieved as living recollections. Does anyone remember the elder's stories as living recollections?

What I learned from listening to the elder's voice in the museum's archive was that it required more than an instrument to be truly understood and preserved. We require multiple forms of contextual guidance—historical, human, ecological, and technological—in order to achieve the cultural dexterity to translate from one language to another.

I was vividly exposed to the limitations of translating archival practice when I used the standard perspective of the West down South in Cuba where I witnessed a motion-picture film archive undergoing deterioration. The beginning of the end of Cuba's moving image heritage had arrived shortly after its "special period" (the economic crisis that ensued after the fall of the Soviet Union in the 1990s) and I was staring its fractured and starving condition right in the face (Ceja Alcalá 2013). In the technical literature the solution to these problems tend to be approached through the use of tools and techniques that afford general claims about technological obsolescence while the political implications are nowhere in sight. Well, there is more to Cuba's story than: "the temperature and humidity in which motion picture film should be housed is . . . and therefore, this collection has a life expectancy of . . ."

As I examined the mold on the motion-picture film stock, saw the decay, smelled the vinegar, endured the heat, and felt the sweat form in my pores, I not only could predict the loss, I also felt it. A knot formed in my stomach because Cuban cinema as envisioned through the eyes of Cuban filmmakers fifty years after the Revolution was in both an oppressed and depressed state. We, in the United States, were not meant to see these images because our foreign policy toward Cuba had from the very beginning sought to deny this country's revolutionary nation-building narrative. The Cuban government chose political and economic independence over dependence on the United States, and Cuba's motion-picture film heritage is ironically one of the most authentic manifestations of this ideology sentenced to expire in this era.

Star

"You're Jewish enough for Hitler." I am sitting at a picnic table, sipping a Coke and half-listening to the full summer birdsong of rural New England. I am at an ecumenical (sic) church summer camp, meant to expose us to other faiths over a two-week period. It is sponsored by my Congregational Church. That winter, I sold over 600 boxes of Girl Scout cookies in order to pay my room and board. Just a couple of months ago, the following encounter had ensued with my mother, my sister, and myself.

"Ma, you have a big nose. Like a Jewish nose," says my sister. I listen, squirming in a big-sister sniffiness at her rudeness, I think, to my mother. I expect my mother to say something reprimand-ish. Instead, in a voice I've never heard from her before, she says, quietly, "We are Jewish. German Jews. Originally from Russia." At the age of thirteen, just when one is trying on various identities, a new, earth-cracking one appears at my feet. A San Andreas Fault of a one, in fact. I am too stunned to speak. But I begin investigating. Quietly. This is not easy. I live in a rural Rhode Island town, before the Internet, sans library, in a house without books.

My conversation with Fred, the only rabbi I have ever met and who in fact is the only person "covering" the Jewish angle at our summer camp, is vitally important to me. I've only begun to study the then-new biological sciences curriculum, where the concept of genes is introduced later on. I have no sense of heredity, except that one is supposed to look like other members of one's family, in a vague sort of way. I turned fourteen just before my conversation with Fred, a kindly, rotund, quiet reform rabbi from New York City. "Jewish enough . . ."

". . . for Hitler." I tell my parents, about a year later, that I wish to study Judaism and participate in her rites and rituals. My mother cries. My father orders me to the room I share with my sister. He threatens to call the police on Fred, with whom I have been carrying on a conversation via letters and brief phone calls. My father comes to the conclusion that he is a pervert out to seduce his oldest daughter.

I do not want the police to bother Fred. I briefly explain, in a letter, what has happened. Several years later, I attend a fancy college on scholarship, and indicate that I want to study religion. There, I am told that the study of religion is "a vocational degree" (clearly A Bad Thing) and that I can only study aspects of the psychology of religion, such as the works of William James. I throw myself into this with a passion. I discover, through some bureaucratic loopholes, that I may take theology classes at a number of nearby institutions by registering at the Divinity School [of one]. One of those places is Boston College, and I take several courses with Mary Daly, a famous feminist philosopher and theologian. For some years, the new women's spirituality takes precedence for me over Judaic studies, dismissed by her as "patriarchal." Nevertheless, in my secret heart, I remain haunted and confused. My worlds within seem irreconcilable.

Decades go by, and my mother passes away at a young age of a very rare cancer, one that is shared by only four hundred people in the United States. She tells me that her grandmother, my great-, had the same thing, she thinks, but nobody had a name for it then. I file this away. Quietly. At the funeral home, my father must fill out her death certificate. The funeral director, a rather awful man with no hair at all whom I call Lurch (I later find out he has a medical condition), asks him, "Name? Religion? Race?" To the latter question, my father says, "Er, English." I see that he is in shock, and decide that this is not the right time to intercede. Certainly, English is not a race, or even an ethnicity. But the funeral director has inscribed something in the book of vital records. Much later, it strikes me that there may be some sort of connection between Ashkenazi Jews and this cancer. Many forms of cancer, including breast cancer, are rampant in my mother's side of the family. In the Internet epidemiological files, my mother becomes a statistic.

Two weeks ago, I stand in my own doctor's office, for my first general exam since starting this new job. I have just read something about the link between Ashkenazi Jews and a number of genetic predispositions, including breast cancer. I ask her, should I get tested for BRCA1 or other genetic possibilities. She questions me about many things, and I tell her, as part of a list, that I am half Jewish. Yes, Ashkenazi. She says to me, "Okay, this is off the record. There is one place in town that does this testing." They are good at the tests, she continues, but more importantly, "They keep only paper records. You know, HIPPA requires that electronic tests as well as paper are confidential. But we both know that isn't really true. At this genetic testing place, they write it down on paper, and put it in a fireproof safe whose combination is known only to the testing MD and the director. That way, it won't make its way onto your insurance information, or the web, nor will it stop you from applying for another job at some future point." She writes, very lightly, the telephone number of the clinic in blue ballpoint pen on a separate piece of paper. I am grateful for the favor, and the risks she is taking. So the paper-web thing goes both ways.

Later, I think:

". . . for Hitler."

This summer, I meet with some new friends from the Ylongu tribe in Arnemland, in north central Australia. Djangal, one of the tribal elders, invites us to her home south of Kakadu National Park, near Gove. Far from anywhere I have ever been. She takes us to a place of Singing, a sacred channel where turtles come to breed, and where they eat small organisms in the water. Her gestures in sharing this place are compact, like a dance. I tell her that one of them reminds me of the Crane form in tai chi, a graceful holding of arms that both points and protects. She laughs and says, yes, I know, my children are of the Crane clan. Later, we visit near her grandmother's sacred tree place, also a part of the local songs. It is surrounded by an iron gate, and backgrounded by the world's longest conveyer belt, all rusty brown in color. It brings bauxite from the hinterland behind her homeland, and puts it on ships heading for industrial ports. Much of the debris from this has poisoned the water. I wonder how she can bear this depredation. But she does.

Bowker

I grew up without race or ethnicity, in Britain and Australia. In Britain (the only country not to put its name on its stamps, since "we" invented them), "races" were terms for other people, not for us. In Australia, we were actively discouraged from any ethnic talk in schools.

Worse than coming from Britain, I was born English. If you ask any English men or women about their ethnicity, they will hedge. Most will refer to some Celtic forefather—the Gaes having had the gall to repress the Roman and Gallic invasions of Julius Caesar and Richard the Conqueror. However, the fact is that most of us are a Roman/Gallic mix.

When I came to this realization, I declared to my friends and others who would listen that I came from the conqueror tradition—I claimed a Roman nose and a French sensibility.

Then I came to the United States and discovered that everybody had an ethnicity—only the morally culpable were missing one. So I'm still hard at work trying to find one. However, I am from an ethnic background which in many ways denies its own ethnicity—much as the Gothic empires following the fall of the Roman empire made no ethnic divides: you were a Goth if you acted like a Goth, full stop. Representing the silent and unspoken is too often seen as the bailiwick of the oppressed; dismantling universalist identities is just as significant. Even though I'd still prefer to live in a world where ethnicity weren't such an identity marker . . . (see, e.g., Reynolds 1999).

A Case Study: How We (as Objects and Subjects) Are Remembered and Forgotten

For decades, theories of "the archive" have been intriguing academic thinkers. Less interesting in academic circles and certainly less understood has been the role of archivists in creating such instruments, objectifying and making subjects of the accumulated information about individuals, activities, and phenomena. Archivists manage recordkeeping systems that account for the ongoing activities of institutions and individuals. They decide which materials make it into archives and which are thrown out as dross. With the exception of a few provocative barbs (archivists, for example, have been called "failed historians") (Starn 2002) the reach of their responsibilities lies unacknowledged for in its concreteness archiving *is* boring to the average humanist. This is more of a reason why archives *should* be studied in all of their dimensions. We address one of them here: provenance.

Archives have historically been organized by their provenance and represented contextually through descriptive practices that validate and preserve the authenticity of the records they contain. This ensures standard arrangement and retrieval mechanisms that adhere to principles known as *respect des fonds* (respect for the source) and *provenienzprinzip* (principle of provenance). The French concept of *respect des fonds* expresses the importance of keeping archival collections intact and in their original order so as to ensure that all critical contextual information is available. Moreover, *provenienzprinzip* is a derivative idea hailing from Germany that establishes the integrity of the records based on their creators (Cook 1992). The aforementioned account of the U.S. Virgin Islands (formerly known as Dutch West Indies) is an example of how these principles can travel through time from one place to another, although not without problems.

The Danish, for instance, considered the provenance of the records they generated and administered on the Virgin Islands as rightfully their own and as a result sent them back to the heart of the nation after they ceased to be in power. U.S. archivists took other records to their own national archives as a result of their new role in the chain of archival custody. In both cases, the records were viewed as possessions and the principle of provenance permitted each country's government to claim them as such. The source location of the records—"the colony"—and the period from which the records originate were no longer respected as a viable place and context of origin even though the records contextually represented the region and society better than they do Denmark or the United States. Yet if the records are interpreted as being a part of more than one social world; that is, if Denmark, the United States, and the U.S. Virgin Islands are viewed as related entities instead of differentiated units, then the principle of provenance breaks down into a community of records that evidences multiple layers of social interaction (Bastian 2003).

More recently, it has been argued that the social milieu of records creation ought to be designated a part of the principle of provenance (Bastian 2001; Nesmith 2006). By being linked to society, the authenticity of archives is far-reaching and complex rather than narrowly based on the procedures used to generate the records in situ by the original administrating body and those implemented while in archival custody. Thus, the notion of a societal provenance is a radical way of thinking about records, their care, ownership, and value.

We will begin to analyze this process by briefly describing some of the foundational work developed in Western archives, and how it has handled alternative archiving proposals made by the first peoples of North America in the United States.

European and U.S. Traditions

It is useful to begin with a statement made by a European archivist about archives in the United States in order to signal the diversity found in Western archival practices and to demonstrate how new archival interpretations can render new traditions. Additionally, the excerpt demonstrates how archival principles travel in time from one geographical location to another and change to be able to *survive*. Thus in describing his impression of annual meetings held in the United States by the Society of American Archivists (SAA), Joan Van Albada, a Dutch archivist commented:

I was filled with astonishment because I was more and more convinced that the SAA annual meetings dealt only partially with my profession, namely the profession of archivist. Most sessions, apart from those related to topics like conservation and user services, dealt with collecting and documenting and not with the core of the profession: archives management, the accessioning, selection and appraisal, and processing of record groups. The sessions did not deal with the essence of archives as defined by Muller, Feith, and Fruin [Dutch archivists] in 1898 or as defined in the 1984 ICA [International Council of Archives] dictionary. (1991, 399)

Van Albada's notice of how archival principles generated in Europe were not faithfully practiced in the United States has much to do with how archival practice develops locally. Archives in the United States were initially shaped by the practices of historical societies with a mission to collect the manuscripts of prominent people, and later of public archives, which administered public records for the sake of all of society. Historical societies, which had existed since the late 1800s, focused on gathering unique and private manuscripts specifically for the use of historians by using organizational schema they initially borrowed from libraries (Gilliland-Swetland 1991). This is why Van Albada mentions the acts of collecting and documenting as different from his profession because it is the task of librarians to collect and of historians to document. By the 1950s, the public archives tradition continued to develop in the United States as the government's administration confronted a growing body of records and borrowed principles from Europe, such as provenance, but also invented homegrown ones like appraisal.

Appraisal was introduced by Theodore R. Schellenberg at the National Archives as a technique for selecting specific records for ongoing maintenance at the cost of disposing of others. It is a concept that grew out of Schellenberg's experience with the vast and speedy accumulation of records from the nation's transformation into a modern bureaucracy. He approached the information overload of his day by conjuring a formula that would permit archivists to appraise records based on their evidential (the evidence of administrative activities from which the records were created) and informational (the content of records as they pertain to the administrative body and as research records for study) values (Schellenberg 1956). This method can be broadly interpreted and depends largely on archivists' formative backgrounds (not to mention subjective judgments) to select records for posterity. This new and divergent principle differed from European archival practice where archivists aimed to intervene minimally in the care of records, if at all, in order for archives to be certified authentic and impartial. Such philosophy hails from positivism and the idea that records can serve as sources to establish truth using scientific methods that go as far back as the seventeenth century. According to Sir Hilary Jenkinson (1944), Deputy Keeper of the British Records Office and a follower of Muller's, Feith's, and Fruin's *Dutch Manual,* the perfect archive was one incapable of lying if it entered the archives in precisely the same manner left behind by the original creators. However, in a recent reiteration of a decades-long debate between Native Americans and SAA, this distinction blurs.

Native American Protocols: Who Is Listening?

Elizabeth Adkins appeared to be listening to Native American archivists when she brought up a story in her presidential address at the 2007 SAA annual meeting about Native American women archivists who "were stopped, thanked and complimented with handshakes and smiles by many Annual Meeting attendees" after they had presented their perspectives concerning the treatment and usage of Native American tribal materials in archives (2008, 49). Adkins commented that such affective attitudes helped lay down the groundwork for diversity in the archival profession—attitudes that were not displayed when the Protocols were submitted for endorsement by the Native American Archives Roundtable section of the SAA to the profession at large that same year. The document was written by a group of nineteen archivists, librarians, museum curators, historians, and anthropologists of Native American and non-Native American heritage to identify the best professional practices for culturally responsive care of Native American archival materials held by nontribal organizations (First Archivists Circle 2007). More specifically, the Protocols were written to open lines of communication regarding the "management, preservation, and transmission of Native American knowledge and information resources" and "to inspire and to foster mutual respect and reciprocity" and encourage the development of the guidelines presented (2).

Shared Stewardship

The Protocols advocate shared stewardship, a concept that would extend the principle of provenance by retooling the manner in which Native American archival materials are cared for in archival environments. Shared stewardship impacts the overall preservation management and transmission of Native American knowledge in nontribal archives and calls for the rethinking and negotiation of how records are organized, described, and made accessible. If applied, shared stewardship requires archivists with records of any kind concerning Native Americans to consult with the communities represented in order to understand how their cultural paradigms bear upon the materials in their custody, as well as to determine how the materials can be best protected. Protection may encompass restricting access to records with religious or ceremonial knowledge that should only be available to certain Native American communities. Furthermore, this type of protection is significant to the health and well-being of Native communities that may suffer from the recorded information's dissemination, or its manifestation in different contexts.

The museum policies in place at the University of British Columbia's (UBC) Museum of Anthropology are a good example of how Canadian archivists have dealt with such archival complexities with the ethnographic archives of Aboriginal peoples of Canada. Some of the archives held by the museum were created during a period when "salvage anthropology" was practiced by anthropologists who believed Aboriginal peoples were dying off and had to be recorded in order to be preserved. Documentary evidence was seen as the most viable option to save their heritage as they underwent forced assimilation. At the UBC, the remnants of this past grew into discussions between Aboriginal peoples and museum archivists with a policy document drafted to express the museum's commitment to "respecting the values and spiritual beliefs of the cultures represented in its collections" (Laszlo 2006, 304). In practice, this statement has meant consulting with Aboriginal communities to identify which culturally sensitive materials should be treated differently from the rest (e.g., sacred expressions); working closely with Aboriginal communities to resolve questions of access to materials that are restricted; creating protocols and informed consent forms for research conducted among Aboriginal communities at the museum (which may include the retention or disposition of their work); and building partnerships by offering internship opportunities to Aboriginal communities in order to help create awareness of the museum's operations. At the center of these relationships, moreover, is a concerted effort to build mutual respect and trust.

Collective Right to Self-Determination

Since the 1950s, the archival as well as the library professions have sought to foster open access and make freedom of access to information a fundamental right. Loriene Roy and Kristen Hogan found that these concepts were introduced at a critical juncture in the nation's pursuit of individualism seen specifically in "librarianship's turn to protecting freedoms of expression" and "the U.S. government['s] increased attempt to disperse and dissolve tribal identity" (Roy and Hogan 2010, 21). During the 1950s, the federal government's Bureau of Indian Affairs (BIA) instituted their relocation program as a way to assimilate Native American communities into mainstream society by removing them from their lands. Additionally, the BIA had put into place a termination policy that removed all protection measures on 109 tribal lands and on the people who lived there (Fixico 2009). Without financial support to maintain a centralized community and public safety measures on their lands, this mechanism created incentives for Native American tribes to abandon them. Approximately one hundred thousand Native American peoples relocated to metropolitan areas under the federal government's relocation program (ibid.). To Roy and Hogan's timeline we can add Schellenberg's implementation of appraisal at the National Archives, which would have coincided with the integration of these one hundred thousand Native American populations into the country's recordkeeping system. Through their integration into U.S society, Native American communities gained the right to openly access information as individuals. However, it was a value that competed with the notion that a Native American community could hold special rights over their traditional knowledge and restrict it depending on the cultural circumstances surrounding its creation. Traditional knowledge is defined by the Protocols as "valued knowledge which is individually or communally owned in accordance with established community rules of ownership; often sacred or sensitive and requiring specialized training or status for inheritance or use; often held in trust for a community by an individual; may include songs, oral traditions, customs, and specialized knowledge" (First Archivists Circle 2007, 22).

Some of the implications the Protocols have for archivists include addressing the right of possession, moral rights, and repatriation.[1] Organizing the right of possession would require archivists to demonstrate evidence that the acquisition of items in their collections was done ethically. It would mean that the integrity of archives in contextualizing materials based on their provenance as well as acquisition policies are put into question. Granting moral rights to Native American communities would give them the right to attribution, as well the possibility of controlling the integrity of the work(s) that were both ethically and unethically acquired by archives. Records that contain sacred rituals or other sensitive expressions could be negotiated for repatriation, restrictions, or duplication. It should be noted that the Protocols would be adopted as deemed appropriate by the archivists and Native American communities, and it does not mean all of these postulations would necessarily apply.

Yet given the apparently serious consequences archivists would have to confront with access restrictions, it is important to point out that in the U.S. such restrictions are not new. As mentioned earlier, since the founding of historical societies in the eighteenth century granting access to records has been an issue in which special group privileges have existed. Further, even in contemporary times when donors of archives request restrictions on their records and they are granted it too produces inequity of access. As much as archivists are against restricting access to archives, the prestigious source of origin of a donation may outweigh the negative outcome because the institution may be endowed with a sense of notoriety.[2] It is not new for archivists to have to mediate between open and restricted access to information through the drafting of policies that set out to balance the competing needs of donors, records creators, and users of archives.

Reacting to the Protocols

Although the Protocols were not meant to be prescriptive and were described as "a work in progress—subject to revision and enhancement" (Boles, Shongo, and Weideman 2008, 32)—many of the comments made about them by archivists did not interpret them as such. Out of a profession of more than five thousand members only thirty-nine people responded to the Protocols when they were sent out for review (some of whom were not archivists). A segment of the Protocols' critics understandably feared the possible violation of federal laws in the justification of intellectual property issues if they were implemented. One section of the SAA stated: "The language in the protocols is very legal in nature—could this be used to support litigation. I don't think it is in the interest of archives to expand their vulnerability to lawsuits" (ibid., 69). However, if it is not in the interest of archivists to expand said vulnerability by openly wrestling with the challenges and possibilities the Protocols afford the profession, the trade-off is that the human rights of Native American communities seeking a right to self-determination are compromised. The aforementioned section of the SAA went on to say: "These protocols appear to be striving to set up private collections with specific rules within open institutional settings based on the legality, accessibility and review elements of the document. I don't believe that is a healthy option, for either the archival or Native American communities" (ibid.).

Their statement is an enactment of Linda Tuhiwai Smith's critical assertion that "Authorities and outside experts are often called in to verify, comment upon, and give judgments about the validity of indigenous claims to cultural beliefs, values, ways of knowing and historical accounts" (1999, 72–73). The comments made by archival professionals about the Protocols sadly reveal the accounts of experts who almost unanimously claim to value cultural sensitivity and diversity, but cannot manage to see how the integration of the worldviews of Native American communities factor into archival work. It has too often been the case with indigenous materials that an outside group of experts fails to acknowledge another viewpoint. Instead of a true negotiation, those in power paternalistically decide the best option for a community. They also fail to recognize that communities themselves change dynamically. In Australia, many young Yolŋu men and women, for example, do not accept the concept of "men's knowledge" and "women's knowledge"—so reifying them into a classification can be seen as supporting a particular conservative section of that society. In turn, this may result in the loss of another form of diversity based on age!

What is at stake with the Protocols is a right to self-determination by Native American communities. Self-determination is a right to collective cultural development (as opposed to an individual's participation in a national culture) and a major tenet for building recordkeeping infrastructures that reflect first peoples' views of the world (Roy and Hogan 2010). As Michael Cook (2006) points out, because a large part of the history of the twentieth century concerns political repression there is also a struggle to recover from it and ethical issues to consider in the process. The repression Native American communities have faced under the direction of the U.S. government is no exception, and records have become an important part of their struggle (see, e.g., Kelly 2009; Boucher Krebs 2012; Underhill 2006). The Protocols have at least raised awareness within the U.S. archival profession of the current issues Native American communities and their allies face in archival settings as creators and users of records, but also as archivists.

In the end and after much discussion, the SAA Council did not endorse the Protocols but acknowledged that, "in a pluralistic society there is a need for ongoing dialog regarding matters of cultural sensitivity among archivists, stakeholders, and the many varied cultures represented within archival repositories" (Greene 2008, 25).

Breaking the Bounds: Rethinking Archives in a Time of Shifting Modalities

The visual aspects of the worlds between offline and online, in the case of Native American knowledge traditions, are always both seen and unseen. They are inscribed under many regimes—sometimes offline, sometimes online. The experience Star describes of an erasure of identity is very common for those identities "we learn to despise" (Cliff 1980). Janet's feeling of gut-wrenching loss at seeing precious films lost to political regimes and enforced poverty occurs commonly in all indigenous communities. Geof's "lack of" ethnicity is erased in an all-too-common claim of "we have no race." How may we see these differently?

There remains a unique form of visualization available if we use the double vision. When Anzaldúa (1987) speaks of *coatlicue* in *Borderlands/La Frontera* she means a kind of affordance available to survivors. This is a state of holding together the torn edges of the border worlds, while staring into the crack between. Her description of this way of looking is at once pedestrian and mystical. But it does offer an alternative to dichotomy. If we can hold that multiple vision, the destruction and the survival, the offline and the online, we may be offered a view of this abyss: Not a picture, or film, or even a word-picture, but the kind of view that transforms the meaning of archives and of memory itself—holding them to the flames of history, and putting ourselves on the line. In the words of T. S. Eliot, perhaps then "the fire and the rose are one," (1971, 59) or, our fragments may be put together again by another self.

Notes

1. The 1992 Native American Graves Protection and Repatriation Action has set a precedent in U.S. museums that hold Native American artifacts (not archival records) with regard to the right of possession, legal rights, and the possibility of both individual and community ownership and repatriation.

2. Such an example is the Freud Archives at the Library of Congress. See Malcolm 1984.

References

Adkins, E. W. 2008. "Our Journey toward Diversity—and a Call to (More) Action." *American Archivist* 71 (1): 21–49.

Anzaldúa, G. 1987. *Borderlands/La Frontera: The New Mestiza*. San Francisco: Aunt Lute Books.

Balka, E. 2009. "Shadow Bodies." Paper presented to the conference on Bodies, Thresholds and Tolerances. Cambridge, UK.

Bastian, J. A. 2001. "A Question of Custody: The Colonial Archives of the United States Virgin Islands." *American Archivist* 64 (1) (Spring/Summer): 96–114.

Bastian, J. A. 2003. *Owning Memory: How a Caribbean Community Lost Its Archives and Found Its History*. Westport, CT: Libraries Unlimited.

Bastian, J. A. 2004. "Whispers in the Archives: Finding the Voices of the Colonized in the Records of the Colonizer." In *Political Pressure and the Archival Record*, ed. M. Procter, M. Cook, and C. Williams, 25–43. Chicago: Society of American Archivists.

Boles, F., D. George Shongo, and C. Weideman. 2008. *Report: Task Force to Review Protocols for Native American Archival Materials*. Chicago: Society of American Archivists. http://www .archivists.org/governance/taskforces/0208-NativeAmProtocols-IIIA.pdf (accessed July 23, 2015).

Boucher Krebs, A. 2012. "Native America's Twenty-First-Century Right to Know." *Archival Science* 12:173–190.

Ceja Alcalá, J. 2013. "Imperfect Archives and the Principle of Social Praxis in the History of Film Preservation in Latin America." *The Moving Image* 13 (1): 66–97.

Cliff, M. 1980. *Claiming an Identity They Taught Me to Despise*. Watertown, MA: Persephone Press.

Cook, M. 2006. "Professional Ethics and Practice in Archives and Records Management in a Human Rights Context." *Journal of the Society of Archivists* 27 (1): 1–15.

Cook, T. 1992. "The Concept of the Archival Fonds in the Post-custodial Era: Theory, Problems and Solutions." *Archivaria* 1:24–37.

Eliot, T. S. 1971. *Four Quartets*. New York: Harcourt Brace Jovanovich.

First Archivists Circle. 2007. "Protocols for Native American Archival Materials." Unpublished manuscript, Salamanca, NY. http://www2.nau.edu/libnap-p/protocols.html.

Fixico, D. L. 2009. "The Federal Indian Relocation Programme of the 1950s and the Urbanization of Indian Identity." In *Removing Peoples: Forced Removal in the Modern World*, ed. R. Bessel and C. B. Haake, 107–129. Oxford; New York: Oxford University Press.

Gilliland-Swetland, L. 1991. "The Provenance of a Profession: The Permanence of the Public Archives and Historical Manuscripts Traditions in American Archival History." *American Archivist* 54 (2): 160–175.

Greene, M. A. 2008. "Protocols for Native American Archival Materials." *Archival Outlook*, March/April, 3, 25.

Irons Walch, V., N. Beaumont, E. Yakel, J. Bastian, N. Zimmelman, S. Davis, et al. 2006. "A*CENSUS (Archival Census and Education Needs Survey in the United States)." *American Archivist* 69 (2): 291–419.

Jenkinson, H. 1944. "Reflections of an Archivist." *Contemporary Review* 165:355–361.

Kelly, G. 2009. "The Single Noongar Claim: Native Title, Archival Records and Aboriginal Community in Western Australia." In *Community Archives: The Shaping of Memory*, ed. J. A. Bastian and B. Alexander, 49–63. London: Facet.

Kling, R. 1996. *Computerization and Controversy: Value Conflicts and Social Choices*. San Diego: Academic Press.

Laszlo, K. 2006. "Ethnographic Archival Records and Cultural Property." *Archivaria* 61:299–307.

Malcolm, J. 1984. *In the Freud Archives*. New York: Knopf.

Nesmith, Tom. 2006. "The Concept of Societal Provenance and Records of Nineteenth-Century Aboriginal–European Relations in Western Canada: Implications for Archival Theory and Practice." *Archival Science* 6 (3–4): 351–360.

Pugh, Mary. 2011. "Educating for the Archival Multiverse: The Archival Education and Research Institute (AERI), Pluralizing the Archival Curriculum Group (PACG)." *The American Archivist* 74 (1) (Spring/Summer): 69–101.

Reynolds, H. 1999. *Why Weren't We Told?: A Personal Search for the Truth about Our History*. Ringwood, Victoria, AU: Viking.

Rich, A. C. [1973] 1994. *Diving into the Wreck: Poems 1971–1972*. New York: Norton.

Roy, L. R., and K. Hogan. 2010. "We Collect, Organize, Preserve, and Provide Access, with Respect: Indigenous Peoples' Cultural Life in Libraries." In *Beyond Article 19: Libraries and Social and Cultural Rights*, ed. J. B. Edwards and S. P. Edwards. Duluth, MN: Library Juice Press.

Schellenberg, T. R. 1956. *The Appraisal of Modern Public Records*. Washington, DC: U.S. Government Printing Office.

Smith, L. T. 1999. *Decolonizing Methodologies: Research and Indigenous Peoples*. London and New York: Zed Books; Dunedin, NZ: University of Otago Press.

Star, S. L. 1983. "Simplification in Scientific Work: An Example from Neuroscience Research." *Social Studies of Science* 13 (2): 205–228.

Starn, R. 2002. "Truths in the Archives." *Common Knowledge* 8 (2): 387–401.

Underhill, K. J. 2006. "Protocols for Native American Archival Materials." *RBM: A Journal of Rare Books, Manuscripts, and Cultural Heritage* 7 (2): 134–145.

Van Albada, J. 1991. "On the Identity of the American Archival Profession: A European Perspective." *American Archivist* 54 (Summer): 398–402.

17   Triangulation from the Margins

John Leslie King

I first got to know Susan Leigh Star when my colleagues and I hired her on to the faculty of Information and Computer Science at UC Irvine. Leigh was not a computer scientist by training (nor was I), but she fit in remarkably well. I kept in touch with her over the years, through her marriage to Geof Bowker and her appointments at Keele, Urbana-Champaign, La Jolla, Santa Clara, and Pittsburgh. She was a close friend and colleague. I learned a great deal from her. I sometimes flatter myself by thinking that Leigh might have learned something from me. But of that I am not sure.

Susan Leigh Star was an interesting character in many ways. For one thing, it was difficult to get a fix on exactly who—or for that matter, what—she was. I remember talking to a long-time friend of hers while referring to her as "Leigh." That was the only name I had known her to use. Leigh's friend seemed puzzled, and then said, "Oh, you mean Susan." Apparently there was a time when Leigh was known as Susan, and not as Leigh. I subsequently heard stories about her radical feminism, her period as a devout lesbian, her varieties of religious experience. At one point I was told that Susan Leigh Star's last name was not originally Star; she'd taken that name in replacement of her own original last name. When I told another mutual friend that I had been at Leigh and Geof's wedding, the response was "Which one?" I felt like I knew Leigh very well, but as time went on I wondered what it meant to know Leigh.

The "deep" Leigh was a person who most intrigued me. She haunts me still. Neil Young has a phrase in the chorus of his song *Cowgirl in the Sand*: "old enough, now, to change your name." That refrain kept running through my mind until it dawned on me that Leigh was an old soul. I kind of always knew this, but I never really thought about it. How many lives and levels did Leigh draw from? I don't think I will ever know, but more than most people.

Once when Leigh and Geof were staying with us in Ann Arbor, I remarked that my teenage daughter had reported that there were spirits living upstairs in the house where my daughter's room and the guest bedroom were located. Leigh immediately said, "Oh yes, there are several," and proceeded to describe them in detail. After that it never occurred to me that the spirits were not really there. I just figured that I had never been able to recognize them, and to this day I have not. While Leigh seemed to disdain most organized religion, she always struck me as an unusually spiritual person. That is perhaps the strongest memory I have of her.

This tribute is focused on Star's work and her intellectual legacy, but it is difficult to separate the personal from the professional with Leigh. More than anyone I've known, her life and work were of one cloth. I've known others who could not separate work and life (and some of them Leigh and I knew together), but the joining of work and life in those people was most noticeable in the imbalance, as exemplified by the expression "get a life." With Star the work was just part of who she was. I often had the feeling that she did not see any clear dividing lines between the two. I found this inspirational. Maybe it had something to do with how old a soul she was. In any case, there is much to relate about her professional side. I will do so in two ways, addressing first her amazingly disarming style of engagement, and then tackling the title of this piece to explain Star's deepest contributions to my thinking.

Readers familiar with higher education, and especially the University of California, will recognize the importance of "diversity" in university policymaking. For many years the University of California was in the vanguard of "affirmative action," which Wikipedia refers to as "policies that take factors including race, color, religion, gender, sexual orientation or national origin into consideration in order to benefit an underrepresented group, usually justified as countering the effects of a history of discrimination."[1] Shortly after moving from the University of California to the University of Michigan I was pulled into the vortex of the University of Michigan's legal fights in favor of affirmative action that led all the way to the U.S. Supreme Court.[2] After the university's narrow victory reaffirming the constitutionality of affirmative action I was appointed to the Executive Committee of the National Center for Institutional Diversity, established at Michigan as a statement about the importance of diversity in the academy. Throughout this time Leigh's voice echoed in my mind from a discussion she and I had in the early 1990s at UC Irvine.

I simply assumed that she was in favor of affirmative action and would decry any criticism of it. I soon discovered that she was adamantly opposed to the whole venture. This conversation occurred well before January of 1996 when UC Regent Ward Connerly was victorious in his effort to ban affirmative action in the University of California. At the time of our discussion, most members of the UC faculty who had any opinion of affirmative action felt it was firmly established in both law and policy. There was no particular reason for her and I to discuss it, but it came up in a conversation.

I was nearly knocked over by the passion that she exhibited on the matter. Those who knew her well will not be surprised to hear that she was in favor of the principles behind affirmative action. Her vehemence grew from deep disappointment with what affirmative action enabled. Leigh felt affirmativaction let "polite" discrimination off the hook, offering a profoundly fake remedy to a serious problem. Affirmative action allowed well-intentioned and generally self-satisfied people to feel they were doing the "right thing" when in fact they were doing nothing at all. I had heard such arguments before, but never from someone like her, and never with such passion. Her words had a big effect on me, but I did not realize how big an effect until later.

After Star and Bowker moved to the United Kingdom, Ward Connerly pushed through his affirmative action ban in UC. The UC faculty was shocked, *shocked* to think that affirmative action could be overturned. The controversy reminded me of her words. When I moved to Michigan in late 1999 I figured that I would not have to think about affirmative action anymore, and the university's victory in the Supreme Court in 2003 seemed to seal the deal. I was wrong. Ward Connerly was not done. Following his victory in UC, he went on to champion California Proposition 209, which in late 1996 banned affirmative action throughout the state (including higher education). He then went on to champion a similar successful effort in Washington State (Initiative 200). In late 2005 he joined forces with Jennifer Gratz, one of the plaintiffs in the 2003 Supreme Court cases, to campaign for passage of the Michigan Civil Rights Initiative, Proposal 2, which passed by a large margin in 2006. Affirmative action was banned in Michigan. My colleagues in the National Center for Institutional Diversity were despondent. I wrote a short memo, inspired largely by the discussion I had with Star from more than a decade earlier. I explained to my colleagues that it was time to leave affirmative action behind and seek new strategies for achieving diversity. I had come to see the wisdom of Star's insights about affirmative action. She was right.

Another surprise for me came around 2004 when I discussed with her recent findings from neurobiology. Her doctoral dissertation research in the 1980s focused on the "localization" movement of the nineteenth century that tied mental activities to regions of the brain.[3] I had expected her to say that brain science was finally catching up with what she was talking about twenty years earlier. Instead, she abruptly informed me that the current thinking in brain research was "all wrong," and that they would find out eventually that they did not understand what they were talking about. By this time I had learned to let her' views find their place in me: in other words, to say, "Oh . . ." and shut up, which I did in this case. She explained to me what would eventually be understood about the relationship between brain and mind. It might take twenty years, but I expect to learn eventually that she was right.

If her's talents were limited to *seeing*—that is, to being a seer—she would have been a profoundly interesting person to know, but it's doubtful that those who knew her would be much more capable for having known her. However, with Leigh the seer part—which was undeniable—was accompanied an extraordinary ability to teach people how to think about things. This brings me to the title of this brief essay, "Triangulating from the Margins." Star taught me how to do this over a period of several years. I do not think she set out to teach me this (or anything else, for that matter), but I learned from her by watching her and listening to her talk about how she saw things.

The principle of triangulation is well established in the history of geometry and trigonometry. It has long been applied to measurement, as with land surveying. It is simple enough: if you can establish two points and measure the distance between them, you can also establish the distance to a remote third point by using precise measurement of the angles between the three points and simple mathematics. She had adopted triangulation as a metaphor for understanding complex social phenomena: if you could establish different viewing angles to examine some phenomenon, and understand the important differences between those viewing angles, you could "triangulate" on the phenomenon and gain considerable understanding of it. This was not new to me; I had read about triangulation and use of multiple methods in the social sciences, and I knew about the importance of "point of view."[4] She made the whole story clearer to me.

What I did not know was the power of combining the triangulation metaphor with her concepts of the "margin." She knew (as opposed to simply believed) that people in marginalized communities actually saw and thought about things differently than people from "within." This shows up in much of her work, and was probably inspired by the fact that she herself had been a marginal in many ways, and could instantly adopt and understand marginal points of view. The thing that most amazed me about her understanding of complex social phenomena was how fast she was: she could cut through the noise and get hold of the signal faster than anyone I knew. This might have been explained by her powers as a seer, but I was convinced that she was also very skillful. After a good deal of watching, I came to see that much of her talent was in her ability to triangulate from the margins.

For Star, it was next to impossible—and therefore foolish—to try to understand something important simply by understanding the master narrative about that thing. Conventional wisdom suggested that the master narrative was provided by those with the most to gain or lose, but she was unconventional. Often the master narrative was dominated by those who had the most to gain, while those who had the most to lose were "outside" in the margins. In any case, those "outside" in the margins saw things—sometimes important things—that those "inside" could not see. If the observer could establish views from the margins, and ideally from different marginal perspectives, the understanding gained by triangulation would reveal insights that were simply impossible to gain otherwise.

In the years since learning this technique from Leigh, I have used it frequently in my work. It has improved what I do, and it has certainly has made my work much more interesting for me. I also have tried to impart this insight to my students, and some of them have used it to great effect.

Triangulating from the margins is embodied in any number of old sayings about the difficulty of understanding how others see things or experience things. Leigh never suggested to me, or to anyone I know that this originated with her. One might argue that she was merely channeling earlier thinkers. But then again, how do we know that they didn't get this from her?

Notes

1. As of April 1, 2012, http://en.wikipedia.org/wiki/Affirmative_action.

2. *Gratz v. Bollinger*, 539 US 244 (2003), and *Grutter v. Bollinger*, 539 US 306 (2003), were brought by Jennifer Gratz, an undergraduate denied admission to the Ann Arbor campus (she attended the university's Dearborn campus), and Barbara Grutter, a graduate student denied admission to the Michigan Law School. Both suits were brought against the University of Michigan, naming Lee Bollinger as president. The cases were combined before being heard by the Supreme Court. The university lost the Gratz case and had to redo undergraduate admissions, but won the Grutter case upholding use of affirmative action in admissions as constitutional.

3. See her book *Regions of the Mind: Brain Research and the Quest for Scientific Certainty* (Palo Alto: Stanford University Press, 1989).

4. Like Star, I had heard Alan Kay's remark to the effect that point of view was worth 80 IQ points, an insight he attributed to the group he was part of at Xerox PARC in its heyday.

18 Reflections on the Visibility and Invisibility of Work

Kjeld Schmidt

I have a vivid and oddly intimidating recollection of Susan Leigh Star at one of the first CSCW (computer-supported cooperative work) conferences. If I'm not mistaken, it was Toronto in 1992. During one of the plenary debates, Leigh took the floor and gave an improvised talk in which she challenged the whole direction of the debate by insisting on the fundamental relevance of *bodies*, human bodies, adding that this is one thing that feminism and Marxism have in common: bodies as a source of creative power and as objects of suffering. Cognitivists were appalled and sociologists were snickering.

I was impressed, not only with the strength of her softly stated message but also with her courage. But I was also intimidated, for as an old Marxist I did my upmost not to engage in discussions about issues of social ethics in a scholarly context. But now that Leigh's softly subversive voice cannot be heard any longer, I can no longer leave it to her (and a few others) to address such issues (issues of repression, exploitation, and so on)—it's time for me to take the floor as Leigh did that day in Toronto.

Leigh's intervention at that event is, in my view, highly characteristic of her, in form and in content. One of the things that characterized Leigh's life and work was a determined effort to have the skills and concerns of ordinary workers respected and taken into account in the design of technical systems, in organizational management, and in public discourse. It's not only a *concern* but also an *indignant insistence*. It runs like a soft but persistent voice through her ethnographic studies and technical papers as well as her more conceptual and philosophical essays.

This voice is perhaps at its most programmatic in the paper entitled "Layers of Silence, Arenas of Voice: The Ecology of Visible and Invisible Work" that Star wrote together with Anselm Strauss in 1999. It is also a paper of great relevance today [see chapter 19, this volume].

The paper is conceptual, but not rigorously so. It rather appears like a number of astute observations that all, in different ways, address or comment on the different ways in which work can be invisible or made invisible—or can be visible or made visible. It is not a coherent exposition of an argument, nor does it claim to be. It is an intricate fabric of mutually illuminating observations. Uniting all this is the authors' clear and strong voice of concern for the dignity of ordinary workers and their work.

In the paper, the concepts of visibility and invisibility of work are explored in different directions. Let me mention just a few.

(1) "Invisibility of work" understood as not seeing, as ignoring, or *depreciating* the value of certain categories of work: their economic, social, and human value. Domestic work, domestic service work, care giving (especially work that involves touching the bodies of others).

This is not just a matter of the ways in which domestic service workers—say, the Hispanic maids and gardeners that make Silicon Valley hum—are being treated by their employers; it's institutionalized in the national economic statistics that exclude work in the domestic arena from being included in the calculation of the national product and in the employment statistics, creating the illusion that so-called service work has virtually exploded over the last half century (and also the illusion that we're well on the way to becoming the Leisure Society).

In short, work is made invisible by being considered valueless, not worth speaking of in polite conversation, something for the "nobodies" of this world (to use a horrific English term).

(2) The other sense in which we can talk of the "invisibility of work," according to Star and Strauss, differs from the first in that the criterion is not one of estimation, or valuation.

In this sense, work is rendered "invisible" in that one denigrates, underestimates, or simply ignores the *skills* involved in ordinary work (witness the other horrific English term "unskilled laborer"); what is denigrated are the local workarounds, the undocumented but essential tricks of the trade, the taken-for-granted practices, the sophisticated skills typically required of articulation work in cooperative work settings, the subtle awareness practices and the unheralded daily innovation of coordinative artifacts, in short, the work to *make work work*. It's of course a widespread—if not ubiquitous—engineering and managerial prejudice (inculcated in technical universities and business schools) to belittle the actual skills involved: workers are routinely conceived of as mindlessly executing the plans of their masters, their practices easily replaced by job cards or robots.

There is of course sweet revenge in the incessantly repeated observation that the organizational schemes and technical systems founded on these prejudices come tumbling down just as frequently. But sweet revenge is a short-lived pleasure: enormous resources are wasted on these harebrained schemes, and the workers who have to shoulder them and make them work in spite of their design are only worse off. They are, for sure, rarely thanked for having saved the day.

(3) There is a third variant of "invisibility" that Star and Strauss only mention in passing, but which now is becoming overwhelming, namely the elimination of *work* from the agenda of respectable intellectual interests.

A colleague of mine, a professor of CSCW and human–computer interaction (HCI) at a technical university, recently told me that her research lab had had to change its name. It used to be called *Interaction Design and Work Practice Lab* but over the years she had come to realize that students no longer were interested in "work" or "work practice." After years of dogged resistance she eventually had to change the name to *Interaction Design and Human Practice Lab*. And then the students were back. That is, for budding designers of digital artifacts, work is *invisible*, in the sense that it's utterly dull and uninteresting. It's just not a sexy topic.

I share my colleague's sadness, not least because her experience is a general one. It's even being promulgated by leading researchers in CSCW, of all places, who argue that CSCW should "move its focus away from work" in favor of "ludic pursuits": entertainment and gossip and the computer-supported presentation of self in public places.

This is not at all confined to CSCW. Take HCI or project design (PD)! Take organizational theory.

Where does this deliberate carefree disinterestedness come from? I'll suggest a couple of sources.

1. It has become generally accepted, as an article of faith, that industrial society is a thing of the past and that we're moving into the age of the postindustrial society. And of course, if you look at the employment statistics of the Organisation for Economic Co-operation and Development OECD countries it would seem correct. But it's a myth of course, in which the work of millions, hundreds of millions of industrial workers, is made invisible.

Here they are: there are 120 million manufacturing workers in China alone, double the number of manufacturing workers in the developed countries (see figure 18.1).

Moreover, the degradation of work, not just *manufacturing* work, which accompanies what is called globalization, seems to have a moral effect in the West comparable to that of slavery in the past. The actual degradation of workers is accompanied by the degradation of the esteem of work as such.

2. It may also be a reflection of the mood of "irrational exuberance," that unbridled speculation and vulgar consumerism that has characterized the West the last two decades or so: The subprime notion of wealth as something miraculous that comes to those who enter the market first. To all this, one can only say what Hesiod said to his wayward younger brother Perses about ten thousand years ago:

But you must work, Perses . . .
Gods and men disapprove of that man who lives without working . . .
Work is no disgrace, it is idleness which is a disgrace . . .
Work, work upon work!
(Hesiod, *Work and Days*, eighth century BCE)

Figure 18.1
Workers at a Foxconn factory in southern Guandong Province. Foxconn employs more than 800,000 workers in China.
Source: New York Times, June 2, 2010

(3) Related to this is, perhaps, the notion, widespread among intellectual middle classes, that we're now *all* self-managing symbol analysts, involved in a weightless economy, the creative class, the key class of the new economy, whose daily life is not really characterized by what is ordinarily understood as *work* but rather by unforced— almost playful—innovative activities interrupted by playing ping-pong. Crudely put, the idea is that *"we" do the creative bit and leave the toil to the Chinese.*

This is of course delusional. It may look like this (as described here) *during* the bubble, but *then comes the crash* and the layoffs. And design, programming, and engineering turn out to be just another kind of work, the work of an extensively trained workforce, of course, but invariably a workforce under the combined pressures of managerial urges to control and the competition from low-cost countries and from machinery. Stress-related depression and burnout is epidemic among computer specialists and similar professions, at least in my part of the world. Psychiatrists are alarmed but organizational theorists and management consultants press on.

We live in troubled and turbulent times. Entire regions and even countries are being de-industrialized, decaying into rustbelts populated to a significant extent by drug addicts and riotous mobs, while workers in the factories of the Pearl River Delta toil, sometimes to the brink of suicide, working up to four thousand hours per year, to produce sneakers and iPods. At the same time, in the West as well as in the East,

Figure 18.2
Workers on strike, Honda Lock factory, Zhongshan, China, June 2010
Source: New York Times, June 16, 2010

cumbersome and often counterproductive coordination technologies are rolled out and imposed.

Unfazed by all this, sociologists and social psychologists, in and outside of CSCW, are irrationally exuberant about Facebook.

Let me end by pointing out that ordinary workers have their own ways of making themselves heard and seen (see figure 18.2). As we saw in China in June of last year, they make themselves visible as follows:

• by turning the Taylorist system around, that is, by using the carefully designed just-in-time workflow so as to identify strategic weak spots, and then paralyzing these by walking out—and thereby showing the essential value of their work;
• by taking photos of themselves and sending them from their mobile phones;
• and by doing that they state that they are not to be considered marginal existences: we are the producers, the salt of the earth,

What I want to suggest by all of this is that a fine way to honor Leigh's work would be to strive to *make work*—the skillful work practices and the problems and concerns of ordinary workers—visible again: make it a legitimate and respectable topic of study.

References

Star, Susan Leigh, and Anselm Strauss. 1999. "Layers of Silence, Arenas of Voice: The Ecology of Visible and Invisible Work." *Computer Supported Cooperative Work: The Journal of Collaborative Computing* 8:9–30.

19   Layers of Silence, Arenas of Voice: The Ecology of Visible and Invisible Work

Susan Leigh Star and Anselm Strauss

Editors' Note: This chapter, written with her former advisor and mentor Anselm Strauss, originally appeared in the journal *Computer Supported Cooperative Work* of which Leigh was a founding coeditor. Here they link silences and voice, distinctive concerns of Leigh's, with the doing of different kinds of work, a major focus of Anselm's writings. Centering on the kinds of work most pertinent in computer and information sciences, they analyze the conditions under which certain kinds of work are rendered invisible while others are extolled. Articulation work—the careful piecing together of a group of tasks that must be coordinated to be accomplished—is a special focus.

No work is inherently either visible or invisible. We always "see" work through a selection of indicators: straining muscles, finished artefacts, a changed state of affairs. The indicators change with context, and that context becomes a negotiation about the relationship between visible and invisible work. With shifts in industrial practice, these negotiations require longer chains of inference and representation, and may become solely abstract.

This chapter provides a framework for analyzing invisible work in computer-supported cooperative work (CSCW) systems. We sample across a variety of kinds of work to enrich the understanding of how invisibility and visibility operate. Processes examined include creating a "non-person" in domestic work, disembedding background work, and going backstage. Understanding these processes may inform the design of CSCW systems and the development of related social theory.

What Counts as Work Is a Matter of Definition: The Big Picture

Much of CSCW is devoted to the support of group work processes. In this, the tension between formal task descriptions and overt work on the one hand, and informal tasks and "behind the scenes" work on the other, has been an important consideration (cf. Schmidt and Bannon 1992; Fjuk, Smørdal, and Nurminen 1997; Robinson 1991, 1993a, 1997; Bowker 1994; Bowker and Star 1994; Bowker et al. 1997; Gasser 1986). Clearly, identification of that work which escapes formal or traditional requirements analysis is crucial in accounting for level of effort, and for representing the subtleties of cooperation. That there are limits to this specification process is also well known. "Less is more" has been widely discussed by many CSCW researchers as a design principle for the field. Some forms of behind the scenes work and of discretionary activity may often be best left unspecified, and not represented in systems requirements. Suchman (1995) provides an elegant analysis of the complex trade-offs involved in making work visible. On the one hand, visibility can mean legitimacy, rescue from obscurity or other aspects of exploitation. On the other, visibility can create reification of work, opportunities for surveillance, or come to increase group communication and process burdens.

As computer use moves from single to group task and communication, the jobs of tuning, adjusting, and monitoring use and users grows in complexity. One type of work is especially important, dubbed "articulation work" by Strauss et al. (1985). Articulation work is "work that gets things back 'on track' in the face of the unexpected, and modifies action to accommodate unanticipated contingencies. *The important thing about articulation work is that it is invisible to rationalized models of work*" (Star 1991a, 275, emphasis in original; Strauss 1985, 1988; Berg 1997). The intuition that articulation work is important for CSCW design has been important since the inception of the field (Gerson and Star 1986; Schmidt and Bannon 1992; Grinter 1996; Suchman 1996; Schmidt and Simone 1996; Kaptelinin and Nardi 1997). The support of articulation work, argue Schmidt and Bannon (1992), means developing a subtle and thorough analysis of the politics and culture of the work to be supported, one which distinguishes the routine from the exceptional, and which does not violate contextual norms. Fjuk, Smørdal, and Nurminen (1997) extend the analysis into the realm of tools, using Leontiev's activity theoretic models of work and development.

Schmidt and Simone made a crucial distinction between articulation work and cooperative work (1996).[1] Cooperative work interleaves distributed tasks; articulation work manages the consequences of the distributed nature of the work. They note the highly complex dynamic and recursive relationship between the two—managing articulation work can itself become articulation work, and vice versa, ad infinitum. The design implications of this important article in light of the topic of invisible work are discussed in what follows.

Throughout this literature, the dynamic *interplay* between the formal and informal has been a focus, as for instance in Robinson's (1991) influential discussion of double-level languages. The relationship between the visible and the invisible needs a similarly rigorous analysis. One point of departure is to ask what exactly *is* work, and to whom it might (or should) be visible or invisible.

However, little is obvious in any general sense about what exactly counts as work. When people agree, it may seem obvious or natural to think of some set of actions as work, or as leisure. But as soon as the legitimacy of the action *qua* work is questioned, debate or dialogue begins.

For example, before the women's movement of the 1970s, much of the activity associated with cleaning houses, raising children, and entertaining families was often defined as an act of love, an expression of a natural role, or even just as a form of being (Kramarae 1988). Feminist movements like the British "Wages for Housework" began a public campaign to define those activities as work—work with real economic value. A recent similar campaign in Canada meant that the May 1996 Canadian Census, for the first time, included questions about unpaid work "taking into consideration everything from bathing children and cutting grass, to caring for seniors and counseling teens" (Clayton 1996). Significantly, one of the participating organizations is the "Canadian Association of Home Managers," not named as housewives (or husbands).

What is at stake in any such movement is an attempt to redefine the relationship between visible and invisible work. When the structural shifts are broad, as with the example above, entire arenas of debate may emerge and be linked with social movements. Ivan Illich, in his provocative essay "Shadow Work" places the creation of invisible work in a broad historical sweep, created in relation to forms of industrial labor, and difficult to define: "In traditional cultures the shadow work is . . . often difficult to identity. In industrial societies, it is assumed as routine. Euphemism, however, scatters it. Strong taboos act against its analysis as a unified entity. Industrial production determines its necessity, extent and forms. . . . To grasp the nature of shadow work we must avoid two confusions. It is not a subsistence activity; it feeds the formal economy, not social subsistence. Nor is it underpaid wage labor; its unpaid performance is the condition for wages to be paid" (1981, 100).

In recent years, there has been such a large-scale shift in how careers, corporations, and the public sector interrelate. On the corporate side, downsizing, reengineering, outsourcing, and other management strategies such as activity-based accounting mean that many people who previously expected lifelong employment in one organization have become freelancers or consultants, without benefits or pensions (Greenbaum 1995). Many more are simply fired. In the public sector, funding cuts and shifting mandates mean a shift toward similar management practices, as for example in many American health maintenance organizations (HMOs),[2] universities, or local government management organizations. These often maximize profit and seek to cut redundant services in a previously nonprofit sector with other goals and values.

These shifts, viewed broadly, form the conditions for a new arena where the relationship between visible and invisible work is challenged. At the extreme, if corporate management chooses to break work down into component tasks and remove it from the biography, job, or career of any particular individual, then it must be fully describable and in some sense rationalized. Features of work which emerged in stable, career-oriented organizational milieux now shift dramatically. The kinds of work especially affected include tacit and contextual knowledge, the expertise acquired by old hands, and long-term teamwork.

Much recent CSCW work research has moved conceptually in resistance to this cultural shift. It analyses the expertise often hidden from view (in even seemingly mindless tasks). If cleaners, file clerks, and other so-called unskilled laborers use significant discretion, and add valuable skills to their jobs, then outsourcing and piecework sold to the lowest bidder may show up down the line as a hidden cost to a firm (Engeström and Engeström 1986). The health-care arena is full of such debates at the moment, as physicians argue that HMO cost restrictions effectively bar them from practicing proper medical care. They argue that such care always relies on practitioner discretion.

Herbert Blumer long ago used the term "cultural drift" to talk about shifts in values and organizations that touch on a multitude of different areas of social life. This present set of circumstances with respect to working life represents a cultural drift concerning the nature of work and professional knowledge. In this chapter, we consider how the question of "what is work" affects invisibility, stepping back a bit from the immediate arguments about how to measure work, or how to design computer systems to support distributed collaboration. We consider varieties of visible and invisible work, some at extremes such as slavery or highest-level corporate decision making, in order to understand better how visibility affects CSCW design considerations and the study of cooperative work. In order to do this, we need to understand how work becomes visible or invisible, and then how negotiations about this status are structured.

The Visibility or Invisibility of Work: Tensions and Negotiations

What exactly counts as work varies a lot. In common parlance, we speak of work as obvious: "work is when you get up in the morning and go to the office, and what you do there *is* working." But as we have seen with the example of wages for housework, there are many kinds of activities that fall into a large, and growing, gray area. Are tasks done in the home to care for a chronically ill spouse really work? No one who has carried bedpans, negotiated with insurance companies, or redesigned a house for wheelchair navigation would deny that it is, indeed, very hard labor in some sense. Yet such work has often been invisible. It may be invisible both to friends and family, and to others in the paid employment workplace. It is squeezed in after hours, hidden as somehow a shameful indicator of a faulty body; it is redefined for public definition as time *away from* work. The recent Family and Medical Leave Act in the United States mandates permission for time off to care for a seriously ill relative. It is an important milestone in the consideration of how family care should be visible with respect to the workplace. It reflects a shift in the structure of the American family, acknowledging the structural changes occasioned by greater longevity and by the epidemics of Alzheimer's disease and other chronic illnesses.

The importance of context in analyzing the visibility of work can be seen in a couple of extreme cases. We include these here to underscore the range of variation in indicators of "real work." Most people would not think of prayer as vital public work; rather, it is a private action. As well, most people think of very elderly people as working less and less. However, Driscoll's (1995) study of a convent inverts both of these common perceptions. Nuns in the convent perform all sorts of duties: washing floors, printing hymnals, ironing priests' robes. One task in particular is important for the nuns' role in the larger community: saying prayers on behalf of sick or troubled parishioners. People from the outside make requests of the nuns to say some number of specific prayers, and there is always a backlog. Saying these prayers is highly valued work. As the nuns age, they become less able to perform physically demanding activities, and may even be bedridden. But as long as their minds are clear, they are ideally suited to reduce the backlog of prayers, a valuable service for both the convent and the wider community. Thus, their work is both visible and valued within this context. In another setting they might be seen as useless (in terms of labor) old women with nothing to occupy them other than prayer and contemplation.

In the case of the nun, the "product" produced by the work is the prayer itself, performed properly, and the indicator used by others in the convent is the eased pressure on the backlog. But the definition of product can also be tricky, certainly situational. For example, in the concentration camps run by the Nazis in World War II, prisoners were often forced by guards to perform meaningless tasks. One task remembered vividly by many inmates was being forced to carry a heavy block of stone up a long stairway in a rock quarry, then carry it down again, and up again, until exhaustion forced the person to drop. The collapse would be an excuse for the guard to shoot the prisoner for being "lazy."

What is the work in this situation, and what is being produced? Clearly, the task is meaningless in the ordinary sense, certainly from the point of view of the prisoner. From the prisoner's view, work in fact shrinks down to a pinpoint of survival—a series of improvised, desperate maneuvers to outlast and outthink the imprisoners. From the Nazi viewpoint, the work of carrying the stone up and down the staircase was not the product—what in fact was being produced was a death. It is clear here that *definition* is all: both in the relative sense of frame of reference, and in the brute sense of who gets to define the work's meaning. On this point, consider the following passage from Toni Morrison's *Beloved*, where a slave, Sixo, is accused of stealing a stoat, an edible animal, from his slaveholder, Schoolteacher:

"You stole that stoat, didn't you?"
"No. Sir," said Sixo, but he had the decency to keep his eyes on the meat.
"You telling me you didn't steal it, and I'm looking right at you?"

"No, sir. I didn't steal it."

Schoolteacher smiled. "Did you kill it?"

"Yes, sir. I killed it."

"Did you butcher it?"

"Yes, sir."

"Did you cook it?"

"Yes, sir."

"Well, then. Did you eat it?"

"Yes, sir. I sure did."

"And you telling me that's not stealing?"

"No, sir. It ain't."

"What is it then?"

"Improving your property, sir."

"What?"

"Sixo plant rye to give the high piece a better chance. Sixo take and feed the soil, give you more crop. Sixo take and feed, Sixo give you more work."

Clever, but Schoolteacher beat him anyway to show him that definitions belonged to the definers—not the defined. (1987, 190)

The concentration camp and American slavery are extremes, of course, in considering relations of power and invisibility. But all the examples given are useful in pointing out the contextual nature of what is work, what may or may not be visible and public, and how radically different conceptions of product and indicator may be. In the film *The Gods Must Be Crazy*, a delightful scene between a Western ecologist studying elephant migration and a !Kung bush tribesman trying to find out what the ecologist does for a living says it all. The !Kung man asks the Western man about his work. The Western man replies that he is an ecologist, a natural historian. Seeing the puzzled look on the !Kung man's face, he translates to the level of activity: "Well, actually, I walk around all day behind elephants and pick up their dung." The !Kung's expression changes to one of pity mixed with thinly veiled amusement. Lacking a mutual context, all that is visible is the unadorned action, meaningful in the wider scientific world, but ludicrous in the world of tribal bush culture.

All modern organizations have some version of these radically varying definitions of the situation. With reorganization and cutbacks, contexts inevitably meet and often clash in precisely this way. "Slimming down" in practice means that someone (a consultant or manager) enters into multiple contexts and judges the necessity of the work being done in any one. They may or may not have a sympathetic understanding of the context. Without that understanding, it may look as though secretaries are often just chitchatting with each other or with clients—surely an activity that indicates lack of real work. However, what gets ignored is the way that the information transmitted between secretaries about their bosses may smooth communication between bosses, speed up unusual requests by building a network of mutual cooperation and favor exchange, or screen out unnecessary interruptions by delaying a troublesome client at the door (Kanter 1977). Deleting such practices through ignoring the context of work may prove expensive in the long run (Suchman 1995; Grinter 1996). Because the design of CSCW systems involves coordination and articulation work, ignoring such invisible work may mean the system is not used, or that it furthers inequities (Grudin 1988).

Forms of Visible and Invisible Work

Definitions belong to the definers, says Morrison. What will count as work does not depend a priori on any set of indicators, but rather on the definition of the situation. It is interesting to think of the gamut of indicators that could indicate work. At one end, there is physical labor, with sweat, hard breathing, and the signs of muscle exertion. At the other end, there is work that involves no movement in the moment at all, such as getting paid for being on retainer, for being available to work whether or not any specific activity takes place in the time period. All along this continuum, the visibility and legitimacy of work can never be taken for granted, as for instance the old nuns' prayers demonstrate. The hardest physical exertion may be defined as sport, as Zen meditation, as punishment, as relaxation. The extremes of physical inertia, such as passive waiting or sometimes just existing, can be defined under some circumstances as legitimate work—being a monarch, lending one's name to an enterprise, even, gory as is the example, having donated one's organs and having them work on behalf of another after one's own death.

Understanding that what counts as work is a matter of definition that allows us to focus on how the relations between indicators appear in different conditions—the matrix of visible and invisible work. We know that it is possible to observe another person sweat and suffer and *not* see exertion as work. We know that it is possible to observe no direct physical action, but to have that lack of movement defined as work, and so paid. This shift moves us away from commonsense and often misleading ideas about work as obvious. In the second section of the chapter, we consider a range of instances along this continuum and the kinds of conversations or silences they engender:

- *Creating a nonperson.* Under some conditions, the act of working or the product of work is visible to both employer and employee, but the employee is invisible (in Goffman's terms, a "nonperson"). Of course, this is linked with power and status differences between employer and employee. We look here at invisibility and struggle among domestic and other service workers, where the defining of legitimate work rests heavily with the employer and where the employee is often invisible. This area has been marked by struggles to raise the status and change the working conditions of domestic workers.

• *Disembedding background work.* In a flip-flop of the preceding scenario, there are circumstances where the workers themselves are quite visible, yet the work they perform is invisible or relegated to a background of expectation. We examine here how nurses, who are very visible as workers in the health-care setting, are struggling to construct an arena of voice to make their work visible. They are attempting to change work previously embedded under a general rubric of "care," and usually taken for granted into work that is legitimate, individuated, and traceable across settings.

• *Abstracting and manipulation of indicators.* There are two cases in which *both* work and people may come to be defined as invisible. (1) Formal and quantitative indicators of work are abstracted away from the work setting, and become the basis for resource allocation and decision making. When productivity is quantified through a series of indirect indicators, for example, the legitimacy of work may rest with the manipulation of those indicators by those who never see the work situation first hand. (2) The products of work are commodities purchased at a distance from the setting of the work. Both the work and the workers are invisible to the consumers, who nonetheless passively contribute to their silencing and continuing invisibility.

In the following section, we explore these conditions and dimensions of invisible work. The first case, "creating a nonperson," uses the example of domestic service work as the quintessential case of deeply invisible work.

Creating a Nonperson: Silences and Struggle among Domestic Workers

The history of domestic service in the United States is a vast, unresolved puzzle, because the social role "servant" frequently carries with it the unspoken adjective *invisible*. (Cowan 1983, 228; emphasis in original)

Domestic service work in the United States—being a maid, cleaner, or child caretaker—is usually private, unrecorded, and unprotected by Social Security, and unregulated. It is usually done by women of color for white women (Romero 1992; Rollins 1985). It is difficult physical work, often to the point of exhaustion. It may also be emotionally exhausting if the employer demands that the employee act as counselor or confidante, or if there are constant assaults on the dignity of the employee (Colen 1989 gives many examples of this). Many of these elements are evident in industrial cleaning and service work as well, although this work may be less lonely and more organized.

Several authors describe a curious mixture of visibility and invisibility in domestic service work. On the one hand, employers usually oversee the work done, sometimes to an astonishing degree of micromanagement. On the other hand, the employees are socially invisible to the employers.

The combination of micromanagement and invisibility is described in some detail in Judith Rollins's extraordinary ethnography, *Between Women: Domestics and Their Employers* (1985). Rollins, an African American sociologist, passed as a maid, working for several employers and interviewed many black maids. She notes: "But seeing me breathing hard, perspiring, and visibly weary never prompted any employer to suggest I take a break—not even when I worked an eight-hour day" (67). This comes from the combination of wanting visible signs of the work and a feeling of ownership of the employees' time.

Other domestics and a few employers also said that employers liked to see domestics working. Employer Margaret Slater described her displeasure at her worker's inactivity this way: "'She really was very good. The only thing that annoyed me was when I would come home and find her sitting down. She'd be just playing with the kids or something. All the cleaning would be finished but it still bothered me to see her sitting'" (Rollins 1985, 65).

"She's a driver. Seems like that's where she gets her therapy from—working you. She likes to *work you*. Seem like the harder she works you the better she feels. She just keeps giving you more and more work, telling you what to do and how to do it. That's the reason today I require a lot of rest" (Rollins 1985, 64–65).

This micromanagement led in extreme cases to electronic monitoring and surveillance. Rollins recalls an incident where the employers put a tape recorder in the bedroom of one of her respondents: "I discovered they had fixed up a tape recorder in my room that recorded my coming in and going out, conversations. Well, you talking about a black woman going off when I found out? What happened was, I had just come in the front door and I noticed him, he went straight upstairs as I came in. And, then one of the little boys [six years old] said, 'Daddy has the tape recorder on'" (Rollins 1985, 145).

A similar instance is noted in Wrigley's study of child care workers, where parents secretly videotaped nannies that they were worried about caring for their children (1995, 96–97). This surveillance is linked with the lack of subjectivity of the person being watched; it is almost as though they are being treated as a machine to be monitored (recalling Foucault 1977; see also Chatman 1987).

Becoming a nonperson follows the hallmarks set out by Goffman in his *Presentation of Self in Everyday Life* (1969) and *Asylums* (1962). In *Asylums*, he defined a sequence through which a person loses their individual identity upon entry to the "total institution." One is stripped of history, family, defining markers. One becomes a number, not a name. Deference behavior is expected.

In the domestic work settings described above, however, the sequence is not as institutionalized as are Goffman's descriptions of indoctrination into mass nonentity status. They are more diffuse. Rollins recounts an incident where she showed up for a job interview at a potential cleaning site, casually but well dressed. Her interviewer expresses concern over hiring her because she's too well educated, which Rollins correctly interprets as being drawn from her being too well dressed and thus not subservient enough. So she shows up the first day in a bandanna and old clothes and "With an exaggeratedly subservient demeanor (standing less erect, eyes usually averted from hers, a tentativeness of movement). Most important, I said almost nothing, asked the few necessary questions in a soft unassertive voice, and responded to her directions with 'Yes, Ma'am'" (1985, 207).

Rollins was shocked at the employer's pleasure in, and acceptance of, this behavior, as the woman had already seen her behaving normally. Rollins analyzes this incident as paving the way toward invisibility. Many of the domestic workers she interviewed learned to hide their abilities and assets from employers. Several spoke of parking their cars down the block from the house to be cleaned, and walking to work so the employer wouldn't see that they owned a car (Rollins 1985, 196).

(Enforced) deference behavior soon gives way to the employer treating the person as a thing. This takes a myriad of nonverbal forms: ignoring the worker, displaying behavior that shows the employer's definition of the domestic servant as a thing. Romero notes that one maid "was expected to share her bedroom with the ironing board, sewing machine, and the other spare room types of objects" (1992, 2). Food, heat, and clothing can often be the symbols for nonpersonhood. Romero notes that employers left them inferior food, even garbage to eat, and forbade them access to the family food (ironically, often the same people who mouthed the cliché, "she's just like one of the family.") Rollins notes one house where when the employers left for the day they turned down the heat—as if the employee wasn't present. Another locked her in, so she was unable to answer the door.

Under these circumstances, the servant becomes "unseeable" by the employer. One says, "To Mrs. Thomas and her son, I became invisible; their conversation was as private with me, the black servant, in the room as it would have been with no one in the room" (Rollins 1985, 209). This leads to the paradoxical experience of: "being there and not being there. Unlike a third person who chose not to take part in a conversation, I knew I was not expected to take part. I wouldn't speak and was related to as if I wouldn't hear. Very peculiar" (208). In an ultimate echo of slavery, one employer became very possessive of "her" cleaning person, even to the point of imposing her last name on the domestic.

Resisting Objectification   Rollins and Romero discuss a number of ways in which the domestic servants fought back against this definition of nonpersonhood. Sabotage, or silent resistance, was common, although in the isolated situation of most domestic workers, was tricky to manage to one's advantage. One maid says:

I did do the washing for her for a long time. But I got tired of it. So I didn't half rinse the soap out. And when you don't half rinse the soap out, the wash turns brown. I got it brown, then she told me she was going to get a woman in to do day work. So when this woman came in and started doing day work, the shirts and things looked so nice. But I could get it nice, you know; but I didn't want to overdo because I had enough work in the house to do without doing all that laundry. So

I knew if I messed up on something, something had to go. So I messed up on the laundry. . . . I had to take care of myself. (Rollins 1985, 143)

A similar resistance, for similar reasons, is documented in the resistance of workers in Silicon Valley against boring and demeaning information work, sometimes by sabotage (Hossfeld 1990). Similarly, an eighty-two-year-old domestic worker says that she too set limits on what tasks she would and would not do, taking the longer view of her own self-worth: "I didn't do everything those folks told me to do. Some I did and some I didn't. They would tell me to get on my knees and scrub the floor and I didn't do it. I didn't mess up my knees. I told one lady, 'my knees aren't for scrubbing. My knees are made to bend and walk on.' I didn't have a lot of bumps and no black knees from scrubbing floors. I took care of myself" (Rollins 1985, 142).

Romero reports that the struggle to redefine the relationship also may take the form of attempting to define a businesslike, or solely contractual relationship between employee and employer. This may be difficult or even impossible to do when employers insist on defining the employee as a pseudo-family member, or the work itself as partially a labor of love. She notes that feminist employers are particularly caught in a status dilemma, as their beliefs in equity are seemingly pulled against by the very concept of having someone clean their home.

At times, the conflicts from these status problems can erupt, as when one of the cleaners discovered her employer had been audiotaping her during working hours: "You tell him he has violated my civil rights and I know he doesn't want me to report this to the NAACP. So you tell him I will not be back. And for him not to contact me, period.' And that was the last of them. I didn't hear from them. I didn't see them. But you see, just like they do that to me, how many other black women are being mistreated and exploited?" (Rollins 1985, 146).

Finally, in a move to define her autonomy that was much more personal, one cleaner wrote about her work, creating a recipe and decor book documenting the creative energy that had gone into the job over a period of many years. She says, "I had created a lot of little decorations for their teas and dinners that I had written in there too. Whenever the ideas came, I'd write them down. And whenever you do, it's like a precious little thing that you do because you want to show your work" (Rollins 1985, 230).

Sadly, underscoring the theme of nonpersonhood, the employer asked to see the book just as she was leaving the job (after many years), and then "lost" it.

Do you know, they moved away and I never got that book back. That was one of the most upsetting experiences I've had. That book was so valuable to me. I wanted my children to read it. I think she took it deliberately. . . . It was a history I would like to have kept. (Rollins 1985, 231)

The circumstances of domestic workers are extreme in their isolation, saturated with race and gender bias. As well, immigrant workers without local networks are often employed in the job. At the same time, there are important links to other lines of work.

Creating a nonperson need not be a monolithic event; every workplace and organization employs people who may give or receive such a status, even if only partially. In academia, some people are judged "dead wood," and their work is ridiculed or they themselves are ignored as people. Janitors, cleaners, physical maintenance workers, and those who work as laboratory technicians are often treated as nonpeople.

Implications for CSCW   Why is this analysis of the processes of invisible work important for CSCW and the design of systems generally? To the extent that such complex, and intimate systems rely on accurate models of working processes, the systematic exclusion of certain forms of work means a displacement of that work and a distortion of the representations of that work (Suchman 1995; Star 1991b, 1992, 1995; Star and Ruhleder 1996). As Illich (1981) would argue, work does not disappear with technological aid. Rather, it is displaced—sometimes onto the machine, as often, onto other workers. To the extent that some people's work is ignored as they are perceived as nonpersons, more "shadow work" or invisible work is generated, as well as the (sometimes) obvious social justice and inequity issues. In the creation of large-scale networked systems, this process may cascade.

Disembedding Background Work
Work may become expected, part of the background, and invisible by virtue of routine (and social status). If one looked, one *could* literally see the work being done—but the taken-for-granted status means that it is functionally invisible. Work in this category includes that done by parents, nurses, secretaries, and others who provide on-call support services for others.

Bowker, Timmermans, and Star (1995) (Bowker, in press) have investigated the world of nursing work, and an attempt by a group of nurses at the University of Iowa to categorize and make visible all the work that nurses do. The Iowa Nursing Intervention Classification spells out a series of nursing interventions, with associated activities and scientific literature (McCloskey and Bulechek 1993, 1994). The interventions were developed through a grassroots network of professional nurses and nursing researchers, and have reached a total of more than three hundred. The book describing the interventions is in its second edition, and has been integrated in several ways in information systems (local software support for medical record keeping, and integration into the National Library of Medicine's *Universal Medical Language System* [*UMLS*] *Thesaurus*).

The impetus for creating the system is several fold, but one primary motive is to disembed what has previously been deeply embedded, invisible work done by nurses, and make it visible to the medical record, for research purposes, and for the legitimation and professionalization of nursing. Historically, nursing work has been lumped in with infrastructural and service work in the hospital, at least in terms of the medical record. Nursing records are even routinely discarded after the patient is discharged, adding another layer of invisibility to the work. With increasing professionalization of nursing, and the rise of nursing research, nurses are entering a vigorous dispute about this set of practices. As one respondent noted, "I am not a bed."

Bowker, Timmermans, and Star detail the ways in which the attempts to make nurses' work visible is fraught with trade-offs and politics (1995). More visibility may mean more surveillance, and an increase in cumbersome paperwork, as Wagner has shown in her powerful study of nursing work (1995). The specification of tasks, as many have noted in the CSCW and requirements analysis literature, requires a light hand if one is not to risk mindless Taylorism and the eradication of discretion from skilled workers. Nurses struggle to be visible, but simultaneously to hold areas of ambiguity and of discretion. It is one thing to note that one has given counseling to a dying patient; quite another to specify the words one would say to that patient.

The trade-offs involved imply that the negotiation between visible and invisible work directly addresses such issues of discretion and Robinson's "double-level languages" (1991). Robinson argues that CSCW systems need to support both explicit and implicit meanings in whatever setting they appear, and those that validate the formal and ignore the informal (or vice versa) are doomed to fail. The risk of such erasure is greater in those settings where the work being modeled is already invisible in the embedded sense discussed previously. Nurses, librarians, parents, and others who are "on call" are particularly vulnerable; but the point is a general one in the modeling of work, in that all work has some of these embedded qualities.

Background work is vulnerable to oversight in the design of CSCW systems, partly as it is diffused through the working process, partly due to the social status of those conducting it, and partly because it requires so much articulation work. Clement (1991, 1993) as well notes that transforming the "invisible" infrastructure of computerized work means recognizing the hidden and devalued skills of routine office workers. For another example, in the design of digital libraries and web indexing systems, the relative lack of attention to the expertise of librarians is evident. This may incur the cost of reinventing the wheel with respect to indexing, bibliographic instruction, and the process of matching patrons to needed reference sources (Bishop and Star 1996).

Going Backstage

There is a special instance of embedded background work, which paradoxically may result in a highly visible public performance. Here Goffman's analysis of "front stage" and "back stage" is particularly compelling (1969). Many performers—athletes, musicians, actors, and arguably, scientists—keep the arduous process of preparation for public display well behind the scenes. Thus the process of trial-and-error in science is less visible than the final published results (Shapin 1989; Star 1989); the hours of practice for a musical performance invisible to the audience observing the recital. Fussell (1991), for example, provides a rare glimpse of the backstage world of body builders, showing the (literal) blood, sweat, and tears involved in creating the perfect body for display, along with the illegal drugs and the isolation. Chambliss (1988), in his observations about the "mundanity of excellence," makes a similar, and more theoretical point, about the training of Olympic athletes: Their seamless performance is built on much invisible backstage work.

The particular aspect of work that is problematic or made backstage in any case is negotiated and historical. Sanjek's remarkable volume (1990) about the taking of field notes in anthropological investigation, for example, carries a number of accounts from fieldworkers who feel that they haven't done it "quite right" or are "too messy" in their practice. Learning to do fieldwork has been a craft skill, often learned by apprenticeship, often conducted in a scientific milieu that looks for rationalization and replicability. Field notes, the stuff of the craft, become a touchstone for one's competence, and at the same time, an embodiment of invisible work. The volume, coming at a historical moment when anthropologists are increasingly self-reflective about the practice of ethnography, goes backstage in the work process in a way that is at first shocking for readers used to the note-taking process as being highly private. It also thus points to a moment where the nature of that privatization is in question.

For any requirements analysis, understanding where in the relationship of visible to invisible work one would locate a given set of work practices is crucial (Goguen 1994, 1997; Jirotka and Goguen 1994). If one attempts to open up backstage practices to scrutiny, where no occupational culture or previous conventions for doing so exist, the analyst or systems developer risks violating people's autonomy, or simply getting no useful information about how work is really done.

Abstracting and Indicator Manipulating

In their paper on the work of cleaners in firms in Finland, Engeström and Engeström note that cleanness is only one kind of motive or measure for doing good cleaning work, and that the measure of success changes over time and with the context of employment: "The 'appropriate level' [of cleanness] is, after all, still based on visible indicators and fixed frequencies of discrete cleaning actions. Both visible indicators and fixed frequencies function badly in environments where there is growing concern about *invisible factors* (bacteria, breathing air, allergies, magnetism, and the like) and *flexibility in responding to fluid functions and needs*" (1986, 8: emphasis in original).

They go on to argue that the definition of the indicators is a prime indication of the developmental aspects of the work being done, and indeed the nature of its specification.

Indicators for productivity, and abstracting away from the process of work being done, are the very stuff of management studies and much of economics. With the introduction of large-scale networked computing, and concomitant changes in how work is tracked and valued, a new ecology of visible and invisible work is being produced. Bannon (1995), Suchman (1995), and Blomberg, Suchman, and Trigg (1997) argue that these representations, often embedded in the neutral language of metrics, are in fact quite political. Invisible work is at the heart of the politics: what will "count" as productive work, creative work, work which cannot be outsourced or replaced in today's new corporation? Boland and Schultze's (1995) impassioned critique of activity-based accounting speaks to this problem. When creative "knowledge work" is highly valued by the firm, and at the same time ever more panoptic and detailed ways of accounting for work are being installed, a kind of double bind emerges. We want to capture creative work, represent it, bill for it. At the same time, such work is notoriously difficult to measure or represent. They sardonically conclude with a suggestion for the invention of an "executive jacket" which electronically would track those moments in which creative thoughts occur by monitoring the executive body: "A fine quality suit coat with microprocessors, data storage, and 'wireless' infrared communications, as well as location sensors, bar code readers, sound recording and high resolution video cameras woven into the lining of the jacket. . . . Each manager. . . . wears an Executive Jacket and in so doing allows much of his or her mental attention to be recorded and allocated as costs of production" (319).

Ten years ago, this might have been merely a science-fiction scenario meant to illustrate an analytic point. But with the advent of "affective computing," smart badges, and other forms of linking presence, thought, and emotion in advanced CSCW applications, it becomes more realistic. The point is thus an instructive one for CSCW. When the relationship between visible and invisible work is solely traded in abstract indicators, both silence and suffering result, to say nothing of inefficiency and obfuscation (Markussen 1995). As CSCW researchers set about modeling work, the use of abstract indicators must be understood in light of the contexts and relationships of visible and invisible work (Star 1995).

Subtle, Positive, and Routine Forms of Invisibility

We by no means wish to imply here that *all* work must be made visible, or else risk social injustice or badly fitting information systems. Much invisible work remains so for good reasons not covered above, and there is no immediate impetus toward change—for good reason. For example, workers—nurses or teachers come to mind as good examples—may quietly carry work reflecting a holistic view of the student or patient, carefully kept out of the range of a more bureaucratic, reductionist set of organizational values. Sometimes positive invisibility comes with discretion, and with not *having* to reveal your work processes to others. Some work is visible only to the worker, as it consists of local workarounds or routines they have developed privately at their desktops or elsewhere. Finally, some invisibility is strategic managing of parts of oneself that are inappropriate or undesirable in the workplace—and this may be positive as in autonomous control of the self, or negative, as in hiding shameful aspects.[3] All of these form important pieces of the visible-invisible matrix as well. We know that it is impossible to define anything inherently as visible or invisible; similarly, it is impossible a priori to say that either are absolutely good or bad, desirable or undesirable.

Some forms of invisibility result from the different perspectives afforded by different points of view in the (always) heterogeneous world of cooperative work. For example, sociologist Everett Hughes was fond of noting that "one person's emergency is another's routine" (1971). We can rarely really see the circumstances of another's world of work—indeed this is one natural consequence of the division of labor and of arm's-length relationship. In every case, it is important to examine the relationship between people and the work that they do, including grounds for trust in how that work is represented. Many studies over the years have cautioned that forced representation of work (especially that which results in computer support) may kill the very processes which are the target of support, by destroying naturally occurring information exchange, stories, and networks (see, e.g., Orr 1996 for a brilliant analysis of this problem in the worlds of photocopier repair technicians). Strübing has developed the notion of "subjective achievement" which is helpful to describe the importance of negotiating the parameters of visibility. He uses the example of computer programming, work organized within organizational structures, but also partly and necessarily against them, partly hidden from view: "It seems reasonable to see negotiation as one of the central subjective achievements of software designers. In programming work the participants have to bring together various dimensions: Every programmer is only capable to act within the framework of his or her subjective dispositions. There is the specific "materiality of the subject" and there is an organization with certain structures, a formal and informal balance of power, explicit and implicit goals. All this has to be taken into account while negotiating" (Strübing 1992, 17).

The invisible and emotional work of negotiating is an indispensable part of design; it is however often ruled out of technical education or design planning (Brooks 1975).

Design Implications

In the examples chosen here, it should be clear that we are not recommending "more visibility" in any simple sense. If we are to accept Grudin's (1988) and Robinson's (1993b) criteria of equity as an evaluation precept for successful CSCW applications, we must look in every circumstance at how the application affects relations of power and the nature of work. We find that in the examples above, there is "good invisibility" and "bad invisibility," often traced to questions of discretion, autonomy, and power over one's resources. So how would one judge the right level of visibility in designing CSCW systems?

The first precept is that the relation between invisible and visible work is a complex matrix, with an ecology of its own. It is relational, that is, there is no absolute visibility, and illuminating one corner may throw another into darkness. For every gain in granularity of description, there may be increased risk of surveillance. In the name of legitimacy and achieving public openness, an increased burden of accounting and tracking may be incurred. The phenomenon is one of trade-offs and balances, not absolutes and clear boundaries.

At the same time, the analysis of Schmidt and Simone suggests a direction that may be helpful for design of complex coordination artifacts. They carefully define a number of qualities of articulation work and cooperative work, and suggest a nomenclature system, Ariadne, for mapping workplace practices to artifact and back (1996). They note that "the distinction between cooperative work and articulation work is *recursive*; that is, an established arrangement of articulating a cooperative effort may itself be subjected to a cooperative effort of re-arrangement which in turn also may need to be articulated, and so forth" (158.)

This process works seamlessly in routine work. But they state that when highly complex artifacts and larger-scale divisions of labor are involved, everyday skills will not ensure continuing seamlessness. Artifacts to manage the complexity of articulation work are also needed. We would add a codicil to this: when serious inequities and divergent frames of reference exist, artifacts meant to decrease the complexity of articulation work will not reduce it, but displace it. That is, if the system does not account for the matrix of visible and invisible work and its questions of equity, those at the bottom will suffer.

How to do this? Schmidt and Simone present "artifactually imprinted protocols," means of articulating activity in work settings, as maps of a sort for coordinating work (1996, 167). They specify that such maps should have certain properties, including malleability and the ability to be locally controlled, and linked with other artifacts. What sorts of maps would we design in order similarly to support an equitable balance of visible and invisible work? In trying to map the invisible, one risks destroying the positive aspects of invisibility—should the map simply be marked, "here be dragons?"

Juggling one's visibility is itself an act of articulation work, and under many circumstances vital to getting work done. Hewitt's notion of the necessity of "arm's length relationships" in open systems and for due process reflects this as well (1985; Gerson and Star 1986). As well, the relationship between visible and invisible work is recursive in the sense meant by Schmidt and Simone. Making visible can incur invisibilities; obscuring may itself become a visible activity. With these caveats in mind, several design criteria suggest themselves.

1. *In making work visible as part of a coordination mechanism, is the reduction of complexity of articulation work equally shared?* For example, a publishing system that helps scientists in labs across geographical distance may reduce duplication and delays in

sharing findings. If at the same time it throws off technicians' reporting schedules and increases data entry for secretaries, this is a hidden cost shared inequitably (Star and Ruhleder 1996).

2. *Is the system temporally flexible as well as locally tailorable?* In managing the balance of visible and invisible work, it may be important to have certain processes become visible for a time, or remain invisible for a time. Using the example above, it may be important for a researcher to be able to time the release of their findings, to declare some work results invisible to others for a time, or to have other things fade or decay after a period of exposure.

3. *Are extant arm's-length relationships preserved?* This is a familiar question for CSCW as framed in terms of privacy and how privacy may be affected by the forms of sharing occasioned in CSCW systems use. Some aspects relevant to the visible/invisible questions mean carefully questioning how different metaphorical "curtains" may be drawn across parts of the work process—or opened up.

4. *Is the requirements analysis and specification of the system understood in terms of trade-offs and balances?* Because CSCW systems may affect the visibility of work processes to others, it is important that the systems development effort account for the pluses and minuses of visibility. System specification creates a complex boundary, not an absolute prescription. Public accessibility to the system itself becomes a condition of the ecology of visibility and invisibility.

The Éclat of Voice: When the Invisible Becomes Visible

The analysis of work as linked with problems in computer science has often produced unexpected views of the work process. For example, the study of the use of expert systems showed that people did not usually use the systems for the designed purposes. However, the knowledge elicitation process became an important occasion organizationally for people to make their work explicit to each other, enhancing communication processes in the result. Beyond being sensitive to the state of relations between visible and invisible work in any given context, it may also be that CSCW design and analysis intervenes in the processes of silence and visibility. CSCW requirements analysis may begin a dialogue between hidden and exiled work, on the one hand, and that which is taken for granted or considered rationalizable on the other. In any event, the stability of undervalued, hidden, shadow, and invisible work should not be assumed—rather, their relations will determine the path of use of CSCW systems.

Acknowledgments

The writings of Geoffrey Bowker, Adele Clarke, Lucy Suchman, and Marc Berg have inspired this article. Jörg Strübing provided very helpful comments on an earlier draft.

A Personal Note

Anselm Strauss's death in September 1996 came in the middle of our coauthoring this article. I am grateful that we had this chance to work together, and deeply saddened by his death. I wrote an appreciation of his work and our friendship (Star 1997) that may be of interest to readers of CSCW.
—Leigh Star

Notes

1. This distinction is in line with Strauss's original distinction between production work and articulation work (Strauss 1988; Strauss et al., 1985).

2. Private health insurance organizations.

3. The authors are grateful to an anonymous referee for clarifying this point.

References

Bannon, Liam. 1995. "The Politics of Design—Representing Work." *Communications of the ACM* 38:66.

Berg, Marc. 1997. *Rationalizing Medical Work.* Cambridge, MA: MIT Press.

Bishop, Ann, and Susan Leigh Star. 1996. "Social Informatics of Digital Library Use and Infrastructure." *Annual Review of Information Science & Technology* 31:301–401.

Blomberg, Jeannette, Lucy Suchman, and Randall Trigg. 1997. "Reflections on a Work-Oriented Design Project." In *Social Science, Technical Systems and Cooperative Work: Beyond the Great Divide,* ed. Geoffrey Bowker, Susan Leigh Star, William Turner, and Les Gasser, 189–215. Hillsdale, NJ: Erlbaum.

Boland, Richard, and Ulrike Schultze. 1995. "From Work to Activity: Technology and the Narrative of Progress." In *Information Technology and Changes in Organizational Work,* ed. Wanda Orlikowski, Geoff Walsham, Matthew Jones, and Janice DeGross, 308–324. London: Chapman and Hall.

Bowker, Geoffrey. 1994. "Information Mythology and Infrastructure." In *Information Acumen: The Understanding and Use of Knowledge in Modern Business,* ed. Lisa Bud-Frierman, 231–247. London: Routledge.

Bowker, Geoffrey. In press. "Lest We Remember: Organizational Forgetting and the Production of Knowledge." *Accounting, Management and Information Technology.*

Bowker, Geoffrey, and Susan Leigh Star. 1994. "Knowledge and Infrastructure in International Information Management: Problems of Classification and Coding." In *Information Acumen: The*

Understanding and Use of Knowledge in Modern Business, ed. Lisa Bud-Frierman, 187–213. London: Routledge.

Bowker, Geoffrey, Stefan Timmermans, and Susan Leigh Star. 1995. "Infrastructure and Organizational Transformation: Classifying Nurses' Work." In *Information Technology and Changes in Organizational Work*, ed. Wanda Orlikowski, Geoff Walsham, Matthew Jones, and Janice DeGross, 344–370. London: Chapman and Hall.

Bowker, Geoffrey, Susan Leigh Star, William Turner, and Les Gasser, eds. 1997. *Social Science, Technical Systems and Cooperative Work: Beyond the Great Divide*. Hillsdale, NJ: Erlbaum.

Brooks, Frederick. 1975. *The Mythical Man-Month: Essays on Software Engineering*. Reading, MA: Addison-Wesley.

Chambliss, D. F. 1988. *Champions: The Making of Olympic Swimmers*. New York: Morrow.

Chatman, Elfreda. 1987. "The Information World of Low-Skilled Workers (Janitorial Workers at a Large University in the South)." *Library & Information Science Research* 9:265–283.

Clayton, Laura Van Tuyl. 1996. "Canadian Census Gives Credit for Unpaid Work." *Christian Science Monitor* (May 8), http://www.csmonitor.com/1996/0508/050896.feat.feat.1.html.

Clement, Andrew. 1991. "Designing without Designers: More Hidden Skill in Office Computerization." In *Women, Work and Computerization: Understanding and Overcoming Bias in Work and Education*, ed. I. Eriksson, B. A. Kitchenham, and K. G. Tijdens, 15–32. Amsterdam: North Holland.

Clement, Andrew. 1993. "Looking for the Designers: Transforming the 'Invisible' Infrastructure of Computerized Office Work." *AI & Society* 5:323–344.

Colen, S. 1989. "'Just a Little Respect': West Indian Domestic Workers in New York City." In *Muchachas No More: Household Workers in Latin America and the Caribbean*, ed. Elsa Chaney and Mary G. Castro, 171–194. Philadelphia: Temple University Press.

Cowan, Ruth Schwartz. 1983. *More Work for Mother: The Ironies of Household Technology from the Open Hearth to the Microwave*. New York: Basic Books.

Driscoll, Mary Kathleen. 1995. "Body Boycotts." Ph.D. dissertation, Department of Sociology, Northwestern University, Evanston, IL.

Engeström, Yrjö, and Ritva Engeström. 1986. "Developmental Work Research: The Approach and an Application in Cleaning Work." *Acta Psychologica Fennica* 11:211–227.

Fjuk, Anita, Ole Smørdal, and Markku Nurminen. 1997. "Taking Articulation Work Seriously—An Activity Theoretical Approach." Unpublished paper submitted to *ECSCW '97*, Department of Informatics, University of Oslo, January.

Foucault, Michel. 1977. *Discipline and Punish: The Birth of the Prison*. Trans. Alan Sheridan. New York: Pantheon Books.

Fussell, Susan R. 1991. "The Coordination of Knowledge in Communication: People's Assumptions about Others' Knowledge and Their Effects on Referential Communication." Ph.D. dissertation, Department of Communication, Columbia University.

Gasser, Les. 1986. "The Integration of Computing and Routine Work." *ACM Transactions on Office Information Systems* 4: 205–2225.

Gerson, Elihu, and Susan Leigh Star. 1986. "Analyzing Due Process in the Workplace." *ACM Transactions on Office Information Systems* 4:257–270.

Goffman, Erving. 1962. *Asylums: Essays on the Social Situation of Mental Patients and Other Inmates.* Chicago: Aldine.

Goffman, Erving. 1969. *The Presentation of Self in Everyday Life.* London: Allen Lane.

Goguen, Joseph. 1994. "Requirements Engineering as the Reconciliation of Technical and Social Issues." In *Requirements Engineering: Social and Technical Issues*, ed. Marina Jirotka and Joseph Goguen, 165–199. New York: Academic Press.

Goguen, Joseph. 1997. "Towards a Social, Ethical Theory of Information." In *Social Science, Technical Systems and Cooperative Work: Beyond the Great Divide*, ed. Geoffrey Bowker, Susan Leigh Star, William Turner, and Les Gasser, 27–56. Hillsdale, NJ: Erlbaum.

Greenbaum, Joan. 1995. *Windows on the Workplace.* New York: Monthly Review Press.

Grinter, Rebecca. 1996. "Supporting Articulation Work Using Software Configuration Management Systems." *Computer Supported Cooperative Work: The Journal of Collaborative Computing* 5:447–465.

Grudin, Jonathan. 1988. "Why CSCW Applications Fail: Problems in the Design and Evaluation of Organizational Influences." In *Proceedings of CSCW 88*, 85–93. New York: ACM Press.

Hewitt, Carl. 1985. "The Challenge of Open Systems." *BYTE*10:223–242.

Hossfeld, Karen J. 1990. "'Their Logic against Them': Contradictions in Sex, Race, and Class in Silicon Valley." In *Women Workers and Global Restructuring*, ed. Kathryn Ward, 149–178. Ithaca, NY: ILR Press.

Hughes, Everett. 1971. *The Sociological Eye, Vol. 2, Selected Papers on Work, Self and the Study of Society.* Chicago: Aldine Atherton.

Illich, Ivan. 1981. *Shadow Work: Vernacular Values Examined.* London: Marion Boyars.

Jirotka, Marina, and Joseph Goguen, eds. 1994. *Requirements Engineering: Social and Technical Issues.* New York: Academic Press.

Kanter, Rosabeth Moss. 1977. *Men and Women of the Corporation.* New York: Basic Books.

Kaptelinin, Victor, and Bonnie Nardi. 1997. "Activity Theory: Basic Concepts and Applications." In *Human Factors in Computing Systems CHI 97 Conference Proceedings*, 74–77.

Kramarae, Cheris. 1988. "Gotta Go Myrtle, Technology's at the Door." In *Technology and Women's Voices: Keeping in Touch*, ed. Cheris Kramarae, 1–14. New York: Routledge and Kegan Paul.

McCloskey, Joanne C., and Gloria M. Bulechek. 1993. *Nursing Interventions Classification.* St. Louis: Mosby Year Book.

McCloskey, J. C., and G. M. Bulechek. 1994. "Standardizing the Language for Nursing Treatments: An Overview of the Issues." *Nursing Outlook* 42:56–63.

Markussen, Randi. 1995. "Constructing Easiness—Historical Perspectives on Work, Computerization and Women." In *The Cultures of Computing*, ed. Susan Leigh Star, 158–180. Oxford, UK: Basil Blackwell.

Morrison, Toni. 1987. *Beloved: A Novel*. New York: Knopf.

Orr, Julian. 1996. *Talking about Machines: An Ethnography of a Modern Job*. Ithaca, NY: ILR Press.

Robinson, Mike. 1991. "Double-Level Languages and Co-operative Working." *AI & Society* 5:34–60.

Robinson, Mike. 1993a. "Design for Unanticipated Use . . ." In *Proceedings of the Third European Conference on Computer Supported Cooperative Work*, ed. G. DiMichelis, C. Simone, and K. Schmidt, 187–202. Milan: Kluwer.

Robinson, Mike. 1993b. "Computer Supported Co-operative Work: Cases and Concepts." In *Readings in Groupware and Computer-Supported Cooperative Work: Assisting Human-Human Collaboration*, ed. Ron Baecker, 29–49. San Francisco: Morgan Kaufmann.

Robinson, Mike. 1997. "'As Real as It Gets . . . ': Taming Models and Reconstructing Procedures." In *Social Science, Technical Systems and Cooperative work: Beyond the Great Divide*, ed. Geoffrey Bowker, Susan Leigh Star, William Turner, and Les Gasser, 257–274. Hillsdale, NJ: Erlbaum.

Rollins, Judith. 1985. *Between Women: Domestics and Their Employers*. Boston: Beacon Press.

Romero, Mary. 1992. *Maid in the U.S.A.* New York: Routledge.

Sanjek, Roger, ed. 1990. *Fieldnotes: The Makings of Anthropology*. Ithaca, NY: Cornell University Press.

Schmidt, Kjeld, and Liam Bannon. 1992. "Taking CSCW Seriously: Supporting Articulation Work." *Computer Supported Cooperative Work: The Journal of Collaborative Computing* 1:7–41.

Schmidt, Kjeld, and Carla Simone. 1996. "Coordination Mechanisms: Towards a Conceptual Foundation of CSCW Systems Design." *Computer Supported Cooperative Work: The Journal of Collaborative Computing* 5:155–200.

Shapin, Steven. 1989. "The Invisible Technician." *American Scientist* 77:554–563.

Star, Susan Leigh. 1989. *Regions of the Mind: Brain Research and the Quest for Scientific Certainty*. Stanford, CA: Stanford University Press.

Star, Susan Leigh. 1991a. "The Sociology of the Invisible: The Primacy of Work in the Writings of Anselm Strauss." In *Social Organization and Social Process: Essays in Honor of Anselm Strauss*, ed. David Maines, 265–283. Hawthorne, NY: Aldine de Gruyter.

Star, Susan Leigh. 1991b. "Invisible Work and Silenced Dialogues in Representing Knowledge." In *Women, Work and Computerization: Understanding and Overcoming Bias in Work and Education*, ed. I. Eriksson, B. A. Kitchenham, and K. G. Tijdens, 81–92. Amsterdam: North Holland.

Star, Susan Leigh. 1992. "Craft vs. Commodity, Mess vs. Transcendence: How the Right Tool Became the Wrong One in the Case of Taxidermy and Natural History." In *The Right Tools for the Job: At Work in Twentieth Century Life Sciences*, ed. Adele Clarke and Joan Fujimura, 257–286. Princeton: Princeton University Press.

Star, Susan Leigh. 1995. "The Politics of Formal Representations: Wizards, Gurus, and Organizational Complexity." In *Ecologies of Knowledge: Work and Politics in Science and Technology*, ed. Susan Leigh Star, 88–118. Albany: SUNY Press.

Star, Susan Leigh. 1997. "Anselm Strauss: An Appreciation." *Studies in Symbolic Interaction* 21:39–48.

Star, Susan Leigh, and Karen Ruhleder. 1996. "Steps toward an Ecology of Infrastructure: Design and Access for Large Information Spaces." *Information Systems Research* 1:111–134.

Strauss, Anselm. 1985. "Work and the Division of Labor." *Sociological Quarterly* 26:1–19.

Strauss, Anselm. 1988. "The Articulation of Project Work: An Organizational Process." *Sociological Quarterly* 29:163–178.

Strauss, Anselm, Shizuko Fagerhaugh, Barbara Suczek, and Carolyn Wiener. 1985. *Social Organization of Medical Work*. Chicago: University of Chicago Press.

Strübing, Jürg. 1992. "Negotiation—A Central Aspect of Collaborative Work in Software Design." In *5ème Workshop sur la Psychologie de la Programmation*, Institut National de Recherche en Informatique et en Automatique, December 10–12, Paris, 31–39.

Suchman, Lucy. 1995. "Making Work Visible." *Communications of the ACM* 38:56–68.

Suchman, Lucy. 1996. "Supporting Articulation Work." In *Computerization and Controversy: Value Conflicts and Social Choices*, 2nd ed., ed. Rob Kling, 407–423. San Diego, CA: Academic Press.

Wagner, Ina. 1995. "Women's Voice: The Case of Nursing Information Systems." *AI & Society* 7:315–334.

Wrigley, Julia. 1995. *Other People's Children*. New York: Basic Books.

IV Infrastructure

20   Steps toward an Ecology of Infrastructure: Design and Access for Large Information Spaces

Susan Leigh Star and Karen Ruhleder

Editors' Note: This chapter was first published in 1996, based on an earlier version presented by Susan Leigh Star and Karen Ruhleder at the 1994 Computer Supported Cooperative Work (CSCW) Conference, and published in its proceedings. At the time, personal computers were far from ubiquitous and those who had home computers connected to the internet via dial-up modems. The World Wide Web barely existed. Although the technologies that now surround us have changed, the analytic points raised in this piece remain highly salient. Popular views of infrastructure described here are just as applicable to views of computing infrastructure predominant in the health care sector today. The dimensions of infrastructure so nicely articulated here remain poorly understood by many—just as they were when this piece was written.

We analyze a large-scale custom software effort, the Worm Community System (WCS), a collaborative system designed for a geographically dispersed community of geneticists.

There were complex challenges in creating this infrastructural tool, ranging from simple lack of resources to complex organizational and intellectual communication failures and trade-offs. Despite high user satisfaction with the system and interface, and extensive user needs assessment, feedback, and analysis, many users experienced difficulties in signing on and use. The study was conducted during a time of unprecedented growth in the Internet and its utilities (1991–1994), and many respondents turned to the World Wide Web for their information exchange. Using Bateson's model of levels of learning, we analyze the levels of infrastructural complexity involved in system access and designer-user communication. We analyze the connection between systems development aimed at supporting specific forms of collaborative knowledge work, local organizational transformation, and large-scale infrastructural change.

"An electronic community system is a computer system which encodes the knowledge of a community and provides an environment which supports manipulation of that knowledge. Different communities have different knowledge but their environment has great similarities. The community knowledge might be thought of as being stored in an electronic library" (Schatz 1991, 88).

"Does virtual community work or not? Should we all go off to cyberspace or should we resist it as a demonic form of symbolic abstraction? Does it supplant the real or is there, in it, reality itself? Like so many true things, this one doesn't resolve itself to a black or a white. Nor is it gray. It is, along with the rest of life, black/white. Both/neither" (Barlow 1995, 56).

What Is Infrastructure?

People who study how technology affects organizational transformation increasingly recognize its dual, paradoxical nature. It is both engine and barrier for change, both customizable and rigid, both inside and outside organizational practices. It is product and process. Some authors have analyzed this seeming paradox as structuration (after Giddens): technological rigidities give rise to adaptations that in turn require calibration and standardization. Over time, structure-agency relations re-form dialectically (Orlikowski 1991; Davies and Mitchell 1994; Korpela 1994). This paradox is integral to large-scale, dispersed technologies (Brown and Duguid 1994; Star 1991a, 1994). It arises from the tension between local, customized, intimate, and flexible use on the one hand, and the need for standards and continuity on the other hand.

With the rise of decentralized technologies used across wide geographical distance, both the need for common standards and the need for situated, tailorable, and flexible technologies grow stronger. A lowest common denominator will not solve the demand for customized possibilities; neither will rigid standards resolve the issue (Trigg and Bodker 1994). It is impossible to have "universal niches"; one person's standard is in fact another's chaos. There are no genuine universals in the design of large-scale information technology (Star 1991a; Bowker 1993).

Furthermore, this simultaneous need for customization and standardization is not geographically based or based on simple group-membership parameters. An individual is often a member of multiple communities of practice that use technologies differently, and that thus have different demands on their flexible standard requirements. There is no absolute center from which control and standards flow; as well, there is no absolute periphery (Hewitt 1986). Yet some sort of infrastructure is needed.

We studied the building of a geographically dispersed, sophisticated digital communication and publishing system for a community of scientists. The system-building effort, which was itself an attempt to enhance and create infrastructural tools for research, took place during a period of immense, even radical change in the larger sphere of electronic information systems (1991–1994). One purpose of the development effort was to transform local laboratory organization, and minimize inefficiencies of scale with respect to knowledge and results. The vision was of a kind of supra-laboratory stretched over the entire scientific community. The needs for both standards and customizable components were equally strong. The system development process also became an effort to bring together communities of practice with very different approaches to computing infrastructure. Designers and users faced two sorts of challenges in developing the system: communicating despite very different practices, technologies, and skills; and keeping up with changes occasioned by the growth of the Internet and tools like Gopher and Mosaic. Trying to develop a large-scale information infrastructure in this climate is metaphorically like building the boat you're on while designing the navigation system and being in a highly competitive boat race with a constantly shifting finish line.

This chapter is about that experience, and about its ultimate failure to produce the expected organizational and infrastructural changes. It offers an analytic framework and vocabulary to begin to answer the question: What is the relationship between large-scale infrastructure and organizational change? Who (or what) is changer, and who changed? We begin with a definition of "infrastructure," and then focus on two aspects of the system development effort: communication and mutual learning between designers and users.

When Is an Infrastructure?

What can be studied is always a relationship or an infinite regress of relationships. Never a "thing." (Bateson 1972)

Yrjö Engeström, in his "When Is a Tool?" answers the implied title question in terms of a web of usability and action (1990). A tool is not just a thing with pre-given attributes frozen in time—but a thing becomes a tool in practice, for someone, when connected to some particular activity. The article is illustrated by a photo of a physician working at a terminal covered with yellow post-it notes, surrounded by hand-scribbled jottings, and talking on the phone—a veritable heterogeneous "web of computing" (Kling and Scacchi 1982). The tool emerges in situ. By analogy, infrastructure is something that emerges for people in practice, connected to activities and structures.

When, then, is an infrastructure? Common metaphors present infrastructure as a substrate: something upon which something else "runs" or "operates," such as a system of railroad tracks upon which rail cars run. This image presents an infrastructure as something that is built and maintained, and which then sinks into an invisible background. It is something that is just there, ready to hand, completely transparent.

But such a metaphor is neither useful nor accurate in understanding the relationship between work/practice and technology. It is the image of "sinking into the background" that concerns us. Furthermore, we know that such a definition will not capture the ambiguities of usage referred to earlier: for example, without a Braille terminal, the

Internet does not work to support a blind person's communication. And for the plumber, the waterworks system in a household connected to the city water system is target object, not background support. Rather, following Jewett and Kling (1991), we hold that infrastructure is a fundamentally relational concept. It becomes infrastructure in relation to organized practices. Within a given cultural context, the cook considers the water system a piece of working infrastructure integral to making dinner; for the city planner, it becomes a variable in a complex equation. Thus we ask, when—not what—is an infrastructure.

Analytically, infrastructure appears only as a relational property, not as a thing stripped of use. Bowker (1994) calls this "infrastructural inversion," a methodological term, referring to a powerful figure ground gestalt shift in studies of the development of large-scale technological infrastructure (Hughes 1983, 1989). The shift deemphasizes things or people as simply causal factors in the development of such systems; rather, changes in infrastructural relations become central. As we learn to rely on electricity for work, our practices and language change, we are "plugged in" and our daily rhythms shift. The nature of scientific and aesthetic problems shifts as well. As this infrastructural change becomes a primary analytic phenomenon, many traditional historical explanations are inverted. Yates (1989) shows how even so humble an infrastructural technology as the file folder is a central factor in changes in management and control in American industry. In the historical analysis, the politics, voice, and authorship embedded in the systems are revealed—not as engines of change, but as articulated components of the system under examination. Substrate becomes substance.

With this caveat, infrastructure emerges with the following dimensions:

• *Embeddedness*. Infrastructure is "sunk" into, inside of, other structures, social arrangements, and technologies.
• *Transparency*. Infrastructure is transparent to use, in the sense that it does not have to be reinvented each time or assembled for each task, but invisibly supports those tasks.
• *Reach or scope*. This may be either spatial or temporal—infrastructure has reach beyond a single event or one-site practice.
• *Learned as part of membership*. The taken-for-grantedness of artifacts and organizational arrangements is a sine qua non of membership in a community of practice (Lave and Wenger 1992; Star 1996). Strangers and outsiders encounter infrastructure as a target object to be learned about. New participants acquire a naturalized familiarity with its objects as they become members.
• *Links with conventions of practice*. Infrastructure both shapes and is shaped by the conventions of a community of practice, for example, the ways that cycles of day-night work are affected by and affect electrical power rates and needs. Generations of typists have learned the QWERTY keyboard; its limitations are inherited by the computer keyboard and thence by the design of today's computer furniture (Becker 1982).

• *Embodiment of standards*. Modified by scope and often by conflicting conventions, infrastructure takes on transparency by plugging into other infrastructures and tools in a standardized fashion.

• *Built on an installed base*. Infrastructure does not grow de novo; it wrestles with the "inertia of the installed base" and inherits strengths and limitations from that base. Optical fibers run along old railroad lines; new systems are designed for backward compatibility; and failing to account for these constraints may be fatal or distorting to new development processes (Monteiro, Hanseth, and Hatling 1994).

• *Becomes visible upon breakdown*. The normally invisible quality of working infrastructure becomes visible when it breaks; the server is down, the bridge washes out, there is a power blackout. Even when there are back-up mechanisms or procedures, their existence further highlights the now-visible infrastructure. The configuration of these dimensions forms "an infrastructure," which is without absolute boundary on a priori definition (Star 1989a,b). Most of us, in speaking loosely of infrastructure, mean those tools which are fairly transparent for most people we know about, wide in both temporal and spatial scope, embedded in familiar structures-like power grids, water, the Internet, airlines. That loose talk is perfectly adequate for most everyday usage, but is dangerous when applied to the design of powerful infrastructural tools on a wide scale, such as is now happening with "national information infrastructures." Most importantly, such talk may obscure the ambiguous nature of tools and technologies for different groups, leading to de facto standardization of a single, powerful group's agenda. Thus it contributes to Kraemer and King's "politics of reinforcement" in computerization (1977). Such talk may also obscure the nature of organizational change occasioned by information technology development.

If we add these dimensions of infrastructure to the dual and paradoxical nature of technology, our understanding deepens. In fact, the ambiguity and multiple meanings of usage mark any real functioning system. An infrastructure occurs when the tension between local and global is resolved. That is, an infrastructure occurs when local practices are afforded by a larger-scale technology, which can then be used in a natural, ready-to-hand fashion. It becomes transparent as local variations are folded into organizational changes, and becomes an unambiguous home—for somebody. This is neither a physical location nor a permanent one, but a working relation since no home is universal (Star 1996).

The empirical data for this chapter come from our work as ethnographers/evaluators of a geographically dispersed virtual laboratory or "collaboratory" system meant to link the work of over 1,400 biologists (Star 1991b). The system itself appeared differently to different groups—for some it was a set of digital publishing and information retrieval tools to "sit upon" already existing infrastructure; for others it supported problem solving and information sharing; for yet others, it was a component of an established set of practices and infrastructural laboratory tools. The target users had vastly differing resources and computing skills and relationships, and these in turn were sharply different from those of the designers.

As well, it is increasingly clear to us that this development effort is taking place at a moment of rare, widespread infrastructural change. With the growth of the Internet/World Wide Web and their utility software (such as Mosaic, Netscape, Gopher, WAIS), as well as the myriad of email uses, electronic bulletin boards, and listservs, the boundaries of system implementation are embedded in the eye of an informational and organizational hurricane of change. For a few of our respondents, the system became a working infrastructure; others, however, turned to Gopher and Mosaic and other Internet tools. And of course, the skill base and learning curve, as well as other factors such as support networks in organizations that help users with such tools, are themselves constantly changing. This changing environment, combined with the complexities of implementation from the user's perspective, contributed to the system's ultimate failure in achieving its original goal of becoming the central information resource and the primary communication conduit within a particular scientific community.

The Worm Community System

Background

The Worm Community System (WCS) is a customized piece of software meant to support the collaborative work of biologists sequencing the gene structure, and studying other aspects of the genetics, behavior, and biology of c. elegans, a tiny nematode (Schatz 1991; Pool 1993). It is one example of a new genre of systems being developed for geographically dispersed collaborative scientific work. WCS is a distributed "hyperlibrary" affording informal and formal communication and data access across many sites. It incorporates graphical representations of the physical structure of the organism; a periodically updated genetic map; formal and informal research annotations (thus also functioning as an electronic publishing medium); directories of scientists; a thesaurus of terms linked with a directory of those interested in the particular subtopic; and a quarterly newsletter, the Worm Breeder's Gazette. It also incorporates an independently developed database, acedb. Many parts of the system are hypertext-linked.

Its principle designers were computer scientists, some with backgrounds in biology. However, WCS was developed with the close cooperation of several biologists; user feedback and requests from those biologists were initially incorporated into the system over the years of development. Its development was part of a broader project to both construct and evaluate the implementation and impact of a scientific collaboratory. Two ethnographers, Star and Ruhleder, were members of the project team, but not part of the technical development effort per se. The ethnographic component of the project is described in more detail as follows.

The community consists, as we have stated, of about 1,400 scientists distributed around the world in some 120 laboratories (as of 1994). They are close-knit and consider themselves extremely friendly, as indeed we found them to be. Until recently, most people were first or second "generation" students of the field's founders. Recently, c. elegans was chosen as the "model organism" for the Human Genome Initiative (HGI), said to be the largest scientific project in history. "Model organism" means both that the actual findings from doing the worm biology and genetics will be directly of interest to human geneticists—for example, when homologs are found between oncogenes (cancer-causing genes) in the worm and in the human (although worms do not get cancer as such, there are developmental analogies). In addition, the tools and techniques developed in the c. elegans mapping effort will be useful for the human project.

Senior biologists are concerned that the impact of the HGI and increasing interest in the worm will adversely affect the close, friendly nature of relationships in the community. Viewing this community as a loosely coupled organization whose members often work in and interact with more formal organizations, these new constraints and opportunities threatened to upset traditional linkages and a collaborative culture heavily dependent on apprenticeship and continued personal contact. Members of the community themselves were willing to become a "model organism" for the ethnographers because they hoped the system would help maintain the community's strong bonds and friendly character in the face of rapidly increasing visibility and growth. In that sense, the goal was not only organizational transformation in terms of available resources and information-sharing opportunities, but also the retention of desired characteristics in the face of transformation for the worse.

The work of c. elegans biologists can be captured by the notion of solving a jigsaw puzzle in four dimensions, across considerable distance (the labs we studied were located in the United States and Canada; input comes from Europe, Japan, and Australia as well). In addition to the four dimensions, the data are structured differently and must be mapped across fields; for example, a behavioral disorder linked with one gene must be triangulated with information from corresponding DNA fragments. Labs working on a particular problem, for example, sperm production, are in frequent contact with each other by phone, fax, and email, exchanging results and specimens.

The worm itself is remarkable both as an organism, and as a component of a complex pattern of information transfer integral to the biologists' work. It is microscopic and transparent (thus easier to work with than opaque creatures such as humans!). It is a hardy creature, and may be frozen, mailed to other labs via UPS, thawed out, and retrieved live for observation. Worms and parts of worms travel from one lab to another as researchers share specimens. Worm strains with particular characteristics, such as a mutation, may be mailed from a central Stock Center to labs requesting specimens.

Tracking the location and characteristics of organisms thus is an important part of recordkeeping and information retrieval.

Computing use and sophistication in the labs varies widely. In the labs most active in trying out WCS, there are one or two active, routine WCS users. In many, computing is confined to email, word processing, or the preparation of graphics for talks. In most labs there is one "computer person," often a student, who is in charge of ordering new programs and designing databases to keep track of strains and other information.

Our role in the project as ethnographers has been to travel to worm labs, interview about and observe both the use of computing and WCS, and other aspects of routine work, as well as to ask questions on topics including careers in the community, competition, routine information-sharing tasks, how computing infrastructure is managed, and so on. We did semistructured interviews and observations at twenty-five labs with more than a hundred biologists over a three-year period (1991–1995) and fed back to designers both specific suggestions ("so-and-so found a bug") and general observations ("such-and-such would violate community norms"), several of which were incorporated into development.[1]

Sociological analysis to support computer design is relatively new (Bucciarelli 1994). The participatory design approach developed in Scandinavia paved the way for workplace studies that inform design (Ehn 1988; Bodker 1991; Anderson and Crocca 1993), usually using a combination of a case study approach and action research, with rapid feedback from users of computing systems. Where possible, we adapted those principles. At the same time, trying to cover a geographically distributed community in aid of complex systems development also meant that neither a strict case study nor rapid prototyping were possible. We covered as much territory as possible with traditional interviewing and observational techniques. We conducted data analysis using a grounded theory approach, beginning with a substantive description of the community and moving to more abstract analytical frameworks as our comparative sites grew in number (Strauss 1986).

On the one hand, most respondents said they liked the system, praising its ease of use and its understanding of the problem domain. On the other hand, most have not signed on; many have chosen instead to use Gopher and Mosaic/Netscape and other simpler net utilities with less technical functionality. Obviously, this is a problem of some concern to us as system developers and evaluators. Despite good user prototype feedback and participation in the system development, there were unforeseen, complex challenges to usage involving infrastructural and organizational relationships. The system was neither widely adopted nor did it have an immediate impact on the field as the resources and communication channels it proffered became available through other (often more accessible) means. However, the WCS itself continues to change and adapt; the latest version is based entirely on web technology, and the web will shortly have enough functionality to reproduce the custom software WCS.[2]

Signing On and Hooking Up

Those working in the emergent field of computer-supported cooperative work (CSCW), of which the collaboratory is a subset, have struggled to understand how infrastructural properties affect work, communication, and decision making (Kraemer and King 1977; Schmidt and Bannon 1992; Malone and Olson 2001). One of the classic CSCW typologies has distinguished important task differences for synchronous/asynchronous systems; proximate/long-distance use; and dedicated user groups vs. distributed groups with fluctuating membership (Ellis, Gibbs, and Rein 1991). This was useful for characterizing an emerging group of technologies; however, it offers no assistance in analyzing the issues associated with implementation or integration (Schmidt and Bannon 1992). It also does not analyze the relational aspects of computing infrastructure and work, either real-time "articulation work" or aspects of longer-term, asynchronous production tasks. We encountered many such issues in the worm community in the process of "signing on" and "hooking up" to WCS tasks related to finding out about the system, installing it, and learning to use it. For most of the worm biologists we interviewed, the tasks involved in signing on and hooking up had preoccupied them, and they had not gotten over the initial hurdle and into routine use.

Consider the set of tasks associated with getting the system up and running. WCS runs on a Sun Workstation as a standalone or remotely, or on a Mac with an ethernet connection remotely over the NSFnet, or, with less functionality, on a PC over the net. Prior to using WCS, one must buy the appropriate computer; identify and buy the appropriate windows-based interface; use a communications protocol such as telnet and/or FTP; and locate the remote address where you "get" or operate the system. Each of these tasks requires that people trained in biology acquire skills taken for granted by systems developers. The latter have interpersonal and organizational networks that help them obtain necessary technical information, and also possess a wealth of tacit knowledge about systems, software, and configurations. For instance, identifying which version of X Windows to use on a workstation means understanding what class of software product X Windows is, installing it, and then linking its configuration properly with the immediate or remote link. Following instructions to "download the system via FTP" requires an understanding of file transfer protocols across the Internet, knowing which issue of the Worm Breeder's Gazette lists the appropriate electronic address, and knowing how FTP and X Windows work together.

All computer users face these common issues of shopping, configuration, and installation in some degree. But solving these "shopping" and informational issues will not always suffice to get work done smoothly. For instance, deciding to buy a SPARC station (one popular UNIX-based workstation) and run it on a campus already standardized on DOS machines may bring you into conflict with the local computer center, and their attempts to limit the sorts of machines they will service. Or there may be enough money to buy the computer, but not enough to support training for all lab staff; in the long term, this disparity may create inequities.

We discovered many such instances, common to a variety of system development efforts and types of users, and all interesting for the design of collaborative systems. With the advent of very large-scale systems such as the U.S. National Information Infrastructure, they become pressing issues of equity and justice and organizational formation, transformation, and demise. At the same time, such instances simultaneously enact technological infrastructure and social order. We encountered a myriad of contexts and tasks surrounding system use. These varied in complexity and consequences, and we borrowed a metaphor from learning theory to characterize these variations.

Levels of Communication and Discontinuities in Hierarchies of Information

The "tangles" encountered in signing on and hooking up occur in many venues, and may inhibit desired organizational transformation; at the least, they inform its character and flavor the growth of infrastructure. We turned to Gregory Bateson as a theorist of communication for a more formal understanding of the ways in which communicative processes are entangled in the development of infrastructure. We rely on his *Steps to an Ecology of Mind* (Bateson 1972). The term "ecology," as adapted to our analysis here, refers to the delicate balance of language and practice across communities and parts of organizations; it draws attention to that balance (or lack of it). It is not meant to imply either a biological approach or a closed, functional systemic one.

Bateson's Model

Bateson (1972), following Russell and Whitehead, distinguishes three levels in any communicative system. At the first level are straightforward "fact" statements, in other words, "the cat is on the mat." A discontinuous shift in context occurs as the statement's object is changed to "I was lying when I said 'the cat is on the mat.'" This second-order statement tells you nothing about the location of the cat, but only something about the reliability of the first-order statement. In Bateson's words: "There is a gulf between context and message (or between metamessage and message), which is of the same nature as the gulf between a thing and the word or sign which stands for it, or between the members of a class and the name of the class. The context (or metamessage) classifies the message, but can never meet it on equal terms" (247).

At the third level, the gulf appears in evaluating the context itself: "There are many conflicting approaches to evaluating whether or not you were lying about the cat and the mat." In this sentence, the listener's attention is forced to a wider and deeper range of possibilities; again, it may classify the message about lying, but is of a different character.

Theorizing the gulf between levels, Bateson and others have gone on to classify levels of learning with similar distinctions and discontinuities. There is a first- and second-order difference in learning something and learning about learning something; and between the second and third order are even more abstract differences between learning to learn, and learning about theories of learning and paradigms of education. As the epigraph to an earlier section indicates, of course the regress upward is potentially infinite.

For our purposes we identify three levels (or orders) of issues that appear in the process of infrastructure development, and discuss each with respect to the worm community and WCS. As with Bateson's levels of communication or learning, the issues become less straightforward as contexts change. This is not an idealization process (i.e., they are not less material and more "mental"), or even essentially one of scope (some widespread issues may be first order), but rather questions of context. Level-one statements appear in our study: "Unix may be used to run WCS." These statements are of a different character than a level-two statement such as "A system developer may say Unix can be used here, but they don't understand our support situation." At the third level, the context widens to include theories of technical culture: "Unix users are evil—we are Mac people." As these levels appear in communication between developers and users, the nature of the gulfs between levels is important.

First-order issues may be solved with a redistribution or increase of extant resources, including information. Examples would be answers to questions such as: What is the email address of WCS? How do I hook up my SPARC station to the campus network?

Second-order issues stem from unforeseen or unknowable contextual effects, perhaps from the interaction of two or more first-order issues. An example here is given above: what are the consequences of my choosing a SPARC station instead of a Mac, if my whole department uses Macs? If I invest my resources in learning WCS, are there other more useful programs I am neglecting?

Third-order issues are inherently political or involve permanent disputes. They include questions about schools of thought of biological theory for designing the genetic map of the organism for WCS. They raise questions such as whether competition or cooperation will prove more important in developing systems privacy requirements, and whether complexity or ease of use should be the main value in interface design. Such questions may arise from an interaction of lower-order issues, such as the choice of computer system and the trade-offs between scientific sophistication and ease of learning.

In this sense, infrastructure is context for both communication and learning within the web of computing (Kling and Scacchi 1982). Computers, people, and tasks together make or break a functioning infrastructure. In Bateson's words: "It becomes clear that the separation between contexts and orders of learning is only an artifact. . . . The separation is only maintained by saying that the contexts have location outside the physical individual, while the orders of learning are located inside. But in the communicational world, this dichotomy is irrelevant and meaningless . . . the characteristics of the system are in no way dependent upon any boundary lines which we may superimpose upon the communicational map" (Bateson 1972, 251).

Information infrastructure is not a substrate that carries information on it, or in it, in a kind of mind–body dichotomy. The discontinuities are not between system and person, or technology and organization, but rather between contexts. Here we echo recent work in the sociology of technology and science, which refuses a "great divide" between nature and artifice, human and nonhuman, technology and society (e.g., Latour 1993).

These discontinuities have the same conceptual importance for the relationship between information infrastructure and organizational transformation that Bateson's work on the double bind had for the psychology of schizophrenia. If we, in large-scale information systems implementation, design messaging systems blind to the discontinuous nature of the different levels of context, we end up with organizations that are split and confused, systems that are unused or circumvented, and a set of circumstances of our own creation that more deeply impress disparities on the organizational landscape.

We apply this typology within the context of "signing on" and "hooking up." Following that application, we discuss the implications of this typology for other forms of information systems development, and the broader implications for understanding the impact of new computer-based media and their integration into established communities.

First-Order Issues

First-order issues are often those that are most obvious to informants, as they tend toward the concrete, and can be addressed by equally concrete solutions (more money, time, training, or support). The first-order issues in this setting center on the installation and use of the system, and include finding out about it, figuring out how to install it, and making different pieces of software work together. First-order issues, however, are not limited to "startup," but recur over time as work patterns and resource constraints shift (and thus perhaps a byproduct of second- or third-order changes).

Informational Issues Potential users needed to find out about the system and determine the requirements for its installation and use. "Shopping" for the system involved decisions about hardware and software, and sometimes also involved agreements with other departments to share resources or funding; at one major lab, the "worm" people had WCS loaded onto a server owned by the "plant" people on the floor above them; establishing this agreement involved finding out about WCS resource needs and the local availability of these resources. This agreement negated the need to find out about system building and maintenance, since the worm people were piggybacking off the original efforts of the plant people to purchase and put the server in place.

Issues of Access   In some labs, physical access was critical. WCS might be located in an overcrowded and noisy room, stuck in the corner of a lounge, on a different floor of the building altogether, or accessible only during certain hours. This was the case in the deal cut between the worm and plant people: "The WCS and acedb are really on a machine upstairs, it belongs to the plant genome project people. . . . We can only use it evenings, weekends" (Brad Thomas, PD).[3]

Other labs experienced time limitations and physical inconveniences: "You can access acedb through the Suns downstairs, but it's not convenient. You can only do it after hours. People just won't use it" (Eliot Red, PD) or: "Our computing is good compared to other labs. I finished up a Ph.D. at UCLA, they had one VAX, some PCs, you had to walk to another building to use the VAX" (Brad Thomas, PD).

When we asked whether lab notebooks would one day be replaced with small palmtop computers or digitized pads, researchers were dubious. Respondents at one cramped lab in an urban high-rise simply noted that there was no place to put another computer, even a small one. They shared their lab with another group, and even lacked space for some necessary lab equipment. Such simple spatial or architectural barriers are crucial for the usability of any system, especially those conceived and designed as integral parts of someone's workflow.

Baseline Knowledge and Computing Expertise   Computing expertise was unevenly distributed within the labs; much equipment seemed out of date or unsophisticated. One senior researcher was not aware that databases were available without fixed-length fields, and a PI made a category error in discussing operating systems and applications (equating "a Mac" and "a UNIX"). In general, PIs thought that the level of knowledge was rising through undergraduate and graduate training, but empirically this did not seem to be the case. This might have constituted a learning level gulf (equating the ability to use online applications with the ability to understand broader systems concepts). Although there were a few highly skilled people, and one or two with advanced computing expertise, these were not clustered in either the graduate student or postdoc categories.

This sort of knowledge may be an access issue just as much as are space or location. First-order issues in this arena certainly include not only learning to use WCS software, but also understanding the platform on which it runs. WCS itself is designed to be extremely user friendly, and can be effectively used without much difficulty. The typical user in our study was a graduate student, postdoc, or principle investigator with enough knowledge about both domain and community to read a genetic map and recognize the importance of the Worm Breeder's Gazette. One user commented: "I just turned it on, pushed buttons" (Ben Tullis, GS).

In fact, at demos and trials at conferences, most users found WCS to be fairly easy and intuitive, once they were on it. However, the platform on which it is based was not transparent (to biologists). WCS runs under UNIX, and both the operating system and software such as X Windows or Suntools requires expertise most biologists didn't have: "UNIX will never cut it as a general operating system. Biologists won't use it, it's for engineers. [Someone in the lab] had a printing question, took him three months to get something to print" (Bob Gates, GS).

Furthermore, many respondents were unclear about carrying out other sorts of networked computer tasks, such as uploading and downloading files from mainframe to terminal. This made it difficult for them to integrate WCS use with email correspondence, word processing files, and other Internet information spaces.

Training often took place in a haphazard way, depending on everything from luck to personal ties: "I learned by using it as an editor. The second time I learned the formatter. A lot of people are comfortable with mail, and a lot of people are now using GenBank and sequencing packages. . . . We get some on the job training. [Two of the grad students in the lab] write up instruction sheets. The person who was the systems administrator until February was a good friend. Got a lot of push and shove from him, a lot of shared ideas" (Jeff Pascal, PD).

No lab offered special training in computing, although some students had taken classes at local computer centers. Several said that they only learn "exactly enough to suit what you have to do" (Carolyn Little, PI).

Addressing First-Order Issues On the surface, these issues may be solved in a fairly straightforward fashion. Effective shopping requires appropriate information, gathered and evaluated by a technically knowledgeable individual. When expertise needed for making computing decisions doesn't reside in the lab itself, it can be brought in from the outside, perhaps by turning to a campus computing facility or hiring a savvy undergraduate in computer science. Proposals can be written for equipment purchases. Issues of physical access can be solved by making the case for additional space. Issues of technical access can be solved by additional training. For instance, just as departments in the humanities are starting to offer tutorials or even certificates in humanities computing, biology departments could offer similar tutorials tailored to the needs of their own communities.

However, first-order selection issues are often intermingled with or converge to form higher-order issues. Shopping, for instance, is not just a matter of getting the right information to the right person, but requires information distribution channels that bridge several academic communities: worm biologists, tool builders, and local computing support centers. Similarly, when shopping and selection raise questions of standards, they become intermingled with questions of organizational and workplace culture ("UNIX is for engineers, not biologists"). This is a particularly salient issue in instances where multiple groups share computing, or where computing support is only available for a limited set of technological choices.

Second-Order Issues

Second-order issues can be analytically seen either as the result of unforeseen contextual effects, such as aversion to UNIX by biologists, or as the collision of two or more first-order issues, such as uncertainty during shopping combined with lack of information about how to hook up the system. These sorts of combinations can mean the person is forced to widen the context of evaluation, and link choices about software packages with best guesses about the direction of the organization. Included in this category are cultural influences on technical choices; paradoxes of infrastructure; "near compatibility" and the "almost-user community"; constraints becoming resources; and understanding the nature of baseline skills and their development. They are second order because they broaden the context of choice and evaluation of the straightforward first-order issues such as obtaining software and access to machines.

Technical Choices and a Clash of Cultures   Shopping and selection interact not only with training and ease-of-use issues, but also with organizational cultural issues. For example, five people independently mentioned being put off by UNIX, usually in the context of comparing it favorably with the Mac. One PI mentioned having no base of UNIX knowledge available from the local computer center, although he had taught himself enough to run a SPARC station (Joe White, PI). Others expressed similar sentiments: "As long as it's easy we'll use anything, like the Macs. So you can do like cut and paste, like you can on the Mac" (Eliot Red, PD); and "We were previously using UNIX but this is much easier. UNIX is impossible. It's a real pain. This is much easier. The Macs, you know . . ." (Linda Smith, PI).

One person who defined himself as a "crossover'" person (between biology and computing) said: "It's a big problem. Biologists are Mac people, and UNIX is an evil word. Most people are afraid of it, and refuse to use it. "If it's not on Mac I don't want it." There are a lot of problems getting people to use it, rather than delegate the use of it" (Harry Jackson, GS).

Yet UNIX, apart from forming a basis for the WCS and the language of its design team, was often also the language of the computer scientists who supported and maintained the local university's computing environment. This apparent gulf between user communities led some biologists to speculate that there are "two types of scientists—love or hate the computer," and that "the only way they'll ever [use] it is by force" (Jeff Pascal, PD). They attributed successful computer use to "some kind of natural affinity"

(Eliot Red, PD). This divergence has important implications for training, as do some other basic "cultural" issues.

Paradoxes of Infrastructure   The uneven spread of computing expertise and resources shows vividly how a simple increase, or lack, of first-order resources cannot fully explain a successful infrastructure. Differences of expertise and local organizational savvy between relatively rich and relatively poor labs may override first-order concerns. One of the poorest labs, for example, still running outdated IBM PC-XT equipment, actively used the system, had developed its own databases, and tracked strain exchange with a level of sophistication unparalleled in the community. The lab's PI loved to "play around" with software and hardware, and loved the challenge of overcoming the limitations of his lab. Second-order problems were thus reduced to first order by his own skill and interest.

The richest lab, on the other hand, had just received a substantial grant from the Human Genome Initiative to completely "hook up" the entire biology infrastructure on campus. However, they were unable to operate the system through a combination of "waiting for the ethernet" and "waiting for the Sun." The PI illustrates the dilemma: "We applied for an ethernet in May. (laughs) It should be here [in a few years] . . . .They'll be independent of the building network, [the people] on the SPARC. The Macs will be on the building network" (Linda Smith , PI).

A graduate student continues the story: "No one will put the wires in, though . . . we made a deal with the network people [network services] that we'd run wires and they'd connect it up. . . . They manage all the campus networks. [Someone else] has dealt with Sun, though" (Steve Grenier, GS).

At the time of the interview, they had strung their own cables and were waiting on the delivery of the SPARC stations. Linda Smith (the PI) then anticipated having to spend a lot of time to "get the software underway."

Even institutions with outstanding technical support had no organizational mechanisms for translating that expertise to highly domain-specific questions, applications, and issues. Campus computer centers were often neither knowledgeable about nor interested in applications packages relevant to biologists and geneticists, nor was there support for independently purchased hardware or software: Computer supports**** at [Research Institution]. I called the center for help with installing WCS on the Sun and they basically told me, find a UNIX guy, buy him some pizza. . . . If we have problems with the network or programs they support, they do it. If you didn't buy your hardware from them, forget it. If they don't support your software, forget it. It's handled on a department by department basis. Biology has no infrastructure" (Bob Gates, GS).

Who "owns" a problem or application was locally determined, and attribution of ownership made a great difference in individuals' ability to get help. Some PIs developed on-campus linkages that would bring computing expertise to bear on their own problems. The PI of one small lab submitted a grant together with a computer science faculty member interested in the visualization of scientific data. Together they planned to develop a tool for visual data representation and analysis; in the process, the PI will get not only a tool to support his research, but a UNIX-based workstation from which he can access WCS.

These issues were of great concern to postdocs looking to start up their own labs with increasingly limited funds. WCS was seen as a tool of the "upper tier" of richer labs (Harry Markson, GS), and described as "a rocket" when "we need a Model-T" (Marc Moreau, GS); a postdoc planning to start his own lab complained that "half a system for everyone is better than a really great system for just a few labs . . . we had to hire [a computer specialist affiliated with another lab]. Even the computer guys here [two graduate students] worked on it three weeks, and they couldn't load the [WCS] system. It's oriented to big labs" (Jay Emery, PD).

He added, "If it's not on a Mac or IBM, it won't get to people," and suggested, "You need a modular system, you need to be able to have parts of the database running on the Mac, *reach the small labs*" (emphasis added).

Tensions between a Discipline in Flux and Constraints as Resources   What might be seen as constraints that could be overcome with technology may become resources from a different perspective. We proposed that it would be trivially easy to make the Worm Breeder's Gazette available on a continual-update basis. On the one hand, continual updates would have served the needs of a very fast-moving community: "The faster the [WCS] update, the better. . . . You do it through the Gazette, you contribute regularly. You're competing [with other labs] on the same gene" (T. Jones, GS). "You need frequent updating, shortly after each Gazette, i.e., within two weeks after a Gazette there should be a new release. . . . The WCS Gazette could replace hard copies, it would be cheaper" (Brad Thomas, PD).

Yet other respondents objected strongly to this option, even though they worked in the same competitive environment. Objections centered on the utility of community-imposed deadlines on structuring work, both in terms of submitting and reading articles: "I would run the newsletter exactly how it's run now. Just leaving it open ended is not good. If there is infinity there is never a time to review things. And no deadlines" (John Wong, GS). "If the WCS were used to publish Gazette articles, what would be optimal? Well, continual would not be so good. There is something to be said for deadlines. Even six times a year, and it becomes background noise . . . it's hard to predict whether a frequency change will change the impact" (Gordon Jackson, PI).

The deadline was simultaneously constraint and resource. The distribution pattern for the Gazette not only affected the work habits of individuals, but it also was integrally linked to communication and coordination within labs and across them: "Do you think the WCS will replace the Gazette? If it replaces it, then we won't read it. I mean, when the Gazette arrives we split it up and each read a part. Then we use it to get into other people's work" (Ed Jones, GS). "What kinds of information do you not keep on the computer? Well, you couldn't have the newsletter on electronically. The constant update would be a nightmare. There would be no referenceable archive" (Paul Green, PI).

This last point is an important one, since the Gazette serves as a reference database containing not only pointers to work being carried out in various labs, but also to protocols, and more. For newcomers to the discipline, it serves as an important teaching tool; the online version would make back issues available more easily. A continual-update format would require a new way of referencing or indexing contributions. One person envisioned a different form of ongoing information service: "something in between a formal and informal database" where "if you have little writeups you could put them in an annotation box" (Alan Merton, PI). As for the Gazette, he suggests, "you could put a more formal thing into an online Gazette format, and keep it as it is" in terms of content, timing, and organization.

"Near-Compatibility" and the "Any Day Now" User   Sometimes the gulf between first and second order appears as a sense that what is happening should be first order. So strong is this sense that it can lead to some seemingly odd behavior. We encountered a persistent idea among respondents that they were "just about to" be hooked up with the system, and that the barriers to hooking up were in effect trivial. Sometimes this even caused them to say that they were using the system, whereas observations and interviews in fact showed that they were not. For instance, when trying to find a site to observe in a large city with several universities and several labs listed as user sites, one of the authors spent almost a week tracking down people who were actually using the system. No one she talked with was using it, but each person knew of someone else in another lab who supposedly was. After following all leads, she concluded that no one was really using the system, though they all "meant to," and figured that it would be available "any day now."

This is not difficult to observe ethnographically, but presents a real difficulty in administering surveys about use and needs. It is clear that this representation is not mendacious, but a common discounting of what seem, from a distance, to be trivial "plug-in" difficulties. The preceding observations of the difficulties associated with hooking up and getting started, coupled with infrastructural limitations would suggest that these issues are not trivial at all. In fact, these issues turn out to be lethal as they become both chronic and ubiquitous in the system as a whole.

Addressing Second-Order Issues   In principal, second-order issues can be resolved by combining an increase in resources with heightened coordination or cooperation between different technical and user communities, such as installing a user support telephone line, hiring a "circuit rider" who can help with hooking up and integration difficulties, and increasing other skill resources locally. However, realistically, biologists are not among the richest of scientists, despite the influx of HGI money. Money for capital expenditures is especially scarce, and decisions made about the purchase of or commitment to a particular system often persist for a decade or more. Thus, second-order issues in system use and development may become third-order issues: "why should this lab get resources, which problem is the most important one?" These issues occur at the level of the broader community and transcend the boundaries of any particular institution.

Third-Order Issues   Third-order issues are those that have been more commonly identified by sociology of science in discussions of problem solving. They have the widest context, involving schools of thought and debates about how to choose among second-order alternatives. These permeate any scientific community, for the reason that all scientific communities are interdisciplinary and contain different approaches and different local histories. They plague communities which are growing rapidly, working in uncharted areas, and which are exceptionally heterogeneous. Community members may not immediately recognize third-order issues as such, as they can be part of the taken-for-granted. Nevertheless, they have long-term implications. With respect to difficulties of signing on and hooking up, they include triangulation and definition of objects, multiple meanings of information, and network externalities.

Triangulation and Definition of Objects   People in different lines of work in the worm community come together in sharing information, including genetics, molecular biology, statistics, and so on. One person explained, "I came from [another lab] where I was working on frogs" (Brad Thomas, PD). Another person described himself as "really a developmental geneticist," and adds that a few years ago, "the field was smaller; . . . now many people are coming from outside, from mammals, protein labs" (Harry Markson, GS). Many people moved into the worm community from other areas after graduate school. Differences sometimes fell along the classical lines of organismal biology vs. molecular or genetic research: "I am more of a wormy person. That's true of the community in general. Sometimes you choose a system that's more organismal" (Jane Sanchez, PD).

Collaboration may take place across disciplinary or geographic boundaries: [*Are you collaborating with anyone?*] "I'm collaborating with people in the worm and nonworm community. Mostly immunologists in the nonworm community, people interested in the immune system. In the worm community, I'm collaborating with [a person in another state], on [a particular gene]" (Harry Markson, GS).

Disciplinary origin and current area of work affected the kinds of information individuals needed, and the tools and data sources with which they are familiar. Those studying the organism for its own sake differed in their information needs from those using it exclusively as a model organism; many informants had very specific expectations for WCS data: "You need more options, especially for sequencing. This will be especially important once the [Human Genome Initiative] gets underway. . . . We need to work with subsets of sequences, examine them in more detail" (Brad Thomas, PD). "What you'd want is a parts list, a list of cells. . . . If it's a neuron, its connections with other neurons. . . . That's for neurobiologists" (Harry Markson, GS).

Identifying the system with a particular subline of work and not as a general utility increased the barriers to usage. System construction was further complicated by another layer of object definition, in that some respondents felt that WCS represented "CS people [computer scientists] . . . building a system only for CS people," and that (WCS) "has a vision that isn't necessarily what biologists want."

Multiple Meanings, Data Interpretation, and Claim Staking   The nature and character of the community was changing as more people entered the "worm world" from other disciplines. During the last decade, it grew from a few hundred to over a thousand members: "It's neat that it's exciting now, but it's also strange to have so many people . . ." (Jane Sanchez, PD/ Research Scientist (RS)).

One goal of WCS is to support communication in a scientific community known for its willingness to share information, but the growth of which has exceeded the ability of informal communication networks to serve as a conduit for this information (Schatz 1991). The issues of developing a collaborative system, however, go far beyond the technical. The multiple meanings or interpretations of particular communications turn out to be important at all levels.

For example, suggestions that it might be useful to have a "who's working on what" directory in the system raised issues of competition and the role of secrecy. Some said they would hesitate to put in certain kinds of information, or wanted announcements delayed until "they had findings":

There's always a problem you're going to get scooped. You always walk a very fine line. There's a lot of people working on my problem. . . . if you publish in the Gazette you can lay claim to it. People would respect it. There have been some clashes, some labs trying to glom on to how much they can. It's going to be a struggle from here on out. . . . It's complex with the claim staking. That's why you want to get into it far enough so you can get ahead—before you announce it. If you could preface it with "wild speculation" (laughs) . . . well, there's a lot of times those can have a big payoff. But then again if five people jump on it, and in the meantime you're scooped . . . that's not so good! (Mike Jones)

Different communication channels also implied different degrees of freedom: "You can be wrong with no stigma" in the newsletter, said one graduate student, and a PI explained: "People are reluctant to do annotations [to the newsletter]. . . . It's the fear of putting yourself on the line. Making a commitment to what you're doing. It means being wrong in the eyes of your colleagues" (Joe White, PI). He suggested delayed publication of annotations, letting them sit locally for a month or so first, and a post-doc at another lab suggested the implementation of a personal level and a public level of annotation (Brad Thomas, PD). Another PI, however, became angry at this idea. He felt that this would work directly against WCS's commitment to community-wide sharing of information and turn the WCS into a local tool rather than a community resource.

Trust and reliability of information is a final concern worth noting: articles in the Gazette, annotations, and so on, all carry some kind of implicit value with them. First of all, information ages; old data is superseded by new, problems are resolved, protocols updated. Neither traditional sources nor the current WCS have any fixed way of marking the relative validity or trustworthiness of a set of data. Annotations with updates or the ability to "grey out" old data in the Gazette might present technical solutions to these problems, yet there are sometimes no clear-cut answers to these questions, especially in a community populated with multiple viewpoints. In general, says one post-doc, "there is no right or wrong, . . . you have to reach consensus on things, you have to look at labs, which labs you trust more" (Brad Thomas, PO). He wanted to use annotations as a means of raising alternate viewpoints: "I'm knowledgeable [in area X]. Sometimes others who clone don't know as much, they write things that are wrong. I would feel entitled to make annotations." Under the scheme he proposes, it would still be up to the reader of the annotation to sort out and make sense of competing information. He noted wryly that people will cite you as a foil when you've said something incorrect in any event, however, and that there's no way to prevent this. All these instances of data meaning different things under different circumstances—who notifies whom and when, what medium is used, who makes an annotation, or why a particular citation is and isn't included—required knowledge of the community that wasn't captured in any formal system (Star 1989a).

Network Externalities and Electronic Participation: Subtleties and Cautions The notion of externalities originates in economics and urban planning; a city may be said to afford "positive externalities" of cultural resources. For an artist, New York's externalities usually outweigh those available in Champaign, Illinois, although other amenities such as cost of housing and safety may be greater in the latter. A network externality means that the more actors actively participate in a system or network, the greater the potential, emergent resources for any given individual; it is distinct from the notion of "critical mass," which focuses on the number of subscribers/users at which system use becomes viable. Externalities may be negative in that, eventually, not being "hooked up" may make it impossible to participate effectively within a given community of work or discourse. For instance, the telephone network became a negative externality for those businesses without telephones sometime in the early twentieth century;

electronic mail has recently acquired a similar status in the academic world. For some purposes, standards (as in information standards) form important aspects of network externalities—in other words, users of nonstandard computing systems are at a disadvantage as network externalities become intertwined with particular operating systems and data interchange protocols (see David 1985 for a cogent analysis and example).

It currently is still difficult to understand the role of network externalities in the worm community, but as electronic access becomes the primary access mechanism for some forms of data, and as participation in all forms of electronic communication rises, they become increasingly important. Let us consider two examples: the WCS as an element of democratization, and, more generally, data repositories as both a means of maintaining openness in the community and a means of providing value to the community.

One goal of the system development is democratization of information—the facilitation of access to critical data through a uniform interface. Yet the more central WCS becomes to the community either as a whole, or as defined by key labs, the more those who cannot sign on along with the others will suffer. The "politics of reinforcement" suggest that the rich labs—either in terms of extant computing infrastructure or in their ability to procure or develop it using internal resources will get richer as network externalities become more dense (Kraemer and King 1977). This issue may be receding in importance as alternatives to WCS emerge via data available at FTP sites and through Gopher and Mosaic; much of the information available via WCS can now be "pulled from the net." Nevertheless, WCS is superior in its possibilities for graphical representation, and some forms of data analysis require such tools.

Issues of participation persist in several venues. For instance, a key repository is the genetic map, which represents the relative positions of genes on the chromosomes; another is the physical map, which represents cloned fragments of worm DNA and how they overlap to form the chromosomes (Schatz 1991). "There's a time problem. You want experts doing this, but you want to do your own stuff, you don't want to maintain a database. If you want this to serve a global community, you have to get the data properly defined" (Brad Thomas, PD). "There are data that should be on the [physical] map, but they are buried in labs all over the world. . . . When it was fragmented, people sent in clones. Now it's filled in, more coherent. The need to communicate back broke down. There used to be a dialogue, now there's a monologue. They don't bother telling Cambridge they've cloned genes. . . . With the genetic map there's still dialogue" (Ben Tullis, PD).

Some of this is an issue of time; two attempts at an electronic bulletin board for the worm community "died out within two weeks due to lack of contributions" (Bob Gates, GS). Annotation and updating take work, and "it's not of immediate profit'" (Sara Wu, PD). However, other reasons were also cited. When asked why the dialogue broke down: "There's a communication pyramid. You've got approximately 600 to 700 people in the community [in 1991]. One third of the community arrived between when [the community] was fragmented and [when it was] cohesive. They know only the cohesive map" (Ben Tullis, PD).

Newcomers weren't there when these repositories were created, and did not share commitment to their upkeep and growth. Competition was also cited as a factor, and was linked to the issue of timing discussed earlier. Someone who overheard the question on dialogue breakdown contributed the following comment: "Yeah, like [one very well-known] lab, . . . not sending in a note [on X]. And [another well-known] lab, they don't publish things when [they] are close to a gene they're working on" (Kyle Jordan, PD).

A graduate student in the same lab echoes a similar view of data sharing: "Instant updating won't go far. People who want an immediate result to be known only want a small number to know. It's more competitive, people are more careful. They don't want everything to be global. By the time it gets into the Worm Breeder's Gazette, it's not critical any more. The people who really need to know already know" (Bob Gates, GS).

WCS does not maintain the databases or publications featured in these discussions, but it would provide uniform access and an easy-to-use interface to them (once the system is up and running). It derives a significant part of its own value from community participation in their upkeep and maintenance. Without community commitment to the maintenance and upkeep of these materials, WCS has neither value nor legitimacy as a system that fosters either communication or collaboration.

Furthermore, if WCS is to develop its own niche within the community, it will also have to develop its own in terms of legitimating, documenting, and disseminating information. Currently, for instance, an annotation published in WCS has uncertain value within the community: ". . . contributing to the WCS, it's not a real publication. You have to send stuff to the Worm Breeder's Gazette if you want to publish widely" (Jane Jones, PD). "You get a better sense of contributing when you send to the Gazette. If you annotate the WCS, you don't know if it's being read" (Morris Owe, GS).

This is an issue that will face similar systems as they try to piggyback on established systems, repositories, and so on, especially when in competition with multiple other avenues for information retrieval and electronic communication. It is also important in the building of digital libraries and publishing systems—will an electronic journal publication "count" as much as a printed one?

All of us, in addition, face paradoxes of efficiency, or information overload and the danger of diverting a successful manual information tracking system over to the computer with a loss of productivity. Many economists have noted the so-called productivity paradox in firms with the introduction of information systems (productivity often declines with investment in IT). Similar paradoxes are a real danger in science with its delicate funding processes, understudied task structures, and fuzzy means of measuring productivity.

Tool Building and the Reward Structure in Scientific Careers   Finally, the role of tool building and tool maintenance may be undergoing a shift as computer-based tools become more prevalent. The tension between traditional notions of work and tool building (and new opportunities for the same) has been observed in other academic communities (Ruhleder 1991, 1995; Weedman 1995). One person was there in the early days of the database acedb, and still contributed regularly, sending email about bugs and suggestions for graphics. Others constructed local tools, such as annotated gene lists (a project carried out part-time over the course of a year), using data from WCS. Yet another person, as mentioned earlier, planned to team up with a computer scientist to develop tools for data visualization. Many of our respondents could list tools (from techniques to compilations of targeted information, to analysis software) that they would have liked to see added or perfected. The difficulty is that there are no clear rewards for this kind of work, except for the contributions the tool makes to one's own work. The biologist working with the computer scientist doesn't get any "credit" for this within his own discipline (he anticipated having tenure by the time this project would begin). As one postdoc put it in a comment appropriate for both sides: "There are a hundred things that are useful, but you don't get a Ph.D. for [them]" (Jay Emery, PD).

Addressing Third-Order Issues   Third-order issues are a feature of complex communities. They may become easier to observe during times of flux because they resist local resolution. Novel technologies, situations, and concerns create immediate resource requirements and gaps in learning that can be addressed locally. Over time, however, the interactions of these first-order and second-order issues combine to raise broader questions that push the magnitude of a "solution" out of the local realm and into the wider community.

Electronic access to data via WCS, for instance, calls into question not only local resource allocations, but also broader institutional alliances and patterns of contributions at the disciplinary level. The resolution of these issues or conflicts (if, indeed, they are resolved) may result in the creation of new subspecialties, new requirements for a discipline or profession, new criteria for the conduct and evaluation of work, and new reward structures. Resolutions or "readjustments" will not only take place overtly (i.e., though a petition to a campus computing committee, or a decision to reallocate travel funds to a lab computing fund). They may also play out on a political level by individuals with high stakes in maintaining stature or controlling resources, or they may be resolved serendipitously, even unconsciously. For instance, as mentioned before, questions of access to the WCS and the maintenance of an open, democratic structure within the community may become moot as other forms of access through the Internet become easier.

Double Binds: The Transcontextual Syndrome on the Net

Until now we have simply followed Bateson's typology for learning in categorizing infrastructural barriers and challenges. Bateson's ideas about levels of learning originated in communication theory and cybernetics; more than a taxonomy, they are an expression of a set of dynamics: "Double bind theory is concerned with the experiential component in the genesis of tangles in the rules or premises of habit. I . . . assert that experienced breaches in the weave of contextual structure are in fact 'double binds' and must necessarily (if they contribute at all to the hierarchic processes of learning and adaptation) promote what I am calling transcontextual syndrome" (Bateson 1972, 276).

The formal statement of the problem is expressed as a logical one, following as we noted earlier Russell and Whitehead's theory of classification. In "The Logical Categories of Learning and Communication," Bateson (1972, 279–308) notes that a category error such as confusing the name of a class and a member of that class will create a logical paradox. In the world of pure logic, this appears as a fatal error, because such logical systems seem to exist outside of time and space. In the real world, particularly the behavioral world, however, people cope by working within multiple frameworks or "worldviews," maintained serially or in parallel.

When messages are given at more than one level simultaneously, or an answer is simultaneously demanded at a higher level and negated on a lower one, there arises a logical paradox or "double bind," an instance of what Bateson terms the "transcontextual syndrome." While Bateson drew his examples from family contexts in the course of his work on schizophrenia, double binds occur in academic and business contexts as well. Middle managers in rapidly changing environments, for instance, are frequently caught between the goals and expectations articulated by senior management and the actions of senior management with respect to budget allocation and performance evaluation (Mishra and Cameron 1991). Companies may formally promote efforts toward "reengineering" and "empowerment," yet offer no mechanisms for employees to participate in decision making, or they may sanction employees for not being active learners while refusing to acknowledge modes of learning and experimentation that fail to conform to very specific models. In the words of Bateson, "There may be incongruence or conflict between context and metacontext" (1972, 245). Over a protracted period of time with many such messages, schizophrenia may result, either literally or figuratively.[4]

People attempting to hook up to complex electronic information systems encounter a similar discontinuity between message types. The rhetoric surrounding the Internet makes "signing on" and "hooking up" sound remarkably straightforward. Furthermore, the benefits sound instantaneous and far reaching. Why, then, do so many problems arise when members of the worm community try to take a similar step? Why are there so many disappointments with accessing information, and how may we understand these?

We identify several varieties of double binds arising across two levels or orders from what we call infrastructural transcontextual syndrome:

1. The gap between diverse contexts of usage;
2. The gap inherent in various computing-related discussions within the worm community itself; and
3. The gulf between "double levels of language" in design and use.

The gap between diverse contexts of usage. What is simple for one group is not for the other, so what appears to be a level-one message to computer scientists posed a level-two problem for users, creating a double bind. For instance, when asked about getting onto the system, designers of WCS might say, "Just throw up X Windows and FTP the file down." The tone of the message is clearly level one, a simple "recipe" for the UNIX literate. For the relatively naive user, however, it requires them to move to a different contextual level and to figure out what type of a thing an "X Window" is, and what it means to FTP a file down. A level-one instruction thus becomes a complex set of level-two questions, closely related to the user's own level of expertise. These kinds of transcontextual difficulties will intensify as collaborative systems and groupware are developed for increasingly nonhomogeneous user communities (Grudin 1991, 1994; Markussen 1994).

Another part of this type of double bind is an infinite regress of barriers to finding out about complex electronic information systems (Markussen 1995). If you don't know already, it's hard to know how to find out, and it isn't always clear how to abstract knowledge from one system to another. What is obvious to one person is not to another; the degrees of obviousness continue indefinitely, forming complex binds. For example, there is no single book that can tell you from scratch about computers or networked computing; the only way in is to switch contexts altogether and work more closely with those who already know, becoming a member of some community (Suchman and Trigg 1991). This may account for the power of the participatory design model popular in Scandinavia, in which designers and users work together to the point of developing a shared context at all levels of interaction (Ehn 1988). It may simultaneously account for the difficulty of explaining or popularizing the model outside of Scandinavia, the working context of which differs greatly from the United States or other parts of Europe.[5]

The gap inherent in various computing-related discussions within the worm community itself. Within the worm community itself there exists a level-two level-three double bind. Just as level-one statements can engender level-two questions, so can level-two discussions open up higher-order issues. Discussions about package or platform choice become discussions about resource allocation, data interpretation, and network externalities. Take, again, the case of "FTPing a file down." Talk of learning about FTP, about alternatives such as gophers, and so on becomes questions of access across labs, of database maintenance and data reliability, and of norms and rewards within the community for contributions to the database.

These issues are particularly poignant ones for "older" members of a fairly new community, who recognize that technical choices and decisions made at the second-level evaluations of the options for responding to level one signals-have the ability to affect dramatically third-order issues. In the worm community, the concerns involve changes in the composition of the community as "outsiders" join, and what this means for data interpretation and tool construction. The concerns also center on the multiple roles that research on the organism plays: "end in and of itself" vs. model organism for the Human Genome Initiative. Tools aimed at second-level problems affect deeply the options open to the discipline when addressing third-order questions and setting broad conceptual directions.

The gulf between "double levels of language" in design and use. There may be double binds in those aspects of the system that are self-contradictory, between formal system properties and informal cultural practices. The language of design centers on technical capacity, while the language of use centers on effectiveness. Robinson (1991) notes that for systems that provide electronic support for computer-supported cooperative work, only those applications that simultaneously take into account both the formal, computational level and the informal, workplace/cultural level are successful. This problem is not unique to this domain. Gasser (1986), for instance, identifies a variety of "workarounds" developed to overcome the rigidity (and limitations) of a transaction-processing system, while users of an insurance claims processing system developed an elaborate and informal set of procedures for articulating alternatives and inconsistencies (Gerson and Star 1986). Other examples abound.

While none of these studies identifies the problems/solutions as evidence of a double bind, each may be expressed in these terms. The "language" of the designer is focused on the technical representation of a particular set of data (i.e., customer records) and the efficiency of processing them to meet a particular goal (i.e., claims processing). The "language" of the user is focused on the need to mediate between conflicting viewpoints (i.e., doctors vs. representatives for large customer groups), and the need to develop effective workflows within their own workplaces. Orlikowski (1993) discusses more narrowly the conceptualization of software design methods and tools as languages and, together with Beath, examines the consequences of nonshared languages or organizational barriers to full participation of users (Beath and Orlikowski 1994).

This double bind is also captured in the discussion of Mac vs. UNIX, and what it means in terms of a clash of cultures between biologists and computer scientists. On one level it is a discussion about operating systems, on another it is representative of two worldviews and sets of values with respect to the relationship between technology and work—the relationship between the tool and its user. In the case of WCS, designers focused on features of technical elegance and sophistication, such as constructing a mechanism for continuous Gazette updating, or fully exploiting hypertext possibilities. Yet constant information updating works against the biologists' informal mechanisms for information distribution, processing, and integration. And biologists were less interested in additional layers of complex hypertext linkages than in simple capabilities, like printouts of parts of the genetic map, which could be taken back to their lab benches, tacked up, pasted into a notebook, and easily annotated in the flow of work.

Summary and Recommendations for Addressing Double Binds

WCS—and the push for collaboratory development that set the stage for this and other projects (Lederberg and Uncapher 1988)—was driven by a desire not only to support collaborative scientific efforts, but also to foster "ideal communities" of rich communication and seamless universal information access. WCS had the advantage of starting with a community in which many of those norms were already in operation, and whose small size made it relatively cohesive. It had a dedicated design team with knowledge of the target domain. It had an interested user population. Yet it never achieved its original goals and, while it does serve as a platform for communication and information access for some, others have found the barriers locally insurmountable, or the system itself superfluous.

When will WCS become infrastructure? The answer is, probably never in its original form, for the reasons outlined. The development of the system and its integration into the community could not overcome the double binds that emerged within the context of system implementation and use. Nor could its development negate the impacts of other technologies such as gopher and Mosaic. Constructed largely as a series of "building blocks" available from other sources, it was easy enough for those building blocks to assemble or duplicate themselves elsewhere. But WCS and other systems development efforts based on this model of collaboratory can benefit from some of the lessons learned or newly illustrated through our analysis. And organizations interested in developing large-scale information and communication infrastructures (whether formal business organizations or loosely coupled academic communities) can become aware of the efforts required on their part to meet developers halfway. Having identified different instances of double binds that predicated the failure of WCS, we are left with the need to suggest positive action; we offer two recommendations below for addressing double binds.

The Role of Multidisciplinary Development Teams.

One of the key difficulties with double binds is recognizing them in the first place: individuals involved in a situation may not be able to identify instances of this transcontextual syndrome. The other key difficulty, once a double bind is identified, is to articulate it such that the other party will recognize it as a problem. Dynamics of power and authority are clearly important here. In family settings, a parent might reject affectionate behavior on the part of the child, then, when the child withdraws, accuse the child of not loving the parent. The child may not always have capacity for analyzing and correcting this inconsistency, just as employees may not really have the power to address problems in a business environment that overtly empowers them. Managers may even subtly sanction the "wrong" kind of empowered behavior. Users are often given computer-based tools that are either cumbersome or ill-explained to them; when they fail to use them, they are labeled as being "resistant" to technology (Markus 1983; Markus and Bjorn-Andersen 1987; Forsythe 1992).[6]

A computing-related analogue would be the denial on the part of developers or system administrators that technical difficulties really mask higher-order conceptual problems centered on work practices and community standards, and a failure on the part of users to recognize the complexity of their work domains, their hidden assumptions, and the various motivations of the stakeholders involved. If we expect designers to learn about the formal and informal aspects of the user domain, to learn to "speak their language," we must ask users to meet designers halfway by learning their language and developing an understanding of the design domain. If designers are at fault for assuming that all user requirements can be formally captured and codified, users are often equally at fault for expecting "magic bullets"—technical systems that will solve social or organizational problems.

The fault may really lie in neither camp. Often miscommunication resulting in the double binds of language, and the context within which the process of design/use occurs are responsible. The emergence of multidisciplinary development teams may help to alleviate aspects of the transcontextual syndrome identified, with ethnographers helping users and designers bridge the contextual divide. "You can FTP that from such-and-such a site" might well give way to "I can give you the FTP address, but the kind of data you'll get won't be detailed enough for what you want to do with it." By sharing an understanding of both the formal, computational level (traditionally the domain of the computer programming and systems analyst) and the informal level of workplace culture, double binds may be more easily identified as all members of the team learn to correctly identify the various orders or levels to which a message might belong. This sharing, however, requires institutional contexts that support and even reward this kind of collaboration.

The Nature of Technical User Education

Many elements of the "computing infrastructures" emerging within the academic and business communities are not custom made. They consist of locally developed applications, off-the-shelf packages or tailored applications, local area networks and the Internet, commercial online services, and "shareware" such as Mosaic and Netscape. They vary greatly in terms of stability, maintainability, interoperability, and access to support. Yet in order to carry out their work effectively in increasingly computer-based environments, individuals must be able to negotiate complex configurations of technical resources. Pentland's (1997) analysis of software help lines attests to this complexity: "Software support is an activity that occurs on the "bleeding edge" of technology, on the boundary between the known and the unknown" (ms 1). Support technicians and customers are often speaking from two disparate viewpoints, and successful support means recognizing and juggling this reality (Heylighen 1991). The emergence of local "tailors" (Trigg and Bodker 1994) and "technology mediators" (Okamura et al. 1994) may provide a bridge between relatively generic technologies and their local interpretation and application.

Individuals are being told that they must adapt to new technologies and become technically literate, yet the type of training and support offered to them rarely gives them the basis necessary to evolve along with the infrastructure. Training sessions, online tutorials, and user manuals focus on a set of skills limited to particular applications, and occur outside of the context of actual work (Bjork 1994). Computer support centers may assist individuals in situ, but tend to be reactive, imparting one solution at a time, without any contextual connection to the kinds of technical problems the user has had before. To apply Bateson's framework, they are aimed at giving people the skills to address first order technical issues, though broader issues—such as whether the implementation of a particular groupware technology is consistent with local, career, or global strategic direction—may require second- or third-order conceptual skills.

Frameworks for various levels of "computer literacy" already exist within the computer science and education communities. What are missing are institutional mechanisms—whether the "institution" is a business enterprise, a university, or a scientific discipline—to support individuals in two ways. First, they do still need to teach specific skills, but they need to place these skills within a technical context that enables users to apply them to the next application, and the next. Second, they need to assist users in developing and maintaining the kind of computer literacy that will allow them to understand and address second- and third-order issues, especially as they unfold over time—a kind of learning that occurs through ongoing dialogue and experimentation. That literacy must thus be coupled with an understanding of emerging work practices (locally and more broadly within their organizations). Finally, organizations also need to develop mechanisms for legitimating and rewarding the work of local tailors and mediators.

These institutional mechanisms can be, in part, consciously constructed. But in order for them to grow dynamically along with emergent user expertise and an emergent base of computing technologies, they must be predicated on the notion that organizations function as complex communities, and that learning takes place within local communities of practice (Lave and Wenger 1992; Star 1995). The creation and use of discrete technologies must occur within a broader context that is constantly reified by participants within and across the various communities of practice that define a particular organization. The success of systems developed to support their work is predicated on the creation of shared objects and practices, boundary objects, and infrastructures (Star and Griesemer 1989; Star 1989b). For instance, use of WCS was and continues to be predicated on the complex interaction between a variety of small and large communities: the WCS development team, a nonhomogeneous target population (the worm community), local systems support groups, and remote data collection and distribution centers. Each of these constitutes extant, partially overlapping communities of practice. Discontinuities in these interactions, unequal participation, and the emergence and continued rise of competing technologies have contributed to the inability of WCS to emerge as boundary object or fully to submerge as infrastructure.

Organizational Environment: Communities and Large-Scale Infrastructure

Using the analysis put forward in the previous section, we would like to understand the nature of the claims about community and the Internet as examples of the complex emergence of infrastructure. We see a number of ways in which the merging of medium and message in the talk about scientific electronic communities is problematic, in addition to the double bind/transcontextual syndrome issues. Scientists do not "live on the net." They do make increasingly heavy use of it; participation is increasingly mandatory for professional advancement or even participation, with a rapidly changing set of information resources radically altering the landscape of information "user" and "provider" (Klobas 1994); and the density of interconnections and infrastructural development is proceeding at a dizzying rate. That development is uneven; it is an interesting mixture of local politics and practices, online and offline interactions, and filled with constantly shifting boundaries between lines of work, cohorts and career stages, physical, virtual, and material culture, and increasingly urgent and interesting problems of scale.

The multiple meanings of WCS for different groups and individuals are useful as exemplars for understanding the challenges posed by the net. From one perspective, the WCS fits well the cognitive map of the scientist with respect to information: links between disparate pieces, graphical representations, layers of detail, and so on. Yet relatively few worm biologists have "signed on" to WCS, even as the community itself is growing rapidly. The seeming paradox of why our respondents chose to use Gopher,

Mosaic, and other simpler, public access systems rather than WCS involves a kind of double bind at larger scale.

To take on board the custom-designed, powerful WCS (with its convenient interface) is to suffer inconvenience at the intersection of work habits, computer use, and lab resources. Its acquisition disrupts resource allocation patterns: ongoing use and support requires an investment in changes of habit and infrastructure. The World Wide Web, on the other hand, can be accessed from a broad variety of terminals and connections, and Internet computer support is readily available at most academic institutions and through relatively inexpensive commercial services.

Yet even within the larger context of infrastructure, there are other ways in which WCS serves its community less well than alternate emerging infrastructures. Science is an integrative and permutable domain, and requires a complementary infrastructure (Ruhleder and King 1991). The construction of the WCS, while it integrates a large number of materials, does so in a constricted fashion. Lab notebooks, by way of contrast, are extremely open and integrative documents (Gorry et al. 1991). At the same time, computing infrastructures, including gophers and FTP sites, while still at a very primitive level fit more closely with this integrative model than relatively closed systems such as the WCS, and these infrastructures are growing at a phenomenal rate. For these reasons, and despite frustrations over the lack of indexing and search capabilities, use of Gopher and Mosaic within the c.elegans world abounds.

Conclusion

Can an organizational support system be developed that allows people to coordinate large-scale efforts, provide navigational aids for newcomers, yet still retain the feeling of an informal, close-knit community or cohesive organizational culture? If structure is not incorporated a priori, then does it emerge, and how? Just as the WCS was intended to bridge geographic and disciplinary boundaries within the worm community, groupware and related technologies are being constructed as technical infrastructures to support members of an organization in bridging physical, temporal, and functional boundaries.

Experience with groupware suggests that highly structured applications for collaboration will fail to become integrated into local work practices (Ruhleder, Jordan, and Elmes 1996). Rather, experimentation over time results in the emergence of a complex constellation of locally tailored applications and repositories, combined with pockets of local knowledge and expertise. They begin to interweave themselves with elements of the formal infrastructure to create a unique and evolving hybrid. This evolution is facilitated by those elements of the formal structure that support the redefinition of local roles and the emergence of communities of practice along the intersection of specific technologies and types of problems. These observations suggest streams of research that continue to explore how infrastructures evolve over time, and how "formal," planned structure melds with or gives way to "informal," locally emergent structure.

The competing requirements of openness and malleability, coupled with structure and navigability, create a fascinating design challenge—even a new science. The emergence of an infrastructure—the "when" of complete transparency—is thus an "organic" one, evolving in response to the community evolution and adoption of infrastructure as natural, involving new forms and conventions that we cannot yet imagine. At the same time, it is highly challenging technically, requiring new forms of computability that are both socially situated and abstract enough to travel across time and space (Eveland and Bikson 1987; Feldman 1987). Goguen (1994) and Jirotka and Goguen (1994) recently referred to these as "abstract situated data types" for requirements analysis, and note that requirements engineering in this view in fact becomes "the reconciliation of technical and social issues."

In the end it seems that organizational change and the resolution into infrastructure are usually very slow processes. Local and large-scale rhythms of change are often mismatched, and what it takes to really make anything like a national or global information space is at the very cutting edge of both social and information science. The mixture of close-in, long-term understanding gained by ethnography and the complex indexing, programming, and transmission tasks afforded by computer science meet here, breaking traditional disciplinary boundaries and reflecting the very nature of the problem: when is an ecology of infrastructure?[7]

Notes

1. Names have been changed to preserve anonymity.

2. Personal communication between Star and Bruce Schatz, September 28, 1995.

3. PI = principle investigator, PD = ostdoc, GS = graduate student.

4. The child insists on seeing the literal level and ignoring context, or inappropriately seeing context literally. The often-noted poetry in schizophrenic language is a result of this refusal—good poets deliberately play with transcontextual double entendres. Formally, this ignores or transgresses the gulf between message and metamessage.

5. Participatory design has its own inherent difficulties (Markussen 1994; Nyce and Löwgren 1995).

6. There is an analogy here with medicine, viz., studies of "patient" compliance that overlook the infrastructural and political features of medicine itself. See, e.g., Strauss 1979 and Strauss et al. 1985.

7. WCS was partially funded by the NSF under grants IRI-90-15047, IRI-92-57252, and BIR-93-19844, and is currently housed within the Community Systems Laboratory (CSL), affiliated with

the Graduate School of Library and Information Science and the National Center for Supercomputing Applications at the University of Illinois, Urbana Champaign. Additional support was provided by the University of Arizona and the University of Illinois. Co-PIs Bruce Schatz and Sam Ward, and developers Terry Friedman and Ed Grossman were extremely generous with their time, comments, and access to data and meetings; we also thank our anonymous respondents for their time and insight. Earlier versions of this paper were presented at the Conference on Computing in the Social Sciences, 1993, CSCW1994, and ASIS 1995. Ruhleder thanks Michael Elmes for an interesting discussion on double binds in organizations, and thanks Sam Politz and the graduate students in his worm lab at Worcester Polytechnic Institute for teaching her how to hook and breed worms, and how to run a gel. Marc Berg, Geof Bowker, Nick Burbules, Tom Jewett, Alaina Kanfer, Rob Kling, Jim Nyce, Kevin Powell, and Stefan Timmermans provided helpful insights and comments. Star's work was also supported by the Program in Culture Values and Ethics and the Advanced Information Technologies Group, University of Illinois; by the Institute for Research on Learning, Palo Alto, and by an NSF Professional Development Grant. The authors would like to thank John Garrett of CNRI; Tone Bratteteig, Pal Seregaard, Eevi Beck, Kari Thoresen, Ole Hanseth, Eric Monteiro, and the Internet Project working group at the Institute for Informatics, University of Olso; Yrjö Engeström, Chuck Goodwin, and Dick Boland for discussions and challenges about the concept of infrastructure. The dimensions of infrastructure outlined were partly developed in an email dialogue between Star and Garrett; we gratefully acknowledge his help and insight. Pauline Cochrane and the students in her information retrieval seminar provided very helpful comments; we acknowledge their help and that of Laura Shoman.

References

Anderson, William L., and William T. Crocca. 1993. "Engineering Practice and Co-development of Product Prototypes." *Communications of the ACM* 3 (6): 49–56.

Barlow, John Perry. 1995. "Is There a There in Cyberspace?" *Utne Reader* 68 (March–April): 53–56.

Bateson, Gregory. 1972. *Steps to an Ecology of Mind: A Revolutionary Approach to Man's Understanding of Himself.* New York: Ballantine Books.

Beath, Cynthia, and Wanda J. Orlikowski. 1994. "The Contradictory Structure of Systems Development Methodologies: Deconstructing the IS-User Relationship in Information Engineering." *Information Systems Research* 5 (4): 350–377.

Becker, Howard S. 1982. *Art Worlds.* Berkeley: University of California Press.

Bodker, Susanne. 1991. *Through the Interface.* Hillsdale, NJ: Erlbaum.

Bowker, Geoffrey C. 1994. "Information Mythology and Infrastructure." In *Information Acumen: The Understanding and Use of Knowledge in Modern Business*, ed. Lisa Bud-Frierman, 231–247. London: Routledge.

Bowker, Geoffry C. 1993. "The Universal Language and the Distributed Passage Point: The Case of Cybernetics." *Social Studies of Science* 23:107–127.

Brown, John Seely, and Paul Duguid. 1994. "Borderline Issues: Social and Material Aspects of Design." *Human–Computer Interaction* 9:3–36.

Bucciarelli, Louis L. 1994. *Designing Engineers*. Cambridge, MA: MIT Press.

David, Paul. 1985. "Clio and the Economics of QWERTY." *American Economic Review* 75:332–337.

Davies, Lynda, and Geoff Mitchell. 1994. "The Dual Nature of the Impact of IT on Organizational Transformations." In *Transforming Organizations with Information Technology*, ed. R. Baskerville, O. Ngwenyarna, S. Smithson, and J. DeGross, 243–261. Amsterdam: North Holland.

Ehn, Pelle. 1988. *Work-Oriented Design of Computer Artifacts*. Hillside, NJ: Lawrence Erlbaurn.

Ellis, C. A., S. J. Gibbs, and G. L. Rein. 1991. "Groupware: Some Issues and Experiences." *Communications of the ACM* 34:38–58.

Engeström, Yrjö. 1990. "When Is a Tool? Multiple Meanings of Artifacts in Human Activity." In *Learning, Working and Imagining*, ed. Y. Engeström, 171–195. Helsinki: Orienta-Konsultit Oy.

Eveland, J. D., and Tora Bikson. 1987. "Evolving Electronic Communication Networks: An Empirical Assessment." *Office: Technology and People* 3: 103–128.

Feldman, Martha. 1987. "Constraints on Communication and Electronic Messaging." *Office: Technology and People* 3:103–128.

Forsythe, Diana. 1992."Blaming the User in Medical Informatics: The Cultural Nature of Scientific Practice." In *Knowledge and Society: The Anthropology of Science and Technology*, vol. 9, ed. David Hess and Linda Layne, 95–111. Greenwich, CT: JAI Press.

Gasser, Les. 1986. "The Integration of Computing and Routine Work." *ACM Transactions on Information Systems* 4:205–225.

Gerson, E. M., and Susan Leigh Star. 1986. "Analyzing Due Process in the Workplace." *ACM Transactions on Information Systems* 4:257–270.

Gorry, G. Anthony, Kevin B. Long, Andrew M. Burger, Cynthia P. Jung, and Barry D. Meyer. 1991. "The Virtual Notebook System™: An Architecture for Collaborative Work." *Journal of Organizational Computing* 1:223–250.

Goguen, Joseph. 1994. "Requirements Engineering as the Reconciliation of Technical and Social Issues." In *Requirements Engineering: Social and Technical Issues*, ed. M. Jirotka and J. Goguen, 165–199. New York: Academic Press.

Grudin, Jonathan. 1991. "Obstacles to User Involvement in Software Product Development, with Implications for CSCW." *International Journal of Man-Machine Studies* 34:435–452.

Grudin, Jonathan. 1994. "Groupware and Social Dynamics: Eight Challenges for Developers." *Communications of the ACM* 37 (1): 92–105.

Hewitt, Carl. 1986. "Offices Are Open Systems." *ACM Transactions on Information Systems* 4:271–287.

Heylighen, Francis. 1991. "Design of a Hypermedia Interface Translating Between Associative and Formal Representations." *International Journal of Man–Machine Studies* 35:491–515.

Hughes, Thomas P. 1983. *Networks of Power: Electrification in Western Society, 1880–1930*. Baltimore, MD: Johns Hopkins University Press.

Hughes, Thomas P. 1989. "The Evolution of Large Technological Systems." In *The Social Construction of Technological Systems*, ed. Wiebe E. Bijker, Thomas P. Hughes, and Trevor Pinch, 51–82. Cambridge, MA: MIT Press.

Jewett, Torn, and Rob Kling. 1991. "The Dynamics of Computerization in a Social Science Research Team: A Case Study of Infrastructure, Strategies, and Skills." *Social Science Computer Review* 9:246–275.

Jirotka, Marine, and Joseph Goguen. 1994. *Requirements Engineering: Social and Technical Issues*. New York: Academic Press.

Kling, R., and W. Scacchi. 1982. "The Web of Computing: Computer Technology as Social Organization." *Advances in Computers* 21:1–90.

Klobas, Jane E. 1994. "Networked Information Resources: Electronic Opportunities for Users and Librarians." *Information Technology & People* 7:5–18.

Korpela, Eija. 1994. "Path to Notes: A Networked Company Choosing Its Information Systems Solution." In *Transforming Organizations with Information Technology*, ed. R. Baskerville, O. Ngwenyama, S. Smithson, and J. DeGross, 219–242. Amsterdam: North Holland.

Kraemer, Kenneth L., and John L. King. 1977. *Computers and Local Government*. New York: Praeger.

Latour, Bruno. 1993. *We Have Never Been Modern*. Trans. Catherine Porter. Cambridge, MA: Harvard University Press.

Lave, Jean, and Etienne Wenger. 1992. *Situated Learning: Legitimate Peripheral Participation*. Cambridge: Cambridge University Press.

Lederberg, J., and K. Uncapher. 1989. *Towards a National Collaboratory: Report of an Invitational Workshop at the Rockefeller University*. Washington, DC: National Science Foundation.

Malone, Torn, and Gary Olson, eds. 2001. *Coordination Theory and Communication Technology*. Mahwah, NJ: Erlbaum.

Markus, M. Lynne. 1983. "Power, Politics, and MIS Implementation." *Communications of the ACM* 26 (6): 430–444.

Markus, M. Lynne, and Niels Bjorn-Andersen. 1987. "Power over Users: Its Exercise by System Professionals." *Communications of the ACM* 30 (6): 498–504.

Markussen, Randi. 1994. "Dilemmas in Cooperative Design." In *Proceedings of the Participatory Design Conference, Computer Professionals for Social Responsibility*, Palo Alto, CA, 59–66.

Markussen, Randi. 1995. "Constructing Easiness: Historical Perspectives on Work, Computerization, and Women." In *The Cultures of Computing*, ed. Susan Leigh Star, 158–180. Oxford, UK: Basil Blackwell.

Mishra, Aneil K., and Kim S. Cameron. 1991. "Double Binds in Organizations: Archetypes, Consequences, and Solutions from the US Auto Industry." Presentation at the Academy of Management Annual Meeting, August 13.

Monteiro, Eric, Ole Hanseth, and Morten Hatling. 1994. "Developing Information Infrastructure: Standardization vs. Flexibility." Working Paper 18 in Science, Technology and Society, University of Trondheim, Norway.

Nyce, James, and Jonas Löwgren. 1995. "Toward Foundational Analysis in Human-Computer Interaction." In *The Social and Interactional Dimensions of Human-Computer Interfaces*, ed. Peter J. Thomas, 37–47. Cambridge, UK: Cambridge University Press.

Okamura, Kazuo, Masayo Fujimoto, Wanda J. Orlikowski, and JoAnne Yates. 1994. "Helping CSCW Applications Succeed: The Role of Mediators in the Context of Use." In *Proceedings of the Conference on Computer Supported Cooperative Work*, 55–65. Chapel Hill, NC: ACM Press.

Orlikowski, Wanda. 1991. "Integrated Information Environment or Matrix of Control? The Contradictory Implications of Information Technology." *Accounting, Management and Information Technology* 1:9–42.

Orlikowski, Wanda. 1993. "CASE Tools as Organizational Change: Investigating Incremental and Radical Changes in Systems Development (Computer Aided Software Engineering)." *MIS Quarterly* 17:309–340.

Pentland, Brian. 1997. "Bleeding Edge Epistemology: Practical Problem Solving in Software Support Hot Lines." In *Between Craft and Science: Technical Work in US Settings*, ed. Stephen Barley and Julian E. Orr, 113–128.

Pool, Robert. 1993. "Beyond Databases and Email." *Science* 261 (April 13): 841–843.

Robinson, Mike. 1991. "Double-level Languages and Co-operative Working." *AI & Society* 5:34–60.

Ruhleder, Karen. 1991. "Information Technologies as Instruments of Transformation: Changes to Work Processes and Work Structure Effected by the Computerization of Classical Scholarship." Ph.D. dissertation, Dept. of Information and Computer Science, University of California Irvine.

Ruhleder, Karen. 1995. "Reconstructing Artifacts, Reconstructing Work: From Textual Edition to On-line Databank." *Science, Technology & Human Values* 20:39–64.

Ruhleder, Karen, Brigitte Jordan, and Michael B. Elmes. 1996. "Wiring the 'New Organization': Integrating Collaborate Technologies and Team-Based Work." Paper presented at the 1996 Annual Meeting of the Academy of Management, Cincinnati, OH, August 9–12.

Ruhleder, Karen, and John Leslie King. 1991. "Computer Support for Work across Space, Time, and Social Worlds." *Journal of Organizational Computing and Electronic Commerce* 1 (4): 341–355.

Schatz, Bruce. 1991."Building an Electronic Community System." *Journal of Management Information Systems* 8: 87–107.

Schmidt, Kjeld, and Liam Bannon. 1992. "Taking CSCW Seriously: Supporting Articulation Work." *Computer Supported Cooperative Work* 1:7–41.

Star, Susan Leigh. 1989a. *Regions of the Mind: Brain Research and the Quest for Scientific Certainty*. Stanford, CA: Stanford University Press.

Star, Susan Leigh. 1989b. "The Structure of Ill-Structured Solutions: Heterogeneous Problem-Solving, Boundary Objects and Distributed Artificial Intelligence." In *Distributed Artificial Intelligence 2*, ed. M. Huhns and L. Gasser, 37–54. Menlo Park, NJ: Morgan Kaufmann.

Star, Susan Leigh. 1991a. "Power, Technologies and the Phenomenology of Conventions: On Being Allergic to Onions." In *A Sociology of Monsters: Essays on Power, Technology and Domination*, ed. John Law. London: Routledge.

Star, Susan Leigh. 1991b. "Organizational Aspects of Implementing a Large-Scale Information System in a Scientific Community." Technical Report. Community Systems Laboratory, University of Arizona, Tucson, November.

Star, Susan Leigh. 1994. "Misplaced Concretism and Concrete Situations: Feminism, Method and Information Technology." Working Paper 11, Gender-Nature-Culture Feminist Research Network Series, Odense University, Odense, Denmark.

Star, Susan Leigh, ed. 1995. *The Cultures of Computing*. Oxford: Blackwell.

Star, Susan Leigh. 1996. "From Hestia to Home Page: Feminism and the Concept of Home in Cyberspace." In *Between Monsters, Goddesses and Cyborgs: Feminist Confrontations with Science, Medicine and Cyberspace*, ed. Nina Lykke and Rosi Braidotti, 30-46. London: Zed Books.

Strauss, Anselm, ed. 1979. *Where Medicine Fails*. NJ: Transaction Books.

Strauss, Anselm. 1986. *Qualitative Methods for Social Scientists*. Cambridge: Cambridge University Press.

Strauss, Anselm, S. Fagerhaugh, B. Suczek, and C. Wiener. 1985. *Social Organization of Medical Work*. Chicago: University of Chicago Press.

Strauss, Anselm, S. Fagerhaugh, B. Suczek, and C. Wiener. 1991. *Social Organization of Medical Work*. Chicago: University of Chicago Press.

Suchman, L., and R. Trigg. 1991. "Understanding Practice: Video as a Medium for Reflection and Design." In *Design at Work*, ed. J. Greenbaum and M. Kyng, 65–89. Hillsdale NJ: Lawrence Erlbaum.

Trigg, Randall, and Susanne Bodker. 1994. "From Implementation to Design: Tailoring and the Emergence of Systematization in CSCW." In *Proceedings of the ACM 1994 Conference on Computer Supported Cooperative Work*, 45–54. New York: ACM Press.

Weedman, Judith. 1995. "Incentive Structures and Multidisciplinary Research: The Sequoia 2000 Project." Paper presented at the American Society for Information Science, Chicago, October.

Yates, JoAnne. 1989. *Control through Communication: The Rise of System in American Management*. Baltimore, MD: Johns Hopkins University Press.

21 Mapping the Body across Diverse Information Systems: Shadow Bodies and How They Make Us Human

Ellen Balka and Susan Leigh Star

As is no doubt the case for many contributors to this volume, writing this chapter is complicated on many levels. It is all at once a story, an intellectual undertaking, and an emotional undertaking. Given Leigh Star's exquisite gift that allowed her to weave life's daily experiences into and through her academic work and her commitment to the personal, it seems fitting to begin with the story of the chapter's origins.

Among the interests Leigh and I shared was a fascination with health indicators (to be discussed in more detail), the information systems that are central to the development of indicators, and the classification systems embedded in those indicators. Ultimately we were concerned with how these indicators were used—often in arguably inappropriate ways, and how they contributed to our understanding of the world around us, and, in doing so, contributed to our own sense of ourselves and our humanness.

We had been talking about the concept of shadow bodies and had decided to write a piece together to flesh out this concept. When we were both invited to present at a workshop in Cambridge at the Centre for Research in the Arts, Social Sciences and Humanities titled Borders, Boundaries and Thresholds of the Body, in June 2009, we saw this as an opportunity to move our writing project forward. We developed our abstracts together, each developed notes or a draft, and then got together in Cambridge before the workshop to talk through our material, finish up our slides, agree on our definition of shadow bodies, and move our paper forward. After the week we spent in Cambridge we continued our discussion by email and phone. At the time of Leigh's death, we were trying to arrange a time to get together to finish our chapter. After she died I was left with my draft paper, both of our abstracts, both sets of slides, and a bucket of grief.

For me, sitting down to finish a paper that I was supposed to finish with Leigh cannot be separated from all the emotions that are evoked at having lost a good friend (not to mention a brilliant theorist). Layered on top of that are intellectual challenges—no doubt the resultant paper would have been different if Leigh and I were working on it, as we did, in the same time and place in Cambridge. And, no doubt, it would have been different had it been finished, as we'd hoped, the year after it was begun. Instead of drawing on the one known published piece where Leigh articulated some of her ideas about shadows (see Star 2010) we would have had the luxury of leisurely discussion and debate. No doubt the piece would have been more nuanced, richer, had Leigh been here in person to finish it. Time and intellect do not stand still, and indeed the ways that I think about the intellectual project of shadow bodies have continued to evolve since her death. Yet, since it is a piece we began work on together, it is written in the collective voice. Having said that, in Leigh's absence, I've had to follow the advice of a mutual friend who suggested that I just sit down and "channel Leigh."

Introduction

In this chapter, we introduce the concept of shadow bodies (Star and Balka 2009). Shadow bodies are partial views of bodies derived from illumination of some aspects of the body, which, like a shadow, leave others underemphasized or less visible. Shadow bodies, like indicators (Balka and Freilich 2008; Balka 2005, 2010; Tolar and Balka 2012; Balka, Mesing, and Armstrong 2006), are representations of the body, and not the real thing. They can be discipline or professionally based, and they can reflect important social constructs in society. In contrast to "the body multiple" (Mol 2003) that is not fragmented, but rather in its multiplicity hangs together (and in which differing versions of the body have to be workably complementary with the others), shadow bodies *are* the fragments which *do not* hang together in their multiplicity, but rather exist, in clouds of indicators, waiting to be woven together in meaningful ways. Shadow bodies are both created by the fragments that are measured, and by the unaccounted for and invisible spaces left in between. Their existence has personal and political consequences.

In the pages that follow, we provide an overview of some of the theoretical and empirical work which gave rise to development of the concept of shadow bodies, and we further elaborate on the concept. We discuss where the notion of shadow bodies can take us, and how it makes us human.

The concept of shadow bodies emerged through multifaceted conversations. Topics included ongoing fieldwork concerned with duplicate medical records, known commonly in healthcare settings as shadow charts (Balka et al. 2012; Balka et al. 2013); the relationship of information systems to data collection and knowledge production in general (Bowker and Star 1999), and health indicators in particular (Balka 2005; Balka, Messing, and Armstrong 2006; Balka and Freilich 2008; Schuurman and Balka 2009); differences in biomedical accounts of health issues and lived experiences of health issues; and what we might learn about science by reflecting on art practices (Balka 2007). Our resulting notion of shadow bodies drew on and from all of these areas, in a progression that was hardly as linear or tidy (or indeed finished) as the material that follows might suggest.

In developing the notion of shadow bodies, we drew on several empirical studies which were undertaken in varied healthcare settings in order to gain insights about both the politics and design of computer based infrastructures in healthcare. The cases yield insights about our choice of terminology, and allow us to describe how some forms of shadow bodies are produced. We then consider the political consequences of shadow bodies, and raise the hopeful possibility that emergent technologies may generate new forms of shadow bodies reflective of more diverse interests and lived experiences.

Theoretical and Methodological Grounding

Theoretically, the notion of shadow bodies derives in part from earlier work undertaken by Berg and Bowker (1997), which outlines how the medical record "mediates the relations that it organizes" (514) along with the bodies that are configured through it. Having pursued a line of inquiry similar in many respects to that outlined by Clarke (2005) in earlier work (e.g., Balka 2003), Balka adapted a similar approach in empirical studies that we draw on here. These included a study of duplicate medical records—known commonly in healthcare settings as "shadow charts" (Balka et al. 2012; Balka et al. 2013), and another study which was concerned with data quality as it relates to epidemiology (Balka et al. 2012; Balka et al. 2013), which was explored through fieldwork concerned with multi-jurisdictional pre-hospital handovers in care. Additionally, we drew on past health issues and challenges we experienced and observed in those around us, which included, among other things, two skiing injuries.

The two case studies we drew on of shadow charts and pre-hospital handovers in care (see Balka et al. 2012 and Balka et al. 2013 for more in-depth accounts) were rooted in infrastructure studies (Star and Ruhleder 1996; Bowker and Star 1999), and reflected an interest in how classification systems (such as patient acuity scoring systems) built into information systems (such as electronic triaging systems) come to bear on the ways we understand our world, as well as our bodies (Balka 2005)—what Berg and Bowker (1997) have referred to as how "the structuring of the record speaks to the structuring of the bodies we investigate" (515). We were interested in and explored the variation and ambiguity in conceptualizations of the body and its processes which occur as notions of the body are created across jurisdictions (healthcare institutions that do not share a single governance structure), through multiple information systems (both paper based and electronic), databases, and data spaces. Additionally, we were interested in tensions that frequently arise between knowledge based on fixed categories, such as specific illnesses or anatomical structures, and the reality of individual bodies and their living processes, and it was here that we engaged in auto-ethnography, reflecting, for example on the differences between x-rays of a knee that yielded inconclusive evidence of a broken bone, and living with the experience of a broken bone in the knee that was only "discovered" after two months of pain, when it was virtually healed.

Methodologically, empirical studies we drew on utilized infrastructure ethnography (Star 1999) and database ethnography (Schuurman 2008). Database ethnographies are a means of using insights from science, technology, and society studies to enhance data analysis. For Schuurman, database ethnographies, which elicit information from data stewards about data in multiple-use databases, are undertaken in order to "provide an archive that describes the context and meaning of the data at a particular point in time" (1529). Where Schuurman has focused on data stewards and the creation of "useful" data, the focus of data collection in the empirical work we have drawn on (Balka et al. 2012; Balka et al. 2013), in thinking through the concept of shadow bodies, included a broader array of data producers and users—the people who collect, coordinate, and use data, both before it gets to and after it has been manipulated by data stewards—than Schuurman's work. In addition, in contrast to Schuurman's (2008) concern with the creation of useful data, we sought to understand the messiness of its production, as well as the myriad factors—human, technological, institutional, and political—that came to bear on its collection, production, and use.

Schuurman (2008) identifies several literatures that have come to bear on the concept of database ethnographies. These include the now substantial literature about boundary objects (Star and Griesemer 1989), writing about semantic heterogeneity (Schuurman and Balka 2009), tacit knowledge in scientific communities (e.g., Collins 2001), and ethnography, as well as science, technology, and society literature (including actor-network theory), and scholarship concerned with ontologies, undertaken from both a social science and computer science perspective (Schuurman 2008). Indeed, material from all of these areas of inquiry influenced data collection in the ethnographic cases we conducted, and draw on as follows.

The ethnography we drew on to articulate the concept of shadow bodies was undertaken within the broader context of challenges arising in contemporary healthcare. In healthcare, the provision of care often requires tracking and managing patients across diverse information systems. Research undertaken in an effort to understand the relationship between health interventions and outcomes also often requires tracking patients across care providers (e.g., from a family physician in a community context to a specialist in an ambulatory care clinic who may then treat the patient as an in-patient in an acute care hospital), and jurisdictions (e.g., from the ambulance service to the emergency department). Typically, each jurisdiction has its own information system, which may be paper based, computer based, or a hybrid that incorporates both kinds of records. The diversity of care providers, jurisdictions, and information systems characteristic of contemporary healthcare systems in developed countries presents challenges for direct patient care, as well as for researchers and administrators whose work depends upon health indicators.

Health indicators measure something that cannot directly be measured. They are "variables linked to the variable (studied), which itself cannot be directly observed" (Chevalier et al 1992, 47). In an era of increased accountability, indicators—which result from performing mathematical operations on data (Moed, Glänzel, and Schmoch 2004)—are increasingly used to evaluate a range of issues, such as whether or not health interventions are resulting in desired health outcomes. Health indicators are an important means through which we both evaluate the success of various health and health system interventions and come to understand health and illness. They are significant in an epistemic sense—they influence the production of biomedical knowledge, how each of us as patients is positioned by care practitioners, and how we must position ourselves—or perform our illnesses—in an effort to successfully maintain our health.

Shadow Charts and Patient Handovers: Overview of Cases

Shadow charts (SCs) are partial records of a patient's medical history, typically maintained by a particular clinical area and widely considered to be a poor medical practice. While working on our shadow chart study, we began a research program concerned with patient handovers and realized there was a relationship between the existence of shadow charts and handovers in care: the "dream of a common infrastructure" (a single, unified electronic patient record) on the one hand seeks to eliminate shadow charts, and on the other hand, can only succeed if considerably greater standardization of handover information occurs than is currently the case. One of the reasons shadow charts exist is precisely because it is not always deemed appropriate to hand over all existing information in a chart to another provider in the care chain, and differences in how different practitioner groups work with information (related to their disciplines as well as the level of care they provide) present challenges for standardization of information exchanged between providers (Balka et al. 2012).

Shadow charts and patient handovers each tell us something about the infrastructure from which shadow bodies emerge.

Case 1: The Pre-Hospital Care (PHC Case)
In 2008, Balka's team began exploring information handovers related to pre-hospital care in the context of ski area injuries (see Balka et al. 2012 and Balka et al. 2013 for a full account). Conducting epidemiological research related to ski area injuries and health outcomes is difficult, partly because of an inability to track patients and the care they receive across multiple jurisdictions (ski area, ambulance, clinic, ambulance,

emergency department) involved in pre-hospital care. After conducting informal interviews with stakeholders we began collecting data in 2009. Subsequently, we conducted several formal interviews with stakeholders involved in ski area operations, and we observed collection and processing of on-hill injury data during the ski hill's summer (bike park and hiking) and winter (skiing and snowboarding) operations. In carrying out this study, we sought to understand work processes and workflows involved in the pre-hospital care chain, and to identify barriers to the production of data that could be used in developing an understanding of health outcomes associated with pre-hospital care in general and ski area epidemiology in particular (see Balka et al. 2012 and Balka et al. 2013 for more detail).

Case 2: The Shadow Chart Case (SC Case)

Balka et al. (2012) carried out a multimethod multisite case study (begun in August 2007, continued in phases through 2010) in order to understand the phenomenon of duplicate medical charts, also known as ghost charts or shadow charts. Shadow charts have long plagued the healthcare profession. They are partial records of a patient's medical history, typically maintained by a particular clinical area. They exist in addition to official medical or hospital records (typically maintained by an institution's health records department), and are a challenge for hospital administrators, who worry that they are exposed to risk (legal action) if a patient is aware that a shadow chart exists and either was not consulted during medical treatment, or was not handed over to a patient in response to a freedom of information request made on behalf of a patient. Accreditation bodies such as Accreditation Canada (responsible for accreditation of Canadian hospitals) have determined that there should be a single patient record for each patient within an institution, however large that institution may be.

We conducted interviews with stakeholders at the hospital which served as our study site, and carried out in-depth ethnographic observation of work in two ambulatory care settings, which included all work that occurs in relation to charts outside of clinical encounters (e.g., chart use by nurses, clerical and administrative staff, researchers, and trainees). We also carried out observation of chart use during clinical encounters (e.g., doctors' use of charts during patient visits) on three clinical units and conducted ongoing observations in the hospital's health records department over a fifteen-month period. Our study was undertaken at a time when the hospital was concerned about the existence of shadow charts (a widely documented phenomenon also viewed as a threat to patient safety) in the context of an upcoming accreditation visit. In addition, the hospital was beginning to plan for computerization of health records and we hoped information gained from our study could help inform decision making related to computerization.

Overview of Findings

In our shadow chart case, we observed a phenomenon that had previously been reported (e.g., that various groups of care providers develop discipline-specific views of the patient). We also observed the local maintenance and preservation of discipline-specific views of the patient in the form of local databases and disease registries in one clinical setting, and maintenance of a physical chart filing and annotation system in another case that was designed to support easy access to clinical data of particular interest to a specialty, for research purposes. At one point during our observations, in all of our SC settings, as patients left the clinical setting and doctors completed their charting, administrative staff in the units were required to complete a form specific to each unit, which had been designed in an effort to build unit-specific indicators that could be used to track the patient flow through each unit over time, and develop a high-level snapshot of the acuity of each unit's patient population, and the mix of ailments that brought patients into the ambulatory care clinics.

Our observations, particularly those in our PHC study, also stressed the importance of context in information handovers. First, each type of care provider (e.g., ambulance personnel vs. medical staff) is trained to construct a view of the patient through a disciplinary lens, and different disciplinary lenses necessitate collection of different constellations of data. While there may be some overlap in data collection across provider groups (e.g., blood pressure), the relative importance of data points changes across types of providers. We had observed a similar phenomenon in our SC case, where each clinical area used a different summary sheet to provide a quick overview of the patient to specialists. In the PHC chain, where each successive handover is generally to a higher level of care, successive care providers may have access to different tests and different problem-solving skills, and hence may require information that differs somewhat from what was handed over by the patient's previous care provider. In addition, as a patient's condition changes, some data may become more relevant (such as a longitudinal view of blood pressure).

Through the PHC case, we also observed the collection of information required to meet the informational needs of numerous agencies such as the workers' compensation boards, insurance companies, and so on (see Balka et al. 2012 and Balka et al. 2013 for a more in-depth discussion). Also, some information was collected about patients for purely administrative purposes (e.g., a patient's address), and at times additional information might be collected to support research and other informational needs that were secondary to care (see Balka et al. 2012 and Balka et al. 2013). We frequently heard that much of the information collected at the point of injury (often mandated by these "off-stage" actors) did not contribute to the delivery of care, and, worse, could lead to delays in care provision, and it was often some of this information which was used to produce indicators. Information collection and display varied with the context of use and the specific substance and format of fields on data collection forms, which often reflected the differing needs of off-stage agencies (such as the regional ski area association) and not those of care providers, patients, or even health researchers.

Although the creation and preservation of data for research purposes is (or should be) secondary to clinical care needs, information collected during healthcare encounters may be used for research as well as clinical purposes (Balka et al. 2012). The usefulness of these data is often enhanced when they are examined in relation to other data sets (e.g., data about interventions performed during pre-hospital care are much more meaningful when examined in relation to the health outcomes of patients who received those interventions). As Schuurman and Balka (2009) have pointed out, preservation of contextual information about how data are collected is also essential to ensuring their usefulness at the point where data are integrated and used for subsequent analysis. Hence, the contextual nature of information is significant in relation to secondary uses of information that is handed over in care encounters (e.g., for subsequent research purposes), as well as to the primary uses of information handed over during care (e.g., to the direct provision of patient care).

Clearly, the contextual nature of information is important in relation to information handover in several ways. Context influences what information is collected in what formats in particular settings (Balka et al. 2012); and it may influence what information is handed over to subsequent care providers, and hence what indicators can be developed. At each stage of data collection, decisions are made about what to measure and how to measure it; classification systems are inscribed into databases that are often subject to the constraints of enterprise-wide or industry-standard reporting systems. The data that result may be integrated and combined with other data sources, subject to other sets of classification systems, data architectures, and so on, to produce indicators. And at each stage, the needs of these off-stage agencies often have a significant influence not only on what information is collected at various points in the pre-hospital care chain, but also on the format in which that information is made available to subsequent care providers. Hence it was often off-stage actors that exerted a powerful—if not explicit—influence on what data were available, and subsequently, on what indicators could be developed.

Through these processes, official mandated accounts of the body are created (which are partial views, but seldom recognized as such), as well as multiple shadow bodies, which result from the byproducts of official, sanctioned data collection and reporting processes—all the data collected but that falls outside of officially sanctioned data collection, as well as those data which are relegated to the residual categories, the "not elsewhere classified." As information is gathered to meet the multi-jurisdictional needs of varied stakeholders, shadow bodies are created, mediated through forms, computing infrastructures, and the classification systems embedded in them. Just as shadow records create partial views of patients, shadow bodies are partial representations and abstractions of lived bodies, which travel into distant social worlds, meeting the informational needs of numerous stakeholders, increasingly distant from our lived experiences and our actual bodies.

The Properties of Shadow Bodies

Shadows are reflections, borne out of partial illumination. Depending upon lighting conditions and angles, shadows can bear a strong likeness to the artifact they reflect, or they can appear as gross distortions or exaggerations of the phenomenon they represent. Shadow bodies are the reflections, illuminations, and impressions of a body created through the illumination of some (but not all) of its processes, as captured in information systems—both paper based and computer supported, and as the byproducts of those systems (the data that fall from view, relegated to the "other" or "not otherwise classified" residual categories). Shadow bodies are constituted of data that fall from view because they are inconsistent with established understandings of bodies, don't fit into existing classification systems (other than "not elsewhere classified," yet somehow persist, precisely because they are rooted in lived experiences.

Shadow bodies are partial views of bodies, derived from illumination of some aspects of the body, which, like a shadow, leave some body parts underemphasized or less visible, and exaggerate others. The specific aspects of a body rendered visible or exposed through a shadow body are reflective of the underlying infrastructure, put in place to meet the informational demands of extra-local actors such as the doctor at a clinic who will eventually see a patient injured on the hill, the worker's compensation board that will eventually adjudicate a claim about a worker injured on the hill, or a provincial or federal government that will eventually need to issue reports about the accessibility of knee surgeries to a regional population.

Shadow bodies, like indicators, are representations, not the real thing. And, like the imaging studies of a knee (an x-ray that wasn't of high enough fidelity to provide a definitive diagnosis of a fractured tibial plateau, the CT scan that ultimately made the fracture visible, and the MRI that made the ruptured ligament visible), each form of representation can be characterized as a series of affordances and constraints, each of which yields some things more visible than other things. And yet neither the initial imaging studies of that knee nor the numerous physical examinations performed on that knee suggested that it was as severely damaged as was eventually found to be the case. Although each of those representations of a knee is subject to the affordances and constraints of the imaging technology used to produce it, and the strengths and limitations of each of these technologies is well known, the patient, faced with imaging studies and physical exam results that were not suggestive of serious injury, bought a knee brace, and carried on as if the initial imaging results and physical exam results had been correct. The lived experience of pain was discounted in favor of these more visible and tangible representations of the knee. Our representations come to bear on the management of our bodies in concrete ways. Shadow bodies have real impacts on our understanding of ourselves, and on how we manage our healthcare.

Shadow bodies are created through the collection of data, through multiple information systems, which can exist in paper or computer-mediated formats. Our work on our shadow chart study and other work that has considered the role of medical charting in the creation of bodies (Berg and Bowker 1997) suggests that shadow bodies can be discipline or professionally based (e.g., gastroenterologists' shadow charts include graphic images of the gastrointestinal (GI) system, while a vascular surgeon's shadow charts are likely to contain vascular maps of the body). Our research to date looking at ski area injuries suggests that shadow bodies can reflect important social constructs in society (e.g., risk as a legal construction; the rights of workers to be compensated when injured at work). Shadow bodies, constructed through paper and computer-based forms, reflect traditions (e.g., the use of small notebooks that fit in the pocket of an ambulance attendant's pants), in addition to the boundaries of academic disciplines and professions. Shadow bodies are often aided in their construction by information systems that yield some things visible (e.g., the mechanisms of injury involved in catastrophic industrial injuries, such as falls), and others invisible (e.g., catching a ski edge, or teaching a snowboard class in a terrain park and landing a jump poorly).

Just as lighting conditions and angles of view come to bear on what an actual shadow looks like, the technologies (broadly defined) and the affordances and constraints of each comes to bear on the representation of the body we can see at any point. At multiple points during the trajectory of an injury or illness, data are collected, and, as the PHC case outlined here suggests, the specific activities of data collection often reflect the needs of slightly off-stage actors (e.g., the National Ski Areas Association, or the workers' compensation board), each firmly entrenched in a social world or arena that has a particular kind of relationship to the injured person, and each of which has a vested interest in how data are collected. The data aren't simply collected—they are collected because they are going to be used *for* something.

The PHC case outlined here suggests that there are numerous social worlds in which shadow bodies—initially injured on a ski hill—are constructed, many of which are not part of the formal health care system (e.g., private corporations, which assume responsibility when people are injured under their watch; industry associations such as the National Ski Areas Association, which may fill multiple and significant roles ranging from protection of its members to marketing). All of these social worlds are involved in the creation of shadow bodies.

The creation of shadow bodies begins with actual bodies—people who get up early on snowy mornings and call their friends from the ski lift to make sure they are out of bed to enjoy the fresh snow. Shadow bodies are rooted in embodiment—presence—one can't ski or ride a mountain bike safely unless one is present. It is these actual bodies, often full of life and full of joy that occasionally are injured, and begin movement along a trajectory where fragments, traces, reflections, and refractions of the lived body oddly remain invisible as data are captured, abstracted, and sent on their way, off in multiple directions, according to embedded ontologies that harken back to the construction of shadow bodies ever so far from the joy of a fresh-powder day or a good string of turns on new telemark skis. The lived experience of joy and physical fitness or the beauty of a sunny day in the alpine remain as elusive through the process of creating shadow bodies as the pain one experiences with an injury. Oddly, the data sets that together constitute those shadow bodies that are most often legitimated yield little insight about the lived experiences that serve as the precursors to their existence, or result when those joyful moments end badly.

Empirical investigations of shadow records and pre-hospital handovers in care yielded insights about how data are collected and moved across and around the healthcare system, and, as the case of pre-hospital handovers in care suggests, related systems such as the legal system, professional associations such as ski area associations, and so on. Those two empirical cases showed us how indicators are used to create bodies (healthy bodies, sick bodies, legal bodies, ambulatory bodies, partial bodies, etc.). Each of these views of the bodies has its own borders, which are often delineated by the data collected, and the broader systems that each data point responded to, or was a part of. For example, the legal view of an injured skier/snowboarder includes information about the frequency with which the injured spent time at the snow sports area where they sustained an injury, because such information may become pivotal to the establishment of prior knowledge of risk, and hence responsibility for injury. But the data that comprise each of these views of the body and delineate its borders, on the surface, tell us little about whose experiences are represented in data, whose bodies are represented, and whose sensoria are rendered invisible or visible. This knowledge remains hidden from view, obfuscated by the layers of data infrastructure, hidden in the unspoken and often unexamined assumptions of multiple classification systems inscribed across paper and electronic data collection systems, many of which span agencies and jurisdictions.

Shadow Bodies and How They Make Us Human

And in the eventual use of data derived from official data-collection activities, the complexities and contingencies of indicators—often accepted like received wisdom, as fact—remain hidden, pushed from view by decontextualization and simplistic assumptions about causality. In the welter of Western scientific folklore resting on single and simple assumptions or chains of evidence, there is a loss of multiplicity and complexity. In our dominant/traditional ways of knowing through numeric indicators, simplicity trumps complexity, and often residual categories are lost. There are a number of reasons for this absence—this silence—in the creation of shadow bodies. Data may be residual because the object to which it refers is unspeakable. As a culture or as individuals or both we may remain silent about phenomena (e.g., living with HIV/ AIDS, or depression). Or, culturally, we may participate in miscounts (e.g., counting of gay couples as single individuals), and "pass," in order to solve problems in some contexts. (A lesbian may pass as straight—a sister rather than partner—to gain access to an area of a hospital restricted to "immediate family" rather than engaging in an argument about definitions of family or the absence of "same-sex partner" on a form.)

Phenomena may be absent or obfuscated from view because the object to which they refer may be too complex or complicated to be easily countable. For example, developing an understanding of how on-hill treatment for dislocated shoulders influences long-term health outcome (including unexpected sequelae) may be beyond the technical capacity of existing information systems. Or, phenomena may remain invisible because the "object" of lived experience is disbelieved by the data collector, and the respondent classed as "crazy" or "disorganized" or sometimes simply "foreign." Experience may be devalued, and stigmatized experiences can and therefore often become residual—(unimportant; too "minority"). Ramsay Hunt syndrome (RHS) provides an interesting example of this.

Caused by reactivation of common viruses (linked to chicken pox and herpes), Ramsay Hunt syndrome has been identified as the second-most common cause of acute peripheral facial palsy, after herpes simplex virus (HSV) in Bell's palsy (BP). Yet, despite its commonality, it has been classified a rare disease by the Office Of Rare Diseases of the National Institutes of Health (United States), because data suggest that it affects fewer than 200,000 people in the United States (which has a population of around 300 million). Because most doctors never encounter a case of RHS in their careers, knowledge about RHS and an understanding of the syndrome are limited, and misdiagnosis as Bell's palsy is a common hazard that can lead to a patient's health deteriorating rapidly (RamsayHunt.org 2013). One patient after several years of misery finally received a diagnosis of RHS from a specialist, having been dismissed on several occasions by a general practitioner who suggested that the patient had a hyper bodily awareness and the symptoms of a worried-well malingerer.

RHS is often misdiagnosed because examination standards are lacking and RHS shares many characteristics with Bell's palsy. Two different viruses can cause RHS, only one of which also causes vesicular eruptions that may be present until long after the onset of the ailment. This complexity—differences in the behavior of the two different RHS causing viruses—also contributes to the misdiagnoses. Additionally, many of the symptoms common to RHS (such as earaches and vertigo) can often easily be explained away as ear infections or resulting from flu or sinusitis. One of the viruses that causes RHS—the herpes simplex -1 virus (HSV-1) typically causes cold sores on the exterior of sufferer's mouths. One patient with RHS was told by a specialist he couldn't possibly have HSV-1 (known to cause RHS), because HSV-1 lesions did not recur inside the mouth near the gum line, where he had experienced a recurrence. The patient's lived experience of illness—a recurrent lesion inside the mouth near the gum line—was inconsistent with common understanding of RHS, and in a search to make sense of this disjuncture between his lived experience (recurrent sores on the inside of the mouth near the gum line) and the dominant understanding of the syndrome (which did not involve recurrent sores in that location), he contributed to the development of a shadow body: he went online and told his story, leaving a trace, an indicator, of his lived experience, that sloshed sloppily outside of the categories that taken together yield the current dominant understanding of the natural history of RHS.

Although lab tests exist that enable a definitive diagnosis of RHS, they are both expensive and nonroutine. Although HSV-1 can—and does—cause blisters in the roof of some sufferer's mouths, this is a comparatively rare occurrence. Another patient with lesions around the gum line and soft palate reported going to varied care providers including general practitioners, dentists, and ear, nose, and throat specialists for seven years, during which she was not tested for HSV-1, presumably because she never got what are commonly called "cold sores," visible on the outside of her mouth. "The discrepancy of these misdiagnoses greatly impact[s] the accuracy of incidence and prevalence statistics for RHS and BP. Furthermore, it implies that there are considerably more RHS patients and considerably less BP patients than previously believed" (RamsayHunt. org 2013).

Building on the example of RHS, if symptoms associated with RHS are not misclassified as symptoms of Bell's palsy, flu, sinus infections, or drug side effects, or dismissed as mental illness, they might easily end up in the "other" category, that residual place where monstrous and complicated "not otherwise classified" remnants are collected and contained. And at the same time that standardized, single indicators (whether they refer to our ethnicity, our marital status, or whether our symptoms are RHS or Bell's palsy) both compose and reduce us, they simultaneously mask all of the complexity of what lives in residual categories (e.g., being one half east Indian Muslim and one half Jewish, or in a long-term same-sex partnership). Complexities (e.g., particular ethnic backgrounds; familial relationships such as one woman and two men who have sex in all possible permutations and have two children together; RHS caused by two different viruses and easily mistaken as many other things) escape the regimes of standardized data collection and quasi-quantification of behaviors and symptoms, leaving our lived experiences somehow outside of the frame, or off the charts.[1] Our lived experiences exist in the shadows, sometimes visible, sometimes not, depending upon how the light falls, and whether or not the richness of the "not otherwise classified" categories has been unpacked, made visible to others, who might then add their experiences to the data that constitute the shadow, adding to its definition.

As the number of standardized indicators increases—across areas such as medicine and intellect (e.g., SATs, GREs, IQ, and tests for specific disorders such as dyslexia and

ADD), as well as other areas (e.g., financial indicators describe creditworthiness)—and as we are defined increasingly in relation to a myriad of standardized indicators, shadow bodies are made, and at the same time bodies of shadows accumulate. Each new set of indicators simultaneously defines norms (substantive forms, accepted constellations of "facts") while relegating other experiences into the shadows where they accumulate to form bodies of shadows, eventually visible in the form of blogs and electronic traces of many sorts such as polls on the RHS web site which aggregate information about individuals. Web sites such as PatientsLikeMe.com (2015), which allow patients to voluntarily contribute data online (which is then sold to varied stakeholders who use it for research purposes), or eHealthMe.com (2015) (which serves as a means of collecting voluntarily supplied patient data concerning symptoms and drug outcomes, and analyzing those data alongside the U.S. Food and Drug Administration data), offer the possibility of illuminating existing phenomena in new ways.

Yet all of this happens with little guidance in the way of moral or sentimental order. The proliferation of these indicators goes unchecked, with documents and traces following, in ever-thickening ways, nearly every domain of life (e.g., book buying habits, health circumstances, where one lives, how "standard" one may be or not be).

Just as the shadow charts we studied were partial and incomplete, so too are shadow bodies. And, just as each practitioner's view of the body instantiated through a shadow chart was partial, so too are shadow bodies. Shadow bodies have an anatomy. They are partial views, which at the same time overlap with or are inhabitants of residual categories. They are the discarded knowing—often the "non-objective" evidence supplied by patients, and complexities that come to comprise our shadow selves, at both an individual and collective level—the RHS sufferers who were misdiagnosed because their HSV-1 didn't cause typical cold sores on the exterior of their mouths, or whose neuropathic pain in other body parts was not understood to be related to their herpes until recently (Kallio-Laine et.al. 2008). They are the bits left out of dominant clinical understandings, left off of data collection forms, not searchable in formal databases. Our shadow bodies remain present in lived experiences, on online discussion forums, and, as they become more visible, first in our residual categories, and then, perhaps later, our shadow bodies may be given legitimacy through forms, data collection, articles, and more. They emerge at the intersection of partial views and the inhabitants of residual categories, which overlap with a context. That context is both created and rendered invisible by clouds of indicators, standards, and classification systems. Unlike the systems of standards/classification systems and indicators that reify some partial views while relegating others to the world of shadow bodies, the nature of that shadow knowledge is layered, complex, and interactional.

We go on living with that which escapes standardized testing, and quantification of our behavior. It is these discarded complexities that comprise our shadow selves that are full of promise and aesthetic surprises. Over time, the discarded knowing comes to comprise our shadow selves. In this world, where we are always being measured by standards, our shadow lives, shadow selves, and shadow communities keep on living in spite of the vigilance and surveillance from officialdom that permeates every part of our lives. We continue to inhabit untold numbers of these residual categories, although these clouds of indicators born from simple categories, standards, and measurements swarm around us, never being able to tell the simplest truth about us directly.

Shadow bodies have a physiology. They exist in the *inbetween of multiple standardized, reified indicators*. They are the unspeakable experiences, those things that are invisible to the formal record. In between these frozen indicators, we live in a forest of misplaced concretism, characterized by notions that there is one place within the body responsible for single complex behaviors such as homosexuality, violence, intelligence, speech recognition, cancer, or genius. Built on notions of relatively simplistic causality, these frozen indicators that together make up the forest of misplaced concretism that we inhabit and that gives rise to our shadow bodies, entail claims of indicators without borders, devoid of context.

How then do we give legitimacy to our shadow bodies, and what does it mean to do so? We can think of indicators as patterns, sequences, clusters, or pathogenic signs. They exist in a world of problems, phenomena and speculations. In science studies—and increasingly many other fields—there is a desire to keep relationships open, and to move beyond notions of simple, singular causality to more complex many-to-one and ultimately many-to-many relationships. Doing so requires opening up pasts, acknowledging multiple selves (and multiple shadow bodies), and seeking creative recombinations in data that allow us to understand not only what resides in residual categories, but also what kinds of processes and relationships exist between those residuals which our current infrastructures and information systems at present may only allow us to see and know in simple ways. We have to overcome data-sharing and infrastructural challenges and seek means of supporting analyses that allow us to recognize creative recombinations of traces, of indicators we may not even today recognize as data.

Exploring the richness of our shadow bodies, those residuals often lost in the forest of misplaced concretism can help restore the importance of context that has been stripped away by our current practices of carving out single-indicator phenomena. Single-indicator phenomena that reduce complex relationships to the simplest plausible explanation leave us vulnerable, as we excise relationships from their context, and complexities from their multiplicity. Though we have many examples of such phenomena—including testing and notions of expertise that suppress (worst case) or make it difficult to discover alternative explanations for a phenomenon (best scenario)—we create vulnerabilities. Each time we presume a relatively simple causal relationship (e.g., facial paralysis means Bell's palsy), we explain away, dismiss, relegate to residual categories precisely that evidence that allows us to make sense of more complex relationships (such as RHS). Indeed, there are real consequences to purifying the chain of "evidence" by uncritically relying on indicators, by failing to acknowledge all of the shadows. Overly simplified indicators support testing, gate keeping, and discarding of anomalies, and leave unexamined questions about who owns the tests, who stands to profit from them, and who gains authority through their use. As chains of causation are simplified and purified, the standard indicators they are built on become substitute theories. We forfeit the infrastructural conditions that afford us the possibility of formulating alternative explanations. Deletion of contexts and simplified causality together narrow the lexicon we have access to in our efforts to make sense of the world around us.

Moving beyond the use of indicators mobilized for purity and simplicity is not a simple undertaking, as many of the indicators we use flow from and reflect the tools through which they are created. There is a kind of tool hegemony, present when technologies such as fMRI (functional magnetic resonance imaging) are used to generate data that are subsequently woven into indicators. Such tools deemphasize ontological and epistemological challenges, and contribute to the burial of histories of multiplicity. Shadows, all the bits that don't fit the existing or dominant categories, find their way into clouds of shadows, often surfacing in the form of patient blogs, patterns in search-engine term use, and other electronic traces. And these clouds of shadows can contribute to new shadow bodies, as well as shadow relationships that challenge hegemony. They represent hopeful "monsters," hopeful residuals, and form resistant bodies. They offer a means through which to challenge diseases without passport—those ailments that have not gained legitimacy in biomedical establishments (some recent examples include multiple chemical sensitivity and chronic fatigue syndrome). These clouds of indicators allow us to uncover some anomalies that are too big to bury, such as the relationship of environmental racism to higher rates of breast cancer in African American women. These clouds of indicators, our emergent shadow bodies allow us to form relationships, to move us a little ahead as we work to articulate the nuances of a more complicated world.

These infrastructure shadows permeate our bodies in action, interaction, and history, where they take the form of absences and presences encoded by all types of information technology—which are not, themselves, yet considered institutionally. Yet these aggregate social forms, these shadow bodies are undertheorized. Their presence marks our humanness, and warrants our concern about the existential and political ways of life that are engendered by the volume and entanglement of multiple infrastructures, and the new forms of indicators they might support.

Note

1. Steven Epstein describes a similar phenomenon in recounting the emergence of AIDS, in his 1996 book *Impure Science: AIDS, Activism, and the Politics of Knowledge*. Epstein describes how

Michael Gottlieb, a young immunologist, began seeing and documenting cases in 1980. He submitted a manuscript about these early cases to the *New England Journal of Medicine*, which rejected the article and referred Gottlieb to the U.S. Center for Disease Control (CDC), in part because the constellation of symptoms the article described had not yet been classified as a disease. The CDC made its first public report about what came to be known as AIDS some months later.

References

Balka, E. 2003. "Getting the Big Picture: The Macro-politics of Information System Development (and Failure) in a Canadian Hospital." *Methods of Information in Medicine* 42 (4): 324–330.

Balka, E. 2005. "The Production of Health Indicators as Computer Supported Cooperative Work: Reflections on the Multiple Roles of Electronic Health Records." In *Reconfiguring Healthcare: Issues in Computer Supported Cooperative Work in Healthcare Environments*, ed. E. Balka and I. Wagner, 67–75. Paris: ATIC Lab, Simon Fraser University.

Balka, E. 2007. Seeing, Being and Knowing: Reflections on Photography, Ethnography and Ways of Knowing. Plenary Address, Society for Social Studies of Science Annual Meeting. Montreal, October 10–13.

Balka, E. 2010. "From Categorization to Public Policy: The Multiple Roles of Electronic Triage." In *By the Very Act of Counting: The Co-construction of Statistics and Society*, ed. A. Rudinow Sætnan, H. M. Lomell, and S. Hammer, 172–190. New York: Routledge.

Balka, E., and J. Freilich. 2008. "Evaluating Nurses' Injury Rates: Challenges Associated with Information Technology and Indicator Content and Design." *Policy and Practice in Health and Safety* 6 (2): 83–99.

Balka, E., K. Messing, and P. Armstrong. 2006. "Indicators for All: Including Occupational Health in Indicators for a Sustainable Health Care System." *Policy and Practice in Health and Safety* (April 1): 45–61.

Balka, E., S. Whitehouse, T. S. Coates, and D. Andrusiek. 2012. "Ski Hill Injuries and Ghost Charts: Socio-technical Issues in Achieving e-Health Interoperability across Jurisdictions." *Information Systems Frontiers* 14 (1): 19–42.

Balka, E., M. Tolar, S. T. Coates, and S. Whitehouse. 2013. Socio-technical Issues and Challenges in Implementing Safe Patient Handovers: Insights from Ethnographic Case Studies.*International Journal of Medical Informatics* 82 (12): e345–357.

Berg, M., and G. Bowker. 1997. "The Multiple Bodies of the Medical Record: Toward a Sociology of an Artifact." *Sociological Quarterly* 38 (3): 513–537.

Bowker, G., and S. L. Star. 1999. *Sorting Things Out: Classification and Its Consequences*. Cambridge, MA: MIT Press.

Chevalier, S., R. Choinière, L. Bernier, Y. Sauvageau, I. Masson, and E. Cadieux. 1992. *User Guide to 40 Community Health Indicators*. Ottawa, Ontario: Health and Welfare Canada, Community Health Division.

Clarke, A. E. 2005. *Situational Analysis: Grounded Theory after the Postmodern Turn*. Thousand Oaks, CA: Sage.

Collins, H. 2001. "Tacit Knowledge, Trust, and the Q of Sapphire." *Social Studies of Science* 31 (1): 71–85.

eHealthMe.com. 2015. http://www.ehealthme.com/ (accessed March 31, 2015).

Kallio-Laine, Katariina, Mikko Seppänen, Marja-Liisa Lokki, Maija Lappalainen, Irma-Leena Notkola, Ilkka Seppälä, Mika Koskinen, Ville Valtonen, and Eija Kalso. 2008. "Widespread Unilateral Pain Associated with Herpes Simplex Virus Infections." *The Journal of Pain* 9 (7): 658–665.

Moed, H. F., W. Glänzel, and U. Schmoch. 2004. *Handbook of Quantitative Science and Technology Research: The Use of Publication and Patent Statistics in Studies of S&T Systems*. New York: Springer.

Mol, A. 2003. *The Body Multiple: Ontology in Medical Practice*. Durham, NC: Duke University Press.

PatientsLikeMe.com. 2015. https://support.patientslikeme.com/hc/en-us (accessed March 31, 2015).

RamsayHunt.org. 2013. http://www.ramsayhunt.org/info/incidence.

Schuurman, N. 2008. "Database Ethnographies: Using Social Science Methodologies to Enhance Data Analysis and Interpretation." *Geography Compass* 2 (5): 1529–1548.

Schuurman, N., and E. Balka. 2009. "Alt.metadata.health: Ontological Context for Data Use and Integration." *Computer Supported Cooperative Work* 18 (1): 83–108.

Star, S. L. 1999. "The Ethnography of Infrastructure." *American Behavioral Scientist* 43 (3): 377–391.

Star, S. L. 2010. "When Shadows Become Complex: Weaving the DANMARRA." *Learning Communities: International Journal of Learning in Social Contexts*, no. 2 (Teaching from Country): 150–158.

Star, S. L., and E. Balka. 2009. "Mapping the Body across Diverse Information Systems: The Anatomy of Cross Jurisdictional Patient Handovers." Linked presentations, Borders, Boundaries and Thresholds of the Body Conference, Centre for Research in the Arts, Social Sciences and Humanities, Cambridge, UK, June 12–13. http://www.crassh.cam.ac.uk/events/22827 (accessed March 21, 2015).

Star, S. L., and J. R. Griesemer. 1989. "Institutional Ecology, 'Translations' and Boundary Objects: Amateurs and Professionals in Berkeley's Museum of Vertebrate Zoology, 1907–39." *Social Studies of Science* 19 (3): 387–420. [See also chapter 7, this volume.]

Star, S. L., and K. Ruhleder. 1996. "Steps toward an Ecology of Infrastructure: Design and Access for Large Information Spaces." *Information Systems Research* 7:111–133.

Tolar, M., and E. Balka. 2012. "Caring for Individual Patients and Beyond: Enhancing Care through Secondary Use of Data in a General Practice Setting. *International Journal of Medical Informatics* 81 (7): 461–474.

22   Thundering Silence: On Death, Fear, Science

Eevi E. Beck

Being

Bodynote[1] from autumn 1996

It was my first time. i felt out of place and had made sure i entered the flat last. i got an eerie feeling—the plain white walls and strange spaciousness of the hallway, the wide doorless openings into more whitewalled rooms—oh the others have finished taking off coats and shoes, they're disappearing through the opening, guess i should follow them . . . a quick glance as i follow them through a room, yes it is empty; from another wide opening sounds & light come. . . .

> *gingerly i turn the corner.................*
> *and freeze.*
> *staring in disbelief*
> *straight into*
> *a*
> *bottom,*
> *naked.*

Someone in bed, turned away from me, the home-helpers busy around her, in charge of events, no consternation, this is everyday stuff.

   what—am—i—do-ing—here!?

   i want to avert my eyes but they are fixed to the sight in front of me. The woman in bed, the softness of the voices asking her questions contrasting strangely with the speed of their measured movements; the sheer routine of the calm busyness around her. No-one takes notice of my presence. i tell myself they need to be here, they're helping her.

   But i—i'm not here to help. An intruder; i'm just adding to the number of people seeing her in this undignified situation, without offering anything.

   She probably hasn't noticed i'm here. Even as she turned and seemed to look this way my presence made no discernible difference to her.

That should surely make it even more important that i take responsibility, to protect some-one's integrity who isn't doing so herself?

i'm not here as a voyeur, i care. I'm here in the name of Science.

—but Science doesn't care, Science is all too happy not to know about this. So how on earth can i defend my intrusion?

The above is from a field study of home help service within a Swedish municipality (Beck 1997, 2002). The overall aim had been to explore implications of municipal "information society" policies, including on marginalisation processes (Beck, Madon, and Sahay 2004). This moment, however, and my struggle to make sense of similar ones turned my priorities: no longer could i sustain focusing my questions on the lack of computer use in the home help service. Bodily decay became of vital importance: how to make sense of the distance between my way of relating to it (awkward, distancing) and that of my hosts—the home-helpers? Despite its pivotal importance, however, in the three publications from the study i never mentioned the event.

Glimpses of academic life such as this are rarely presented when scientific studies are written up.[2] Their invisibility to most science and to predominant notions of scientific truths, of scholarly work, hides worlds. Sitting in the middle of the daily experiences of many researchers as they do, why have they not been subject to more interest? Behind lies frustration at the poor ability of science to address issues or experiences such as the one that opens this chapter (most strikingly when there is an expectation to "formulate researchable questions")—and curiosity as to why this has come to be. Maybe it is due to lack of interest—lack of relevance to *theory*,[3] lack of theory of relevance?

In developing this chapter i kept returning to questions troubling me, such as how these nonaddressed and conventionally perhaps nonaddressable questions reflect and co-constitute science.[4] This is inspired by Leigh Star's ability to open for investigation areas that had not previously been thought noteworthy.[5]

An aim is to heighten awareness of boundaries of science, a means transgressing a number of them. The form, then, is an experiment in locating the building of theory in the heart-minds of researchers.[6] And the chapter is thus intensely interested in vague areas of science; where the solidity of theory seems to peter out; where scientists emerge as eminently human in our splendor and our fallibility. Why would i want to write this way? Because . . . i want to contribute to expanding the sense of the world. i believe, or have experienced, that some people—academic researchers and lay—will at times see only that which science can encompass. The world is so much more interesting, mysterious, and unfolding, than that which can be explained or fathomed by science. My yearning is for more modesty among those of us who produce and consume science. Not that modesty is never present, it is, but often it gets lost in translation, in abbreviation, hardened. Why this interest in the shortcomings of science?[7] Why would examples far away from science even be of relevance? It is because of the dominance of scientific ways of posing answers and the questions that would fit them.[8]

My heart and mind are drawn to attempting to articulate a nameless gap between experiences as a scientist and the ideals of my scientific training. The gap is particular, unspeakable, yet a product of some very ordinary cultures, including the sciences. Many scientists *in fact* habitually delimiting ourselves from swathes of empirical non/fact, including our own direct experiences. While this is not particular to science, my interest is in its particular expression as science, as part and parcel of predominant notions about how to come to know things in such ways that they can be legitimately presented as knowledge.[9] "To be radical, an empiricism must never admit into its construction any element that is not directly experienced, nor exclude from it any element that *is* directly experienced" (William James, n.d., cited in Myskja 2012, 5).

Sixteen years later

Sixteen years after the above visit, images are still vivid in my mind. Not necessarily correct, but vivid. The bedroom's minimalist elegance (lacking from other equally sparsely furnished flats in the block of purpose-built flats). Her young age, 42 or something. The framed photo of her son on the bedside table; the only one she showed reaction to (. . . as the home-helpers demonstrated for me by speaking his name, to which she duly reacted—a passing aliveness coming over her—which made me only more uneasy about the ethics). The softly spoken "he rarely comes to visit her;" the vast, silent pain behind. i'm convinced they did not wish to mock her but gently teach me something. Not sure i did learn . . . ?

i never found an answer to my burning question of ethics; not an articulable one. Rather, the purity of decontextualized "Ethics" dissolved into the practice of revisiting her and other recipients of daily home-help; each with their stories. i learnt that in any new situation my best action was to wait and watch, see what the quiet wisdom of experienced home-helpers produced, born from the marriage of experience with caring, strained through the sieve of strict time-keeping. After a few visits this woman motioned interest in me, asking the others who i was. So, she had noticed. But an experienced help-recipient, she'd wasted no attention on me till i turned out to be a returner. "Oh, she's just here for a look" would've been their response; easy, apparently forthcoming, but uninviting of further questions.

A Culture of No Culture

Critique of the claim to uniquely represent truth that can be read into much scientific rhetoric has come from various places. Feminist thought, science and technology studies (STS), interdisciplinary/cross-faith dialogues on consciousness (e.g., Wallace 2000;

the Dalai Lama's conferences), examination of method including from disability research/special needs education.

In Beck 1997, i articulated the possibility that computer systems—plastic as they are—might have been developed from different rationalities, addressing different kinds of needs than they commonly do. The question *why don't they?* stayed with me, taking me into the backwaters not only of computer science and nonscience, but of science, of the academy.[10] In contributing to an archeology of the lived lives of science, of "knowing-work," the promises and fulfilled wishes of research are blatantly highly visible, at times literally mass broadcast, and most of the time revered. My need for excavation, then, is of the non-advert stuff, the unpleasant mess of some everyday lives and work as academics. In STS many have looked at such matters in terms of the makings of "scientific" "truths" (among numerous examples, see Akrich 1992; Star and Ruhleder 1994). From these, i have learned looking beyond received boundaries of what "science" is about. Pushing further, my interest is in exploring the workings of "non-truths" in science and the rest of the academy—that which is excluded, nonseen. Thus i am following the lead and the inspiration of Donna Haraway and especially Susan Leigh Star.

Shapin and Schaffer present an origin of experimental science in Boyle's seventeenth-century vacuum pump (cited in Haraway 1997, 23). Extending this, Haraway 1997 discussed the deletion of voices that contributed to the contraption and to establishing it as an instrument of science. Boyle's *approach* (and Boyle) came out victorious from, in this depiction, a power struggle with elder, more introspection-oriented approaches.[11] Within the STS research area, studies have discussed contemporary practices among scientists of apparent deletions of their own values. This is at times referred to as the "Culture of No Culture"—in other words, a culture that professes and believes in its own 'neutrality" regarding beliefs, values, and so on. As a young academic, i learned (usually by reproachful comments) about the strict limits of the genre of academic discourse, including subtle ways that the researcher's voice (or chemistry lab discoveries having happened by accident) was to be concealed. Yet, studies such as those by Bruno Latour, Susan Leigh Star, and many others document networks, power struggles, the pivotal role played by infrastructure, and so on, in science. Donna Haraway thus stands on firm (well-documented) ground when exploring Boyle's legacy of deletion of the self as alive and well today.

Un-learning [*shifting levels -consider-reducing/deleting-this-section-*]
The chapter is a patchwork, or a weave. Both metaphors involve threads not visible (hidden stitches, and the warp, respectively) and a pattern that emerges through repetition of parts that themselves do not carry the greater image (bits of cloth, and the weft). A complex rhythm emerges from simpler ones repeated. One rhythm is that of seeing, facing, disconnection. By speaking the true name(s), the positive and negative, of

science; merging, till the positive/negative dichotomy holds (or is given) no power. Ursula K. Le Guin and Thích Nhất Hạnh—writers from America and Europe/Asia respectively—both speak of the power of the true name that is beyond opposites (Le Guin 1968; Nhất Hạnh 1993).

The culture of no culture is under pressure, at least at the fringes. And this chapter wants to contribute by challenging—not least myself. A process of un-learning. Example:

Leigh had invited me to write a chapter for *The Cultures of Computing*. i did, based on my Ph.D. work on collaborative writing practices across distance among researchers. The first i knew of the other contributions was when a book review was forwarded to us. The reviewer described the chapters in order of appearance, stopping after mine, which was briefly described as "scholarly." After initial relief that my writing was good enough, on second thought i wondered why the reviewer had stopped reading. Could this term be a polite way of saying "dry," "providing more of the same kind of stuff that scholars keep pouring out"? Irrespective of the reviewer's actual intentions, the lesson hit hard: That other ways of contributing in the academy may exist (as Star had been trying to convey to me), ways other than empirical traceability of claims, clarity of logic of argument, and so on. The very possibility shook my ground, i dared not fully take it in, i barely dared breathe.

i had no words then. Because in my culture of no culture, i should strive to be a Modest Witness (Haraway 1997), show no self. Early in my Ph.D. work i was exasperated to discover that if i had an idea, i was not supposed to write it except by citing an existing publication. Experience and Truth existed on separate planes. i lived and learned. A number of years and contemplative practice has been needed to un-learn this.

In Fact . . . Deletion of the Self in Science and in Buddhism

Science and Buddhism appear to share an ideal of deletion of the self, yet my experiences of the two do not overlap. How come?

When we are part- or full-time Modest Witnesses we are not meant ever to ask how practices of deletion of the self from science functions. Or what it may be like to live, love, suffer from, and practice the "culture of no culture." This is a beam of light i wish to help diffract (Haraway 1997); to explore some of the constituents through the prism of my experience. "Experience" includes reading academic papers—how could this be exempt?—that gnaw at the distinction between experience and truth. Diffractions might include:

• What are places where fear enters the practices of science? Fear of fact, fear of non-fact? Fear, even, *of our students*. While not every teacher fears, it's a familiar aspect of teaching—known as nervousness (Palmer 2007 writes beautifully of this). STS research has not to my knowledge studied the passing on of the "culture of no culture" through teaching. Yet, is not the ordinariness of Modest Witnessing, of the negation of non-culture, itself a feat? It is the passing on of a specific form of non-self, of deletion of experiencing, through ordinary practices of teaching at university and elsewhere.

• When considering two phenomena, Buddhism and science share an ideal of equanimity: not valuing the one as higher than the other. When exploration of non-self is to be practiced (not necessarily completed), could gentle self-observation of Buddhist practice coexist with the "compare and contrast" of the scientific? Could techniques of comparison, splitting into parts, examining bit by bit what's similar and what different, somehow be brought into merger with deeply nondichotomous practices?

~~At the time of writing, doubt seeps in in the shape of comparing my intentions with advice i've been given to handle the genres of academic writing and presenting a paper: make one point only. This is such a strong imperative that if i find myself with more than one main point, i am to drop all but one and/or turn them into more than one paper (each with one easily identifiable main point).~~[12]

A leap into faith: what would it be like to write the dregs, permitting stuff that didn't fit into previous well-disciplined writings?

Some of the Dregs (aka Proto-analysis)

Being, embodying. The work to make inconveniently visible fluids of the body stay invisible so the person in receipt of home help can be socially presentable. Such work is done either in the privacy of the family (Isaksen 1994) or by home-help workers. One of the ways that social life gets ordered is through the ordering of our bodies, in this case bodies that fail the norm of discharging waste only with the permission of the social/mind. My own immense confusion upon initial and subsequent encounters with this work: how to make sense of the distance between what to my untrained eye and heart seem to be boundary transgressions with other people's bodily functions, and the plain fact of this being in the service of the adults whose bottoms were being wiped and diapers donned? The work at times involved beautiful moments of mutual recognition, gratitude, while it also could include visiting places daily where the troubled person one was helping would hurl verbal abuse. The distance between, on the one hand, the provision of institutionalized care—with sensitivity and way beyond personal comfort; the importance of this work to keeping social life looking as "neat" as i was accustomed to; and, on the other hand, the low status, pay, and say of those doing the work and the high-status objects, the computers, that stood there dumb with plenty of expectations. How to absorb and transform my pain in attempting to make sense of that immense gap, theoretically and as a human? While Donna Haraway's notion of *diffracted rationalities* (1997) provided one lens (Beck 1997) and lack of participation another (Beck 2002), the enormity of the misfit between what i believed i was meant to be looking at—being, after all, a Computer Scientist studying the creation of an Information

Society—and what i saw staggered me. What kind of society do we have that renders invisible such basic issues, experienced by huge numbers of those fortunate enough to live into old age? And while plenty of attention, money, and so on, goes into technical means of ordering? Later, the issue came full circle as reports started appear arguing that there was no need for home help workers to shop for their clients as this could be done more efficiently through a centralized phone order service and delivery companies. How many elderly with an already poor appetite experienced received the wrong kind of bread because they remembered what the wrapping looked like but not the precise, decontextualized info needed by the person at the end of the phone with their huge list of available foods in front of them? One home-helper asked, peering into someone's fridge, "the usual treat?"—the relation implied in the question being for some as important as the help in retrieving the object. Who would now be asking? Another helper asked, into the air, who would now gently encourage someone to eat a little more, as the doctor had prescribed? She used to sit down by a help recipient's table at meals, initially sharing food, and when no longer allowed to share she had started bringing her own. Now she was no longer allowed to sit down.

A further detail: During the three years of my periodic visits to the home-helpers, those recipients who could be granted the time for someone to spend time with them just talking went from "only a few, those above 80" to "almost none."

And when i started my fieldwork, when visiting a home, we would take off our shoes in the hallway as is customary in Sweden. A few years later we no longer were permitted time for such luxury as tying our shoelaces, slipping instead blue plastic covers over our shoes.

This catalog of gradual erosion of relations, with and without the support of computer systems, and in the service of the efficiency of enumerable actions, is far too common. While in the world some die from lack of food, others die in front of food from lack of love, lack of connection.

Erosion of relations was evident also in the relations between the workers and their superiors, as i analyzed in Beck 2000. Most of this, however, i've been holding for years. Holding back the pain of witnessing this as a daily reality, releasing only what i was able to construct a scientific discourse around. Thus i've not only been a well-behaved scientist, but have also passed on what to speak and how, and what to silently retain in my body-mind.

~~What kinds of permissions to be am i holding back from granting other people, both those considered my juniors and my peers? How does holding shape my thoughts/speech/action?~~

~~Can this be brought into Eevi the Academic?~~

~~("Når hodet er kropp~~

~~ikke sinn,~~

~~drønner fossen~~

~~rett inn")~~

=>Silence

Among ways in which those of us who work as researchers constitute a "culture of no culture," originally brought to my attention by Leigh's writings—is silence. Silence as a vehicle of power, silencing, but also *Thundering Silence*: facing up to theory, terms, words, categorizations, by employing something beyond their grasp.

Pastry

One home-help recipient was not satisfied with the usual, evasive answers to who i was. She was the only recipient i ever told i was a researcher. Back at the group rooms—the space for home-helpers to meet, rest, and talk—she was considered one of the "difficult" ones, due to her behaviour especially before medication took effect. Yet she was the one who often had bought a pastry for the visit and invited us to sit down for a coffee. Futile, as she always received the same answer: Sorry, we don't have the time. After visiting her a few times, i stopped appearing as my fieldwork visit was over. A year later i was back. Unannounced, i walked into her flat—and looking into my face, she remembered even before medication took effect who i was, including details of my research interest and family health situation. i believe she told me i was the only visitor who ever had time to talk with her.

Death follows on the heels of home-help service. She was the one i mourned upon receiving news of her death. The pastry always on her kitchen counter, always ready to be served: the gift we never let her offer us. Silent pain, screams unscreamed.

September 2011: "This Is Not a Performance." Dancing an Academic Paper

Waking up early, wide awake. Sore-throated. Have a drink of water. No better. Try sleeping more. Nope; too restless. Next "suspect": nervousness. Bingo. Or fear, as my heart-mind knows it as: the quality of the feeling is fear, its quantity small, the context such that we commonly label it "nervousness."

This fear is fine. We have been companions for a while, have become friends. Still, the wave of nervousness surprises me, but less so as i let go of the idea that i shouldn't be nervous for this talk.

Yet my sore throat means i cannot speak. Well, if i can't talk—or if giving the talking will make me ill—i won't. An impulse to go ahead even so flashes, a heavy resigned "what i really feel like doesn't make a difference anyway, i just have to go through with what i'm meant to do." Just as clearly i allow a response to arise "No. This is not what i came here for. This is not Leigh's legacy which i travelled thousands of miles to explore. i will not push myself."

Someone could read my slides. Fear objects: But there is no "red thread" in the written form. . . . So be it. Hey, this means i can move, i don't have to stay at the roster (remembering a camerawoman disciplining the speakers yesterday), i'll be free to point to parts of the screen, free to move . . . FREE TO MOVE. That's it, i shall move! i will dance.

i find someone to be my voice, whom i trust with my vulnerability and who can carry some silence, some dancing. We get busy in the breaks, i show her the talk slides, write a few changes. Respecting my nervousness i ask for an additional smiling face in the audience, a gentle presence. Fear and connection in & out.

Lunch break. Stefan Timmermans reads Leigh's poem aloud (Star 2010). i read my response. A dear friend of Leigh's knows Leigh's poem and takes in how i end mine. That "glow" from stillness is the deepest preparation for dancing. Inside and outside in harmony. Nothing to hide. Deep connection within; non-fear, wholeness.

Judith offers my words gently to the audience. Inner connection expanded to include her. And dance i do.

First, move the title, moving with the "Thundering Silence" book. Sensing the audience the thought comes that no one will understand the relevance, "probably stop doing that, just move." Dialogue with present: connection with perception of immediate outside.

Then the first section; pain and healing and how they go together, the silent lifting of flowers to the sky, i am the flower, the uplift, the sky itself, the people calling for and enacting healing, all people yearning for and willing healing from fear. Connection with the good will of humans.

Second section, on the screen an image i am intimate with: i am the old woman. Her hands in her lap. i've known for eons—minutes—i'd dance her half-standing, crouching down, clutching hands; then a quick smile upon the mention of happiness; then back in the down, down, down; the heavy "down." Next, preparing to move again, i give her, me, permission to open up, upward, outward; then as my body starts letting go. . . . Suddenly i sense a shadow of . . . it's a cry, devastation; instant recognition: "Oh it's you?"—a decision to permit an old knot to be untied: "have i been holding you, hiding you? Come, join us!"—and the scream splits the air: aaaAAAAAAAAArgh! The pain, the despair wells up from the centre of my being, rolls towards my mind then turns outwards, out of my gaping mouth, the vehicle of the outward expression; the years-old despair. (In passing, my mind recognises it as familiar, but not whether it is mine or non-mine—it is both.) Connection with pain of humans; despair.

In the corner of my eye a conference participant on the front row jumps: "i should be a little compassionate with my audience too." Dialogue with present.

Next, with deliberate movements i pick up a woolen shawl, wrap it neatly around me; i am now the pained person who performs normalcy in public. Walking as if nothing had happened across the 'stage' set by the projector's light beam i glimpse the effort of performing "public normalcy" and "long for" a "home" to retreat to. i enter the space behind the rostrum, unwrap my shawl, and—making the choice of emotional clarity rather than physical accuracy—i "half collapse." This i hold for a moment . . . then let go of the scene.

Third section onwards i illustrate with no sound, only gentle movements, for the people on the front row. i move up the aisle, creating more 'front rows;' now being with the audience, the spoken words receiving centre stage.

Performing science

This was not a performance. It had none of the distance of a professional performer (had not been rehearsed and could not be repeated), and all of the performance distance of a professional academic. It was me, a scientist, doing science; doing inward work (listening inward and dancing that; also, permitting the greater purpose to become more important than instants of fear); doing connections between past relations ("observation") and an idiosyncratic distillation of importance (creating "findings" from "data") based on long familiarity with the reasoning processes of science (i.e., an informed following of the data, including self-awareness of breaking with conventions); presenting it for an audience of peers and thus placing the data and findings open for alternative opinions. Only the minor detail of nonverbal expression—silent and at one point voiced—made this unusual. Yet it made a difference. Testifying to the narrowness of scientific conventions, both the silence and the scream em-bodied more than em-intellectualizing/em-thinking/em-categorizing.

The glimpses of home-help work seemed paradoxical as long as i was looking for straight categories of either/or; yet opening new worlds when viewed through lenses that transcend dichotomous thought: In the home-help work, while emotion plays a part, there is a merging of emotion with intellect, a heart-mind. i see a careful balancing of structures (e.g., faithfulness to the system of strictly allocated time, a sense of responsibility to recipients, colleagues, management, etc.—so many layers of meaning) and relations (which can include emotions, and which despite and because of the highlighting of connection presupposes distance and disconnection). Attempting to make sense of this, i once labeled it *rationality of care* (Beck 1997). Yet . . . "rationality" and "caring" both miss that which transcends labeling, which resists words.

Home-helpers and those receiving help, together hide pain (both physical and emotional) from public view. This is an important function, yet poorly acknowledged in terms of status, salary, or being listened to. The pain and suffering of the situation stays with those nearest, they in turn are forced to dehumanize their actions, creating further suffering.[13] Consider the following example:

Back at the group rooms i had inquired about what seemed the legitimate needs of a lonely person: at 60–70 she was too young to be granted "X mins of conversation" on the list of services from the home-help service. Such decisions were made not by the workers but by the manager, who sat in a different building and rarely visited the group rooms. The manager made home calls to assess needs and grant activities according to the regulations. She had been my main contact into this fieldwork, and talking with her i had no difficulties seeing her point of view. But the distance to the world of her immediate subordinates i found increasingly difficult to fathom, as its implications dawned on me. Yet they all were parts of a system and despite its weaknesses the system was better there than not there. People did receive attention, at low cost. The manager was only doing as she should. Such spoke my rational voice. But the deep pains remained untouched, the layering of insulation against the vast pain that was the life-blood of this system and which it was meant to heal, yet (literally) systematically distanced itself from. Insulation between the everyday lives of the recipients and those making the decisions that structured their key contact with others. Sys-tem-a-tic silencing, deletion of pain, from experiences and records.

Why the Shortcomings of Science?

If science/the academy includes a project of building, maintaining, fortifying, extending the skeleton that holds the body of society in shape, which enables dramatic movement, and which above all holds the head-mind high and encased in a hard shell, then my interest is in its soft underbelly. That texture which enables something else to seem hard, clearly delineable, dependable, predictable.

As Leigh taught me, softness is more complex. Less repeatable. Does not lend itself to experimentation and measurement (plenty of evidence to which can be found in a century or more of research into educational methods, uncomfortably poised between the rock of scientific approaches and the hard place of political interests in short-term applicability. Very little that is precise, enduring, has come from this, other than that things are complex and that simplistic ideas don't get us very far, because, i would contend, its nature is more that of the water flowing metaphorically between the rocks on its way somewhere else, and never twice the same . . .)

This "soft underbelly" can be thought of as a counterpoint to a "hard" surface of science. My interest is, however, in the places where these melt into each other: the softness within science as a practice. This is what seems to me to be the gift of STS to the academy: to see our work as just that. Our own work. STS studies are not necessarily soft. But STS has provided an opening—a Trojan horse—for examining softness in science. Leigh Star and Geoffrey Bowker studied what (hardened) systems such as the ICD (International Classification of Diseases) do with the stuff that doesn't fit (the residual category) and is a case in point. Their work demonstrates both how there always is some stuff that doesn't fit and how this necessitates some management or other for the system to exist. Therefore, residual categories co-constitute categorizations as presentably neat, with in Leigh's words "hardened categories." In their 2007 paper they take this further, examining their personal experiences of residuality (Star and Bowker 2007).[14] Here, their writing is pushing the boundaries of what can be considered an academic paper by "outing" (exposing) themselves as living beings, academics (scientist in Leigh's case) with living, thinking, feeling bodies, with pain and with uncertainty. To me, the question posed between the lines is what does this do to their (Leigh's) scientistness . . . approaching the soft underbelly as a necessary constituent of hard science, within ourselves as scientists.

Fear and Beyond

One rhythm underlying this chapter is fear; facing fear. Academics' fear as we go about our trade.[15] What might it be like to live in fear? Ursula Le Guin has given a description in her *A Wizard of Earthsea:* Ged, known as *the Sparrowhawk*, is chased by an unknown, evil being he calls "unlife." A monster, which even knows the *True Name* of Ged and thus has power over him—while Ged does not know the True Name of the monster. Ged is almost caught on several occasions, flees to exhaustion, this way, then that, till he ends up at his first master, Ogion the Silent. "You must turn around . . . [I]f you keep running, wherever you run you will meet danger and evil, for it drives you, it chooses the way you go. You must choose. You must seek what seeks you. You must hunt the hunter" (1968, 120).

So, being driven by fear is a consequence of not facing up to it. A question then is: What might it mean to hunt the hunter of fear in science? My journey goes its uncharted ways. Hunting that which hunts could be looking at non-fear where fear was expected and massively present, such as with public mass murder.

Publicly causing death

The deep significance of willed murder, conducted in numbing numbers in full view of the public with the purpose of shocking us, its audience—as took place in New York City on September 11, 2001, and in Norway on July 22, 2011—lies not so much in the fear it arouses, nor in the perceived need for restoring "order" or "safety," through incarceration and punishment (or ostracizing) of the offender. Neither does it lie in debating the extent to which such measures are really needed, important as those debates are. The greater significance lies in focusing the meaning of life. And not just any life. While "surviving affirms the value of life," it may merely reproduce what was before that produced the events in the first place. Reaffirming the political position of others' death. Rather, life-affirming, spontaneous, generosity of heart; love; giving without expectations—these go by many names, choose one that speaks to you—by infusing life with meaning; with relations worth living.

Walter Benjamin ([1955] 1999, 73) wrote: "All purposeful manifestations of life, including their very purposiveness . . . have their end not in life, but in the expression of its nature, in the representation of its significance." And many help create community after atrocities, taking part in the generosity, exercising of love, softening of hearts and voices . . .

This is from a note i wrote for friends outside Norway in 2011:

The summer's dramatic events in Norway affected us in more beautiful ways than anyone could've foreseen. On 22nd July, we were in Plum Village, the meditation centre in France where we go each summer. . . . We heard of the killings, but also of the silence in the shops. Of people gathered to show their silent "no," of openness and inclusion as a response. When we arrived in Oslo a week later, the atmosphere was different from ever before. Palpable love, between strangers. Bus drivers didn't want to take our money. People of colour and of little colour talking with each other as if previous barriers had evaporated.[16] Flowers hung everywhere in the city. Listening to friends and colleagues retell their experiences of the 25th July gathering of 200.000 people made my heart sing. At short notice this had been turned into a flower gathering, and at one point, spontaneously, the flowers were lifted to the sky in a massive, silent wave. The sheer softening of their faces as they relived those moments, the infinite gentleness with which the word "healing . . ." was spoken, will stay with me for a long time. . . . Born from deep pain, it remains one of the treasures of this year.

Other golden moments included hearing of the Prime Minister, whose colleagues and friends had been killed, broadcasting the wisdom of survivors, such as a young woman's "if one person's hatred can be this powerful, how much more cannot our combined love achieve?" The Mayor of Oslo, from a different political party, spoke to a gathering in a mosque, saying "i'm a white man from the West of Oslo [the affluent part, from where the killer came], yet you don't look upon me as a killer. For this i'm grateful." The Crown Prince, saying to the crowd "this evening, the streets are filled with love." Each showing leadership in picking up on and reflecting back some of the beauty on the streets. When friends outside Norway gently wished us that things would be returning to normalcy soon, i was hoping that'll take a long time. That we won't forget the message of love. . . .

My awe has deepened as i was told of similar spontaneity among e.g. New Yorkers after 11th September 2001. They also went gentler, more appreciative of each other, more willing to help. (Unfortunately, collective ways of expressing and nourishing this seemed not to be supported by the leadership.) i come out of this summer, then, with strengthened faith in humanity.

Science may ask about faith, about fear, about silence; may tally the numbers of people who report various effects. This is a long way from allowing ourselves to be touched, in our heart-minds. It is not only a matter of emotion, but of intelligent caring, of choosing the humane, and of speaking through silence when words do not suffice. *"Being human is the cracks between"* (Star).

Earthsea: fear re-enters Ged

As advised by Ogion, Ged has started chasing the unknown monster. He doesn't know where it is but follows his hunches. Here he is on a sea journey: "At the sight of [the land] fear had come into him again, the sinking dread that urged him to turn away, to run away. And he followed that fear as a hunter follows the . . . tracks of the bear, that may at any moment turn on him from the thickets. For he was close now: he knew it" (Le Guin 1968, 135–136).

So, he knew fear, felt it, felt its power, wanted to turn away—and used that fear as his guide.

death

floating outside the perimeters of science with a stronger power over its dominant notions than those notions would grant it, is the hard fact of death. the ultimate challenger of perspectives, of vision in the first place, of thought. in more than one sense, death[17] produces life as finite. death is the ultimate marker of what we do not know, of nonfiniteness and therefore of the finiteness of projects such as science.

by touching the infinite (whether conceived of as infinite existence, e.g., in relation to some God, or as infinite nonexistence), the issue of death breaks all scientific conceptions and laws. death looms over each and every scientist as much as over anyone else, and though some may try for glory to achieve intellectual infinity, the illusive comfort upon which such a project is built would be shattered by the demands of the decaying body of anyone who does live into old age. what good is a great idea that made others admire your mind if that same mind can no longer provide the most basic control of the discharges of your body? or if the mind cannot remember the names of your nearest and dearest? such matters we pass by in silence. not relevant to scientists, unless made objects for research, e.g., into an intellectual or chemical cure for some detail in the picture. hushed messages passed between colleagues of someone's exit from the fold (i.e., no longer to be reckoned with, often years before their death), speaks tons of silence. such experiences do not even belong in lunchtime conversation. yet, the silence roars.

why would science have difficulties dealing with death? do i appeal only to cultural reasons? its methods build on dichotomizing; an approach so fundamental to all the academic subjects i have visited. throughout, i see categorizations, delineations, sometimes definitions . . . there are exams, grades, and the assessment of submitted research papers; review procedures, quality assessment exercises with and without financial implications for the department/individual . . . and nervousness not to make it past some milestone among members of faculty as among students and administrators.

while tracing his fear, ged later stops by a friend's house and is asked: "'What of death?" . . . 'For a word to be spoken,' Ged answered slowly, 'there must be silence. Before, and after'" (Le Guin 1968, 152). so the silence which is a prerequisite for speech and naming, illuminates death.

"'I think this voyage he is on leads him to his death,' the girl said, 'and he fears that, yet he goes on'" (ibid.).

mourning and ethnographies of science (Star 2010). how do they intersect? the more i look the less i know. what is mourning? mourning the love of my life, my late husband? who was always there, receiving me at home, and who helped me find my own home inside, which among other things has given me great courage as an academic? whom my deeper habits and longings at times still refuse to admit the head's insistence that he really is dead? mourning the mentor who showed me that life in the academy could include being vibrantly alive: leigh? with whom i'd not been in touch for a long time when the sad news reached me? mourning = daring. diving into mourning = freeing, into pain, into liberty. no hesitation. no longer waiting, putting off until the imaginary day when conditions will be right for . . . (what? living as a free person?). no putting off. death—a teacher with clarity.[18]

"no death, no fear" (nhất hạnh 2002)—and sometimes, "death, no fear."

do i ever fear scientific death? or bodily decay, the nonnegotiable signal of approaching bodily death?

(when) do i allow fear to accompany me as ged continued *with* awareness of death, *accompanied by* his fear? (when) do i turn away, busily pretending it's not present? ged, remembering words spoken to him by ogion: "'To hear, one must be silent . . .'" (Le Guin 1968, 157).

the infinite silences marked (to the living) by death . . .

. . . are the ultimate example of thundering silence. the prospect of death speaks back at our multitude of thoughts, ideas, theories, with a silence beyond descriptions, beyond comprehension, beyond breathing. attempts by artists and religious people to provide words or images fall short even of the experience of witnessing the death of a loved one, so certainly these cannot capture the actual experience of death. death has until recently remained the provenance of the spiritual, of religion. these have touched the academy through the clergy. a glossy research magazine included humanist/theological studies of ancient views of death. in it, an unnamed editorial voice: "Death is one of the great mysteries of humanity—where academic research falls short: There is no empirical data to relate to" (Apollon 2015) indeed. working with the dying has in recent years emerged as an area of interest beyond that delegated to clergy (e.g., work pioneered by elizabeth kübler-ross ([1991] 1998), also stephen levine (1982) and others). one journalist (Shroder [1999] 2001) has collected stories of remembering beyond death—stories from contested-ignored research. and what of foucault's interest in death . . . it resonates with me. writing obituaries gave value to friends' deaths. also, his fascination with a story of a young aristocrat whom the soldiers saved, and whose brush with death must have left a mark on the rest of his life in, it seems that foucault insists, important but inexpressible ways . . . [in that small library book with the red and gold cover and the text gently layed out dual-page in english and French—(i remember it was classified as literary critique, but not the details i need to cite it; too many hits for foucault in the library search and can't remember the key author's name—the pressure is on/in me, therefore, to delete this section)]. nothing can convince me foucault didn't know something in the depths of his heart, hadn't experienced something.

Earthsea spoiler warning: Ged stops and turns

As Ged braves falling off the end of the earth and finally catches up with the evil being, "breaking the old silence, Ged spoke the shadow's name . . . [who at the same time spoke his]: 'Ged' (. . .) Light and darkness met and joined and were one" (Le Guin 1968,

164). "Ged had neither lost nor won but, naming the shadow of his death with his own name, had made himself whole" (165–166).

Dis-connectedness. Connectedness.

The Ged character illuminates how seeking out a fear that drives me, drives cultures i live by and shape, may not constitute working for their demise but for their wholeness.

Looking at unspoken pains of silence at the mercy of caregivers who themselves lack permission to spend enough time to be merciful. Looking at unspoken pains of teaching . . . the soft underbelly of which includes the passing on in teaching of caring and noncaring, of chopping up the world according to a tradition. Connecting with, becoming, part of a tradition, by dis-connecting smaller and smaller pieces of the world.

J'accuse

J'accuse: my accusation against science—and therefore against myself, against a part i love and care about, and people whose hopes and aspirations touch and impress and scare me—is, in the end, this: that science, as a system of thought, does not include its own limit. Granted, its practices (and some meta-theory) do include the shedding of old truths, but in favor of new ones.[19] Its theory doesn't allow for its end, has no hint of worlds outside.[20]

So my yearning for a meaningful relation with death, at a first glance perhaps naïve or just quirky, includes examining the limitations of scholarship: because when keeping the reality of death in mind, in body, in motions and emotions, in fact: considering academics habits of structuring what it is permissible to ask questions about, *how could it become possible/imaginable for us/the academy to point us toward the end of all categorizations?*

Of the (Deeper) Purpose of Science

Based on the purpose of life quoted earlier, Benjamin saw the purpose of translation (he was a literary critic) as expressing, in embryonic form, the relationship between languages—as i understand it by enacting a fraction of it. What, then, is the purpose of science, of academic scholarship? It proceeds by establishing distinctions—between legitimate and illegitimate ways of arriving at knowledge; between person and facts/actions. Separation; *between.* For example, between a natural scientist and facts she may legitimately establish; a social scientist and his critique of theory; a philosopher and ways she may establish a question as relevant and legitimately investigate it. Thus, science creates knowers and known.[21] The institution of the university is geared toward this creation as a never-ending process: a rotation of knowers is maintained (initiating new ones through teaching while established knowers decay through sickness and age or changing the focus of their efforts whether voluntarily or through being discarded as inferior in selection processes); means of disembodying knowledge are established and patrolled (tearing knowledge apart from the knower, forcing externalization through systematically valuing the semi-permanent media (print and its lookalikes) above personal means of transmission). This much is clear. The question is: what for? What purpose does this separation, the establishment of knower and known serve? Is it, like translation, an embryonic expression of a hidden relationship? Is its purpose to manifest the importance of such a relationship?

On translation ultimately serving the purpose of expressing the central reciprocal relationship between languages, Benjamin continues: "It cannot possibly reveal or establish this hidden relationship itself; but it can represent it by realizing it in embryonic or intensive form." What then is the significance of science, that makes me stay, makes others stay, makes us want to be part of it? How do i represent, for, with, and through science, "hidden significance through an embryonic attempt at making it visible"? What would a science based less in adherence to genre and more on, for example, awe, be like?

Attempts at Living with Less Separation, Less Category Hardening

Language relies on categorization, and categories can as Leigh would remind listeners, be more and less hardened. Academic language i often find hardened, in at least two senses: specifically delineated/defined, and in terms of importance. i engage in practices that uphold this when i react to others' usage of certain terms. When i put energy into this, i co-construct the importance of such ways of perceiving the world. i also have experienced glimpses of what existence beyond categorizations can be like. The following are attempts at communicating some of these. Whether or not such moments can extend to the entire life i don't know. And i have no need to know; an answer could be closure, apparently settling an issue that is beyond closures and finalities.

Letting Be, Letting Go, Letting Come

At times, i have no need to sort things out. Whatever arises in me is fine. It may not be comfortable; that's part of the package. Sadness is ok when i see it for what it is and don't resist its presence. The mind spinning off on some strange tangent is ok when i see it for what it is. The impulse to do or say something unhelpful is ok when i allow it as just an impulse of the mind, which i do not have to act on. Joy is ok when i just enjoy it as it presents itself and don't try to hold on to it when it doesn't. A deep sense of community, of non-separation, is beautiful when i don't damage it by trying to make it last.

There are times when the dancer fills all of my experiencing. Contained by a sense of responsibility toward myself and others, allow all. Trust. Let be, let come, let go. There is only movement—big, small, wide, narrow, fast, slow, subtle, dramatic. Whirling movement. Still movement. Flowing movement, soft movement, sharp movement; movement on the floor, high up, in between; moving old pain, moving new joy. Familiar movement, repetitive movement, movements i didn't know could live in this body. Fear. Joy. Anger. Bottomless pain. Lightness. The falling away of all labeling, just being. With others, with myself, with-myself-and-others—no separation. An experience of following. Following. Following. Each moment afresh. No prediction, no knowing; the start of one movement turns into another, then gives birth to the next. At some point thoughts appear and the movements appear as separate, from each other, from myself. . . . Returning to following, movement re-melts into a flow of muscles tensing and relaxing, one arm giving way to the shoulder giving way to the spine giving way to the lower back and hips giving way to the legs giving way to a foot giving way to the head.

Postlude: Acknowledgments

For readers[22] who have read this far, my gratitude.

May words that don't seem to make sense be forgotten; words that do, likewise.

Because my embryonic attempts are at touching—in fact, in confusion, in peace, or otherwise—

awareness of silences beyond what i can see on my own.

For those who have contributed reading along the way, given the gift of feedback whether verbalized or not, struggling or a softening in the voice—thank you.

For the unliving, thank you for reflecting space beyond.

Notes

This chapter is part of a continuing conversation with S. Leigh Star, written with Leigh-in-me after her death for creators and users of science-with-a-heart: we are many. It is by intention "unfinished," as life is.

1. As a "headnote" is a fieldnote that seems lodged in one's memory, to me a *bodynote* is one that got lodged in my body-mind.

2. Early work by Anselm Strauss and others comes close, where repeated observation of how specific kinds of work were actually conducted helped build "grounded theory." The observer was still outside. Feminist approaches highlighted own experience. Leigh Star developed her own ways of writing theory from experience: articulation work and invisibility (initially with Anselm Strauss; later, famously, her "onion paper" [Star 1991]) and residuality (with Geof Bowker; e.g., Star and Bowker 2007).

3. i remember yearning to understand what theory *was* so I'd be a real academic ("not just a computer scientist"). Yet, the more i looked, the more it meant different things depending on context. Now i remind myself not to play too easily with the term, or people get upset. *In fact.*

4. *Science* in this chapter is used in its broad sense (*vitenskap*, or the creation of knowledge), not limited to the natural sciences, medicine, etc. At times i use *scholarship* as a reminder of this broad reading, and the *academy*.

5. Leigh Star's intelligent and caring heart-mind is between the lines all over this chapter. She offered goggles i dared not touch as a young researcher. To honor her encouragement to do as i thought right even when that did not conform to expectations, this chapter is an attempt to take it all the way, an expérience in courage. The form—namely, working through a mosaic approach in which various angles leave impressions without necessarily being explained or the red thread being drawn out—is born from the silent subversion of my once-naïve trust in science, which crept under my skin from her.

In this footnote, i attempt to sketch what meetings with Leigh and her work has nourished in me. Claims as to what she "is"/"was"/"thought" should be taken lightly.

To Leigh, while science often presented itself as self-sufficient, with boundaries strongly patrolled, it was neither impenetrable nor uncontested. She contributed influential theory, often with others, in computer science and the research area of computer-supported cooperative work (CSCW) by seeing "what was not being seen": While there was discussion of "cooperation vs. collaboration," Leigh called attention to conflict. While others were studying contained groups (small distributed groups or single-location workplaces), Leigh Star (with Karen Ruhleder) highlighted ecologies, infrastructure. Her ways of working included personal stories contributing to theory (a phenomenology of conventions from her allergy to onions to the state of her/their desktop to "Sorting Things Out"), poems, and snippets. Networking people for their benefit.

The significance of Leigh's contribution—opening that which in our minds and research fields gets closed, expanding the boundaries of what is acceptable—may sound simple. Like a Columbi egg, the answer is easily repeated once revealed. Yet to think of it when no one else does is a feat of her interest in openness itself.

Nonduality: Paradox need not be a wall but can be a window. Example: passage in Star 1994 (and chapter 6, this volume), a (feminist) paper by Leigh on how people used to debate whether or not they kept birthing so many boys because they traveled back home with semen in turkey basters. Leigh commented that maybe there weren't more boys; it just seemed like it at the time.

From nonduality, sensitivity to the workings of dualist thinking, its hardening, and the consequent suffering caused. Avenues of exploration in Leigh's writing include the following:

a. Pointing to the existence of such suffering (e.g., tuberculosis sanatorium; though note: not by Leigh described as one-sided "evil").
b. Exploring the expression and experience of the suffering itself. Explored from the marginalized position: for example, allergy to onions (Star 1991); and the residual category (Bowker and Star 2000, Star and Bowker 2007).
c. Analyzing causes of such suffering. The notion of "hardening of the categories" (personal communication—maybe also printed somewhere); also of how the categories come to exist and continue existing (by the necessary aid of the residual); asking *cui bono?*

This field is vast. Language itself (as intimated in Star 2007) relies on categorizations. Yet systems differ as to how hard the categories are and to their consequences.

d. Demonstrating a possible end to this specific kind of suffering. The "contents" of the argument in terms of "intellectual" ideas are expressed also through Leigh's way of writing and talking. The 2007 paper by Leigh and Geof (Star and Bowker 2007) vividly shows not only the relevance of personal stories, but also how these differ, and must differ, because of the matter itself, the alternative to hard categorization at times, to categorization at all.

6. In other words, do-it-yourself—but be aware that you are. The dream is touching not only reader curiosity but also reactions including attraction, neutrality, and rejection of all or parts of the chapter. Yet, provoking reactions is no aim; clarity about them is.

7. For an argument why not to engage in "science-bashing," cf. Star's insistence that she was a scientist. Also, see Latour 2004 (a helpful call for precision but does not go far enough toward alternatives.

8. At an extreme end, some claim the omniscience of science, the impossibility of anything existing of which science is not aware (Wallace 2000).

9. I'm doing science, as commenters helpfully have pointed out, a disservice by at times writing with the hardened notions in mind, thus perhaps upkeeping hardening. Yet, this is the pain this chapter sets out to explore and at present this is how i am capable of doing so. May i later mature (following the lead of Leigh and others) to write different kinds of work.

10. Including my nine-year experience as a "seasonal laborer." Zinged with the shame of not having been chosen, i had to live on short-term contracts not only due to the lack of long-term ones or a sufficiently substantive CV—as beforehand i might have imagined—but due also to unwholesome employment practices. The belly of the beast: i knew (a part of) it, and found little to help understand why being in that belly hurt for a lover of theory, with plenty of feedback on her skills and understanding, with items on her CV supposedly in high demand. With time, bruises cushioned by a tenured position, it became less a matter of survival and more one of curiosity: What are we—the academics, members of faculty, scientists—doing to each other and therefore ourselves? What cultures are we building, how, and why?

11. Note that Wallace (2000), taking interest as he does in the deletion of God/nonmaterial aspects from the discourse of science proper, is less interested in Boyle than in one of his contemporaries who, through writing the history of the Royal Society, evicted God from this discourse on science. In Wallace's account, the eviction of the self (in Wallace's terms, of subjectivity) was built on the eviction of God.

12. For this chapter, then, the question is whether to choose a single point now (saves time, simplifies the writing and the reading, makes it easier to complete the chapter, maybe i can get two papers so more publication points and the Department will be happier . . . [aaargh!—Oh hello scream]) or stay with the multiple, permitting the possibility of surprise. The choice to go safe vs. trust. This time, i choose faith. I'm writing for what Leigh inspired in me, which i didn't

dare to meet in myself and thus could not offer to others. Once again, sister death has brought me the gift of non-fear, of not postponing my own growth. Of challenge.

13. i am using here a distinction from Buddhist terminology between pain—that which cannot be helped, for example, from ill health—and suffering—the worry, regrets, blame, greed, sorrow, etc., which often builds around pain. The terms are relative and can be used recursively.

14. As discussed above, I'm also interested in the softness that resides less demonstrably in relation to science. "In fact," that on the edge which helps define what is science by being "non-science," or even more interestingly [and more closely inspired by Leigh, as Sisse F reminded me]: that which resides in the uncertain terrain of the borderlands (Gloria Anzaldúa?).

15. Does such fear exist? It seems to, looking at book titles on specific subjects. These are chiefly science applied upon some topic (e.g., used probably deliberately as suggested by a title on viruses and fear; or perhaps inadvertently, as in a title on clients' fear of dental work).

Details: Looking up "fear" in the book title on the joint database for Norwegian university libraries, the number of hits comes close to 1,400: 500 of these since 2000; 67 during 2010–1011. Fear is a topic made relevant to medicine (e.g., virus, science, fear); christianity; much in business (e.g., *MarketPsych : How to Manage Fear and Build Your Investor Identity,* and *The Fear Index* of 2011); filmmaking; regional national concerns (America, Norway, etc.); globalization; design; literature; computer programming (C++ *without fear*); dentistry (clients' fear); psychology (Ph.D.: "Guided by Fear: Effects on Attention and Awareness"); history; political analyses (*Entertaining Fear: Rhetoric and the Political Economy of Social Control* [edited volume]); philosophy. Interestingly, substantially more of these recent titles seem to relate to business than to psychology. Two older volumes on fear i was already aware of, the Batesons' *Angels Fear* (which as far as i am aware is beautiful though not particularly about fear) and Furedi's *Culture of Fear* (sharp, unhappy). Searching for "fear" and "academic" (any year or medium) yields no hits, nor "fear"—"scholar," but "fear"—"scien?" gives three. One looks relevant: *The Inertia of Fear: And the Scientific Worldview* by Valentin Turčin; trans. Guy Daniels (Oxford: Martin Robertson, 1981).

16. Later, i heard that sadly, during the initial hours of uncertainty, Muslims had been the target of accusations.

17. i will write of death in small letters.

18. in a song attributed to st. francis of assisi, he offers gratitude to, among others, sister death, "the final helper" in our difficulties.

19. For example, what of the love in the streets, love in the hearts, discussed above—is science love-free? i think not. A slow learner perhaps, but not immutable. A beauty of science is our ideal of and ability to change. To accept more of what was previously hidden. But whenever new topics are made to fit familiar molds of clear categories, we are moving in circles. Could love be acceptable without being dissected, classified?

20. "But it is impossible," i hear the objection raised, "for a system to include something outside of it." Depends on what the claims of the system are. For (dominant notions of) science it is.

21. Also created is the category of those who do not know (in the correct way), namely, not scientists/scholars, and the category of that which is not "known," namely, not seen with the eyes of science/scholarship. The former is not explored here. The latter category poses science—especially people with hardened categories about science—with an undercommunicated problem. This gets dealt with in a range of ways, including to ignore but permit it, to posit it as a soon-to-be-addressed and (scientifically) resolved problem, or to render the whole problem nonexistent, for example, by defining the nonscientifically measurable as nonexistent. For more on this, see, for example, Wallace 2000.

22. An image i have of you is that of creators and/or consumers of science, well aware of soft underbellies; also people i cannot imagine or categorize; in total, therefore manifestations of Leigh's legacy.

References

Akrich, Madeleine. 1992. "The De-Scription of Technical Objects." In *Shaping Technology/Building Society,* ed. W. Bijker and J. Law, 205–224. Cambridge, MA: MIT Press.

Apollon. 2015. *Apollon.* In Norwegian, on death. http://www.apollon.uio.no/tema/doden/ (accessed March 31, 2015).

Beck, Eevi E. 1997. "Managing Diffracted Rationalities: IT in a Home Assistance Service." In *Gender, Technology and Politics in Transition? Proceedings from Workshop 4,* ed. Ingunn Moser and Gro Hanne Aas. TMV Report no. 29. Series on Technology and Democracy. Conference on Technology and Democracy—Comparative Perspectives, Oslo, January 17–19, Oslo.

Beck, Eevi E. 2002. "Mediation, Non-Participation, and Technology in Care Giving Work." In *PDC 02 Proceedings of the Participatory Design Conference,* ed. T. Binder, J. Gregory, and I. Wagner, 204–214. Palo Alto CA: CPSR.

Beck, Eevi E., Shirin Madon, and Sundeep Sahay. 2004. "On the Margins of the 'Information Society': A Comparative Study of Mediation." *The Information Society* 20 (4): 79–290.

Benjamin, Walter. [1955] 1999. "The Task of the Translator." In *Illuminations,* ed. Hannah Arendt, 70–82. London: Pimlico.

Bowker, Geoffrey C., and S. Leigh Star. 2000. *Sorting Things Out: Classification and Its Consequences.* Cambridge, MA: MIT Press.

Haraway, Donna. 1997. *Modest_Witness@Second_Millennium.FemaleMan©_Meets_OncoMouse™: Feminism and Technoscience.* New York: Routledge.

Isaksen, Lise Widding. 1994. "Den tabubelagte kroppen. Kropp, kjønn og tabuer i dagens omsorgsarbeid" [The taboo-ridden body. Body, gender, and taboos in contemporary care work]. Ph.D. dissertation, Sociology, University of Bergen.

Kübler-Ross, Elizabeth. [1991] 1998. *Døden er livsviktig* [Death is of vital importance]. Norwegian ed. Oslo: Ex Libris.

Latour, Bruno. 2004. "Why Has Critique Run Out of Steam? From Matters of Fact to Matters of Concern." *Critical Inquiry* 30 (2): 225–248.

Le Guin, Ursula. 1968. *The Wizard of Earthsea*. Berkeley, CA: Parnassus Press.

Levine, Stephen. 1982. *Who Dies? An Investigation into Conscious Living and Conscious Dying*. New York: Anchor.

Myskja, Audun. 2012. *Kunsten å dø: livet før og etter døden i et nytt lys* [The art of dying]. Oslo: Stenersen.

Nhất Hạnh, Thích. 1993. *Call Me by My True Names: The Collected Poems of Thích Nhất Hạnh*. Berkeley, CA: Parallax Press.

Nhất Hạnh, Thích. 2002. *No Death, No Fear: Comforting Wisdom for Life*. London: Rider/Random House.

Palmer, Parker. 2007. *The Courage to Teach: The Inner Landscape of a Teacher's Life*. San Francisco: Jossey-Bass.

Shroder, Tom. [1999] 2001. *Old Souls: The Scientific Evidence for Past Lives*. New York: Fireside.

Star, S. Leigh. 1991. "Power, Technologies and the Phenomenology of Conventions: On Being Allergic to Onions." In *A Sociology of Monsters: Essays on Power, Technology and Domination*, ed. J. Law, 26–56. London and New York: Routledge.

Star, S. Leigh. 1994. "Misplaced Concretism and Concrete Situations: Feminism, Method and Information Technology." Working Paper 11, Gender-Nature-Culture. Odense: Department of Feminist Studies, Odense University, Denmark.

Star, S. Leigh. 2010. "Mourning Light: The Ethnography of Science and Love." *Science, Technology, and Human Values* 35 (5): 618–619.

Star, S. Leigh, and Geoffrey Bowker. 2007. "Enacting Silence: Residual Categories as a Challenge for Ethics, Information Systems, and Communication." *Ethics and Information Technology* 9:273–280.

Star, S. Leigh, and Karen Ruhleder. 1994. "Steps toward an Ecology of Infrastructure." In *CSCW '94: Proceedings of the Conference on Computer Supported Cooperative Work*, ed. Richard Furuta and Christine Neuwirth, 253–264. New York: ACM.

Wallace, B. Allan. 2000. *The Taboo of Subjectivity: Toward a New Science of Consciousness*. New York: Oxford University Press.

23 Those Who Are Not Served? Exploring Exclusions and Silences in Transport Infrastructures

Jane Summerton

Infrastructure is both relational and ecological—it means different things to different groups and it is part of the balance of action, tools, and the built environment, inseparable from them. . . . Struggles with infrastructure are built into the very fabric of technical work. . . . Study a city and neglect its sewers and power supplies (as many have), and you miss essential aspects of distributional justice and planning power.
(Star 1999, 377–379)

What might it mean to take Star's passionate call to study issues of "distributional justice" in infrastructures as a point of departure to understand exclusions and silences specifically in transport infrastructures? How can concepts and approaches from Star's work within science and technology studies (STS) be used to identify "those who are not served" (Star 1999, 380) by existing transport infrastructures? How might we understand the experiences, silences, and struggles of excluded, marginalized, or "nonstandard" subjects in transport infrastructures? Finally, what are some of the linkages between Star's influential work and other theoretical and empirical work that explores issues related to exclusions and their implications for the lived experiences of various individuals and groups specifically in transport infrastructures?

These questions are the focus of this chapter. My purpose is to explore the ways in which approaches and concepts from Star's work on infrastructures can be extended to understand some of the complex linkages between exclusions, standards, and silences, specifically in transport infrastructures. The chapter will discuss recent work by scholars within STS and other areas who explicitly draw upon insights from Star's work to address these themes, as well as work by scholars whose perspectives on exclusions in transport resonate with core themes from her research.

Within science and technology studies, there has been relatively little critical work that has examined the workings of transport infrastructures and their implications for various groups. As Wyatt (2007, 822) notes, "Mobility and temporality ha[ve] received far more attention in mainstream social theory than they have in STS." Similarly, Sheller and Urry (2006, 208) argue that within the social sciences, travel has been seen as "a black box, a neutral set of technologies and processes." These observations make the admonition of Star's teacher and friend Anselm Strauss to "study the unstudied" particularly relevant as a point of departure for exploring interlinkages between mobilities and politics in transport.

The chapter is organized as follows. First I will briefly present relevant concepts and approaches from Star's work that can be used as theoretical tools for understanding ways in which exclusions, silences and suffering are expressed in transport infrastructures. I will then explore examples of recent work in STS and other areas that explicitly apply Star's concepts or in various ways embody her concerns. The text will be organized around discussions of selected work within three themes: exclusions and travelers' articulation work, inscribed narratives in the configuration of transport infrastructures, and linkages between disability, obduracy, and exclusions in transport. Rather than a comprehensive review, the chapter will focus on selected examples of relevant theoretical and conceptual work on these themes. Finally, I will suggest that there is currently a neglect of an important dimension of exclusions and silences in such infrastructures that is linked to victims of traffic accidents. The chapter concludes by proposing some topics for possible future research in this area.

Star's Work on Standardization, Exclusions, and Lived Experiences of Infrastructures

Star (1999) and Bowker and Star (1999) propose an explicitly relational approach to infrastructures: infrastructures are layered, complex, and mean different things to individuals with divergent life situations, identities, and contingencies. Bowker and Star conceptualize infrastructures as embedded in complex sociotechnical arrangements and structures that are both transparent in daily use and extensive in scope and scale. Infrastructures are also highly standardized with regard to institutionalized practices, norms, and modes of operation, where Bowker and Star's work reflects a strong and long-standing commitment to critically explore the ways in which such standardized practices work to marginalize, exclude, or silence certain individuals and groups. Their concern is particularly for those who do not "fit in," who are residual categories, or whose needs and lived experiences are not compatible with the standards and classifications that guide the institutionalized practices and ongoing operations of infrastructures (Star 1991; Bowker and Star 1999; Star and Bowker 2007).

An important part of this work is thus to trace how multiple exclusions are enacted and how these exclusions shape the lived experiences and sufferings of "non-standard" individuals and groups. In her widely cited 1991 essay, Star uses her own experiences (as a person who is allergic to onions) as a point of departure for discussing the personal costs that are often incurred when individuals' needs clash with the norms and conventional modes of operations of standardized forms of infrastructures (Star 1991).[1]

Similarly, in her work with Anselm Strauss (Star and Strauss 1999), Star points to the invisible work, or *articulation work*, that individuals or groups often must do in their everyday practices in order to adapt to—or maneuver in—such standardized orderings. Articulation work means the multiple ongoing, real-time adjustments that users of infrastructures often must perform to compensate for gaps in existing institutionalized work practices (Star 1999). Star urges researchers to pay attention to the ways in which the obduracy of seemingly small barriers to use in infrastructures (Star 1999, 386) can prevent people from using standardized systems without considerable articulation work. Here Star uses an example of the challenges that people with disabilities often face when attempting to maneuver in infrastructures of the built environment, a theme that I will return to later in this chapter. Star observes:

People commonly envision infrastructure as a system of substrates—railroad lines, pipes and plumbing, electrical power plants, and wires. . . . This image holds up well enough for many purposes—turn on the faucet for a drink of water and you use a vast infrastructure of plumbing and water regulation without usually thinking about it.

The image becomes more complicated when one begins to . . . examine the situations of those who are *not* served by a particular infrastructure. For a railroad engineer, the rails are not infrastructure but topic. For the person in a wheelchair, the stairs and door-jamb in front of a building are not seamless subtenders of use, but barriers. One person's infrastructure is another's topic, or difficulty. (Star 1999, 380)

As a metaphor to illustrate such tensions between standardized infrastructural configurations and the situations or experiences of those who are not accounted for in these standardizations, Bowker and Star (1999) suggest the concept of *torque* or stress. Torque refers to the often-painful misalignments between standardized infrastructural orders and the lived experiences of those individuals and groups whose identities, capabilities, and life situations are not compatible with these orders.[2] Finally, Star argues that an urgent agenda for social science researchers is to examine the perspectives of such marginalized, deleted, or unnamed individuals as a means to expose the *inscribed, often hidden narratives* in infrastructures that fail to recognize diversity in experiences and identities (Star 1999, 384–385).

How then might Star's work serve as a point of departure for exploring exclusions in transport infrastructures, the articulation work that often must be performed by those who do not fit in to such standardized infrastructures, the torque or misalignments that often results, and the inscribed narratives in transport infrastructures as seen from the perspective of marginalized others? These issues will be discussed in the following sections.

Exclusions and Articulation Work in Transport Infrastructures

Similarly to the infrastructures studied by Star, transport infrastructures (that is, infrastructures that provide goods and services that enable movements of people and things) are highly standardized in their designs, material configurations, and operations. For example, roads and highways are configured in accordance with specific standards with regard to the size of lanes, formats for road signs, designated speed limits, and driving rules; similarly, rail systems are based on standards for the interconnecting of rails, trains, signal systems, timetables, stations, ticket systems, and passengers. Because these infrastructures are typically tightly coupled and extensive in scope and scale, their extensive standardization as reflected in the configuration of technical artifacts (e.g. buses, subways, trains, planes, boats), coordinated modes of operation (e.g., travel routes, frequency of buses and trains, scope of services) and institutionalized practices has significant implications for the everyday mobility of various groups.

Within social science work on transport, there is a relatively extensive literature that explores various dimensions of the linkages between the dynamics of transportation infrastructures—their planning, design, and modes of operation—and the implications of these dynamics for access, mobility and "social exclusions" among various groups.[3] My purpose here is not to provide an overview of work in this area. Instead I will provide a few examples of recent theoretical and empirical work that has linkages to Star's work on the intertwinings of lived experiences, struggles and articulation work among heterogeneous users of infrastructures.

First, Hine and Mitchell (2001) discuss concrete contingencies that often make the ordering of everyday travel a difficult task for various travelers. Based on qualitative, in-depth studies of travelers' transport needs and lived experiences of transport provision, Hine and Mitchell discuss the everyday struggles that follow from inaccessible design of buses, trains, and stations (which influence not only elderly and disabled persons but also persons with many small children and lots of shopping bags), the often distant location of bus stops and stations (which can make it a physical ordeal to get to a station), the "unpredictable long wait" for buses or trains (often in cold, dark, or threatening environments), and deficiencies in customer care and travel information (which is sometimes lacking when it is most needed). In Star's terms, these dynamics underscore the material obduracy of barriers to use in transport infrastructures and the articulation work that many travelers must perform as part of their routinized practices of everyday mobility.

Articulation work is also a core theme in a second example of recent work on implications of travelers' lived experiences in transport that has linkages to Star's work. Peters, Klobbenburg, and Wyatt (2010) discuss the "coordination work" that is done by multiple users in order to achieve spatio-temporal order among heterogeneous entities in transport infrastructures. As Wyatt also observes (2007, 822), this concept is very similar to the concept of articulation work/invisible work as developed by Star and Strauss (1999). Specifically, Peters, Klobbenburg and Wyatt (2010) emphasize the work that is called for when travelers make connections between people, people and objects, and people and place. In order to understand the fluid and relational character of mobility practices and the required coordination work on the part of travelers, the authors develop the concepts of *projects* and *passages*. Projects ("the *why* of travel") are intertwined with passages ("the *how* of travel"). In other words, projects are what people do in their everyday lives in ways that entail both material dimensions and movements in space and time, while passages refer to the ways in which heterogeneous entities such as people, buses, trains, schedules, and clocks are ordered in the situated practices in which travel is embedded. Thus specific projects (such as getting to work on time, or visiting a relative) call for specific types of passages (such as traveling by car or bus, cycling). Peters, Klobbenburg, and Wyatt (2010) argue that projects and passages require extensive, and often demanding, coordination work on the part of users of transport infrastructures—work that is largely invisible. Conceptualizing projects and passages is thus useful because it allows us to ask what kinds of articulation work various groups must perform in order to carry out specific, situated actions in their everyday lives (Peters, Klobbenburg and Wyatt 2010).

A third example of recent work in this area that resonates with Star's insights on invisibility and lived experiences of exclusion is Jocoy and Del Casino's (2010) work on homeless adults' experiences of mobility and the sociocultural discourses that frame expectations for their mobility. Jocoy and Del Casino show the ways in which institutional practices and policies with regard to containment and mobility often rendered homeless people both spatially excluded and invisible. In Star's terms, their accounts are a vivid expression of the type of *torque* that can result from misalignments between infrastructural orderings and the lived experiences of those who "don't fit" easily into transport infrastructures. Based on the nuances of *varying levels* of inclusion and exclusion that homeless people themselves expressed, the authors reject homogeneous categorizations of binary relations of mobility/power and immobility/powerlessness.

These and other studies provide a fruitful point of departure for exploring the experiences of those who are not served (or not served well) by existing infrastructural orders in transport infrastructure, the articulation work that many individuals and groups must perform in their everyday travel practices, as well as the ways in which certain groups may be rendered invisible. As the work of Star and others such as Jocoy and Del Casino (2010) reminds us, however, lived experiences of marginalization do not neatly reflect dichotomies of exclusion/inclusion, and essentialist categorizations of broad groups tend to mask the heterogeneities of user identities, experiences, and practices. As Farrington and Farrington also point out, "There is not just one account, either of what, in terms of accessibility, people 'need' or what they experience" (2005, 3–4). Following Star and Bowker, every categorization tends to valorize some point of view while silencing another (Bowker and Star 1999).

Inscribed Narratives in the Configuration of Transport Infrastructures

As noted earlier, a core theme in Star's work concerns the importance of identifying and analyzing the narratives that are inscribed in the work of configuring infrastructures, while paying particular attention to those subjects or entities that have been deleted or marginalized in these narratives (Star 1999, 387). Specifically in relation to transport infrastructures, Jensen (2011) proposes an interesting approach to understanding such inscribed narratives within urban transport planning and politics in particular. Jensen explores the ways in which specific "imagined" mobile subjects and patterns of mobility are inscribed in the logic that contributes to the ordering of urban spaces in particular ways. Drawing on Foucault's work on governmentality as well as work by Deleuze and Huxley, Jensen argues:

> Perceptions and daily practices of mobility rest on a special gaze that produces particular social aspects of mobility while simultaneously silencing others, and which is embedded in governing and planning a city/region. It is a matter of seeing mobility in particular ways, thus also delineating how future mobility can be imagined. Such imagined mobilities and e.g. imagined mobile subjects emerge and take shape also through the aspects of the city and urban life that its users and designers see mobility as part and parcel of, assume or neglect. This is for example the case when new regions are pushed through representations of moving people as cosmopolitan, knowledge intensive workers who desire easy access to opera houses in Berlin and Oslo and shopping in New York. . . . In this sense, "seeing mobilities" also defines what is actually within the realm of mobilities and for whom." (Jensen 2011, 262)

Jensen (2011) thus draws attention to the "silencing" of others through particular ways of imagining mobile subjects and mobilities that are embedded in both the design and use of urban spaces. Representations of imagined mobile/immobile subjects tend to reinforce certain mobilities while excluding, neglecting, or silencing others.

Similarly, Graham and Marvin's (2001) influential work examines the changing narratives that are inscribed in starkly reconfigured infrastructures for transport, telecommunications, energy, and water in the modern metropolis. While traditionally the provision of services in such infrastructures was guided by an ethos of universal service on equal terms for all users within monopoly markets, neoliberal policies of privatization and liberalization have led to a radical departure from this ethos in recent decades. The authors argue that configurations of urban infrastructures are increasingly being driven by systematic attempts to design, commercially exploit, and expand privileged, high-service enclaves, zones, and spaces that are configured to fit the perceived needs of the most lucrative network users in urban areas. While the privileged groups in such spaces are offered premium infrastructural goods and services, other "less valued and less powerful" groups in "network ghettoes" are often marginalized or excluded from access to even basic infrastructural services (Graham and Marvin 2001, 382, 387). Similar to Star's commitment to exposing master narratives from the perspective of the unnamed other (1999), Graham and Marvin provide a rich account of the ways in which inscribed narratives in infrastructures may privilege some groups while marginalizing others, as well as how these narratives are materialized in the built environment.

Disability, Obduracy, and Exclusion in Transport

A core insight from Star's work is that the provision of services within standardized infrastructures is primarily designed and configured for those users whose bodies, identities, and needs are compatible with the standards, classifications, and categories that guide the operation of these infrastructures, while those with "nonstandard" bodies often experience torque or misalignments as a result of not fitting in. As noted earlier, Star also points out the ways in which the obduracy of specific material configurations in infrastructures can prevent people from using standardized systems without considerable articulation work. These themes are expressed in different ways in recent work at the intersection of disability studies, STS, and studies of transport.

Like Star, Imrie (2000) notes the assumptions about a "universal" mobile body that are embedded in transport planning and politics:

In particular, political and policy assumptions about mobility and movement are premised on a universal, disembodied subject which is conceived of as neutered, that is without sex, gender, or any other attributed characteristic. . . . The hegemony of what one might term the mobile body is decontextualized from the messy world of multiple and ever-changing embodiments, where there is little or no recognition of bodily differences or capabilities. The mobile body, then, is conceived of in terms of independence of movement and bodily functions; a body without physical and mental impairments. (1643)

Imrie (2000) thus emphasizes the hegemony of discourses about the *non-impaired-mobile body* in transport planning and politics. Imrie argues that inequities in mobility often reflect sociocultural values and practices among transport planners and operators that privilege certain bodies. He notes that planners often tend to *categorize* disabled people in broad classifications or groups (e.g., a diversity of impairments can be conflated to constructions of disabled travelers as "individuals in wheelchairs"). Such practices enact particular modes of mobility while excluding others, which means that actual impaired bodies are largely invisible in transportation planning and policy (Imrie 2000; Aldred and Woodcock 2008). Like Star, Imrie also reiterates that struggles for mobility are integral to the everyday lives of people with "impairments" (Imrie 2000).

Working at the intersection of STS and disability studies,[4] Galis (2006) also focuses on the dynamics by which disability issues become included or excluded in transport planning processes, as well as the struggles of organized groups of disabled travelers to intervene in these processes. Specifically, Galis analyzes the planning, design, and construction of a large infrastructural project, namely the Athens metro. Prior to the building of the Athens metro, persons with disabilities and the organizations that represented them were largely excluded from access to transport opportunities in Athens due to multiple material barriers in the urban built environment. Galis's work traces the processes by which representatives of disability organizations evolved from being weak actors to powerful spokespersons who successfully articulated and problematized their lived experiences of limited access, formulated transport-related disability issues and made these issues visible to others, and constructed their own agenda –all of which eventually resulted in the construction of an accessible metro. Galis's study has linkages to Star's work in pointing to the obduracy of material barriers in the built environment of infrastructures that work to exclude "nonstandard" subjects, as well as the articulation work that must be performed in order to give voice to (and intervene on behalf of) those who experience these barriers.

Finally, Moser (2003) provides a compelling account of the ways in which disability is experienced and enacted in the everyday lives and experiences of victims of traffic accidents. Moser's work is concerned with exploring the relations among embodiment, subjectivity, and disability. What does it mean to become disabled in a road traffic accident? How do victims of traffic accidents constitute, reconstitute, and order their identities and lives? Moser traces the multiple modes by which victims of traffic accidents handle and "order" their disability in their everyday practices. While some of these subjects enacted a "normalization" mode of ordering their lives as disabled, others performed an ordering that centered on disability as fate or disability as passion. An example of the latter was one victim's passionate engagement in extreme sports such as downhill wheelchair racing. The various modes were not, however, mutually exclusive. In practice, these subjects moved between multiple, coexisting modes of ordering disability and reconstituting their identities in everyday life.

In making this argument, Moser makes explicit linkages to Star's work (1991) about the ways in which infrastructural standards exclude or disable non-standard bodies and subjectivities, often producing their own "monsters" that do not fit. As Moser (2005) points out, a key contribution of Star's work is showing the ways in which the particularity of some bodies is "materially produced" (Moser 2005, 677) when nonstandard bodies interact with standardized environments.

Understanding the Suffering and Silences of Deaths on the Road

Moser's work is a unique,[5] in-depth account of the ways in which victims of severe traffic accidents may interpret, reconstitute, and reorder their identities and lives in the aftermath of an accident, as well the implications of these reorderings for our understandings of the intertwinings between materialities and subjectivities. With the exception of Moser's work, there is, however, little research within STS that focuses on the experiences, silences and suffering of those who are *victims* of traffic accidents or those who are *professionally or personally implicated* in various ways by collisions on the road. This lack of work is somewhat surprising in light of the continued ubiquity of traffic accidents and the extent of the individual suffering incurred in such accidents: traffic accidents constitute a significant cause of death in most areas of the world, and these deaths are differentially distributed among different groups.[6] The loss of lives as represented in the statistics of traffic deaths is daunting in scope and scale. At the same time, it can be argued that we lack approaches that could capture the kind of data that would allow us to better understand—and indeed more fully grasp—the meanings and consequences of death in traffic for various individuals and groups. I argue that it is important from an STS perspective to break this silence, give voice to victims and implicated others, and intervene in this issue as a "matter of care" (Puig de la Bellacasa 2011). Hence, we might ask, what kinds of concrete work might be considered in the spirit of Star's scholarship?

First, such work might further explore *traffic victims' lived experiences, meanings, and everyday practices* in dealing with the torque and altered trajectories of their everyday lives, as well their struggles in handling the obduracy of multiple small barriers in the built environment that they more or less suddenly have to deal with. How do victims of traffic accidents juggle their lives in daily negotiations between their bodies, their imagined (or aspired) mobilities and flows, and the material configurations of transport infrastructures that are perhaps only partially "accessible" and only in relation to some types of impaired bodies while excluding others? What are the assemblages of material and nonmaterial entities that constrain/enable/facilitate their daily lives?

A second area of research might examine the *interpretations, negotiations, and practices of the many professional teams* that respond to, interpret, and handle severe accidents on the road. The organizational and institutional infrastructure by which traffic accidents are handled and afflicted travelers taken care of—both on an emergency basis and more long-term—is extensive. For example, the professional emergency personnel include the ambulance workers who are called to the scene of the accident, the police officers who are charged with restoring order and literally "clearing the streets," the emergency room attendants and other medical professionals who receive and treat accident victims in emergency rooms, and the medical teams who provide short-term, follow-up care in hospital wards. On a more long-term basis, professional teams include rehabilitation specialists, therapists, psychologists, and other professionals in public and private health services who in various ways support the ongoing lives of victims of severe traffic accidents. How do various individuals and groups absorb, interpret, and give meanings to their work, and what professional routines and structures are in place to support these practices? How are "gaps in work processes" (Star 1999, 387) expressed and negotiated in their articulation work, and what further interventions might be possible?

Finally, STS research on the implications of traffic deaths for sufferings and silences among various groups might focus on the *lived experiences and everyday practices of caretakers and nonprofessional others* (that is, families, friends, relatives, neighbors, colleagues and others in the victim's extended social networks) in their ongoing work of supporting, caring, and perhaps serving as spokespersons for victims of traffic accidents. The articulation work of these individuals and groups in "filling the gaps" between damaged bodies and care is significant but remains largely invisible.[7]

Concluding Remarks

This chapter has shown the multiple ways in which approaches, concepts, and insights from Star's work have been fruitfully extended to areas of scholarship that seek to explore how exclusions, silences, and suffering related to victims of traffic accidents are expressed specifically in transport infrastructures. The scope and saliency of Star's work is reflected in both the theoretical heterogeneity of these accounts and the new understandings that they have provided. It is clear that Star's rich and multifaceted work offers a fruitful space for posing a number of critical questions about the workings of standardized transport infrastructures and their implications for the work of individuals and groups in negotiating meanings of mobility, identities, and exclusions/inclusions in everyday practices of mobility.

The new articulations and extensions of Star's work to transport issues that have been discussed here also show the ways in which her work can provide useful tools that can contribute to making the politics of transport infrastructures more visible. Understanding more fully the politics by which such infrastructures are imagined, translated into planning approaches and tools, configured and reconfigured—and again, with what concrete implications—is a potentially fertile area for further extensions of Star's work in STS, mobility studies, urban studies, disability studies, and other areas of social science.

As Star often noted, there is much work to be done.

Notes

1. This piece by Star (my personal favorite among her work) is particularly useful as a resource both for scholars who are new to STS and for teachers of STS courses.

2. For a fuller discussion of this concept, see Bowker and Star 1999, 27.

3. See, for example, Church, Frost, and Sullivan 2000; Kenyon, Lyons, and Rafferty 2002; Rajé 2003; Hodgson and Turner 2003; Cass, Shove, and Urry 2005; Farrington and Farrington 2005; Lucas 2006; Martens 2006; Sheller and Urry 2006; Preston and Rajé 2007; Sen 2008; Martens 2006; Beyazit 2011.

4. See also Galis 2011 and Blume, Galis, and Pineda 2014, the latter part of a special section "STS and Disability" in *Science, Technology & Human Values*.

5. See also Cuttler and Malone 2005 as another example.

6. More specifically, it is estimated that 1.3 million people are killed in traffic annually, which is more than the deaths caused by malaria or diabetes. Expressed another way, about 3,000 people die from traffic accidents every day. The risk of dying in a traffic accident varies among groups, as expressed in particular vulnerabilities among youth (aged 10–24) and individuals from low- and middle-income countries. See, for example, United Nations n.d., Peden et. al. 2004, Dahl 2004, and Mundell 2008.

7. I would like to acknowledge that Leigh Star and I had several inspiring talks about some of these research ideas.

References

Aldred, Rachel, and James Woodcock. 2008. "Transport: Challenging Disabling Environments." *Local Environment* 13 (6): 485–496.

Beyazit, Eda. 2011. "Evaluating Social Justice in Transport: Lessons to Be Learned from the Capability Approach." *Transport Reviews* 31 (1): 117–134.

Blume, Stuart, Vasilis Galis, and Andrés Valderrama Pineda. 2014. "Introduction: STS and Disability." *Science, Technology & Human Values* 39 (1) (January): 98–104.

Bowker, Geoffrey, and Susan Leigh Star. 1999. *Sorting Things Out: Classification and Its Consequences*. Cambridge, MA: MIT Press.

Cass, Noel, Elizabeth Shove, and John Urry. 2005. "Social Exclusion, Mobility and Access." *The Sociological Review* 53 (3) (August): 539–555.

Church, A., M. Frost, and K. Sullivan. 2000. "Transport and Social Exclusion in London." *Transport Policy* 7:195–205.

Cuttler, Sasha J., and Ruth E. Malone. 2005. "Using Marginalization Theory to Examine Pedestrian Injury: A Case Study." *Advances in Nursing Science* 28 (3): 278–286.

Dahl, Richard. 2004. "Vehicular Manslaughter: The Global Epidemic of Traffic Deaths." *Environmental Health Perspectives* 112 (11) (August): A628–A631.

Farrington, John, and Conor Farrington. 2005. "Rural Accessibility, Social Inclusion and Social Justice: Towards Conceptualization." *Journal of Transport Geography* 13:1–12.

Galis, Vasilis. 2006. "From Shrieks to Technical Reports: Technology, Disability and Political Processes in Building Athens Metro." Ph.D. dissertation, Linköping Studies in Arts and Sciences, no. 374, Linköping, Sweden.

Galis, Vasilis. 2011. "Enacting Disability: How Can Science and Technology Studies Inform Disability Studies. *Disability & Society* 26 (7): 825–838.

Graham, Stephen, and Simon Marvin. 2001. *Splintering Urbanism: Networked Infrastructures, Technological Mobilities and the Urban Condition*. London and New York: Routledge.

Hine, Julian, and Fiona Mitchell. 2001. "Better for Everyone? Travel Experiences and Transport Exclusion." *Urban Studies* 38 (2): 319–332.

Hodgson, F. C., and J. Turner. 2003. Participation Not Consumption: The Need for New Participatory Practices to Address Transport and Social Exclusion. *Transport Policy* 10:265–272.

Imrie, Rob. 2000. "Disability and Discourses of Mobility and Movement." *Environment & Planning A* 32:1641–1656.

Jensen, Anne. 2011. "Mobility, Space and Power: On the Multiplicities of Seeing Mobility." *Mobilities* 6 (2): 255–271.

Jocoy, Christine L., and Vincent J. Del Casino Jr. 2010. "Homelessness, Travel Behavior, and the Politics of Transportation Mobilities in Long Beach, California." *Environment & Planning A* 42:1943–1963.

Kenyon, Susan, Glenn Lyons, and Jackie Rafferty. 2002. "Transport and Social Exclusion: Investigating the Possibility of Promoting Inclusion through Virtual Mobility." *Journal of Transport Geography* 10:207–219.

Lucas, Karen. 2006. "Providing Transport for Social Inclusion within a Framework for Environmental Justice in the UK." *Transportation Research Part A, Policy and Practice* 40:801–809.

Martens, Karel. 2006. "Basing Transport Planning on Principles of Social Justice." *Berkeley Planning Journal* 19 (1): 1–17.

Moser, Ingunn. 2003. "Road Traffic Accidents: The Ordering of Subjects, Bodies and Disability." Ph.D. dissertation, Faculty of Arts, no. 173, University of Oslo.

Moser, Ingunn. 2005. "On Becoming Disabled and Articulating Alternatives: The Multiple Modes of Ordering Disability and Their Interferences." *Cultural Studies* 19 (6): 667–700.

Mundell, E. J. 2008. "U.N. Seeks to Curb World's Traffic Deaths." *The Washington Post*, April 1. http://www.washingtonpost.com/wp-dyn/content/article/2008/04/01/AR2008040101507.html, accessed March 23, 2015.

Peden, Margie, et al., eds. 2004. *World Report on Road Traffic Injury Prevention*. Geneva: World Health Organization.

Peters, Peter, Sanneke Klobbenburg, and Sally Wyatt. 2010. "Co-ordinating Passages: Understanding the Resources Needed for Everyday Mobility." *Mobilities* 5 (3): 349–368.

Preston, John, and Fiona Rajé. 2007. "Accessibility, Mobility and Transport-Related Social Exclusion." *Journal of Transport Geography* 15:151–160.

Puig de la Bellacasa, Maria. 2011. "Matters of Care in Technoscience: Assembling Neglected Things." *Social Studies of Science* 41 (1): 85–106.

Rajé, Fiona. 2003. "The Impact of Transport on Social Exclusion Processes with Specific Emphasis on Road User Charging." *Transport Policy* 10:321–338.

Sen, Siddhartha. 2008. "Environmental Justice in Transportation Planning and Policy: A View from Practitioners and Other Stakeholders in the Baltimore–Washington, D.C. Metropolitan Region." *Journal of Urban Technology* 15:117–138.

Sheller, Mimi, and John Urry. 2006. "The New Mobilities Paradigm." *Environment & Planning A* 38:207–226.

Star, Susan Leigh. 1991. "Power, Technology and the Phenomenology of Conventions: On Being Allergic to Onions." In *A Sociology of Monsters: Essays on Power, Technology and Domination*, ed. J. Law, 25–26. London: Routledge.

Star, Susan Leigh. 1999. "The Ethnography of Infrastructure." *American Behavioral Scientist* 43:377–391.

Star, Susan Leigh, and Geoffrey Bowker. 2007. "Enacting Silence: Residual Categories as a Challenge for Ethics, Information Systems, and Communications." *Ethics and Information Technology* 9:273.

Star, Susan Leigh, and Anselm Strauss. 1999. "Layers of Silence, Arenas of Voice: The Ecology of Visible and Invisible Work." *Computer Supported Cooperative Work: An International Journal* 8: 9–30.

United Nations. n.d. "Global Plan for the Decade of Action for Road Safety 2011–2020." http://www.who.int/roadsafety/decade_of_action/plan/en (accessed March 23, 2015).

Wyatt, Sally. 2007. "Making Time and Taking Time: Review of P. F. Peters' *Time, Innovation and Mobilities: Travel in Technological Cultures*." *Social Studies of Science* 37 (5) (October): 821–824.

24   The Ethnography of Infrastructure

Susan Leigh Star

> *Editor's Note:* This chapter, originally published in 1999, builds on work introduced in chapter 20. It extends our understanding of infrastructure as relational, and pays particular attention to the methodological challenges that emerge as a result. Although the use of ethnographic methods to study infrastructures has gained currency since this chapter was originally published, many of the issues raised in the chapter—for example, how to scale ethnographic studies—continue to challenge ethnographers of infrastructure today.

This chapter asks methodological questions about studying infrastructure with some of the tools and perspectives of ethnography. Infrastructure is both relational and ecological—it means different things to different groups and it is part of the balance of action, tools, and the built environment, inseparable from them. It also is frequently mundane to the point of boredom, involving things such as plugs, standards, and bureaucratic forms. Some of the difficulties of studying infrastructure are how to scale up from traditional ethnographic sites, how to manage large quantities of data such as those produced by transaction logs, and how to understand the interplay of online and offline behavior. Some of the tricks of the trade involved in meeting these challenges include studying the design of infrastructure, understanding the paradoxes of infrastructure as both transparent and opaque, including invisible work in the ecological analysis, and pinpointing the epistemological status of indictors.

Resources appear, too, as shared visions of the possible and acceptable dreams of the innovative, as techniques, knowledge, know-how, and the institutions for learning these things. Infrastructure in these terms is a dense interwoven fabric that is, at the same time, dynamic, thoroughly ecological, even fragile. (Bucciarelli 1994, 131)

Tell that its sculptor well those passions read
Which yet survive, stamped on these lifeless things.

(Shelley 1817)

General Methodological Problems

This chapter is in a way a call to study boring things. Many aspects of infrastructure are singularly unexciting. They appear as lists of numbers and technical specifications, or as hidden mechanisms subtending those processes more familiar to social scientists. It takes some digging to unearth the dramas inherent in system design creating, to restore narrative to what appears to be dead lists.

Bowker and Star (1999) note the following about the International Classification of Diseases (lCD), a global information-collecting system administered by the World Health Organization: "Reading the ICD is a lot like reading the telephone book. In fact, it is worse. The telephone book, especially the yellow pages, contains a more obvious degree of narrative structure. It tells how local businesses see themselves, how many restaurants of a given ethnicity there are in the locale, whether or not hot tubs or plastic surgeons are to be found there. (Yet most people don't curl up with a good telephone book of a Saturday night)" (Bowker and Star 1999, 56).

They note that aside from this direct information, indirect readings of such dry documents can also be instructive. In the case of phone books, for instance, a slender volume indicates a rural area; those that list only husbands' names for married couples indicate a heterosexually biased, sexist society.

Historical changes are important in reading these documents. Names and locations of services may change with political currents and social movements. Bowker and Star (1999, 57) note:

In the Santa Cruz, California, phone book, Alcoholics Anonymous and Narcotics Anonymous are listed in emergency services; years ago they would have been listed under "rehabilitation" if at all. The changed status reflects the widespread recognition of the organizations' reliability in crisis situations, as well as acceptance of their theory of addiction as a medical condition. Under the community events section in the beginning, next to the Garlic Festival and the celebration of the anniversary of the city's founding, the Gay and Lesbian Pride Parade is listed as an annual event. Behind this simple telephone book listing lies decades of activism and conflict—for gays and lesbians, becoming part of the civic infrastructure in this way betokens a kind of public acceptance almost unthinkable thirty years ago . . . excursions into this aspect of information infrastructure can be stiflingly boring. Many classifications appear as nothing more than lists of numbers with labels attached, buried in software menus, users' manuals, or other references.

Much of the ethnographic study of information systems implicitly involves the study of infrastructure. Struggles with infrastructure are built into the very fabric of technical work (Neumann and Star 1996). However, it is easy to stay within the traditional purview of field studies: talk, community, identity, and group processes, as now mediated by information technology. There have been several good studies of multiuser dungeons (MUDs), or virtual role-playing spaces, distance-mediated identity, cyberspace communities, and status hierarchies. There are much fewer on the effect of standardization or formal classification on group formation, the design of networks and their import for various communities, or on the fierce policy debates about domain names, exchange protocols, or languages.

Perhaps this is not surprising. The latter topics tend to be squirreled away in semi-private settings or buried in inaccessible electronic code. Theirs is not the usual sort of anthropological strangeness. Rather, it is an embedded strangeness, a second-order one, that of the forgotten, the background, the frozen in place.

Studies of gender bending in MUDs, of anonymity in decision making, and new electronic affiliations are important; they stretch our understanding of identity, status, and community. The challenges they present are nontrivial methodologically. How does one study action at a distance? How does one even observe the interaction of keyboard, embodied groups, and language? What are the ethics of studying people whose identity you may never know? When is an infrastructure finished, and how would we know that? How do we understand the ecology of work as affected by standardization and classification? What is universal or local about standardized interfaces? Perhaps most important of all, what values and ethical principles do we inscribe in the inner depths of the built information environment (Goguen 1997; Hanseth and Monteiro 1996; Hanseth, Monteiro, and Hatling 1996). We need new methods to understand this imbrication of infrastructure and human organization.

As well as the important studies of body snatching, identity tourism, and transglobal knowledge networks, let us also attend ethnographically to the plugs, settings, sizes, and other profoundly mundane aspects of cyberspace, in some of the same ways we might parse a telephone book. My teacher Anselm Strauss had a favorite aphorism, "study the unstudied." This led him and his students to research in understudied areas: chronic illness (Strauss 1979), low-status workers such as janitors, death and dying, and the materials used in life sciences including experimental animals and taxidermy (Clarke and Fujimura 1992). The aphorism was not a methodological perversion. Rather, it opened a more ecological understanding of workplaces, materiality, and interaction, and underpinned a social justice agenda by valorizing previously neglected people and things.

The ecological effect of studying boring things (infrastructure, in this case) is in some ways similar. The ecology of the distributed high-tech workplace, home, or school is profoundly impacted by the relatively unstudied infrastructure that permeates all its functions. Study a city and neglect its sewers and power supplies (as many have), and you miss essential aspects of distributional justice and planning power (Latour and Hermant 1998). Study an information system and neglect its standards, wires, and settings, and you miss equally essential aspects of aesthetics, justice, and change. Perhaps if we stopped thinking of computers as information highways and began to think of them more modestly as symbolic sewers, this realm would open up a bit.

Defining Infrastructure

What can be studied is always a relationship or an infinite regress of relationships. Never a "thing." (Bateson 1978, 249)

People commonly envision infrastructure as a system of substrates—railroad lines, pipes and plumbing, electrical power plants, and wires. It is by definition invisible, part of the background for other kinds of work. It is ready-to-hand. This image holds up well enough for many purposes—turn on the faucet for a drink of water and you use a vast infrastructure of plumbing and water regulation without usually thinking much about it.

The image becomes more complicated when one begins to investigate large-scale technical systems in the making, or to examine the situations of those who are not served by a particular infrastructure. For a railroad engineer, the rails are not infrastructure but topic. For the person in a wheelchair, the stairs and doorjamb in front of a building are not seamless subtenders of use, but barriers (Star 1991). One person's infrastructure is another's topic, or difficulty. As Star and Ruhleder (1996) put it, infrastructure is a fundamentally relational concept, becoming real infrastructure in relation to organized practices (see also Jewett and Kling 1991). So, within a given cultural context, the cook considers the water system as working infrastructure integral to making dinner. For the city planner or the plumber, it is a variable in a complex planning process or a target for repair: "Analytically, infrastructure appears only as a relational property, not as a thing stripped of use" (Star and Ruhleder 1996, 113).

In my own research, this became clear when I did fieldwork over three years with a community of biologists, in partnership with a computer scientist who was building an electronic shared laboratory and publishing space for them (Schatz 1991). I was studying their work practices and traveling to many laboratories to observe computer use and communication patterns. Although we were following the principles of participatory design-using ethnography to understand the details of work practice, extensive prototyping, and user feedback, and testing the system in laboratories and at conferences, few biologists ended up using the system. It seemed the difficulty was not in the interface or the representation of the work processes embedded in the system, but rather in infrastructure-incompatible platforms, recalcitrant local computing centers, and bottlenecked resources. We were forced to develop a more relational definition of infrastructure, and at the same time, challenge received views of good use of ethnography in systems development.

We began to see infrastructure as part of human organization, and as problematic as any other. We performed what Bowker (1994) has called an "infrastructural inversion"—foregrounding the truly backstage elements of work practice. Recent work in the history of science (ibid.; Edwards 1996; Hughes 1983, 1989; Summerton 1994; Yates 1989) has begun to describe the history of large-scale systems in precisely this way.

Whether in science or in the arts, we see and name things differently under different infrastructural regimes. Technological developments move from either independent or dependent variables, to processes arid relations braided in with thought and work. In the Worm Community Study (see chapter 20, this volume), Ruhleder and I came to define infrastructure as having the following properties, with examples following each dimension:

Embeddedness. Infrastructure is sunk into and inside of other structures, social arrangements, and technologies. People do not necessarily distinguish the several coordinated aspects of infrastructure. In the worm study, our respondents did not usually distinguish programs or subcomponents of the software—they were simply "in" it.

Transparency. Infrastructure is transparent to use, in the sense that it does not have to be reinvented each time or assembled for each task, but invisibly supports those tasks. For our respondents, the task of using FTP to download the system was new and thus difficult; for a computer scientist, this is an easy, routine task. Thus, the step of using FTP made the system less than transparent for the biologists, and thus much less usable.

Reach or scope. This may be either spatial or temporal—infrastructure has reach beyond a single event or one-site practice. One of the first things we did in system development was scan in the quarterly newsletter of the biologists so that one of the long-term rhythms of the community could be emulated online.

Learned as part of membership. The taken-for-grantedness of artifacts and organizational arrangements is a sine qua non of membership in a community of practice (Bowker and Star 1999; Lave and Wenger 1991). Strangers and outsiders encounter infrastructure as a target object to be learned about. New participants acquire a naturalized familiarity with its objects, as they become members. Although many of the objects of biology were strange to us as ethnographers, and to the computer scientists, and we made a special effort to overcome this strangeness, it was easy to overlook other things that we had already naturalized, such as information retrieval practices over networked systems.

Links with conventions of practice. Infrastructure both shapes and is shaped by the conventions of a community of practice (e.g., the ways that cycles of day-night work are affected by and affect electrical power rates and needs). Generations of typists have learned the QWERTY keyboard; its limitations are inherited by the computer keyboard and thence by the design of today's computer furniture (Becker 1982). The practices of reporting quarterly via the newsletter could not be changed in the biologists' system—when we suggested continual update, it was soundly rejected as interfering with important conventions of practice.

Embodiment of standards. Modified by scope and often by conflicting conventions, infrastructure takes on transparency by plugging into other infrastructures and tools in a standardized fashion. Our system embodied many standards used in the biological and academic community such as the names and maps for genetic strains, and photographs of relevant parts of the organism. But other standards escaped us at first, such as the use of specific programs for producing photographs on the Macintosh.

Built on an installed base. Infrastructure does not grow de novo; it wrestles with the inertia of the installed base and inherits strengths and limitations from that base. Optical fibers run along old railroad lines; new systems are designed for backward compatibility, and failing to account for these constraints may be fatal or distorting to new development processes (Hanseth and Monteiro 1996). We partially availed ourselves of this in activities such as scanning in the newsletter and providing a searchable archive; but our failure to understand the extent of the Macintosh entrenchment in the community proved expensive.

Becomes visible upon breakdown. The normally invisible quality of working infrastructure becomes visible when it breaks: the server is down, the bridge washes out, there is a power blackout. Even when there are back-up mechanisms or procedures, their existence further highlights the now-visible infrastructure. One of the flags for our understanding of the importance of infrastructure came with field visits to check the system usability. Respondents would say prior to the visit that they were using the system with no problems—during the site visit, they were unable even to tell us where the system was on their local machines. This breakdown became the basis for a much more detailed understanding of the relational nature of infrastructure.

Is fixed in modular increments, not all at once or globally. Because infrastructure is big, layered, and complex, and because it means different things locally, it is never changed from above. Changes take time and negotiation, and adjustment with other aspects of the systems is involved.[1] Nobody is really in charge of infrastructure. When in the field, we would attempt to get systems up and running for respondents, and our attempts were often stymied by the myriad of ways in which lab computing was inveigled in local campus or hospital computing efforts, and in legacy systems. There simply was no magic wand to be waved over the development effort.

Infrastructure and Methods

The methodological implications of this relational approach to infrastructure are considerable. Sites to examine then include decisions about encoding and standardizing, tinkering and tailoring activities (see, e.g., Gasser 1986; Trigg and Bødker 1994), and the observation and deconstruction of decisions carried into infrastructural forms (Bowker and Star, 1999). The fieldwork in this case transmogrifies to a combination of historical and literary analysis, traditional tools like interviews and observations, systems analysis, and usability studies. For example, in studying the development of categories as part of information infrastructure, I observed meetings of nurses striving to categorize their own work (Bowker, Timmermans, and Star 1995), studied the archives of meetings at the World Health Organization and its predecessors arguing about establishing and refining categories used on death certificates, and read old newspapers and law books recording cases of racial recategorization under apartheid in South Africa (Bowker and Star 1999). In each case, I brought an ethnographic sensibility to the data collection and analysis: an idea that people make meanings based on their circumstances, and that these meanings would be inscribed into their judgments about the built information environment.

I have also worked with computer scientists designing complex information systems. I began this work as a kind of informant about social organization. At first, the computer scientists sought examples of real organizational problem solving in order to model large-scale artificial intelligence systems. They identified problems from the realm of complex system development, and asked me to investigate their analog in organizational settings, primarily of scientists and engineers (Hewitt 1986; Star 1989). For example, when designers tried to model how a smart system would determine closure for a complex problem, I investigated how this was managed in nineteenth-century England by a group of neurophysiologists debating the functions of the brain (Star 1989), and made formal models of the processes that were fed back to the computer scientists.

This early work began in the 1980s, before the current development in information systems partnering ethnographers with computer scientists for the purpose of improving usability (as, for example, in the Worm Community Study). During the last decade, some ethnographers have created durable partnerships with system developers in many countries, especially in the areas of computer-supported cooperative work (CSCW) and human–computer interaction (Bowker et al. 1997). This work has emerged from a number of intellectual traditions, including ethnomethodology, symbolic interactionism, labor process research, and activity theory (cultural-historical psychology), among others.

All of us doing this work have begun to wrestle with questions of scalability that inherently touch on questions of infrastructure. It is possible (sort of) to maintain a traditional ethnographic research project when the setting involves one group of people and a small number of computer terminals. However, many settings involving computer design and use no longer fit this model. Groups are distributed geographically and temporally, and may involve hundreds of people and terminals. There have always been inherent scale limits on ethnography, by definition. The labor-intensive and analysis-intensive craft of qualitative research, combined with a historical emphasis on single investigator studies, has never lent itself to ethnography of thousands.[2]

At the same time, ethnography is a tempting tool for analyzing online interaction. Its strength has been that it is capable of surfacing silenced voices, juggling disparate meanings, and understanding the gap between words and deeds. Ethnographers are trained to understand viewpoints, the definition of the situation. Intuitively, these seem like important strengths for understanding the enormous changes being wrought by information technology. The scale question remains a pressing and open one for methodological concerns in the study of infrastructure. It is an ironic and tempting moment—we have the promise of a complete transcript of interactions, almost ready-made "field notes" in the form of transaction logs and archives of email discussions. At the same time, reducing this volume of material to something both manageable and analytically interesting is a tough nut to crack, despite the emergence of increasingly sophisticated tools such as Atlas/ti for qualitative analysis. Yet, I know of no one who has analyzed transaction logs to their own satisfaction, never mind to a standard of ethnographic veridicality (see Spasser 1998 for a good discussion of some of these problems).

And we are still stuck with the problem of where online interactions fit with people's lives and organizations offline. In the Worm Community Study, I tried simply to scale up traditional fieldwork techniques—and I and my research partner ended up traveling to dozens of labs, doing entree work for each one, interviewing more than a hundred biologists, and exhausting myself in the process. In the Illinois Digital Library Project, our social science evaluation team found that we had to transform our original study of "emergent community processes in the digital library" (via fieldwork and transaction logs) to a linked set of interviews with potential users and ethnographies of the design team while we waited for the system testbed to emerge, some two years behind schedule (Bishop et al. 2000; Neumann and Star 1996). We had to invent new ways of triangulating and bootstrapping along with the systems developers. These new ways of working broke old forms both for our respondents and for us.

Tricks of the Trade

The following section examines several tricks I have developed in the previously mentioned studies, helpful for "reading" infrastructure and unfreezing some of its features.[3]

Identifying Master Narratives and "Others"

Many information systems employ what literary theorists would call a master narrative, or a single voice that does not problematize diversity. This voice speaks unconsciously from the presumed center of things. An example of this encoding into infrastructure would be a medical history form for women that encodes monogamous traditional heterosexuality as the only class of responses: blanks for "maiden name" and "husband's name," blanks for "form of birth control," but none for other sexual practices that may have medical consequences, and no place at all for partners other than husband to be called in a medical emergency. Latour (1996) discusses the narrative inscribed in the failed metro system, Aramis, as encoding a particular size of car based on the presumed nuclear family. Bandages or mastectomy prostheses labeled "flesh colored" that are closest to the color of White people's skin are another kind of example.

Listening for the master narrative and identifying it as such means identifying first with that which has been made other, or unnamed. Some of the literary devices that represent master narratives include creating global actors, or turning a diverse set of activities and interests into one actor with a presumably monolithic agenda ("the United States stands for democracy"); personification, or making a set of actions into a single actor with volition ("science seeks a cure for cancer"); passive voice ("the data have revealed that"); and deletion of modalities. The latter has been well described by sociologists of science—the process by which a scientific fact is gradually stripped of the circumstances of its development, and the attendant uncertainties, and becomes an unvarnished truth.

In the previously mentioned study of the International Classification of Diseases, Bowker and I discovered many moments when the master narrative in the making became visible. One such deconstructive moment occurred when a committee of statisticians attempted to codify the "moment of life": How can you tell, for the purposes of filling out a birth certificate, when a baby is alive? Religious differences (as, for example, between Catholics and Protestants) were argued about, as well as phenomenological distinctions such as the number of breaths a baby would draw, try to draw, or fail to draw (Bowker and Star 1999). In studies we read of the actual practices of filling in death certificates, how the distinctions made by the "designers" upstream did not match the ways that attending doctors saw the world. We came to understand how the blanks on the forms were both heteropraxial (different practices according to region, local constraints, beliefs) and heteroglossial (inscribing different voices in the seemingly monotonous form).

Surfacing Invisible Work

Information systems encode and embed work in several ways. They may directly attempt to represent that work. They may sit in the middle of a work process like a rock in a stream, and require workarounds in order for interaction to proceed around them. They also may leave gaps in work processes that require real-time adjustments, or articulation work, to complete the processes.

Finding the invisible work in information systems requires looking for these processes in the traces left behind by coders, designers, and users of systems (Star and Strauss 1999 discuss this in relation to the design of CSCW systems). In some instances, this means going backstage, in Goffman's (1959) terms, and recovering the mess obscured by the boring sameness of the information represented. It is often in such backstage work that important requirements are discovered. For example, in the Worm

Community Study, we discovered that there were crucial moments in a biologist's career—especially during the postdoc period, just before getting one's own lab—when secrecy and professional smoothness are valued over the usual community norms of sharing preliminary results in semiformal venues.

With any form of work, there are always people whose work goes unnoticed or is not formally recognized (cleaners, janitors, maids, and often parents, for instance). Where the object of systems design is to support all work, leaving out what are locally perceived as "nonpeople" can mean a nonworking system. For example, with the biologists, I had originally wanted to include secretaries in the publication and communication stream, as they were so obviously (to me) part of the community communications. This was strongly resisted by both biologists and systems developers, as they did not see the secretaries as doing real science, and thus the idea was dropped. There is often a delicate balance of this sort between making things visible and leaving things tacit. With the nurses previously mentioned, whose work was categorizing all the tasks done by nurses, this was an important issue. Leave the work tacit, and it fades into the wallpaper (in one respondent's words, "we are thrown in with the price of the room"). Make it explicit, and it will become a target for hospital cost accounting. The job of the nursing classifiers was to balance someone in the middle, making their work just visible enough for legitimation, but maintaining an area of discretion. Without the fieldwork at their sessions where they were building the classification system, Bowker, Timmermans, and Star (1995) would never have known about this conflict.

Paradoxes of Infrastructure

Why does the slightest small obstacle often present a barrier to the user of a computer system? One of the findings of our studies of users in the Illinois Digital Library Project (Bishop et al. 2000) is that seemingly trivial alterations in routine, or demands for action, will act to prevent them from using the system. This can be an extra button to push, another link to follow to find help, or even looking up from the screen. The obduracy of these "tiny" barriers presents, at first glance, a puzzle in human irrationality. Why would someone not punch a couple of buttons rather than walk across campus to get a copy of something? Why do people persist in using less functional, but more routine actions when cheaper alternatives are nearby? Are people so routinized, so rigid in their ability to adapt to change that even such a slight impediment is too much?

Rather than characterize human nature with such broad strokes, I return to a fieldwork example for an explanation of this phenomenon. At a phenomenological level, what has happened is that these slight impediments have become magnified in the flow of the work process. An extra keyboard stroke might as well be an extra ten push-ups. What is going on here? One way to explain this magnification process is to understand that in fact two processes of work are occurring simultaneously: Only one is visible to the traditional analysis of user-at-terminal or user-with-system. That is the one that concerns keystrokes and functionality. The other is the process of assemblage, the delicate, complex weaving together of desktop resources, organizational routines, running memory of complicated task queues (only a couple of which really concern the terminal or system), and all manner of articulation work performed invisibly by the user.

Schmidt and Simone (1996) show that production/coordination work and articulation work (the second set of invisible tasks previously described) are recursively related in the work situation. Only by describing both the production task and the hidden tasks of articulation, together and recursively, can we come up with a good analysis of why some systems work and others do not. The magnification we encountered in our studies of users concerns the disruption of the users' articulation work. This system is necessarily fragile (as it is in real time), depending on local and situated contingencies, and requires a great deal of street smarts to pull off. Small disruptions in the articulating processes may ramify throughout the workflow of the user, causing the seemingly small anomaly or extra gesture to have a far greater impact than a rational user-meets-terminal model would suggest.

The Thorny Problem of Indicators

One of the difficulties in studying infrastructure is distinguishing different levels of reference in one's subject matter. This is a difficulty shared by all interpretive studies of media. For instance, suppose one wishes to understand the relationship of scientific advertising to cultural values about science. At one level of reference, one could count the frequency of ads, their claimed links with sales, and the attendant budget without even reading a single ad. In this case, the ads are indicators of resources spent promoting scientific products. Taking a step into the content of the advertisements, one could trace the emphases placed on certain types of activity, or the gender-stereotyped behavior embodied in them, or what sorts of images and aesthetics are used to display success. Here, one is required to assess the stylistics of the advertisements' creators—including ironic usage, multiple levels of meaning, psychological strategies employed, and thus their meanings. Finally, one could simply take the advertisements as a literal transcript about the process and progress of science, to be read directly for their claims, as indicators of scientific activity. To generalize this, one can read information infrastructure either as

- a material artifact constructed by people, with physical properties and pragmatic properties in its effects on human organization. The truth status of the content of the information is not relevant in this perspective, only its impact; or as
- a trace or record of activities. Here, the information and its status become much more relevant, if the infrastructure itself becomes an information-collecting device. Transaction logs, email records, as well as reading things like classification systems for evidence of cultural values, conflicts, or other decisions taken in construction fall into this

category. The information infrastructure here sits (often uneasily) somewhere between research assistant to the investigator and found cultural artifact. The information must still be analyzed, and placed in a larger framework of activities; or as

• a veridical representation of the world. Here, the information system is taken unproblematically as a mirror of actions in the world, and often tacitly, as a complete enough record of those actions. Where Usenet groups' interactions replace field notes entirely in the analysis of a particular social world, for example, one has this sort of substitution.

These three sorts of representations are not mutually exclusive, of course. There is, however, an important methodological point to be made about where one's analysis is located. I have several times advised students on theses that elide these functions of indicators, and it is a difficult and painful process to disentangle them. Films about rape may say a great deal about a given culture's acceptance of sexual violence, but they are not the same thing as police statistics about rape, nor the same as phenomenological investigations of the experience of being raped. Films are made by filmmakers who work within an industry, constrained by budgets, conventions, and their imaginations. Similarly, as an example from information infrastructure, people send email according to certain conventions and within certain genres (Yates and Orlikowski 1992). The relationship between email and the larger sphere of lived activity cannot be presumed, but must be investigated.

The processes of discovering the status of indicators are complex. This is partly due to our own elisions as researchers, and partly due to sleights of hand undertaken by those creating them. A common example is the substitution of precision for validity in the creation of a system of indicators or categories. When large epistemological stakes are at issue in the development of a system, one political tactic is to focus away from the larger question, and instead to seize control of the indicators. Kirk and Kutchins (1992), in their study of the DSM, show precisely this set of tactics at work between psychoanalysts and biologically oriented psychiatrists in the construction of that category system. Rather than (as they had in fact been doing for years) focus on the larger questions of mind and psychopathology, the designers of the DSM reframed the indicators, including how to frame requests for reimbursement from third parties, into a set of numbers that gradually squeezed out psychoanalytic approaches. I noted a similar set of activities by brain researchers at the turn of the century (Star 1989).

Bridges and Barriers

At least since Winner's (1986) classic chapter "Do Artifacts Have Politics?" the question of whether and how values are inscribed into technical systems has been a live one in the communities studying technology and its design. Winner used the example of Robert Moses, a city planner in New York, who made a behind-the-scenes policy decision to make the automobile bridges over the Grand Central Parkway low in height.

The reason? The bridges would then be too low for public transport—buses—to pass under them. The result? Poor people would be effectively barred from the richer Long Island suburbs, not by policy, but by design.

Whether or not one takes the Moses example at face value (and it has been a controversial one), the example is an instructive one. There are millions of tiny bridges built into large-scale information infrastructures, and millions of (literal and metaphoric) public buses that cannot pass through them. The example of computers given to inner-city schools and the developing world is an infamous one. The computers may work fine, but the electricity is dirty or lacking. Old floppy disks do not fit new drives, and new disks are expensive. Local phone calls are not always free. New browsers are faster, but more memory hungry. And one of those now popular will not support the most popular web browser for blind people in text-only format.

In information infrastructure, every conceivable form of variation in practice, culture, and norm is inscribed at the deepest levels of design. Some are malleable, changeable, and programmable—if you have the knowledge, time, and other resources to do so. Others—such as a fixed-choice category set—present barriers to users that may only be changed by a full-scale social movement. Consider the example of choice of race in the U.S. Census forms. In the year 2000, for the first time, people may check more than one racial category. This simple infrastructural change took a march on Washington, years of political activism, and will cost billions of dollars. It is opposed by many progressive social justice groups, on the grounds that although it is biologically correct to say that most of us are multiracial, the effects of discrimination will be lost in the count by those who claim multiple racial origins.

Applying the insights, methods, and perspectives of ethnography to this class of issues is a terrifying and delightful challenge for what some would call the information age. The effort to date has linked historians, sociologists, anthropologists, philosophers, literary theorists, and computer scientists. The methodological side of the questions posed is underdeveloped by contrast with the power of the findings of this "invisible college." Thus, the articles in this issue[4] are a most welcome addition to a literature of growing importance.

Notes

The author thanks Howie Becker, Geof Bowker, Jay Lemke, Nina Wakeford, and Barry Wellman for helpful comments. This chapter is for the other members of the Society of People Interested in Boring Things, especially cofounder Charlotte Linde.

1. I am grateful to Kevin Powell for this point. This modularity is formally similar to Hewitt's (1986) open systems properties (see also Star 1989).

2. At least, that is, when those thousands are heterogeneous, distributed over many sites, and perhaps anonymous. Becker (personal communication, February 25, 1999) points out that some

ethnographies of thousands have been done in large organizations (see, e.g., Becker, Geer, and Hughes 1968).

3. This title is stolen from Becker's (1998) invaluable *Tricks of the Trade*, a handbook for conducting good social science research. The stealing, of course, is one of the key tricks of the trade. To quote Latour (1987), "les deux mamelles de la science sont peage et bricolage" (the twin teats of science are petty theft and bricolage).

4. In this sentence, issue refers to a special issue of *American Behavioral Scientist* that had as its aim shaping the then emergent interdisciplinary field for the study of digital and networked technologies. Journal editors in particular sought to explore a methodological approach that critically examines the cultural and social structures within which technologies are embedded.

References

Bateson, G. 1978. *Steps to an Ecology of Mind.* New York: Ballantine.

Becker, H. S. 1982. *Art Worlds.* Berkeley: University of California Press.

Becker, H. S. 1998. *Tricks of the Trade: How to Think about Your Research While You're Doing It.* Chicago: University of Chicago Press.

Becker, H. S., B. Geer, and E. C. Hughes. 1968. *Making the Grade: The Academic Side of College Life.* New York: John Wiley.

Bishop, A. P., L. J. Neumann, S. L. Star, C. Merkel, E. Ignacio, and R. J. Sandusky. 2000. "Digital Libraries: Situating Use in Changing Information Infrastructure." *Journal of the American Society for Information Science* 51 (4) (Special Issue: Digital Libraries: Part 2): 394–413.

Bowker, G. 1994. "Information Mythology and Infrastructure." In *Information Acumen: The Understanding and Use of Knowledge in Modern Business*, ed. L. Bud-Frierman, 231–247. London: Routledge.

Bowker, G., and S. L. Star. 1999. *Sorting Things Out: Classification and Its Consequences.* Cambridge, MA: MIT Press.

Bowker, G., S. L. Star, W. Turner, and L. Gasser, eds. 1997. *Social Science, Information Systems and Cooperative Work: Beyond the Great Divide.* Hillsdale, NJ: Lawrence Erlbaum.

Bowker, G., S. Timmermans, and S. L. Star. 1995. "Infrastructure and Organizational Transformation: Classifying Nurses' Work." In *Information Technology and Changes in Organizational Work*, ed. W. Orlikowski, G. Walsham, M. Jones, and J. DeGross, 344–370. London: Chapman and Hall.

Bucciarelli, L. L. 1994. *Designing Engineers.* Cambridge, MA: MIT Press.

Clarke, A. E., and J. H. Fujimura, eds. 1992. *The Right Tools for the Job: At Work in Twentieth-century Life Sciences.* Princeton, NJ: Princeton University Press.

Edwards, P. N. 1996. *The Closed World: Computers and the Politics of Discourse in Cold War America.* Cambridge, MA: MIT Press.

Gasser, L. 1986. "The Integration of Computing and Routine Work." *ACM Transactions on Office Information Systems* 4:205–225.

Goffman, E. 1959. *The Presentation of Self in Everyday Life*. Garden City, NY: Doubleday.

Goguen, J. 1997. "Requirements Engineering as the Reconciliation of Technical and Social Issues." In *Requirements Engineering: Social and Technical Issues*, ed. M. Jirotka and J. Goguen, 27–56. New York: Academic Press.

Hanseth, O., and E. Monteiro. 1996. "Inscribing Behavior in Information Infrastructure Standards." *Accounting, Management & Information Technology* 7: 183–211.

Hanseth, O., E. Monteiro, and M. Hatling. 1996. "Developing Information Infrastructure: The Tension between Standardization and Flexibility." *Science, Technology and Human Values* 21:407–426.

Hewitt, C. 1986. "Offices Are Open Systems." *ACM Transactions on Office Information Systems* 4:271–287.

Hughes, T. P. 1983. *Networks of Power: Electrification in Western Society, 1880–1930*. Baltimore: Johns Hopkins University Press.

Hughes, T. P. 1989. "The Evolution of Large Technological Systems." In *The Social Construction of Technological Systems*, ed. W. E. Bijker, T. P. Hughes, and T. Pinch, 51–82. Cambridge, MA: MIT Press.

Jewett, T., and R. Kling. 1991. "The Dynamics of Computerization in a Social Science Research Team: A Case Study of Infrastructure, Strategies, and Skills." *Social Science Computer Review* 9:246–275.

Kirk, S. A., and H. Kutchins. 1992. *The Selling of the DSM: The Rhetoric of Science in Psychiatry*. New York: Aldine de Gruyter.

Latour, B. 1987. *Science in Action: How to Follow Scientists and Engineers through Society*. Milton Keynes, UK: Open University Press.

Latour, B. 1996. *Aramis, or the Love of Technology*. Cambridge, MA: Harvard University Press.

Latour, B., and E. Hermant. 1998. *Paris: Ville invisible*. Paris: La Decouverte.

Lave, J., and E. Wenger. 1991. *Situated Learning: Legitimate Peripheral Participation*. Cambridge, UK: Cambridge University Press.

Neumann, L., and S. L. Star. 1996. "Making Infrastructure: The Dream of a Common Language." In *Proceedings of the PDC '96*, ed. J. Blomberg, F. Kensing, and E. Dykstra-Erickson, 231–240. Palo Alto, CA: Computer Professionals for Social Responsibility.

Schatz, B. 1991. "Building an Electronic Community System." *Journal of Management Information Systems* 8:87–107.

Schmidt, K., and C. Simone. 1996. "Coordination Mechanisms: Towards a Conceptual Foundation of CSCW Systems Design. Computer Supported Cooperative Work (CSCW)." *Journal of Collaborative Computing* 5:155–200.

Shelley, P. B. 1817. "Ozymandias." *The Examiner*, January 11, 1818.

Spasser, M. A. 1998. "Computational Workspace Coordination: Design-in-Use of Collaborative Publishing Services for Computer-Mediated Cooperative Publishing." Ph.D. dissertation, University of Illinois, Urbana, IL.

Star, S. L. 1989. *Regions of the Mind: Brain Research and the Quest for Scientific Certainty.* Stanford, CA: Stanford University Press.

Star, S. L. 1991. "Power, Technologies and the Phenomenology of Conventions: On Being Allergic to Onions." In *A Sociology of Monsters: Essays on Power, Technology and Domination*, ed. J. Law, 26–56. London: Routledge.

Star, S. L., and K. Ruhleder. 1996. "Steps toward an Ecology of Infrastructure: Design and Access for Large Information Spaces." *Information Systems Research* 7 (1): 111–134.

Star, S. L., and A. L. Strauss. 1999. "Layers of Silence, Arenas of Voice: The Ecology of Visible and Invisible Work." *Computer Supported Cooperative Work: The Journal of Collaborative Computing* 8:9–30.

Strauss, A., ed. 1979. *Where Medicine Fails.* New Brunswick, NJ: Transaction Books.

Summerton, J., ed. 1994. *Changing Large Technical Systems.* Boulder, CO: Westview.

Trigg, R., and S. Bødker. 1994. "From Implementation to Design: Tailoring and the Emergence of Systematization in CSCW." In *Proceedings of the ACM 1994 Conference on Computer Supported Cooperative Work*, 45–54. New York: ACM Press.

Winner, L. 1986. "Do Artifacts Have Politics?" In *The Social Shaping of Technology: How the Refrigerator Got Its Hum*, ed. J. Wacjman and D. Mackenzie, 26–37. Milton Keynes, UK: Open University Press.

Yates, J. 1989. *Control through Communication: The Rise of System in American Management.* Baltimore: Johns Hopkins University Press.

Yates, J., and W. J. Orlikowsi. 1992. "Genes of Organizational Communication: A Structurational Approach to Studying Communication and Media." *Academy of Management Review* 17:299–326.

Envoi: When Shadows Become Complex: Weaving the Danmarra

Susan Leigh Star

Editors' Note: In August 2009, Susan Leigh Star participated in a "teaching from country" program, part of a symposium sponsored by Charles Darwin University in Northern Australia. The chapter is based on conversations that took place in this unique environment where Leigh found herself among Yolŋu Aboriginal Australians who own land in Australia's northeast Arnhem Land. The symposium used a "blended learning model" wherein participants—teachers and learners alike—were in a diversity of locations in Northern Australia, communicating via computer and video links. Thus the chapter reads at times like a conversation, and at times as though Leigh was addressing an audience, which she also was. Published after her death, it originally appeared in a journal with other pieces from the Learning from Country symposium. Demonstrating her sensitivity and reflexivity, as well as her openness and analytic acuity, it is a most apt final chapter for this volume commemorating Leigh's legacies.

Every girl growing up in North America faces her relationship with quilts. A quilt is a bed covering composed of a colorful top, a filling for warmth, and a plain bottom, sewn together to create a blanket. Quilts are home art, a way of remembering, art in use, part of culture and family and for friends. There are stages in the making of a quilt—first a vision of what you want to say or achieve, perhaps a sketch. Then, you collect fabric, usually from places where that fabric has been worn or saved, sometimes pieces of clothing, or many other eclectic forms. The pieces can be put together as a simple checkerboard or grid of squares; or matched up in no particular pre-made form (a "crazy quilt"); or they can be pieced together in the most elaborate of forms and patterns, either those invented by others or by oneself. The final stage of making a quilt is considered the most important, technically, by many quilters. This is the almost-invisible stitching that will run across the fabric, making little distinct areas, and forming its own pattern on top of the big pattern. It might take the form of a spiral, or branches, trails, stripes, according to what the quilter envisions. In old times, and often even today, many of these steps, especially the last, were collective. Women would gather around a big frame that held the pieces, and together do the last, nearly invisible stitches (called a "quilting bee"), as well as, sometimes, the larger pattern.

Quilts are important for North Americans, although no one region can be said to "own" the form. They reach complex, collective art forms that only recently (1970s) became public "high" art. The best ones are astonishing blends of color, form, quilting stitches, imaginations, or stories. African American and Appalachian women in particular are known for their quilting art, from slavery days until now, both embedding stories and history, lovely in their variety and artistic sophistication (see, for example, hartcottagequilts.com, and the essay by Fellner 2009). I myself have my great-grandmother's quilt, made in the 1930s, of creamy squares bordered with lavender, and each square containing a different color and shape of butterfly, with little black embroidered antennae coming out of each cotton body. I slept under it as a child at my grandmother's, as did my father as a boy, and now it lives in my adult bedroom. When I sleep under it, I am reminded of the goat-hair mattress at my grandmother's house on the bed, how squeaky it was, and of the plain pine wood boards of the floor, smells of wax, and her old-fashioned bathtub, with its animal form feet and prodigious depth.

I have coordinated quilt making two times, for what are called "memory quilts," one for my mother, and another for my dear friend Alison Harlow, both for their fiftieth birthdays. I asked each family member or close friend to create a square that remembered a happy day, or that somehow captured the essence of the person being honored. People varied so much in how they responded to this call: some took photographs and made digital representations of a good time, and then transferred the image to the fabric. Others, skilled in embroidery, made pieces of complex needlepoint that took months to complete. Others made collages and sewed on trinkets, souvenirs, or symbols of their times with Mom or Alison.

So many times I have meditated before those quilts, seeking to understand the wide web of relationships, skill, and care involved. I knew I was not in control of the process, but simply myself a kind of frame. Yet, somehow, those quilts evolved into a whole in both cases, with conversations whispering across the fabric squares, bringing layers of memory and life together.

When I first came to Yolŋu land last summer, and listened to the delicate and strong philosophies of how differences come to be honored and negotiated, I realized that I had stepped into a living quilt. Everywhere pieces talked to each other, made wholes and then separated; the stitches were songs and dreams like the last, nearly invisible quilting step. A group of us traveled into East Arnhem Land before the conference. Dhäŋgal very kindly took us to places near her home, sacred song places and also to the tree honoring her grandmother, encased in a tiny plot of land, surrounded by an iron fence. At the beach near her house, I found myself enchanted by a channel full of sparkling light where turtles swam and fed, and also by her tales of the small islet across the channel that she knew every inch of. She said there was a spring there that Balanda could never find, laughing to herself. She made the landscape come alive to me, and I will never forget it. Perhaps I will never leave it—because I am a nomad, I don't have a sense of one land that is my core, that holds all my stories. Perhaps quilts—and those quilts I carry in my body of images, stories, and friends from all over the world—are my land. Perhaps my writing and keeping in touch are my invisible stitches.

So this piece of writing is a way of talking about what I learned and shared during that visit.

Simplicity and Creating the Shadowy Other

The modern industrial world is saturated with categories and systems of classification (Bowker and Star 1999). Bureaucracies that deal with human beings use classifications such as race, income, and geographical location to identify what bureaucrats see as essential to the person. Often, such categories are used to simplify a complex human into a single thing. When you are reduced to one of "them," say, a "mere" girl, or "just" an Aborigine, it feels as if a piece of your soul gets stolen from you. When you are simplified by the powers that be, you are severed from history and from your own body. I believe that everyone has some experience of this, whether from childhood or from being an adult outcast of some sort, somewhere. At the same time, it is clear, too, that some forms of simplification are more life threatening or pervasive than others. If the simplification includes hatred, war, genocide, or slavery, those who are simplified are in extreme danger.

At times it feels to me that complexity and the respect that seeing people as complex means, are vulnerable to simplicity. The slogan beats out the essay. The name calling beats out the teachings of tolerance. War against "them" (whoever they are) beats out peace. When I feel this sadness, I fear for our world. Over the last several years, however, I have also found hope in looking more deeply into how the discarded complex is formed, lives, and carries the shadows as a living entity.

Most classification systems have spaces in them designated as "not elsewhere classified." These residuals have other names, such as "none of the above" or "not otherwise specified." There is an important connection between simplifying other people, and the use of residual categories. This is that simplifications reduce people to one thing, and residual categories are often a way of forming the discarded complex. So I envision this as a stream where some sort of screen takes things out, filters us and our multiple selves. This is the exact opposite of the Yolŋu system of thinking, where the flow is from apparently simple into more heterogeneous, more complex forms of living together. I feel that I am accepted into the Yolŋu as an entity filled with differences. This is a rare enough experience in my country of origin, the United States of America. So I learn here of a space where so many kinds of people have a chance to be central, not residual.

The reasons for why people become not elsewhere classified, none of the above, may be for reasons of keeping secrets, of silence, or perhaps of trying to pass as one kind of person while feeling like another. Sometimes one's own experience is unspeakable where there are no words for it. At the Teaching from Country International Seminar I told the story of discovering in my teens that my mother was Jewish. Now, I grew up in a rural place that seemed to have almost no Jews, African Americans, Hispanics, or Native Americans (I now know better than to take that for granted). So when I first found out this secret Jewishness, I didn't even know what it meant. I had no cultural context as I had been raised Christian in a rural place—I may as well have been told that I was part Venusian. So in a visceral way, sometimes experiences are much more simple and complex than current knowledge can maintain. Perhaps this is always true.

In school, or when applying for a job, people are often subjected to standardized tests. This is a stressful moment, one that severely tests the relationship of the simplifying imposed by the test, and the experience of being somehow residual. Sometimes, if you are taking standardized tests, it can become almost like a game. That is, it is not whether you know the "right" answer; it is whether you know why the designer put the question that way in the first place. So you have to double guess not just the answer but also the motivations of the test-setters. And in this process, our knowledge becomes twisted. We become objects instead of our own subjects—and in the process, create a whole class of shadowy strategies that will not appear on the test or in any statistics concerning us as a group. Let me give an example from the research of Rogers Hall et al. (1989) on how children solve algebra problems. These math problems take the form of something like: "If a train leaving Darwin travels at 200 km per hour and carries 50 passengers, and passes another train leaving from Sydney for Darwin traveling at 125 km/hour, carrying 200 passengers, but 40 of the Sydney passengers get off at Katoomba and take the bus to Darwin instead, and the bus is traveling at 70 km/hour, then how many passengers will arrive in Darwin and Sydney and at what times?" (I'm making this up, these problems always drive me crazy.)

Hall found that students devised ingenious ways of solving such problems, using metaphors for speed and number of passengers. One student substituted the rate at which a swimming pool would fill with water and the size of the hoses instead of the speed of trains and buses. She came up with the right answer. But when Hall, as a teacher, went to check on her work, she put her arm over the metaphor she had devised, and then attempted to erase it. When he questioned her about it, she said, deeply ashamed, that she had done it the "dirt way," not the right way. He reassured her that all ways are fine, and that she in fact had done something creative.

I think that the dirt way, the discarded complex, is full of promise and aesthetic surprises. Over time, the discarded knowing comes to comprise our shadow selves (both individual and collective). And the nature of that shadow knowledge is layered, complex, and interactional. We go on living with that which escapes standardized testing, quantification of our behavior.

In this world, where we are always being measured by standards, our shadow lives and shadow selves and shadow communities keep on living and keep on, you know, in spite of this kind of vigilance and surveillance from officialdom on every part of our behavior. We are inhabiting untold numbers of these residual categories. I see this as a kind of cloud of indicators: standards, measurements, simple categories. These indicators swarm around us, never being able to tell the simplest truth about us directly. At the same moment, in between these frozen indicators, we have our own experience that is unspeakable except to others who speak from that truth. We form the relationship, we have each other and we are able to keep moving with that, just a bit ahead, hopefully, of all these ways that are trying to freeze us and make us into a nothing or a thing that can be mowed down.

I am trying to figure out, first of all, what happens when we lose out on that, the simplicity wins and the standard, frozen standard way of viewing human beings wins. How does that work out? So, a thought is that it begins with these indicators, patterns, things you buy, things you do, grades you have in school, all that kind of stuff, and there are always different problems and phenomena and speculations on the part of those who would be governing us or claim to. How do we fight such deep bureaucracy and seeming craving to simplify us? In my brief time in Australia, I learned something about this. I felt that we started with respectful differences, with respect for each other's pasts. I was especially aware of how deep the currents ran for respect of each other's multiplicity. I am always forgetting and learning again how it is never just one past either, it is multiple futures, multiple pasts, multiple selves (Bowker 2006). There are always new combinations, new kinds of possibility within these differences that we bring to each other.

So in the time of questions during the seminar, Yiŋiya noted:

I actually wanted to make a comment and I think it is to, I think it attends to a long silence or a long secret. . . . Leigh has mentioned her teacher Howie Becker several times. Howie Becker is a sociologist. The person who taught Howie Becker who taught Leigh was William Lloyd Warner . . . who was taught by Gupupunyu in the 1930s by Makarrwala. So I think it is no accident that Leigh is coming back here at this point.

I didn't know this, and later Yiŋiya came to me and said, "Well, then, you are my grandmother!" I have thought much about that since then, because to my slow ears I couldn't exactly take that on myself at that time. But now, I have talked with Howie about this, and I have thought about the complex workings of knowledge and kinship. It is certainly true that for many of us scholars, the genealogy that matters most as kinship is that of teacher to student, student to her students, and so on. This is a very complex genealogy, too, one that criss-crosses time and many other sorts of kinship. So I am now understanding that a little better, and will honor it.

Another person asked me:

. . . to see at least how working with, going backwards and forwards between Yolŋu knowledge authorities and the governments that are giving this money for consultancies or for programs, it is part of that ongoing process where maybe we're never gong to come up with the answer but we might come up with some good ways of remembering how to do it next time. So, I think that is one of the things that we can keep thinking about and we'll hear more . . .

Like how we're making a big ŋanmarra, or mat. It starts from the centre, and it starts to, like Yolŋu and Balanda weaving together. The fibres are thin, thin and bright so it's going to happen like a big ŋanmarra so as to accommodate everbody to make that forum of the mat, the ŋanmarra. These two sisters, Djaŋ'kawu sisters, with their coming from the east to west, they brought the law and gave the riŋgitj names and the likan names, power names, they gave water, they gave the law, and the Wagilak sisters from central Arnhem way walking across the land, Yolŋu lands, giving the Yolŋu people their skin names, the subsections that come from the Wagilak, so there was like two sisters bringing water, and bringing the law.

Yes, this to me is very much what I mean by making quilts and especially the stitches. The stitches are the thin, bright fibers of our weaving together.

Governments, organizational structures, and people who use them use this testing and experience when they try to suppress the memory of multiple selves and multiple beings. They isolate and they capture one indicator, I think that is the thing that they would love to have, just one little indicator that could pin you down, that would be it and it would be wonderful (or so they say!).

We have also to respect multiple futures with multiple pasts. When I was a child, my great uncle had an ulcer. At this time, the standard treatment was to drink a lot of milk, take nothing spicy and try to be calm. I did not keep track of this knowledge. But then, a few years ago, my teacher Howie Becker had an ulcer. He went to the hospital and expected to hear a similar prescription. But the doctors said, that was then, we thought we knew everything about ulcers. But now, no, that it is actually not the case. Ulcers are because of an infection and they can be cured by antibiotics. Then we thought, now we know. Howie found this to be hilarious and so did I. I went to my doctor after that and told her the story. She began to laugh, too, and said, no, then we thought, and now we think.

Then We Thought, Now We Know. Then We Thought, Now We Think.

I do prefer the second way of being. It is more modest. Geof Bowker and I worked for a long time to understand and compare how different classifications emerge all the time, and a lot of times for many, many reasons. I remember finding an old manual that classified the causes of death deemed "acceptable" by the World Health Organization, one that changed every decade. There is a page in there from the 1930s, where the World

Health Organization was collecting causes of death and old age. It was possible then to die of old age—this is no longer permitted formally in filling out certificates of death. Also, then, one could die of being worn out—too bad you can no longer officially die of that!

There are many other interesting ways to die here but unfortunately, they have gone, replaced by other, more biomedical ways of thinking about it. So in this process, in this battle of sorts, there is always this tendency to simplify something as complex as death.

Whose Authority?

It is often profitable to simplify human experience. Tests that simplify are popular, easily processed, and the profits from their manufacture and use are often gained by those who develop them.

But I think that actually in the "not elsewhere classified world" there is this kind of hope. Donna Haraway, a theorist who has been important to me, used to speak of hopeful monsters. She spoke exactly of our selves that are outside of formal categories. They are classified as monsters or weird or Other. But there are helpful residuals, too. If you are not elsewhere classified, you can also move around a bit, it might even be for just a second, in which you can revert out there, you can resist with your bodies and we can share some hope. We're sharing our own stories and our own knowledge so that we remain more complex. We refuse to be universal singleton indicators. Where we have time and space and relationship and multiplicity moving, like we used the metaphors quilting and weaving above—dancing together in kind of a fullness that is not comprehensible, it is a way of escaping from these traps for a time.

I'm not naive about the risks for those who have been especially simplified and been at the end of receiving horrible violence. There are risks too in being blended together in some sort of phony "family of man" sort of talk. How can we resist, especially groups that are the targets of violence and prejudice, just becoming a bland mixture? How to keep our identities and not run the risk of being a default (and all too often, North and White default)? I'm not thinking that we are going to change the world overnight, but when we do speak out, and join together, it becomes an available way of thinking, and an available resource. The Other, the complex "not otherwise classified" is a resource. It can offer ways to creatively make tensions that begin the complexity (again).

Teaching from Country is making wonderful and important steps in keeping the heterogeneity as well as the commonality between people going on. I am honored to pay respect to the Yolŋu of Arnhem Land, and to take the lessons I have learned and try to make them more accessible to many others.

It is perhaps "Seeing like a City" as Canadian legal scholar Mariana Valverde puts it, rather than "Seeing like a State" as James Scott has it. One of the things that she points out is that the city is not decomposed into this sort of hierarchical standard process. It is instead a patchwork of many different kinds of simplifications—police precincts, school catchment areas, tax assessment zones, all these sorts of things, none of which line up with each other. There is this horrible, fascinating, wonderful mess. And there is maybe, partially, an organism there. Actually that classificatory gaze can be your friend when it is brought in alongside your "creating the cacophony" that allows for multiplicity and the kinds of ways you are arguing. So, all these many singular gazes, there are opportunities in that.

It is almost a trickster element—when we gather in the shadows, we sort of learn to make a different niche out of place. We resist and examine these simplifications, both. We even use them ourselves sometimes, but perhaps a bit more carefully. We can work with a kaleidoscope, but carefully, noting limits.

Shadows and Lights

I was wondering about this. If there are shadow spaces where there are many possibilities, including possibilities to think of those light spaces of the orderly? There, in those light spaces maybe we have conventions and it is considered to be the sort of mainstream, or the dominant paradigm, and so on. Often we recognize that they don't really exist as such but there is actually a multiplicity there.

Still the shadows must be quite a precarious spot from which to actually try to defend and protect those kinds of other categorizations. And so I don't know, I am just wondering about your thoughts on that, because we are talking about the academy as a unit or the university as a unit, and how that exchange between shadows and light is happening. What does it mean for those places that are in the light?

We are always working back and forth. We struggle to keep the tension going, and can kind of stay "on guard," and use what power we have to protect complexity. This is no easier in the academy than in any other institution. However, when the simplifiers get power and are imperialist about it and want to make everything the same essential thing, then those are the times when the power to be other that we still have in the academy can be helpful, for example, as in this program at Charles Darwin University. Again to quote Yiŋiya: "We just worked this kinship of the academy, and that is the kinship of the academy making its presence felt, it is no accident that she comes here."

I agree! As Native Americans in the Southwest of the United States say, when we respect this time-weaving, we respect Spider Grandmother, who spins fate. She weaves the creation again and again through Her body, and keeps us all complex.

References

Bowker, G. C.2006. *Memory Practices in the Sciences*. Cambridge, MA: MIT Press.

Bowker, G. C., and S. L. Star. 1999. *Sorting Things Out: Classification and Its Consequences*. Cambridge, MA: MIT Press.

Fellner, L.2009. "Betsy Ross Redux: The Underground Railroad 'Quilt Code.'" hartcottagequilts.com (earlier print article cited, 2003, in *Traditional Quilting*).

Hall, R., D. Kibler, E. Wenger, and C. Truxaw. 1989. "Exploring the Episodic Structure of Algebra Story Problem Solving." *Cognition and Instruction* 6 (3): 223–283.

Afterword: On the Distributedness of Leigh

Helen Verran

The distributedness of the figure of the analyst who inhabits an ecology of knowledge practices does not come easily to some of us. Others experience it as a welcome relief from the seemingly relentless pressure that moderns face to "clot themselves" as more or less coherent. Susan Leigh Star fell into this second category, and what's more she had over many years cultivated it not only as an analytic figuration, but as a mode of existence. This well-developed capacity served Leigh well when toward the end of her life she spent a little time among Yolŋu Aboriginal Australians who own land in Australia's northeast Arnhem Land. Leigh was ailing, and it is costly to be distributed when merely going on takes effort. Nevertheless the alacrity with which she seeped into the knowledge milieu and to which it in return began to inhabit her, was at times alarming for those of us who felt responsible for her. We need not have worried, and indeed after a while our concern dissipated. It was clear that Leigh drew in life, receiving as much as she contributed.

Yolŋu Aboriginal Australians grow up through a formalized kinship system of ordering people and places. Rather than face a relentless pressure to clot themselves as singular beings, Yolngu children face an implacable insistence that they distribute their being through the world. Leigh unexpectedly found herself among equals in the matter of distributedness, and responded in kind.

Along with other scholars, Leigh had come to Northern Australia for a symposium associated with the Charles Darwin University's Teaching from Country program—ecology of knowledges in practice in the academy, so to speak. This program has Yolŋu Aboriginal Australian elders teaching students located in Darwin, by distance education mode from their remote home places (remote by most Australians' standards) from which they draw inspiration.

It took a day or so, but Leigh's evident capacity in this matter of exchanging properties with the milieu became explicable—at least to our Yolŋu friends and colleagues. To everyone's delight—not least Leigh's—she was recognized as the intellectual granddaughter of a man adopted by Yolŋu near the beginning of the twentieth century.

Lloyd Warner, trained by Yolŋgu, in turn trained Howard Becker who in turn trained Leigh.

Why offer this anecdote as an afterword to *Boundary Objects and Beyond: Working with Leigh Star*? Susan Leigh Star had an extraordinary capacity as an analyst to let herself be inhabited by the world, and in turn to insinuate her being into many of the world's nooks and crannies. Leigh has left something with me, and naming that feels like an appropriate expression of respect. I am privileged to be one of the world's nooks Leigh insinuated herself into.

Obituaries, Memorial Events, and Selected Publications of Susan Leigh Star

Obituaries

Balka, Ellen. 2010. "Susan Leigh Star, 1954–2010." *Social Studies of Science* 40 (4): 647–651.

Clarke, Adele E. 2010. "In Memoriam: Susan Leigh Star (1954–2010)." *Science, Technology & Human Values* 35:581–600. [Includes bibliography of her publications and see below.]

http://en.wikipedia.org/wiki/Susan_Star (accessed March 23, 2015).

Conferences

"The State of Science and Justice: Conversations in Honor of Susan Leigh Star." 2011. Sponsored by the Science & Justice Training Program, UC Santa Cruz, June. http://scijust.ucsc.edu/wp-content/uploads/2013/04/Leigh-Star-Rapp-2011-Final.pdf (accessed March 23, 2015).

"Honoring the Intellectual Legacies of Susan Leigh Star." 2011. Sponsored by the STS Program, NSF; Stefan Timmermans (Sociology, UC Los Angles); Geoffrey C. Bowker (Computer Science, UC Irvine); and Adele E. Clarke (Social and Behavioral Sciences, UC San Francisco). At UC San Francisco, September. http://www.sscnet.ucla.edu/soc/LeighStar/ (accessed March 23, 2015).

Review

Timmermans, Stefan, and Steven Epstein. 2010. "A World of Standards But Not a Standard World: Toward a Sociology of Standards and Standardization." *Annual Review of Sociology* 36:69–89.

Chronological Listing of Books and Journal Special Issues

Star, Susan Leigh. 1984. *Zone of the Free Radicals*. Berkeley, CA: Running Deer Press.

Star, Susan Leigh. 1988. "Guest Editor's Introduction to Special Issue: The Sociology of Science and Technology." *Social Problems* 35 (3): 197–205.

Star, Susan Leigh. 1989. *Regions of the Mind: Brain Research and the Quest for Scientific Certainty*. Stanford, CA: Stanford University Press.

Star, Susan Leigh, ed. 1995a. *Ecologies of Knowledge: Work and Politics in Science and Technology.* Albany: SUNY Press.

Star, Susan Leigh, ed. 1995b. *The Cultures of Computing. Sociological Review Monograph.* Oxford: Basil Blackwell.

Star, Susan Leigh. 1995c. "Listening for Connections: Introduction to the Symposium on the Work of Anselm Strauss." *Mind, Culture, and Activity* 2 (1): 12–17.

Bowker, Geoffrey, Susan Leigh Star, William Turner, and Les Gasser, eds. 1997. *Social Science, Information Systems and Cooperative Work: Beyond the Great Divide.* Mahwah, NJ: Lawrence Erlbaum Associates.

Bowker, Geoffey, and Susan Leigh Star, eds. 1998. "Special Issue: How Classifications Work: Problems and Challenges in an Electronic Age." *Library Trends* 47 (2): 185–190.

Clarke, Adele E., and Susan Leigh Star, eds. 1998. "Introduction to Anselm Strauss Memorial Issue: On Coming Home and Intellectual Generosity." *Symbolic Interaction* 21 (4): 341–464.

Star, Susan Leigh, ed. 2000. "Introduction: Making Music with Cases: Improvisation and the Work of Howard Becker." *Mind, Culture, and Activity* 7:167–170.

Bowker, Geoffrey, and Susan Leigh Star. 2000. *Sorting Things Out: Classification and Its Consequences.* Cambridge, MA: MIT Press.

Lampland, Martha, and Susan Leigh Star, eds. 2009. *Standards and Their Stories: How Quantifying, Classifying and Formalizing Practices Shape Everyday Life.* Ithaca, NY: Cornell University Press.

Chronological Listing of Key Articles and Chapters

Star, Susan Leigh. 1979a. "The Politics of Right and Left." In *Women Look at Biology Looking at Women*, ed. R. Hubbard, M. S. Henifin, and B. Fried, 61–74. Cambridge, MA: Schenkman.

Star, Susan Leigh. 1979b. "Sex Differences and Brain Asymmetry: Problems, Methods and Politics in the Study of Consciousness." In *Genes and Gender II*, ed. Marian Lowe and Ruth Hubbard, 113–130. New York: Gordian Press.

Star, Susan Leigh. 1979c. "Feminism and Consciousness." *Science/Technology and the Humanities* 2:303–308.

Star, Susan Leigh. 1981. "I Want My Accent Back." *Sinister Wisdom* 16: 20–23.

Star, Susan Leigh. 1983. "Simplification in Scientific Work: An Example from Neuroscience Research." *Social Studies of Science* 13: 208–226.

Star, Susan Leigh. 1985. "Scientific Work and Uncertainty." *Social Studies of Science* 15: 391–427.

Star, Susan Leigh. 1986. "Triangulating Clinical and Basic Research: British Localizationists, 1870–1906." *History of Science* 24: 29–48.

Star, Susan Leigh, and Elihu M. Gerson. 1987. "The Management and Dynamics of Anomalies in Scientific Work." *Sociological Quarterly* 28: 147–169.

Fujimura, Joan, Susan Leigh Star, and Elihu Gerson. 1987. "Méthodes de recherche en sociologie des sciences: Travail, pragmatisme et interactionnisme symbolique" [Research methods in the sociology of science: Work, pragmatism and symbolic interactionism]. *Cahiers de Recherche Sociologique* 5: 65–85.

Star, Susan Leigh. 1988a. "The Structure of Ill-Structured Solutions: Heterogeneous Problem-Solving, Boundary Objects and Distributed Artificial Intelligence." In *Proceedings of the 8th AAAI Workshop on Distributed Artificial Intelligence*, Technical Report, Department of Computer Science, University of Southern California. Reprinted in *Distributed Artificial Intelligence* 2, ed. M. Huhns and L. Gasser, 37–54. Menlo Park: Morgan Kaufmann, 1989.

Star, Susan Leigh, and James R. Griesemer. 1989. "Institutional Ecology, 'Translations' and Boundary Objects: Amateurs and Professionals in Berkeley's Museum of Vertebrate Zoology, 1907–39." *Social Studies of Science* 19:387–420. Excerpted and reprinted in *The Science Studies Reader*, ed. Mario Biagioli, 505–524. New York: Routledge, 1999.

Star, Susan Leigh. 1989. "Layered Space, Formal Representations and Long-Distance Control: The Politics of Information." *Fundamenta Scientiae* 10:125–155.

Star, Susan Leigh. 1990. "What Difference Does It Make Where the Mind Is? Some Questions for the History of Neuropsychiatry." *Journal of Neurology and Clinical Neuropsychology* 2:436–443.

Hornstein, Gail A., and Susan Leigh Star. 1990. "Universality Biases: How Theories about Human Nature Succeed." *Philosophy of the Social Sciences* 20:421–436.

Star, Susan Leigh. 1991a. "Power, Technologies and the Phenomenology of Conventions: On Being Allergic to Onions." In *A Sociology of Monsters: Essays on Power, Technology and Domination*, ed. John Law, 26–56. London: Routledge.

Star, Susan Leigh. 1991b. "The Sociology of the Invisible: The Primacy of Work in the Writings of Anselm Strauss." In *Social Organization and Social Process: Essays in Honor of Anselm Strauss*, ed. David R. Maines, 265–283. Hawthorne, NY: Aldine de Gruyter.

Star, Susan Leigh. 1991c. "Invisible Work and Silenced Dialogues in Representing Knowledge." In *Women, Work and Computerization: Understanding and Overcoming Bias in Work and Education*, ed. I. V. Eriksson, B. A. Kitchenham, and K. G. Tijdens, 81–92. Amsterdam: North Holland.

Star, Susan Leigh. 1992a. "Craft vs. Commodity, Mess vs. Transcendence: How the Right Tools Became the Wrong One in the Case of Taxidermy and Natural History." In *The Right Tools for the Job: At Work in the Twentieth-Century Life Sciences*, ed. Adele E. Clarke and Joan H. Fujimura, 257–286. Princeton, NJ: Princeton University Press.

Star, Susan Leigh. 1992b. "The Trojan Door: Organizations, Work, and the 'Open Black Box.'" *Systems Practice* 5:395–410.

Star, Susan Leigh. 1992c. "The Skin, the Skull, and the Self: Toward a Sociology of the Brain." In *So Human a Brain: Knowledge and Values in the Neurosciences*, ed. Anne Harrington, 204–228. Boston: Birkhäuser.

Star, Susan Leigh. 1993. "Cooperation without Consensus in Scientific Problem Solving: Dynamics of Closure in Open Systems." In *CSCW: Cooperation or Conflict?*, ed. Steve Easterbrook, 93–105. London: Springer-Verlag.

Star, Susan Leigh. 1995a. "Epilogue: Work and Practice in Social Studies of Science, Medicine and Technology." *Science, Technology & Human Values* 20 (4): 501–507.

Star, Susan Leigh. 1995b. "The Politics of Formal Representations: Wizards, Gurus and Organizational Complexity." In *Ecologies of Knowledge: Work and Politics in Science and Technology*, ed. Susan Leigh Star, 88–118. Albany: SUNY Press.

Star, Susan Leigh. 1996. "From Hestia to Home Page: Feminism and the Concept of Home in Cyberspace." In *Between Monsters, Goddesses and Cyborgs: Feminist Confrontations with Science, Medicine and Cyberspace*, ed. Nina Lykke and Rosi Braidotti, 30–46. London: ZED Books. Reprinted in *Oxford Readings in Feminism: Feminism and Cultural Studies*, ed. Morag Shiach, 565–582. Oxford, UK: Oxford University Press, 1999. Also reprinted in *The Cybercultures Reader*, ed. David Bell and Barbara Kennedy, 632–643. London: Routledge, 2000.

Star, Susan Leigh, and Karen Ruhleder. 1996. "Steps toward an Ecology of Infrastructure: Design and Access for Large Information Spaces." *Information Systems Research* 7 (1): 111–134. Reprinted in *IT and Organizational Transformation: History, Rhetoric, and Practice*, ed. JoAnne Yates and John Van Maanen, 305–346. Thousand Oaks, CA: Sage Publications, 2001.

Star, Susan Leigh. 1997a. "The Feminism(s) Question in Science Projects: Queering the Infrastructure(s)." In *Technology and Democracy: Gender, Technology and Politics in Transition?*, ed. Ingunn Moser and Gro Hanne Aas, 13–22. Oslo: Center for Technology and Culture TMV Skrift-serie 29.

Star, Susan Leigh. 1997b. "Working Together: Symbolic Interactionism, Activity Theory and Information Systems." In *Communication and Cognition at Work*, ed. Yrjö Engeström and David Middleton, 296–318. Cambridge: Cambridge University Press.

Star, Susan Leigh. 1997c. "Anselm Strauss: An Appreciation." *Studies in Symbolic Interaction* 21: 39–48.

Bowker, Geoffrey, and Susan Leigh Star. 1997. "Problèmes de classification et de codage dans la gestion internationale de l'information." In *Cognition et Information en Société 8*, ed. B. Conein and L. Thévenot, 283–310. Paris: Éditions de l'École des Hautes Études en Science Sociales Raisons Pratiques.

Bowker, Geoffrey, and Susan Leigh Star. 1998. "Building Information Infrastructures for Social Worlds: The Role of Classifications and Standards." In *Community Computing and Support Systems: Social Interaction in Networked Communities*, ed. Toru Ishida, 231–248. Berlin: Springer-Verlag.

Kling, Rob, and Susan Leigh Star. 1998. "Human Centered Systems in the Perspective of Organizational and Social Informatics." *Computers & Society* 28 (1): 22–29.

Star, Susan Leigh. 1998a. "Experience: The Link between Science, Sociology of Science and Science Education." In *Thinking Practices*, ed. Shelley Goldman and James Greeno, 127–146. Hillsdale, NJ: Lawrence Erlbaum.

Star, Susan Leigh. 1998b. "Grounded Classifications: Grounded Theory and Faceted Classifications." *Library Trends* 47:218–232.

Star, Susan Leigh, and Anselm L. Strauss. 1999. "Layers of Silence, Arenas of Voice: The Ecology of Visible and Invisible Work." *Computer Supported Cooperative Work: The Journal of Collaborative Computing* 8:9–30.

Timmermans, Stefan, Geoffrey Bowker, and Leigh Star. 1998. "The Architecture of Difference: Visibility, Controllability, and Comparability in Building a Nursing Intervention Classification." In *Differences in Medicine: Unraveling Practices, Techniques and Bodies*, ed. Marc Berg and Annamarie Mol, 202–225. Durham, NC: Duke University Press.

Star, Susan Leigh. 1999. "The Ethnography of Infrastructure." *American Behavioral Scientist* 43:377–391.

Star, Susan Leigh. 2002. "Commentary: 'Betweeness' in Design Education." In *Computer Supported Cooperative Learning*, ed. T. Koschmann, 259–262. Fairfax, VA: TechBooks.

Star, Susan Leigh. 2003. "Computers/Information Technology and the Social Study of Science and Technology." In *International Encyclopedia of Social and Behavioral Sciences*, ed. Niel Smelser and Peter Baltes, 13638–13644. Amsterdam: Elsevier.

Clarke, Adele, and Susan Leigh Star. 2003. "Science, Technology and Medicine Studies." In *Handbook of Symbolic Interaction*, ed. Nancy Herman and Larry Reynolds, 539–574. Walnut Creek, CA: Alta Mira Press.

Star, Susan Leigh. 2004. "Infrastructure and Ethnographic Practice: Working on the Fringes." *Scandinavian Journal of Information Systems* 14 (2): 107–122.

Star, Susan Leigh, Geoffrey Bowker, and Laura Neumann. 2004. "Transparency Beyond the Individual Level of Scale: Convergence between Information Artifacts and Communities of Practice." In *Digital Library Use: Social Practice in Design and Evaluation*, ed. Ann P. Bishop, Barbara P. Buttenfield, and Nancy Van House, 241–270. Cambridge, MA: MIT Press.

Star, Susan Leigh. 2005. "Categories and Cognition: Material and Conceptual Aspects of Large-Scale Category Systems." In *Interdisciplinary Collaboration: An Emerging Cognitive Science*, ed. Sharon J. Derry, Christian D. Schunn, and Morton Ann Gernsbacher, 167–186. Mahwah, NJ: Lawrence Erlbaum Associates/Psychology Press.

Bowker, Geoffrey, and Susan Leigh Star. 2006. "Infrastructure." In *Handbook of New Media and Communication*, ed. L. Lievrouw and S. Livingstone, 151–162. London: SAGE.

Clarke, Adele E., and Susan Leigh Star. 2007. "Social Worlds/Arenas as a Theory-Methods Package." In *Handbook of Science and Technology Studies*, 2nd. ed., ed. Edward Hackett, Olga Amsterdamska, Michael Lynch, and Judy Wacjman, 113–137. Cambridge, MA: MIT Press.

Star, Susan Leigh. 2007a. "Living Grounded Theory: Cognitive and Emotional Forms of Pragmatism." In *The SAGE Handbook of Grounded Theory*, ed. Anthony Bryant and Kathy Charmaz, 75–94. Thousand Oaks, CA: SAGE.

Star, Susan Leigh. 2007b. "Interview on the Information Society." *Daedalus*. [In Italian.]

Star, Susan Leigh. 2007c. "Five Answers." In *Philosophy of Technology: Five Questions*, ed. J. B. Olsen and E. Selinger, 223–231. Copenhagen, Denmark: Automatic/VIP.

Star, Susan Leigh, and Geoffrey Bowker. 2007. "Enacting Silence: Residual Categories as a Challenge for Ethics, Information Systems, and Communication Technology." *Ethics and Information Technology* 9:273–280.

Star, Susan Leigh. 2009. "Susan's Piece: Weaving As Method in Feminist Science Studies: The Subjective Collective." *Subjectivity* 28:344–346.

Star, Susan Leigh. 2010a. "This Is Not a Boundary Object: Reflections on the Origin of a Concept." *Science, Technology & Human Values* 35:601–617.

Star, Susan Leigh. "When Shadows Become Complex, Weaving the Danmarra." 2010b. http://www.cdu.edu.au/centres/spill//journal/LC_Journal_Issue2_2010.pdf (accessed March 23, 2015).

Contributors

Ellen Balka
Simon Fraser University, School of Communication

Eevi E. Beck
University of Oslo, Department of Education

Dick Boland
Case Western Reserve University, Weatherhead School of Management

Geoffrey C. Bowker
University of California, Irvine, Department of Informatics

Janet Ceja Alcalá
The University of Arizona, School of Information

Adele E. Clarke
University of California, San Francisco, Department of Social and Behavioral Sciences

Les Gasser
University of Illinois Graduate School of Library and Information Science

James R. Griesemer
University of California, Davis, Philosophy Department

Gail A. Hornstein
Mount Holyoke College, Department of Psychology

John Leslie King
University of Michigan, School of Information

Cheris Kramarae
University of Oregon, Center for the Study of Women in Society

Maria Puig de la Bellacasa
University of Leicester, School of Management

Karen Ruhleder
University of Illinois at Urbana-Champaign, Graduate College

Kjeld Schmidt
Copenhagen Business School, Department of Organization

Brian Cantwell Smith
University of Toronto, Faculty of Information and Coach House Institute

Susan Leigh Star*
University of Pittsburgh, School of Information Sciences

Anselm L. Strauss*
University of California, San Francisco, Department of Social and Behavioral Sciences

Jane Summerton
Swedish National Road and Transport Research Institute, Research Group on Mobility, Actors, Planning Processes

Stefan Timmermans
 University of California, Los Angeles, Department of Sociology

Helen Verran
Charles Darwin University, Northern Institute

Nina Wakeford
Goldsmiths, University of London, Department of Sociology

Jutta Weber
University of Paderborn, Faculty of Cultural Sciences

*Contributor is deceased.

Index

Discovery of Grounded Theory, The (Glaser and
Strauss), 88, 123
Distributed artificial intelligence (DAI),
240
Distributedness, 499
Distributed problem solving, 8
boundary objects and, 243–257
combinatorial implosion and, 248–249
due process and, 243, 247–250
Durkheim and, 243–244, 246, 256
environment and, 248–249
frame problem and, 247–250
heterogeneous information characteristics
and, 251
information systems and, 250, 257
lesbianism and, 265
metaphor and, 243–250, 257
negotiations and, 245–250
networks and, 246
open systems and, 243–251, 255, 257n3
participant systems and, 247
physiology and, 252, 254
sociology and, 251
truth and, 250–251
Turing test and, 243–247, 249, 255
"Do Artifacts Have Politics?" (Winner),
484–485
Doing Fieldwork: Warnings and Advice (Wax),
126
Driscoll, Mary Kathleen, 355
Drones
bug splat and, 304–306
Bush administration and, 308, 310, 314
capabilities of, 315–316
categorizing life and death in robot wars
and, 303–317
CIA and, 304, 306, 309, 311–312, 316
CIVIC and, 312–314
civilians and, 303, 306–317
control teams and, 304–306, 309, 315
effectiveness of, 310–311
Geneva Convention and, 313
Global Information Grid (GIG) and, 308

Hellfire missiles and, 304
increased U.S. reliance on, 304–306
international humanitarian law and, 306,
311
law of war and, 306, 311
moral choices and, 316–317
Obama administration and, 310–312
Pakistan and, 309–315
Predator, 308, 311, 315
Reaper, 315
reconnaissance and, 315–316
remembering and, 324
Revolution in Military Affairs (RMA) and,
307–309, 317
silences and, 303, 306
Duchamp, Marcel, 79
Due process
boundary objects and, 193, 243, 247–250
distributed problem solving and, 243,
247–250
frame problem and, 247–250
visibility and, 367
Dumit, Joseph, 65n1, 91, 100
Durfee, E., 256
Durkheim, Emile, 136, 243–244, 246, 256
Durkheim test, 243–247, 256
Dutch Manual (Muller, Feith, and Fruin), 331
Dyslexia, 429

East Arnhem Land, 490
Ecological thinking, 2
anthroporganizationalism and, 56–57
bioinfrastructure and, 54–58
marginalities and, 48
modes of attention and, 47
networks without voids and, 50–54
poetics of infrastructure and, 47–65
power of metaphor and, 54–58
relational form and, 51
soil and, 54–61
spaces between and, 37, 48–53, 55, 62, 64,
92, 324
speculative commitment and, 49–50